REINFORCED CONCRETE DESIGN

Third Edition

S.U. Pillai
Director
Cooperative Academy of Professional Education (Kerala)
Trivandrum, India
Formerly Visiting Professor of Civil Engineering
Royal Military College of Canada

D.W. Kirk
Professor Emeritus of Civil Engineering
Royal Military College of Canada

M.A. Erki
Professor of Civil Engineering
Royal Military College of Canada

Toronto Montréal New York Burr Ridge Bangkok Bogotá Caracas Lisbon London
Madrid Mexico City Milan New Delhi Seoul Singapore Sydney Taipei

REINFORCED CONCRETE DESIGN 3/E

Copyright © McGraw-Hill Ryerson Limited, 1999, 1988, 1983. All rights reserved. No part of this publication may be reproduced, stored in a retrieval system, or transmitted, in any form or by any means, electronic, mechanical photocopying, recording, or otherwise, without prior written permission of McGraw-Hill Ryerson Limited.

Any request for photocopying, recording, or taping of any part of this publication shall be directed in writing to CANCOPY.

ISBN 0-07-560829-4

1 2 3 4 5 6 7 8 9 0 GTC 8 7 6 5 4 3 2 1 0 9

Printed and bound in Canada

Care has been taken to trace ownership of copyright material contained in this text. The publishers will gladly take any information that will enable them to rectify any reference or credit in subsequent editions.

Canadian Cataloguing in Publication Data

Pillai, S.U.
Reinforced concrete design

3rd ed.
First ed. (1983) published under title: Reinforced concrete design in Canada.
Includes index.
ISBN 0-07-560829-4

1. Reinforced concrete. 2. Reinforced concrete construction – Canada.
I. Kirk, D.W. (Donald Wayne), 1934- II. Erki, Marie-Anne. III. Title: Reinforced concrete design in Canada.

TA683.2.P54 1999   624.1'8341   C99-931585-4

COVER PHOTO CREDIT
McGraw-Hill Ryerson Ltd. would like to thank Boily Photo of Summerside, Prince Edward Island, for their cooperation in obtaining the cover photograph. Special thanks is expressed to Dr. Gamil Tadros, P.Eng., of Speco Engineering Limited, Calgary, Alberta, for his assistance in obtaining the photograph of the construction view of the conical pier base of the Confederation Bridge.

CONTENTS

PREFACE

CHAPTER 1	**INTRODUCTION AND MATERIAL PROPERTIES**	1
1.1	Introduction	1
1.2	Cement and Water-Cement Ratio	2
1.3	Aggregates	2
1.4	Design and Control of Concrete Mixes	3
1.5	Compressive Strength and Stress-Strain Curve	5
1.6	Behaviour Under Tension	9
1.7	Shear Strength	10
1.8	Combined Stresses	11
1.9	Triaxial Strength	13
1.10	Stress-Strain Relationships for Multiaxial Stress States	14
1.11	Creep	14
1.12	Shrinkage and Temperature Effects	17
1.13	Steel Reinforcement	19
1.14	Codes and Specifications	22
1.15	International System of Units	25
	References	26
CHAPTER 2	**REINFORCED CONCRETE BUILDINGS**	27
2.1	Introduction	27
2.2	Loads	28
2.3	Structural Systems	35
2.4	Structural Analysis and Design	44
	References	46
CHAPTER 3	**STRUCTURAL SAFETY**	47
3.1	Introduction	47
3.2	Safety Considerations	47
3.3	Limit States Design	49
3.4	Partial Safety Factors	51
3.5	Limit States Design by Canadian Codes	52
3.6	Load and Resistance Factors	55
3.7	Notation	56
3.8	Tolerances	57
	References	57
CHAPTER 4	**BEHAVIOUR IN FLEXURE**	59
4.1	Theory of Flexure for Homogeneous Materials	59
4.2	Elastic Behaviour and Transformed Sections	61
4.3	Transformed Steel Area	64
4.4	Flexure of Reinforced Concrete Beams	65

4.5		Analysis and Design	70
4.6		Service Load Stresses	71
4.7		Strength in Flexure	79
4.8		Code Recommendations	85
4.9		Maximum Rinforcement Ratio	85
4.10		Minimum Reinforcement in Flexural Members	93
4.11		Analysis for Strength by CSA Code	94
		Problems	111
		References	114

CHAPTER 5   DESIGN OF BEAMS AND ONE-WAY SLABS FOR FLEXURE   115

5.1	Introduction		115
5.2	Concrete Cover and Bar Spacing Requirements		116
5.3	Selection of Member Sizes		119
5.4	Design of Rectangular Sections for Bending –Tension Reinforcement Only		121
5.5	Design Aids for Rectangular Beams		128
5.6	Design of One-Way Slabs		132
5.7	Arrangement of Reinforcement in Continuous One-Way Slabs		138
5.8	Design of Rectangular Beams with Compression Reinforcement		138
5.9	Design of T-Beams		142
5.10	Cut-Off of Flexural Reinforcement		148
	Problems		161
	References		163

CHAPTER 6   DESIGN FOR SHEAR   164

6.1	Introduction	164
6.2	Nominal Shear Stress	168
6.3	Effect of Shear on Beam Behaviour	168
6.4	Shear Strength without Shear Reinforcement	174
6.5	Shear Strength at Inclined Cracking	174
6.6	Behaviour of Beams with Shear Reinforcement	177
6.7	Action of Web Reinforcement	179
6.8	Shear Strength of Beams with Web Reinforcement	181
6.9	Shear Design of Beams by Simplified Method	183
6.10	Shear Design Near Supports	189
6.11	Anchorage of Shear Reinforcement	190
6.12	Additional Comments on Shear Reinforcement Design	192
6.13	Shear Design Examples	192
6.14	Shear Strength of Flexural Members Subjected to Axial Loads	202
6.15	Structural Low Density Concrete	202
6.16	Members of Varying Depth	203
6.17	Interface Shear and Shear Friction	204
	Problems	209
	References	211

CHAPTER 7	DESIGN FOR TORSION	212
7.1	General	212
7.2	Torsion Formulas	213
7.3	Strength and Behaviour of Concrete Members in Torsion	216
7.4	Combined Loadings	223
7.5	CSA Code Approach to Analysis and Design for Torsion	224
	Problems	234
	References	235
CHAPTER 8	SHEAR AND TORSION DESIGN - GENERAL METHOD	237
8.1	Introduction	237
8.2	Stress-Strain Relationhip for Diagonally Cracked Concrete	239
8.3	Analysis Based on Modified Compression Field Theory	241
8.4	Combined Shear and Torsion	253
8.5	Design Using Strut-and-Tie Model	254
	Problems	265
	References	265
CHAPTER 9	BOND AND DEVELOPMENT	266
9.1	Bond Stress	266
9.2	Flexural Bond	266
9.3	Anchorage or Development Bond	267
9.4	Variation of Bond Stress	268
9.5	Bond Failure	270
9.6	Factors Influencing Bond Strength	272
9.7	Bond Tests	274
9.8	Development Length	275
9.9	CSA Code Development Length	277
9.10	Hooks and Mechanical Anchorages	281
9.11	CSA Code Development Requirements	284
9.12	Factored Moment Resistance Diagrams	285
9.13	Design Examples	286
9.14	Splicing of Reinforcement	296
	References	299
CHAPTER 10	CONTINUITY IN REINFORCED CONCRETE CONSTRUCTION	300
10.1	General	300
10.2	Loading Patterns for Continuous Beams and Plane Frames	301
10.3	Approximations Permitted by Code for Frame Analysis	306
10.4	Analysis Procedures	309
10.5	Stiffness of Members	311
10.6	Use of Centre-to-Centre Span and Moment at Support Face	315
10.7	Inelastic Analysis and Moment Redistribution	331
10.8	Preliminary Design	338
	Problems	339
	References	340

CHAPTER 11 SERVICEABILITY LIMITS — 341
- 11.1 Introduction — 341
- 11.2 Deflections — 341
- 11.3 Deflections by Elastic Theory — 342
- 11.4 Immediate Deflections — 344
- 11.5 Time-Dependent Deflection — 355
- 11.6 Deflection due to Shrinkage — 356
- 11.7 Shrinkage Curvature — 357
- 11.8 Computation of Shrinkage Deflections — 358
- 11.9 Deflection Due to Creep — 358
- 11.10 Long-Time Deflection by Code — 360
- 11.11 Deflection Control in Code — 366
- 11.12 Control of Cracking — 366
- Problems — 370
- References — 372

CHAPTER 12 ONE-WAY FLOOR SYSTEMS — 374
- 12.1 One-Way Slab - Beam and Girder Floor — 374
- 12.2 Design Example — 376
- 12.3 One-Way Joist Floor — 396
- Problems — 398
- References — 399

CHAPTER 13 TWO-WAY SLABS ON STIFF SUPPORTS — 400
- 13.1 One- and Two-Way Action of Slabs — 400
- 13.2 Background of Two-Way Slab Design — 402
- 13.3 Design of Two-Way Slabs Supported on Walls and Stiff Beams — 403
- 13.4 Difference Between Wall Supports and Column Supports — 420
- Problems — 421
- References — 421

CHAPTER 14 DESIGN OF TWO-WAY SLAB SYSTEMS — 422
- 14.1 Introduction — 422
- 14.2 Design Procedures in CSA Code — 425
- 14.3 Minimum Thickness of Two-Way Slabs for Deflection Control — 429
- 14.4 Transfer of Shear and Moments to Columns — 431
- 14.5 Direct Design Method — 433
- 14.6 Shear in Two-Way Slabs — 443
- 14.7 Design Example for Direct Design Method — 454
- 14.8 Design Aids — 464
- 14.9 Elastic Frame Method — 464
- 14.10 Slab Reinforcement Details — 474
- 14.11 Design Example for Elastic Frame Method — 477
- Problems — 497
- References — 498

CHAPTER 15		**DESIGN OF COMPRESSION MEMBERS - SHORT COLUMNS**	**499**
	15.1	Compression Members	499
	15.2	Design of Short Columns	504
	15.3	Strength and Behaviour Under Axial Compression	504
	15.4	Compression with Bending	508
	15.5	Strength Interaction Diagrams	518
	15.6	Interaction Diagrams as Design Aids	521
	15.7	Analysis of Sections Using Interaction Diagrams	523
	15.8	Design of Sections Using Interaction Diagrams	528
	15.9	Columns of Circular Section	533
	15.10	Combined Axial Compression and Biaxial Bending	537
		Problems	544
		References	545
CHAPTER 16		**SLENDER COLUMNS**	**546**
	16.1	Behaviour of Slender Columns	546
	16.2	Effective Length of Columns in Frames	552
	16.3	Design of Slender Columns	556
		Problems	571
		References	571
CHAPTER 17		**FOOTINGS**	**570**
	17.1	General	570
	17.2	Types of Footings	570
	17.3	Allowable Soil Pressure	572
	17.4	Soil Reaction under Footing	573
	17.5	General Design Considerations for Footings and Code Requirements	576
	17.6	Combined Footings	591
	17.7	Design of Two-Column Combined Footings	593
		Problems	602
		References	602
CHAPTER 18		**SPECIAL PROVISIONS FOR SEISMIC DESIGN**	**606**
	18.1	General	606
	18.2	Role of Ductility in Seismic Design	607
	18.3	Major Design Considerations	610
	18.4	Recent Advances	619
		References	619

PREFACE

This 3rd edition of *Reinforced Concrete Design* includes the latest information from the standards and codes pertaining to the design of reinforced concrete structures, principally the CSA Standard A23.3-94, *Design of Concrete Structures* of the Canadian Standards Association and the *National Building Code of Canada* 1995 (NBC 1995) of the National Research Council. Additional referenced documents include the provisions of CSA Standards A23.1, A23.2, and other Canadian and American codes for concrete materials, structures, and steel reinforcement. The 1994 version of the CSA Standard A23.3-94 has some major changes and rationalisations compared its previous 1984 edition. The most far-reaching of these are the changes in the ultimate concrete strain from 0.003 to 0.0035 and the reformulation of the equivalent rectangular stress block, wherein the factors of 0.85 and β_1, for the average concrete stress and depth of the stress block, are replaced by an α_1 and a new β_1. Almost equally far-reaching are the changes to the expressions for the modulus of elasticity of concrete, the minimum amount of flexural reinforcement, the minimum amount of shear reinforcement, and the development length of reinforcing bars. The *Simplified Method* for calculating shear reinforcement for beams has been further simplified, and the *General Method* for shear and torsion design has been revised, based on the modified compression field theory. The *Strut-and-Tie* or *Truss* model of structural behaviour has been given explicit recognition in the design procedures. The provisions have also been simplified for the *Direct Design Method* for slabs and for the calculations for slender columns and transmission of column loads through floors. The 3rd edition of *Reinforced Concrete Design* presents updated load factors in CSA A23.3-94 that correspond to those of the NBC 1995, and the new factors in the NBC 1995 for calculating earthquake forces on structures. The 3rd edition has been entirely reformatted, with newly drawn figures, for improved clarity. All the examples have been reworked to incorporate the many changes to the design standards and codes.

The target audience of this book is mainly the undergraduate Civil Engineering student, taking a first or second course in design of reinforced concrete structures. However, the book also addresses itself also the practising engineer, who may wish to have a companion textbook to use with the CSA A23.3-94. Currently, the *Concrete Design Handbook*, published by the Canadian Portland Cement Association, is widely used in Canadian design offices and universities. This handbook contains numerous design aids, all of which cannot be provided in a general textbook, and it is a desirable supplement to any textbook on the design of reinforced concrete structures. Similarly, the *Reinforcing Steel Manual of Standard Practice*, published by the Reinforcing Steel Institute of Canada, is a resource of standard practices in reinforced concrete design and construction. It, too, is an excellent reference for designers and students.

The 3rd edition of *Reinforced Concrete Design* has kept its original eighteen chapter format, that covers beam, slab, column, and footing design, as well as special considerations in the analysis of reinforced concrete structures, for both the ultimate and serviceability limit states. The final chapter of the book gives an introduction to seismic design. Appendix A contains suggestions for maintaining systematic bookkeeping, and, for convenience, all design aids contained in the book are listed in Appendix B.

The authors gratefully acknowledge the outstanding professional competence of Ms. Marie-Claude Leblanc, Department of Civil Engineering, Royal Military College of Canada, who organised, typed, and prepared the camera-ready copy for this 3rd edition. A graphic arts team, headed by Mr. P. Beljith, drew the figures herein, and their meticulous work and exceptional skill are greatly appreciated by the authors. The authors thank Dr. Khaled Soudki, Department of Civil Engineering, University of Waterloo, for undertaking an independent check of many of the example solutions. The authors also thank Dr. Gamil Tadros, conceptual designer of the Confederation Bridge and now head of Speco Engineering Limited of Calgary, Alberta, by whose intermediary the authors received from Boily Photo of Summerside, Prince Edward Island, the permission to reproduce the cover photograph. The authors are very grateful to Boily Photo for their generosity. Finally, the authors acknowledge with gratitude the co-operation of the Canadian Portland Cement Association, the CSA International, and the Reinforcing Institute of Canada for their permission to reproduce the copyright material presented herein.

S.U. Pillai
D.W. Kirk
M.A. Erki

CHAPTER 1 Introduction and Material Properties

1.1 INTRODUCTION

The following sections provide a brief introduction to the materials in, and the properties of, concrete. For a detailed study of concrete technology, reference should be made to any of the standard works on the subject (Refs. 1.1, 1.2, 1.3).

When cement, aggregates (sand, gravel, crushed rock) and water are mixed together in the proper proportions and allowed to harden, a solid rocklike mass is obtained which is called *concrete*. The cement and water in the paste react chemically (*hydration*), and the paste hardens binding the aggregates. The process of hydration continues as long as the concrete remains saturated, the temperature remains favourable, and any unhydrated cement is present. To aid hydration, the freshly placed concrete is cured by keeping it continually moist and under controlled temperature, until the desired quality is attained.

The gain in strength of concrete is very rapid in the first few days, after which the rate of gain decreases steadily as shown in Fig. 1.1. Concrete is generally designed on the basis of its strength at 28 days, by which time it has acquired most of its strength. Normal concrete under proper curing acquires about 75 percent of its design strength (28-day strength) by the end of two weeks. This strength is usually adequate to carry the dead loads and construction loads, so that the formwork can be removed about this time. Considerable variations in the properties of concrete can be achieved

Fig. 1.1 Variation of compressive strength with age (Ref. 1.2)

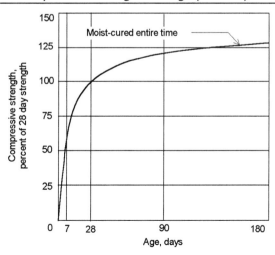

by variations in mix proportions, type of cement used, type of aggregates, methods of curing, and by the use of special *admixtures*. Plain concrete has a high compressive strength, but a very low tensile strength. Therefore, when areas of a concrete structural member are subjected to tensile stresses (either due to direct tensile forces or indirectly due to flexural tension, diagonal tension, temperature changes or shrinkage, and volume changes) it is *reinforced* with steel bars embedded in these locations while casting the member. The resulting composite is termed *reinforced concrete*.

1.2 CEMENT AND WATER-CEMENT RATIO

Hydraulic cements, most notably *Portland cements*, which set and harden by chemically reacting with water, are used to make structural concrete. CSA Standard A5 provides for five types of Portland cements as follows:

    Type 10, Normal
    Type 20, Moderate
    Type 30, High-early-strength
    Type 40, Low-heat-of-hydration
    Type 50, Sulphate-resisting

    Type 10 Normal cement is used in general concrete construction where the special properties of the other types are not required. Other types of special cements, such as air-entraining Portland cement, white and coloured cements are also available.
    The ratio, by mass, of water to cement in the mix is termed the *water-cement ratio* (w/c), and is the major factor that controls the strength and many other properties of the concrete. The minimum w/c ratio required in a concrete mix to ensure complete hydration of the cement is about 0.35 to 0.40 (even though the purely chemical requirement is somewhat lower). However, higher w/c ratios are usually used in order to improve the *workability*; that is, to make the mixture more plastic and workable, so that it can be placed in the formwork and compacted without difficulty. A reduction in the water content will result in increased compressive strength, bond strength, and resistance to weathering, and lower porosity, creep, absorption, and volume changes in the hardened concrete.

1.3 AGGREGATES

The aggregates make up 60 to 80 percent of the total volume of concrete. Aggregates are usually divided into *fine aggregate*, consisting of sand with particle sizes up to 10 mm, and *coarse aggregate* composed of gravel, crushed rock, or other similar materials. Coarse aggregate has particle sizes in excess of 1.25 mm, but usually the bulk of it has sizes in excess of 5 mm. Commonly used aggregates, such as sand, gravel and crushed stone, produce *normal density concrete* with a mass density of about 2150 to 2500 (average 2320) kg/m^3. For design computations, normal density concrete reinforced with steel is commonly assumed to have a mass density of 2400 kg/m^3. Both the size and grading of the aggregates are important considerations in proportioning the mix. These aggregates should meet the requirements of CSA Standard A23.1-94 – *Concrete Materials and Methods of Concrete Construction*.

Occasionally, structural low density aggregates are used to produce low density concrete for special applications. Such aggregates consist of expanded shale, clay, slate, and slag, and the resulting concrete has a density of about 1350 to 1850 kg/m^3, with compressive strengths comparable to those of normal density concrete. Requirements to be met by structural low density aggregates are given in American Society for Testing and Materials (ASTM) C330-89 – *Lightweight Aggregate for Structural Concrete*. Low density concrete having a 28-day compressive strength in excess of 20 MPa and an air dry density not exceeding 1850 kg/m^3 is termed *Structural Low Density Concrete*, and low density concrete having a 28-day compressive strength in excess of 20 MPa and an air dry density between 1850 and 2150 kg/m^3 is termed *Structural Semi-low Density Concrete*. Similarly, high density aggregates such as limonite, barite, ilmenite, magnetite, haematite, iron (shots and punchings) and steel slugs are used occasionally to produce high density concrete for special applications such as radiation shielding and counterweighting.

1.4   DESIGN AND CONTROL OF CONCRETE MIXES

For given materials, the w/c ratio and the proportion of the ingredients most significantly influence the quality of the hardened concrete. Figure 1.2 shows typical variation of compressive strength with w/c ratio for normal concrete. Maximum permissible w/c ratios for different types of structures and exposure conditions are recommended in CSA Standard A23.1-94, Clause 14.

The mix is designed on the basis of the desired mean compressive strength, workability (slump), availability of materials, and economy. CSA Standard A23.1-94, Clause 14, recommends ranges of slump to be used for different types of construction. The concrete mix proportion is usually expressed as the ratio, by volume, or more accurately by mass, of cement to fine aggregate to coarse aggregate, in that order (for example, 1:2:4) together with a w/c ratio.

Concrete is normally specified by the required *compressive strength at 28 days*, f_c', as determined by tests on standard cylindrical specimens 150 mm diameter by 300 mm long. A concrete is considered satisfactory (CSA Standard A23.1-94) if the averages of all sets of three consecutive strength tests equal or exceed the specified

Fig. 1.2 Typical variation of 28-day compressive strength with water-cement ratio

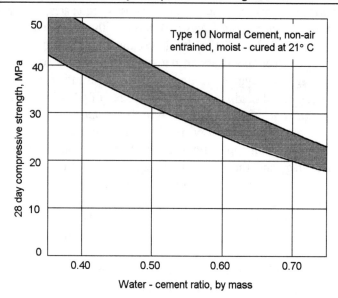

strength, with no individual test more than 3.5 MPa below the specified strength. Thus, the specified *compressive strength of concrete*, f_c', is the strength that will be equalled or exceeded by the average of all sets of three consecutive strength tests, with no individual test result failing more than 3.5 MPa below f_c'. Results of standard strength tests on cylinders cast from concrete of identical proportions show inevitable scatter, the amount of scatter depending on the degree of control. To account for these variations in strength test results, it is necessary to proportion the concrete mix for a *mean strength* higher than the specified strength, f_c'. The minimum magnitude of this over-allowance is determined from a statistical analysis of test results. Methods and procedures for proportioning of concrete mixes are not included in this book. Several such methods are given in detail in Refs. 1.1-1.6.

In addition to cement, aggregates, and water, other materials termed collectively as *admixtures* are sometimes added to the mix in order to impart special properties to the concrete. For instance, admixtures are often used to entrain air, to improve workability, and to accelerate or retard setting and rate of strength gain. Air-entraining admixtures, which entrain microscopic air bubbles into the concrete, are widely used particularly in cold environments as they improve the resistance of concrete to freezing and thawing and to surface scaling caused by de-icing agents. CSA Standards A266.1, A266.2, and A266.6, as well as ASTM C494, give specifications for chemical admixtures. Guidelines for the use of admixtures are contained in CSA Standard A266.4 and the American Concrete Institute report ACI 212.3R-91 (Ref. 1.7).

Proper mixing, placing, compacting, and curing are all essential for the production of quality concrete. A very concise and pertinent presentation of all the major aspects relevant to the manufacture of good quality concrete, including examples of mix design, with reference to the requirements of Canadian Standards, is given in Ref. 1.3.

1.5 COMPRESSIVE STRENGTH AND STRESS-STRAIN CURVE

The compressive strength of concrete is one of its most important and useful properties. Structural concrete is primarily used to withstand compressive stresses. Many important properties which influence the strength and deformation of concrete structures, such as tensile strength, shear strength, and elastic modulus, can be approximately correlated to the compressive strength and are, therefore, normally determined using the compressive strength. However, all such correlations are not precise, but only approximate or empirical in nature, and are generally applicable to normal concrete with nominal compressive strength in the range of 20 to 45 MPa.

The determination of compressive strength is very simple and is generally done by the standard compression test (CSA Test Method A23.2-9C) on 150 mm by 300 mm cylinders at 28 days. The standard cylinder has a length-to-diameter ratio of two. For cylinders with this ratio appreciably less than two, ASTM C42-90 provides the correction factors given in Table 1.1. The size of the cylinders also influences the strength test results. A variation of compression test results, with the cylinder diameter expressed as a percentage of the strength of the standard 150 mm by 300 mm cylinder, is shown in Fig. 1.3 (Refs. 1.8, 1.9).

Table 1.1 Correction Factors for Non-Standard Test Cylinders (ASTM C42-90)

Ratio of length of cylinder to diameter	Strength correction factor
2.00	1.00
1.75	0.98
1.50	0.96
1.25	0.93
1.00	0.87

In usual structural applications, the concrete strength specified is in the range of 20 to 40 MPa for reinforced concrete and 40 to 55 MPa for prestressed concrete. Typical stress-strain curves for normal concrete in this range, obtained from standard compression tests, are shown in Fig. 1.4. The initial part of the curve, up to a stress level equal to about one-half of the maximum, is nearly linear. As the stress reaches

Fig. 1.3 Effect of diameter of cylinder on compressive strength (Ref. 1.9)

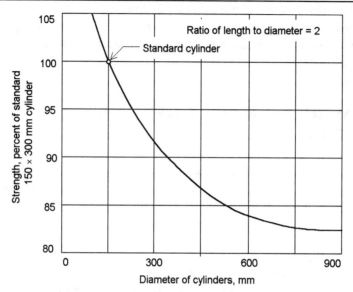

about 70 to 90 percent of the maximum, minute internal cracks parallel to the direction of the applied load are initiated in the mortar throughout the concrete mass (Ref. 1.10). The lateral tensile strain (due to the Poisson's ratio effect) at this loading may exceed the failure strain of concrete in tension and may be the cause of this cracking. As a result of the associated larger lateral extensions, the apparent Poisson's ratio starts to increase sharply (Ref. 1.11). The maximum stress, f_c', is reached at a strain of

Fig. 1.4 Typical stress-strain curves for concrete in compression

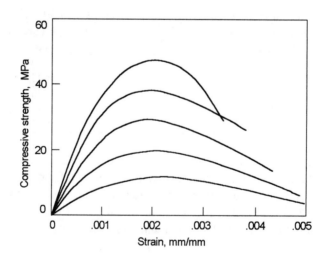

approximately 0.002, and beyond this an increase in strain is accompanied by a decrease in stress. This is believed to be the result of extensive microcracking in the mortar. With most testing machines the concrete cylinder falls abruptly soon after the maximum load is attained. However, the descending branch of the curve can be obtained by using a suitably stiff testing machine or special techniques (Ref. 1.12). For the usual range of concrete strengths, the strain at failure is in the range of 0.0035 to 0.0045. The higher the strength of concrete, the less the failure strain. For design purposes the CSA Standard A23.3-94(10.1.3) limits the usable strain at the extreme concrete compression fibre to a maximum of 0.0035. Tests (Ref. 1.12) have confirmed that the stress-strain relationship for the compression zone of a flexural member is nearly identical to that obtained for a standard cylinder test.

The *modulus of elasticity* of concrete is necessary for all computations of deformations and for the design of sections by the limit states design procedure. The term *Young's modulus of elasticity* has relevance only in the linear elastic part of a stress-strain curve. For a material like concrete, whose stress-strain curve is nonlinear throughout the range, an elastic modulus may be defined in any of the four ways shown in Fig. 1.5. Amongst these the secant modulus at a stress of 40 percent of the compressive strength would represent approximately the average modulus of elasticity in compression throughout the normal range of applied loads. ASTM C469 describes a test for determining the static chord modulus between a lower point A (Fig. 1.5) corresponding to a strain of 50×10^{-6} and an upper point D corresponding to a stress equal to 40 percent of the concrete strength. The lower point is set far enough from the origin to be free of the influence of any initial errors in the setup.

Fig. 1.5 Various definitions of static modulus of elasticity

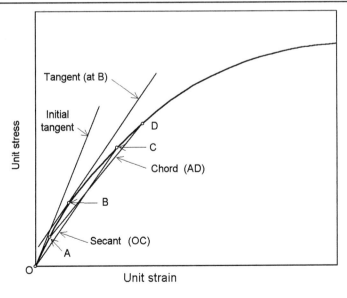

8 REINFORCED CONCRETE DESIGN

CSA Standard A23.3-94(8.6.2) gives an approximate expression for the modulus of elasticity of concrete, in terms of its strength and mass density as:

$$E_c = (3300\sqrt{f_c'} + 6900)(\gamma_c/2300)^{1.5} \qquad (1.1)$$

where E_c is the modulus of elasticity in MPa, γ_c is the mass density of concrete in kg/m³, and f_c' is the specified compressive strength in MPa. This empirical equation was determined from test results on concrete with γ_c in the range of 1500 to 2500 kg/m³, and represents the secant modulus at a stress of approximately $0.4f_c'$. For normal density concrete ($\gamma_c = 2400$ kg/m³), having a compressive strength between 20 and 40 MPa, E_c may be taken as $4500\sqrt{f_c'}$.

Poisson's ratio, which is the ratio of the transverse strain to the axial strain under uniform axial stress, is generally in the range of 0.15 to 0.20 for concrete and is usually taken as 0.20 for design. Observed variations of axial, transverse, and volumetric strains in a concrete prism tested to failure in uniaxial compression are shown in Fig. 1.6 (Ref. 1.11). At a stress equal to about 80 percent of the compressive strength, there is a point of inflection in the volumetric strain curve. As the stress is increased beyond this point, the rate of volume reduction decreases, and soon after the volume begins to increase. It is believed that this inflection point coincides with the initiation of major microcracking in the concrete, leading to large lateral extensions. Poisson's ratio appears to be essentially constant for stresses below the inflection point. At higher stresses, the apparent Poisson's ratio begins to increase sharply.

Fig. 1.6 Strains in concrete prism under uniaxial compression (Ref. 1.11)

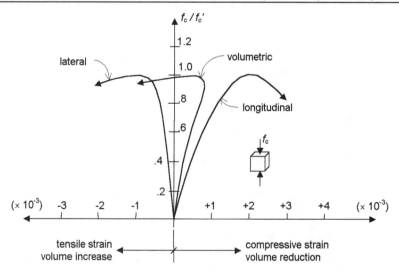

1.6 BEHAVIOUR UNDER TENSION

Concrete is not normally designed to resist direct tension. However, tensile stresses do develop in concrete members as a result of flexure, shrinkage, and temperature changes. Principal tensile stresses may also result from multiaxial states of stress. Cracking in concrete is a result of tensile failures.

Concrete is weak in tension, the direct tensile strength usually being only about 8 to 15 percent of the compressive strength (Ref. 1.8). A direct tension test on a concrete specimen, where a purely axial tensile force must be applied, free of any misalignments and secondary stresses at the grips, is difficult to perform. Therefore, the tensile strength of concrete is generally determined by indirect tension tests, most commonly by *flexure tests* or *cylinder splitting tests*.

In the flexure test most commonly used (CSA Test Method A23.2-8C), a standard, plain concrete, simple beam of square or rectangular cross-section is loaded to failure with third-point loading. Assuming a linear stress distribution across the cross-section, the theoretical maximum tensile stress reached in the extreme fibre, termed the modulus of rupture, f_r, is computed with the flexure formula as

$$f_r = M/Z \qquad (1.2)$$

where M is the maximum bending moment, and Z is the section modulus. The actual stress distribution is not linear, and the modulus of rupture so computed is greater than the direct tensile strength, usually by about 60 to 100 percent (Ref. 1.8). However, the modulus of rupture, f_r, will be the appropriate limiting tensile strength to be used in the computation of the cracking moment, M_{cr}, of a beam using the flexure formula $M = fZ$, as the same assumptions are involved in calculations for both f_r and M_{cr}.

In the splitting test (CSA Test Method A23.2-13C), a standard cylinder of the same type as used for the compression test is loaded in compression on its side along a diametral plane until failure by splitting along the loaded plane (Fig. 1.7). In an elastic homogeneous cylinder, this loading produces a nearly uniform tensile stress across the loaded plane, as shown in Fig. 1.7c. The splitting tensile strength, f_{ct}, may be computed as:

$$f_{ct} = \frac{2P}{\pi dL} \qquad (1.3)$$

where P is the maximum applied load, d is the diameter, and L the length of the cylinder. The splitting test is the easiest to perform and gives more uniform results than the other tension tests. For normal density concrete, the splitting strength is close to the direct tensile strength, being 5 to 12 percent higher, and about two-thirds of the modulus of rupture (Ref. 1.13). The splitting tensile strength, f_{ct}, for normal density concrete varies between 0.5 to 0.58 times $\sqrt{f'_c}$ and an average is taken as:

10 REINFORCED CONCRETE DESIGN

Fig. 1.7 Cylinder splitting test for tensile strength

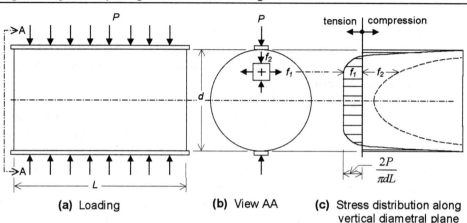

(a) Loading   (b) View AA   (c) Stress distribution along vertical diametral plane

$$f_{ct} = 0.56\sqrt{f_c'} \text{ MPa} \tag{1.4}$$

The failure strain of concrete in uniaxial tension is in the range of 0.0001 to 0.0002 (average 0.00015), and the stress-strain curve in tension can be approximated as a straight line to failure. The modulus of elasticity in tension does not appear to differ from its value in compression. Since the tensile strength of concrete is low and often ignored in design, the tensile stress-strain relationship is of little practical value.

The tensile strength of concrete is found to be approximately proportional to $\sqrt{f_c'}$. On this basis, tensile strength is often taken as a constant times $\sqrt{f_c'}$ for design purposes. Thus CSA Standard A23.3-94 takes the modulus of rupture, f_r, of concrete as:

$$f_r = 0.6\lambda\sqrt{f_c'} \text{ MPa} \tag{1.5}$$

where f_c' is in MPa and λ is a factor to account for the density of concrete. Recommended values for factor λ are:

$\lambda = 1.00$ for normal density concrete;
$\lambda = 0.85$ for structural semi-low density concrete; and
$\lambda = 0.75$ for structural low density concrete.

1.7 SHEAR STRENGTH

The strength of concrete in pure shear is of little practical relevance in design. The case of pure shear rarely exists in actual structures. Furthermore, a state of pure shear is accompanied by principal tensile stresses of equal magnitude on a diagonal plane, and, since the tensile strength of concrete is far less than its shear strength, failure invariably

CHAPTER 1 INTRODUCTION AND MATERIAL PROPERTIES 11

occurs in tension. This makes it difficult to determine experimentally the resistance of concrete to pure shearing stresses. A reliable assessment of the shear strength of concrete can be obtained only from tests under combined stresses. The strength of concrete in pure shear has been reported as approximately 20 percent of the compressive strength (Ref. 1.9). In normal design practice, the shear stress in concrete has to be limited to values well below the shear strength to control the associated principal tensile (diagonal tension) stresses and cracking.

1.8 COMBINED STRESSES

Structural members are usually subjected to various combinations of axial force, bending moments, transverse shear forces, and twisting moments. Concrete is therefore subjected to combinations of normal (tensile or compressive) and shearing stresses. Such a general three-dimensional state of stress acting on an element can be transformed into an equivalent set of three normal stresses (principal stresses) acting in three orthogonal directions. When one of these three principal stresses is zero, the state of stress is termed *biaxial*. The failure strength of materials under combined stresses is normally defined by appropriate failure criteria. As yet, there is no

Fig. 1.8 *Failure stress envelope - biaxial stress*

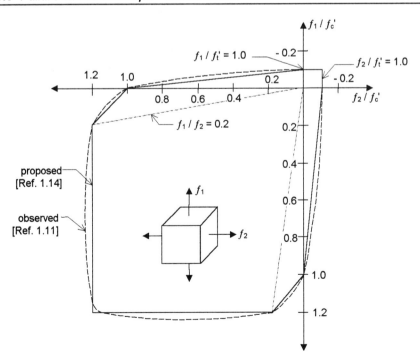

12 REINFORCED CONCRETE DESIGN

Fig. 1.9 Mohr's failure envelope for concrete

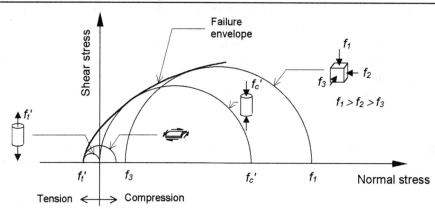

universally accepted criterion for failure of concrete.

Concrete subjected to a biaxial state of stress has been studied extensively, due to its relative simplicity in comparison with the triaxial case, and because of its common occurrence in flexural members, plates and thin shells. Figure 1.8 shows the general shape of the biaxial strength envelopes for concrete, obtained experimentally (Refs. 1.11, 1.14), along with proposed approximations. In general, the strength of concrete in biaxial compression is greater than in uniaxial compression by up to 27 percent. The biaxial tensile strength is nearly equal to its uniaxial tensile strength. However, in the region of combined compression and tension, the compressive strength decreases nearly linearly, with an increase in the accompanying tensile stress. The failure strains under biaxial states of stress depend on the nature of the stress states - tensile or compressive. Failure itself occurs generally by tensile splitting along a plane at right angles to the maximum tensile strain. For biaxial compression, as in the case of uniaxial compression, the volumetric strain begins to increase as the peak stress is approached. This is again due to the progressive increase of microcracking in the mortar. Observed failure modes suggest that tensile strains are of vital importance in the failure criteria and failure mechanism of concrete for both uniaxial and biaxial states of stress (Ref. 1.14).

Another commonly used, but approximate, failure criterion for concrete is the Mohr's failure envelope (Fig. 1.9). Here, the stresses at failure for any given test under uniaxial stress or combined stresses are represented by the corresponding Mohr's circle. Figure 1.9 presents the results of a direct tension test, a standard compression test, and triaxial compression test, of which only the smallest and largest principal stresses are considered. Applying torsion to a thin-walled hollow cylinder can create a state of pure shear. However, this creates direct tensile and compressive stresses of equal magnitude on diagonal planes, and failure is inevitably in tension along the diagonal plane. The result of such a test is also shown in Fig. 1.9. Since the circles correspond to conditions at failure, an envelope drawn to them is termed a *failure envelope*. Points on the failure envelope define limiting states of stress that can exist in the material prior to failure. For any combination of stress for which the Mohr's circle is tangent to the failure envelope, failure will occur. The intercept of the envelope on the vertical axis, gives the apparent strength of concrete in pure shear. The strength criteria presented in Figs. 1.8 and 1.9 are not in complete mutual agreement. The former is obtained by curve fitting of biaxial test data, while the latter is an extension of a standard failure criterion for concrete. For biaxial compression, Mohr's theory gives more conservative results, because the confining influence of the intermediate principal stress is neglected. However, for the cases of biaxial compression-tension and tension-tension, the influence of the intermediate principal stress on the failure strength is negligible.

1.9 TRIAXIAL STRENGTH

Strength and behaviour of concrete under triaxial stress systems has received considerable attention, and a large volume of experimental data is available (Ref. 1.15). Based on this, several strength criteria for concrete subjected to short-term multiaxial stresses have been proposed. A generalised strength criterion for triaxial states of stress, proposed in Ref. 1.16, is shown as an ultimate strength surface in Fig. 1.10. This criterion can be expressed mathematically in terms of the magnitude of the non-dimensionalized (with respect to f_c') octahedral normal stress component and the magnitude and direction of the octahedral shear stress component. The orthogonal axes in Fig. 1.10 represent the three principal stresses. This ultimate strength surface was found to provide a close fit for extensive biaxial and triaxial strength test results. The trace of the surface on a coordinate plane, which gives the strength envelope for the biaxial case, is in very close agreement with the experimentally obtained envelope in

Fig. 1.10 *Schematic representation of the ultimate strength surface (Ref. 1.16)*

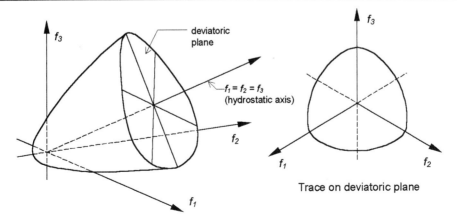

Fig. 1.8. This criterion shows promise as a unified strength criterion common for uniaxial, biaxial, and triaxial stress states.

Tests on concrete under triaxial compression have shown that both the compressive and the shear strength increase with increase in confining pressures.

1.10 STRESS-STRAIN RELATIONSHIPS FOR MULTIAXIAL STRESS STATES

For the use of nonlinear computer-based methods in the analysis of concrete structures subjected to multiaxial stress states, an essential input is a realistic stress-strain relationship. Although a variety of models characterising the stress-strain relations and failure criteria for concrete under multiaxial stress states have been proposed in recent years (Refs. 1.17, 1.18, 1.19), none of them has received general acceptance. A critical evaluation of some of these proposals is given in Ref. 1.19.

1.11 CREEP

Concrete under sustained loads shows an increase in strain with time (Fig. 1.11). The time-dependent part of strain resulting from stress is termed *creep*. In Fig. 1.11, ε_{inst} is the instantaneous elastic strain on the application of stress. If the stress is maintained at this level, the strain will continue to increase with time along the solid curve, although at a progressively decreasing rate. At any given time, the increase in strain above the initial elastic strain is termed the creep strain, ε_{cp}. Measurements indicate that small increases in creep strain take place after as long as 30 years (Ref. 1.20, Fig. 1.12). The 30-year creep may be on the average about 1.36 times the one-year creep. The ultimate creep strain may be as much as 1.5 to 3 times the instantaneous elastic strain, ε_{inst}. If the sustained load is removed, the strain follows the curve shown by the broken lines in Fig. 1.11. There is an instantaneous recovery of strain by an amount equal to the elastic strain due to the load removed at this age. This is followed by a gradual decrease in strain, which is termed creep recovery. The exact mechanism of creep in concrete is as yet not fully understood, but creep is generally attributed in varying degrees to internal movement of adsorbed water, viscous flow or sliding between the gel particles, moisture loss, and the growth in microcracks.

There are several independent and interacting factors related to the material properties and composition, curing and environmental conditions, and loading conditions that influence the magnitude of creep strains. The major factors related to the composition of the mix that influence creep are the type and amount of cement, w/c ratio, aggregate type and content, and admixtures. Creep increases with increases in w/c

Fig. 1.11 *Typical strain - time curve for concrete in uniaxial compression*

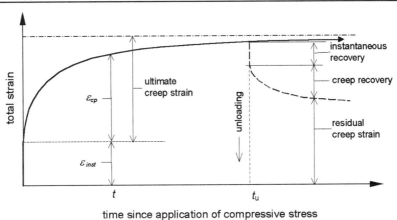

Fig. 1.12 *Variation of creep with time (Ref. 1.20)*

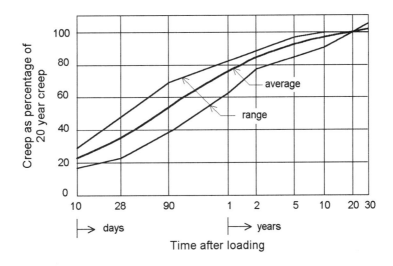

ratio, cement content, and air entrainment. Since creep is associated primarily with the cement paste portion of the concrete, an increase in aggregate content reduces the creep. Physical properties of the aggregates, such as the modulus of elasticity and porosity, also influence the creep.

The ambient relative humidity and temperature are significant factors related to the environment that influence creep. A decrease in relative humidity, as well as an increase in temperature, increases the creep strain. Both these factors increase the loss of moisture. For unsealed specimens exposed to a drying atmosphere, there is also a size effect associated with the moisture gradient in the specimen. The effect of size on creep can be expressed in terms of the ratio of volume to surface area or in terms of an average thickness of the specimen. An increase in this factor results in a decrease in creep.

In usual structural applications, creep is proportional to the applied stress. Furthermore, creep strain in general is inversely proportional to the strength of concrete at the time of application of stress. Thus, creep can be expressed as a linear function of the stress/strength ratio. Creep is also dependent on the age of the concrete at the time of loading. At higher stresses creep increases with applied stress at an increasing rate, and at a stress in the region of 0.8 to $0.9f_c'$ creep leads to failure in the course of time. Cyclic loading produces higher creep strain than a static load of equal intensity acting for the same total time.

Measurement of creep is usually made on specimens loaded in uniaxial compression (ASTM C512). Creep under multiaxial compression is less than that associated with uniaxial stress, and may be as low as 50 percent of the latter for a hydrostatic state of stress (Ref. 1.21).

Except under very high stress (0.75 to $0.9f_c'$), creep does not affect the strength of concrete. However, it often influences the stress distribution in members and increases the strains and deflections. Thus, in columns, the creep of concrete under stress results in a gradual transfer of load from concrete to the reinforcement, so that the stress distribution will be different from that indicated by elastic theory. When the stress in the concrete is the result of an imposed deformation, as with the settlement of a support in a continuous structure, the effect of creep is to reduce the induced stresses. Similarly, creep leads to a loss of prestress in prestressed concrete. In reinforced concrete, however, the major concern for creep is with regard to the increased deformations.

A review of different equations proposed for predicting creep, along with comparisons with experimental data and numerical examples, is presented in Ref. 1.22. In one widely recommended empirical method (Refs. 1.22, 1.23), a creep coefficient, C_t (defined as the ratio of creep strain ε_{cp} to the initial elastic strain ε_{inst}), is computed for "standard conditions" using Eq. 1.6. For departures from standard conditions, correction factors are applied to the ultimate (in time) creep coefficient, C_u, in Eq. 1.6. Standard conditions mean a relative humidity of not more than 40 percent, a loading age of seven days for moist-cured concrete and one to three days for steam-cured concrete, an average thickness of 150 mm (6 in.), a slump of 102 mm (4 in.) or less, and an air content of 7 percent or less. For standard conditions

$$C_t = \frac{t^{0.6}}{10 + t^{0.6}} C_u \tag{1.6}$$

where $C_t = \varepsilon_{cp}/\varepsilon_{inst}$ and t is the time after application of load, in days. The ultimate creep coefficient, C_u, varies in the range of 1.30 to 4.15 and has an average value of 2.35. Since the value of C_t predicted using Eq. 1.6 depends primarily on the value taken for C_u, it is preferable that C_u for local conditions be established based on tests or local experience, and the average value of 2.35 used only in the absence of such data. Recommended correction factors for variations from "standard conditions" are presented in Ref. 1.22. The two major correction factors (CF) are given below:

1) for age of loading later than seven days for moist cured and one to three days for steam cured concrete,

$$(CF)_{AL} = 1.25 \, t_a^{-0.118} \text{ for moist cured}$$
$$(CF)_{AL} = 1.13 \, t_a^{-0.095} \text{ for steam cured} \tag{1.7}$$

where t_a is the age of concrete when load is applied, in days

2) for relative humidity greater than 40 percent,

$$(CF)_H = 1.27 - 0.0067H \tag{1.8}$$

where H is the ambient relative humidity in percent

1.12 SHRINKAGE AND TEMPERATURE EFFECTS

Slight changes in the volume of concrete occur during and after hardening (Ref. 1.2). The decrease in volume, other than that due to externally applied forces and temperature changes, and resulting mostly from moisture loss during drying, is broadly termed *shrinkage*. Unlike creep, shrinkage is not related to an applied stress and is reversible to a greater extent. Shrinkage and creep are not independent phenomena; however, for convenience it is normal practice to treat their effects as separate, independent, and additive. All the factors related to constituent material properties, composition of mix, curing and environmental conditions, member size, and age that affect creep also affect shrinkage. Typical variations of shrinkage with time are shown in Fig. 1.13.

Major concerns for shrinkage in concrete structures are related to its potential for inducing cracking and deformations. When shrinkage is restrained, as it often is in concrete structures, tensile stresses develop which, if excessive, may lead to cracking. Similarly, a differential shrinkage due to a moisture or thermal gradient or due to a differential restraint to shrinkage (for example, the reinforcement being placed nearer to one side of a beam) will result in internal stresses, curvature, and deflections. Shrinkage, like creep, also leads to a loss of prestress in prestressed concrete.

Since the primary cause of shrinkage is moisture loss from the cement paste phase of the concrete, it can be minimised by keeping the unit water content in the mix as low as possible and the total aggregate content as high as possible.

Shrinkage is usually expressed as a linear strain (mm/mm). ASTM C157 provides a test procedure for the measurement of length change of cement mortar and concrete. Shrinkage measurements in the range of 200×10^{-6} to 1200×10^{-6} mm/mm have been reported (Ref. 1.2) for small plain concrete specimens, 127 mm square in cross section, stored for six months at 50 percent relative humidity and 21°C. Shrinkage is less in reinforced concrete due to the restraint offered by the reinforcing and is ordinarily in the range of 200×10^{-6} to 300×10^{-6} mm/mm.

Because of the several interacting variables involved, prediction of shrinkage, as for creep, can only be approximated. An empirical method recommended by ACI Committee 209 (Refs. 1.22, 1.23) uses an approach similar to that for creep discussed in the previous section. The shrinkage strain at time t, $(\varepsilon_{sh})_t$, for "standard conditions" (relative humidity 40 percent or less, average thickness of part under consideration 152 mm or less, and slump 100 mm or less) is given by:

1) for shrinkage after seven days, for moist curing,

$$(\varepsilon_{sh})_t = \frac{t}{35+t}(\varepsilon_{sh})_u \qquad (1.9)$$

2) for shrinkage after age one to three days, for steam curing,

$$(\varepsilon_{sh})_t = \frac{t}{55+t}(\varepsilon_{sh})_u \qquad (1.10)$$

where t is the time in days, from age seven days for moist curing and from age one to three days for steam curing; and $(\varepsilon_{sh})_u$ is the ultimate shrinkage strain.

Ref. 1.22 suggests an average value for $(\varepsilon_{sh})_u$ of 780×10^{-6} mm/mm for both moist and steam cured concrete for use in the absence of more precise data. Correction

Fig. 1.13 Typical variation of shrinkage with time

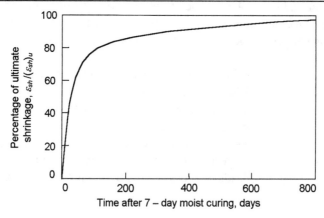

factors to be applied to the ultimate shrinkage, to account for variations in humidity, thickness of member, slump, cement content, percentage of fine aggregate, and air content are given in Ref. 1.22. Except for the first, which is given below, the influence of these correction factors is only marginal and can be neglected under normal conditions.

For relative humidity, H, greater than 40 percent:

$$\text{Shrinkage correction factor} = 1.40 - 0.01 H \text{ for } 40\% < H < 80\% \quad (1.11)$$
$$= 3.00 - 0.03 H \text{ for } 80\% < H < 100\%$$

Concrete expands with a temperature rise and contracts as the temperature drops. An average value of the coefficient of thermal expansion is 10×10^{-6} mm/mm/°C. This is close to the value for steel (which is about 12×10^{-6} mm/mm/°C) so that there is little likelihood of any differential thermal expansion and associated relative movements between the steel and surrounding concrete. Thermal contraction has effects similar to shrinkage.

1.13 STEEL REINFORCEMENT

1.13.1 General

Concrete is strong in resisting compressive stresses, but weak in tension. For this reason it has to be "reinforced" with a material strong in tension if tensile stresses are to be resisted. Usually this is done by embedding steel in the form of bars, wires, or welded wire fabric, at the required locations. Because steel is much stronger than concrete even in compression, it can also be introduced into concrete to carry compressive stresses.

To be effective, the reinforcement and the surrounding concrete must deform under load as an integral unit, without any relative movement or slip between them. To assist the bond between the steel bars and concrete, and mechanically inhibit longitudinal movement of the bar relative to the concrete around it, reinforcing bars are provided with lugs or protrusions (called "deformations") on their surface. Plain bars are normally permitted only as spiral reinforcing in columns and, in sizes smaller than 10 mm diameter, as stirrups and ties.

Requirements for concrete reinforcing steel are given in CSA Standard A23.1-94. The primary requirements are a minimum specified yield strength, f_y, (specified by "Grade"), minimum ultimate tensile strength, f_u, minimum percentage elongation and cold bending capability. The mass density of reinforcing steel is normally taken as 7850 kg/m^3 and the coefficient of thermal expansion as 12×10^{-6} per degree Celsius.

1.13.2 Types, Sizes, and Grades

The two types of reinforcing steel generally in use are hot-rolled round bars and cold-drawn wires. The former is the most widely used type in reinforced concrete. Specifications for deformed and plain bars used for concrete reinforcement are given in CSA Standard G30.18. A summary of the standard sizes and dimensions of deformed bars and their number designations is given in Table 1.2. Reinforcing bars are classified into three grades based on minimum specified yield strength: 300, 350, and 400 (MPa). Grade 400 bars (with $f_y = 400$ MPa) are the most frequently used type of reinforcement.

Table 1.2 Metric Deformed Bar Size and Designation Numbers - CSA G30.18

Bar Designation No.[1]	Nominal Mass kg/m	Nominal Dimensions[2]		
		Diameter mm	Cross-sectional Area mm^2	Perimeter mm
10	0.785	11.3	100	35.5
15	1.570	16.0	200	50.1
20	2.355	19.5	300	61.3
25	3.925	25.2	500	79.2
30	5.495	29.9	700	93.9
35	7.850	35.7	1000	112.2
45	11.775	43.7	1500	137.3
55	19.625	56.4	2500	177.2

1. Bar designation numbers approximate the nominal diameter of the bar in millimetres.
2. The nominal dimensions of a deformed bar are equivalent to those of a plain round bar having the same mass per metre as the deformed bar.

There are situations where it is necessary to ensure that the designed strength is not exceeded by much, and it may be undesirable to have a yield strength much higher than the specified value. For instance, in seismic design, a ductile flexural failure with considerable yielding is desired. For use in such cases an upper limit on the yield strength and a minimum ratio of actual tensile strength to actual yield strength may be specified. Steel conforming to CSA G30.18 and ASTM A706 generally meets these superior ductility requirements, as well as chemical composition restrictions for improved weldability.

All deformed bars are identified by a distinguishing set of marks rolled on the bar surface. The marking system, shown in Fig. 1.14, is given in the following order: a letter or symbol designating the producer mill, the bar designation number, the type of steel, and the number 400 or a continuous line through five or more spaces for 400 grade steel only.

Cold-drawn wires used as reinforcement may be smooth (CSA G30.3) or deformed (CSA G30.14). Smooth wire is denoted by the letter W followed by a number indicating the cross-sectional area in mm^2. Similarly, deformed wire is designated by the letter D followed by the number representing the area as for the smooth wire. In both cases for metric size wires a letter M is prefixed to the above designations (example: MW25.8, MD22.6). Cold-drawn wires are used mostly (in slabs and similar situations) in the form of welded wire fabric, which consist of a series of wires arranged in orthogonal patterns (forming square or rectangular openings) and welded together at all intersections. These are denoted by the letters WWF followed by spacings of longitudinal and transverse wires in mm and the designations of the longitudinal and transverse wires, in that order (example: WWF 152 × 152 - MW25.8 × MW25.8). Welded wire fabric must conform to CSA G30.5-M1983 (R1991) if made of smooth wires and to CSA G30.15 if made of deformed wires. A summary listing of all the standard reinforcing steels, giving the type of steel, specification number, bar sizes, grades and major tensile strength properties is presented in Table 1.3.

1.13.3 Stress-Strain Curves

Typical stress-strain curves for reinforcing steel are shown in Fig. 1.15. For all steels there is an initial linear elastic portion with constant slope which gives the modulus of elasticity, E_s. For hot rolled bars in the lower grades (300 and 350 MPa) the elastic part is followed first by a yield plateau, where the strain increases at almost constant stress (yielding), and then a strain hardening range, in which the stress once again increases with further increase in strain (although at a decreasing rate) until the peak stress (tensile strength, f_u) is reached. Finally, there is a descending portion of the curve with the nominal stress decreasing until fracture occurs. In general, the higher the strength, the shorter the length of the yield plateau. High-strength high-carbon steels and cold-drawn wires may not exhibit a distinct yield plateau, and the elastic phase may be

Fig. 1.14 Marking system for reinforcing bars

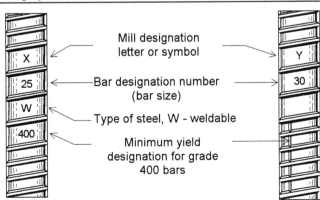

22 REINFORCED CONCRETE DESIGN

followed by a strain hardening phase.

The structurally significant properties of reinforcing steel are its yield strength, f_y, and modulus of elasticity, E_s. The yield strength for reinforcing steel is taken as the stress corresponding to a specified strain, as indicated in Fig. 1.16. The modulus of elasticity varies very little and is generally taken as 200 GPa for all reinforcing steels.

For design calculations, CSA A23.3-94(8.5) does not permit the use of a value of f_y in excess of 500 MPa except for prestressing steel. Furthermore, for compression reinforcement having a specified f_y in excess of 400 MPa, the value of f_y assumed in design calculations shall not exceed the stress corresponding to a strain of 0.35 percent. In lateral load resisting systems, frames and walls, designed with modification factors R greater than 2.0 (Chapter 18), reinforcements must be weldable grade conforming to CSA Standard G30.18. In lateral load resisting systems designed with force modification factor R of 2.0 or less, reinforcement must conform to CSA Standard G30.18, but need not be weldable grade (CSA A23.3-94(21.2.4.1)).

1.14 CODES AND SPECIFICATIONS

This book is written primarily with reference to the National Building Code of Canada 1995 (NBC 1995, Ref. 1.24). The NBC requires that the design of plain, reinforced and prestressed concrete conform to CSA Standard A23.3-94, *Design of Concrete Structures* (Ref. 1.25). CSA has several other standards applicable to concrete construction and dealing with areas such as materials, construction methods and bridge

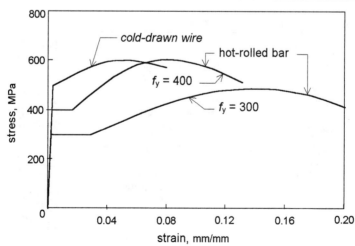

Fig. 1.15 *Typical stress-strain curves for reinforcing steels*

Table 1.3   Standard Reinforcing Steel Data

Type of Reinforcement	Type of Steel and CSA Specification	Size No. Inclusive	Grade	Minimum Yield Strength f_y MPa	Minimum Tensile Strength f_u MPa	Minimum Percent elongation for 200 mm Gauge Length[2]
Bars						
Deformed and plain	Billet G30.18	10-55	300	300	450	10-12
		10-55	350	350	550	7-9
		10-55	400	400	600	7-9
Deformed weldable	low alloy G30.18	10-55	400	Min. 400 Max. 540	Min. 550 but ≥ 1.25 (test f_y)	12-13
Cold-drawn wires[3]						
Smooth	G30.3	MW12.9-MW87.3		485	550	
Deformed	G30.14	MD12.9-MD87.3		517	585	
Welded wire fabric[3]						
Smooth	G30.5	(steel conforming to G30.3)		$\{450^*$ $385^\ddagger$	$\{515^*$ $485^\ddagger$	
Deformed	G30.15	(steel conforming to G30.14)		485	550	

(1) Plain bars in sizes form 50 mm² (8 mm diameter) to 500 mm² (25 mm diameter).
(2) Specified minimum elongation depends on the bar size.
(3) At present time the metric wire sizes and dimensions have been "soft" converted from imperial units.
 *   For size MW7.7 and larger.
 ‡   For size smaller than MW7.7.

Fig. 1.16 Initial portion of stress-strain curves for reinforcing steels

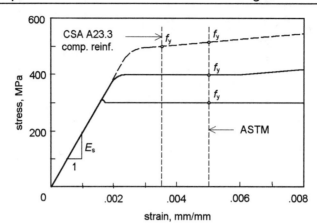

design. The important standards, which control the design of reinforced concrete structures in Canada, are given below:

1. National Building Code of Canada 1995
2. CSA Standard A5   *Portland Cements*
3. CSA Standard A23.1-94   *Concrete Materials and Methods of Concrete Construction*
4. CSA Standard A23.2-94   *Methods of Test for Concrete*
5. CSA Standard A23.3-94   *Design of Concrete Structures*
6. CSA Standard A23.4-94   *Precast Concrete - Materials and Construction*
7. CSA Standard G30.18-M92   *Billet-Steel Bars for Concrete Reinforcement*
8. Canadian Highway Bridge Design Code (in press)

The design of reinforced concrete buildings in the United States is mostly governed by the *Building Code Requirements for Structural Concrete* (ACI 318-95) of the American Concrete Institute (Ref. 1.26). The requirements in the ACI Code are for the most part very similar to those of CSA Standard A23.3. To avoid lengthy repetitions, the simple reference to Code herein will mean CSA A23.3-94.

1.15 INTERNATIONAL SYSTEM OF UNITS

The International System of Units (SI Units) is used throughout this book. With this system, all loads must be converted from their mass (kilogram) into force (Newton kg·m/s^2) by multiplying with the acceleration due to gravity ($g = 9.80665 \approx 9.81$ m/s^2).

REFERENCES

1.1 Mindess, S. and Young, J.F., *Concrete*, Prentice-Hall, Englewoods Cliffs, N.J., 1981, 671 pp.

1.2 Neville, A.M., *Properties of Concrete*, 4th ed., John Wiley and Sons, Inc., New York, 1996, 844 pp.

1.3 *Design and Control of Concrete Mixtures*, Sixth Canadian Edition, Canadian Portland Cement Association, Ottawa, Ontario, 1995.

1.4 ACI Standard 211.1-91, Standard Practice for Selecting Proportions for Normal, Heavyweight, and Mass Concrete, *ACI Manual of Concrete Practice, Part 1: Materials and General Properties of Concrete*, Detroit, Michigan, 1994, 38 pp.

1.5 ACI Standard 211.2-92, Standard Practice for Selecting Proportions for Structural Lightweight Concrete, *ACI Manual of Concrete Practice, Part 1: Materials and General Properties of Concrete*, Detroit, Michigan, 1994, 14 pp.

1.6 ACI Standard 211.3-75, Revised 1987, Reapproved 1992, Standard Practice for Selecting Proportions for No-Slump Concrete, *ACI Manual of Concrete Practice, Part 1: Materials and General Properties of Concrete*, Detroit, Michigan, 1994, 11 pp.

1.7 ACI 212.3R-91, Chemical Admixtures for Concrete, *ACI Manual of Concrete Practice, Part 1: Materials and General Properties of Concrete*, Detroit, Michigan, 1994, 31 pp.

1.8 Raphael, J.M. *Tensile Strength of Concrete*, ACI Journal, Proceedings, Vol. 81, No. 2, 1984, pp. 158-165.

1.9 Kesler, C.E., *Hardened Concrete-Strength*, Tests and Properties of Concrete, ASTM Special Technical Publication No. 169-A. Am. Soc. for Testing and Materials, 1966, pp. 144-159.

1.10 Hsu, T.T.C., Slate, F.O., Sturman, G.M. and Winter G., *Microcracking of Plain Concrete and the Shape of the Stress-Strain Curve*, J. ACI, Vol. 60, Feb. 1963, pp. 209-223.

1.11 Kupfer, H., Hilsdorf, H.K. and Rüsch, H., *Behaviour of Concrete Under Biaxial Stresses*, J. ACI, Vol. 66, Aug. 1969, pp. 656-666.

1.12 Hognestad, E., Hanson, N.W., and McHenry D., *Concrete Stress Distribution in Ultimate Strength Design*, J. ACI, Vol. 52, Dec. 1955, pp. 455-479.

1.13 Wright, P.J.F., *Comments on an Indirect Tensile Test on Concrete Cylinders*, Magazine of Concrete Research, Vol. 7, No. 20, 1955, p. 87-96.

1.14 Tasuji, M.E., Slate, F.O., and Nilson, A.H., *Stress-Strain Response and Fracture of Concrete in Biaxial Loading*, J. ACI, Vol. 75, Jul. 1978, pp. 306-312.

1.15   Wang, C-Z., Guo, Z-H., Zhang, X-Q., *Experimental Investigation of Biaxial and Triaxial Compressive Concrete*, ACI Materials Journal, Vol 84, No. 2, 1987, pp. 92-96.

1.16   Kotsovos, M.D., *A Mathematical Description of the Strength Properties of Concrete under Generalised Stress*, Magazine of Concrete Research, Vol. 31, No. 108, Sept. 1979, pp. 151-158.

1.17   Kotsovos, M.D., and Newman, J.B., *A Mathematical Description of the Deformational Behaviour of Concrete under Complex Loading*, Magazine of Concrete Research, Vol. 31, No. 107, June 1979, pp. 77-90.

1.18   Gerstle, K.H., *Simple Formulation of Biaxial Concrete Behaviour*, J. ACI, Vol. 78, Jan.-Feb. 1981, pp. 62-68.

1.19   Chen, W.F., and Ting, E.C., *Constitutive Models for Concrete Structures*, Proc. ASCE, Vol. 106, No. EM1, Feb. 1980, pp. 1-19.

1.20   Troxell, G.E., Raphael, J.M., and Davis, R.E., *Long Time Creep and Shrinkage Tests of Plain and Reinforced Concrete*, Proc. ASTM, Vol. 58, 1958, pp. 1101-1120.

1.21   McDonald, J.E., *Creep of Concrete under Various Temperature, Moisture, and Loading Conditions*, Douglas McHenry International Symposium on Concrete and Concrete Structures, ACI Publication SP-55, American Concrete Institute, Detroit, 1978, pp. 31-53.

1.22   Branson, D.E., *Deformation of Concrete Structures*, McGraw-Hill, Inc., New York, 1977, 546 pp.

1.23   ACI 209R-92, Prediction of Creep, Shrinkage and Temperature Effects in Concrete Structures, *ACI Manual of Concrete Practice, Part 1: Materials and General Properties of Concrete*, Detroit, Michigan, 1994, 47 pp.

1.24   *National Building Code of Canada*, National Research Council of Canada, Ottawa, 1995.

1.25   CSA Standard A23.3-94 C Design of Concrete Structures, Canadian Standards Association, Rexdale, Ontario, 1995.

1.26   ACI Standard 318-95, *Building Code Requirements for Structural Concrete* (ACI 318-95), and *Commentary* (ACI 318R-95), American Concrete Institute, Detroit, Michigan, 1995, 369 pp.

CHAPTER 2 Reinforced Concrete Buildings

2.1 INTRODUCTION

Reinforced concrete is extensively used for the construction of various types of structures such as buildings, bridges, retaining walls, water tanks, grandstands, etc. The basic principles involved in the *design of sections* for all these structures are essentially the same. In general, the most common type of reinforced concrete structure is the building, and, therefore, it will be considered in some detail in this chapter. There are many types of reinforced concrete systems that are used in building construction, such as precast, lift slab, post-tensioned, composite, and modular systems. The most common construction is cast-in-place. The two main categories of building structures are office buildings and residential type structures, such as apartments and hotels.

Office buildings (Fig. 2.1) are characterised by service cores and flexible floor

Fig. 2.1 Floor plan of a typical office building

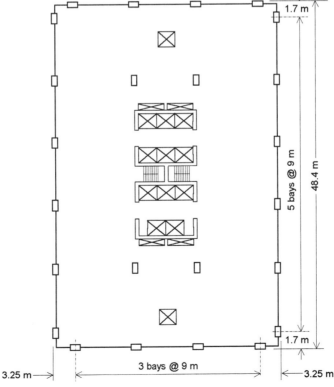

spaces. Service cores contain elevators, stairwells, mechanical and electrical supply ducts, equipment rooms, janitors' closets, and washrooms. Floor slabs may span the complete distance from the exterior walls to the core or they may be supported on intermediate columns. The distance to the core is generally no greater than 12 m, because that represents an acceptable maximum distance from a natural light source. For clear spans, structural economy also tends to limit this distance.

More expensive structural schemes with column-free spaces between the exterior walls and the core are sometimes desired by office tenants, who wish to provide flexibility in floor layouts by using the open office concept. Other tenants, who require individual offices, can generally accept less expensive structural layouts with intermediate columns on 7.5 to 9 m grids, which can be conveniently concealed by partitions.

Apartment buildings or hotels (Fig.2.2) can accommodate most architectural layouts with smaller structural spans of 6 to 7.5 m. Structural shear walls are commonly used as partitions because of their acoustic properties. Floor to floor heights are less for apartment buildings, because there is generally no requirement to provide dropped ceilings to conceal mechanical and electrical services, and the soffits of flat plate floor systems can be used directly as ceilings for the floor level below.

2.2 LOADS

Loads, which must be considered in the design of structures, are (Ref. 2.1):

    D – *dead load*;
    E – *live load* due to earthquake;
    L – *live load* due to intended use and occupancy;

Fig. 2.2 Floor plan of a typical hotel building

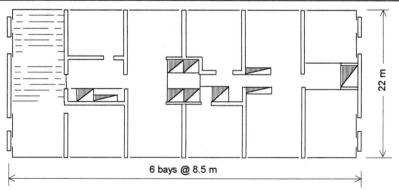

W – *live load* due to wind;

T – loads due to contraction or expansion caused by temperature changes, shrinkage, moisture changes, creep, movement due to differential settlement or combinations thereof.

Where a structure or member is likely to be subjected to loads other than those listed above, such loads are also be taken into account in the design. Where it is known from experience, or it can be shown theoretically, that the neglect of some or all of the effects due to loads, T, does not adversely affect the strength and serviceability of the structure, they need not be considered in the design calculations. For design purposes it is often convenient to group loads into gravity loads and lateral (horizontal) loads.

Dead loads consist of the weight of the structure itself and all other permanent installations, such as partitions, ceiling and floor finishes, piping and lighting fixtures, permanent equipment, and the forces due to prestressing. The magnitude and distribution of these loads are fixed in relation to time and can be estimated with reasonable accuracy from known or assumed dimensions. When dimensions are assumed, their accuracy must be verified, and they must be revised if necessary as the design progresses. Average mass densities of the more commonly used engineering materials are given in Table 2.1. For ceilings, fixtures, irregularly spaced partitions, etc., it may be convenient to reduce their dead load to a uniformly distributed load (kN/m^2) over the entire floor area. For ordinary reinforced concrete buildings, the average gross dead load considering the whole building may be in the range of 5 to 7 kN/m^2 of floor area. Such approximate figures may be helpful for the preliminary

Table 2.1   *Average Mass-Densities of Building Materials*

Material	kg/m^3
Brick masonry	1600-2400
Concrete, normal density	2400
Concrete, light weight	1440-1920
Granite	2250-2650
Steel	7850
Aluminium	2640
Water	1000
Snow, fresh	100
Snow, old	200-400
Snow, wet	640-800
Sand	1600-1920
Glass	2560
Asphalt	1280-1600
Mortar	1600
Timber, redwood	415
douglas fir	513
pine	560-640
oak	865

choices of foundation type and structural layout. In computing the load effects on the structure, dead loads are considered to act at all times, except that during the construction period these loads may be somewhat different from their values expected on the finished structure.

Live loads are the loads on the structure resulting from its use and occupancy (including vertical loads due to cranes); snow, ice and rain; earth and hydrostatic pressure; and horizontal components of static or inertia forces. They are more or less of a temporary duration and can vary considerably in magnitude and distribution. It is possible to have very heavy loads concentrated on small areas, as with a bookshelf or a safe box. However, the average load over larger areas may be much smaller. Furthermore, the worst load effect on a structure or member may not in all cases be when the entire structure is fully loaded. Therefore, the conditions of live load being absent and of live load acting only on part of the structure (pattern loading) must also be considered for design.

Minimum specified live loads on floors and roofs in buildings for various uses and occupancies are given in sentence 4.1.6 of the National Building Code of Canada (NBC) 1995 (Ref. 2.1). This consists of a uniformly distributed load (kN/m^2) applied over the entire area or on any portion of the area (Table 2.2), or a concentrated load applied over an area of 750 mm by 750 mm (Table 2.3), whichever produces the most critical effects. When a member supports a large tributary area of floor and/or roof (e.g., the columns in the lower storeys of a multi-storey building), it is highly unlikely that this entire area would be loaded uniformly with the full specified live load. Therefore, for large tributary areas a reduction in the specified live load intensity is allowed.

For roofs, the live loads due to the accumulation of snow, ice and rain may be more critical than the specified load based on use and occupancy. Specified snow load is taken as the product of the ground snow load specified in Appendix C, "Climatic Information for Building Design in Canada" of Ref. 2.1 for various locations in Canada and the snow load factors detailed in sentence 4.1.7 of Ref. 2.1. Live loads on structures also include, where applicable, the lateral pressure due to liquid and lateral earth pressure.

Live loads, even if moving or dynamic in type, are generally treated as static loads. However, when their dynamic, vibratory, and impact effects are significant, as with heavy moving equipment such as cranes, elevators, etc., these effects are accounted for by increasing the live load by an impact factor. Specified impact factors for building structures are given in sentence 4.1.10 of Ref. 2.1. Specified live loads on bridge structures due to traffic and their impact effects are given in Ref. 2.3.

Wind loads Wind pressure on the surface of a structure results from the differential pressures caused by the obstruction to the free flow of the wind. Such loads, therefore, depend on the wind velocity; the orientation, area and shape of the surface;

Table 2.2 Minimum Specified Distributed Live Loads for Floors and Roofs (Abstracted from National Building Code of Canada-1995(Sen. 4.1.6)*

Use of Area of Floor or Roof	Minimum specified load, kN/m^2
Assembly areas with fixed seats that have backs over at least 80 percent of the assembly area and used for churches, courtrooms, lecture halls, theatres and classrooms with or without fixed seats	2.4
Assembly areas other than those listed above	4.8
Balconies, exterior and interior	4.8
Corridors, lobbies and aisles over 1200 mm wide (except upper floor corridors of residential areas of apartments, hotels and motels)	4.8
Equipment areas and service rooms	3.6
Exits and fire escapes	4.8
Factories	6.0
Garages for	
Passenger cars	2.4
Unloaded buses and light trucks	6.0
Loaded buses and trucks and all other trucking spaces	12.0
Kitchens (other than residential)	4.8
Libraries	
Stack rooms	7.2
Reading and study rooms	2.9
Office areas in office buildings and other buildings (not including record storage and computer rooms) located in	
Basement and first floor	4.8
Floors above first floor	2.4
Operating rooms and laboratories	3.6
Residential areas	
Sleeping and living quarters in apartments, hotels, motels, boarding schools and colleges	1.9
Retail and wholesale areas	4.8
Roofs	1.0
Sidewalks and driveways over areaways and basements	12.0
Storage areas	4.8
Toilet areas	2.4
Warehouses	4.8

*Abstracted from Ref. 2.1

and the density of the air, which itself depends on the atmospheric pressure and air temperature. Within the height range of usual structures, the wind speed increases with with height, and is influenced by the roughness of the terrain. Wind speed also fluctuates constantly, and the peaks in the wind speed represent gusts. Because of their limited spatial extent, wind gusts are unlikely to act simultaneously over the full extent of the structure.

Table 2.3 Minimum Specified Concentrated Live Loads for Floors and Roofs (Ref. 2.1)

Area of Floor or Roof	Minimum specified load, kN
Roof surfaces	1.3
Floors of Classrooms	4.5
Floors of offices, manufacturing buildings, hospital wards and stages	9.0
Floors and areas used by passenger cars	11.0
Floors and areas used by vehicles not exceeding 3600 kg gross weight	18.0
Floors and areas used by vehicles exceeding 3600 kg but not exceeding 9000 kg gross weight	36.0
Floors and areas used by vehicles exceeding 9000 kg gross weight	54.0
Driveways and sidewalks over areaways and basements	54.0

Although the determination of wind forces on a structure is a dynamic problem, for buildings of moderate proportions which are not susceptible to wind induced vibrations, it has been the usual practice to treat such forces as equivalent static loads expressed in kN/m² of exposed surface area. Wind loads are derived from pressures resulting from a mean wind velocity and amplified for the effect of gusts. A reference wind speed, $\overline{V}$, based on mean wind speeds taken at a standard elevation of 10 m above ground in open exposure is used to determine a reference velocity pressure (dynamic pressure of moving air), q, as:

$$q = \frac{1}{2}\rho \overline{V}^2 = CV^2 \tag{2.1}$$

where ρ is the mass density of air. For Canadian conditions, with q in kN/m² and $\overline{V}$ in km/h, a value of 50×10^{-6} is recommended for C. Values of q, with probabilities of exceedance in any one year of 1/10, 1/30 and 1/100, for various selected locations in Canada are given in Appendix C of Ref. 2.1. Values specified for q seldom exceed 1.5 kN/m². Reference 2.1, sentence 4.1.8, recommends that the specified external static pressure (or suction), p, due to wind acting normal to the surface be calculated as:

$$p = q\, C_e\, C_g\, C_p \tag{2.2}$$

In Eq. 2.2, C_e is an exposure factor, which allows for the variation of the wind speed with height and the effects of the variations in the surrounding terrain (shielding) and ranges between 0.9 and 2.0; C_g is a gust effect factor, with recommended values of 1.0 or 1.0 for internal pressures, 2.0 for the building as a whole and main structural members, and 2.5 for small elements including cladding; and C_p is the external pressure coefficient and allows for the effects of shape, wind direction, and the profile of wind velocity. These are generally determined from wind tunnel experiments. Recommended values for C_p are given in Commentary B of Ref. 2.2. Equations similar to 2.2 above are also recommended for specified internal pressure (or suction).

The net specified wind pressure on the surface of a building is the algebraic difference between the external and internal pressures. Similarly, the algebraic difference between the external wind pressures on the windward and leeward sides of the building is the specified net wind pressure for the design of the whole structure.

Exposed elements such as cladding, walls, and roof must be designed to carry the net load due to the wind distributed over the exposed area. However, for the design of the structural frame, it is normal practice to assume the wind load as concentrated at the frame joints. The concentrated wind load at a joint is taken as the total wind force acting on the exterior surface tributary to that joint. The area tributary to a joint is the portion of the exterior surface bounded by the horizontal centrelines between the joint and the floors above and below, and the vertical centrelines between the joint and the frames on either side (Fig. 2.3b).

For design of cladding, as well as for design of structural members for deflection and vibration, the reference velocity pressure, q, is based on a probability of being exceeded in any one year of 1/10. For design of structural members for strength, this probability is reduced to 1/30, except for post-disaster buildings for which the probability is kept as low as 1/100. Post-disaster buildings, such as hospitals, fire stations, and power stations, are those which are expected to be functional in the event of a disaster.

An example of wind load calculations by the static approach is given in Example 2.1. Buildings which are likely to be susceptible to wind load vibrations, such as tall and slender buildings, must be designed for wind load effects either by experimental methods, such as wind tunnel tests, or by a dynamic approach to the action of wind gusts. Details of such a dynamic approach are given in Commentary B of Ref. 2.2.

Earthquake loads The sudden horizontal translation of the foundation of a structure due to earthquake ground motion leads to horizontal inertia forces distributed throughout the mass of the structure. Seismic loads on structures depend on the peak horizontal ground acceleration, the peak horizontal velocity, and the characteristics of the structure and foundation. For simplicity in analysis, the loading due to earthquake motion is generally specified in Ref. 2.1, sentence 4.1.9, as equivalent static loads acting horizontally. Specified earthquake loads are based on the requirements that the structure be able to resist moderate earthquakes without significant damage and to resist major earthquakes without collapse.

Based on a statistical analysis of earthquake data, seismic zoning maps for Canada have been prepared giving contours of peak horizontal ground accelerations (in units of g, the acceleration due to gravity) and contours of peak horizontal ground velocities in m/s, both having a probability of exceedance of 10 percent in 50 years (Commentary J, Ref. 2.2). The peak ground motion parameters that have a probability of exceedance of 10 percent in 50 years (that is, 0.0021 per annum) are the basis for design specifications in the NBC 1995. Based on the ranges of the ratio, a, of peak horizontal ground acceleration to the acceleration due to gravity (taken nominally as 10 m/s^2) and of the ratio, v, of peak horizontal ground velocity to a velocity of 1 m/s, seven seismic zones are defined, and for each zone a, v, Z_a and Z_v are specified:

where $a =$ zonal acceleration ratio, from 0.00 to 0.40;
$v =$ zonal velocity ratio, from 0.00 to 0.40;
$Z_a =$ acceleration-related seismic zone, from 0 to 6; and
$Z_v =$ velocity-related seismic zone, from 0 to 6.

Values for a, v, Z_a and Z_v for selected locations across Canada are tabulated in Commentary J of Ref. 2.2. The specified minimum lateral seismic force at the base of the structure, V, is given as:

$$V = (V_e / R)U \qquad (2.3)$$

where, R is a force modification factor that reflects the capability of a structure to dissipate energy through inelastic behaviour; U is a factor representing a level of protection based on experience and is equal to the value 0.6; and V_e is the equivalent lateral force at the base of the structure representing elastic response and is given by

$$V_e = vSIFW \qquad (2.3a)$$

where, v is the zonal velocity ratio, equal to the specified zonal horizontal ground velocity expressed as a ratio of 1 m/s; S is a seismic response factor for the structure which reflects the dependence of seismic force on the fundamental period of the structure as well as the contributions of the higher modes for tall buildings; I is an importance factor of the structure; F is a foundation factor; and W is the weight of the structure. A portion of the total V (up to 25 percent) is assumed to be concentrated at the top of the structure, and the remainder distributed along the height of the building (including the top level) in proportion to the product of the weight at each level and its height above the base.

As an alternative to this, the distribution of the total lateral seismic force, V, over the height of a building may also be determined by a dynamic analysis, and guidelines for such a procedure are presented in Commentary J of Ref. 2.2.

Other loads Temperature variations, as well as shrinkage and creep of materials, cause dimensional changes in members which, if restrained by other parts of the structure, give rise to internal strains and stresses. Similarly, differential settlement of the foundation of continuous structures also produces internal forces. Structures must be designed to withstand such loads. These forces may, however, be minimised by making provisions in the structure (expansion joints, construction joints, discontinuity, etc.) to accommodate differential movements.

2.3 STRUCTURAL SYSTEMS

2.3.1 Load Transfer Schemes

Reinforced concrete buildings are three-dimensional structures; however, they are generally constructed as an assemblage of more or less two-dimensional or planar subsystems lying primarily in the horizontal and vertical planes (e.g., floors, roofs, walls, plane frames, etc., Fig. 2.3). As an integrated system, the structure must carry and transfer to the foundation and the ground below all gravity loads on it, as well as all horizontal loads and associated overturning moments resulting from wind, earthquake, or asymmetry. The structural system can, for the sake of convenience, be separated into gravity load resisting schemes and lateral load resisting schemes, although these two are mutually interacting and complementary.

Gravity load resisting schemes in building structures consist of a horizontal subsystem (floors and roofs) and a vertical subsystem (columns, walls, shafts, transfer girders, hangers, etc.).

2.3.2 Floor Systems

The floor systems (horizontal subsystems) in buildings pick up the gravity loads acting on them and transfer them to the vertical elements (columns, walls, etc.). In this process, the floor system is primarily subjected to flexure and transverse shear and the vertical elements are subjected to axial force (usually compression), bending and shear (Fig. 2.3a). The floor also functions as a horizontal diaphragm, connecting together and stiffening all the vertical elements. When these connections are rigid, "frame action" is also developed which greatly improves the efficiency of the system. In addition, the floor system, acting as diaphragms, picks up and distributes the horizontal loads generated in its tributary area (from wind or earthquake loadings) to the vertical elements of the main lateral load-resisting system (Fig. 2.3b). The diaphragms further help to maintain the overall cross-sectional geometry of the structure.

In cast-in-place reinforced concrete construction, the floor system usually consists of one of the following: (1) slabs supported on bearing walls, (2) flat plates, (3) flat slabs, (4) joist floor, and (5) beam and slab systems.

Wall supported slabs A bearing wall and slab system is used mainly in low-rise residential type buildings. Slabs are generally 100 mm to 200 mm thick with spans from 3 in to 7.5 m. When the slab is supported only on opposite sides (Fig. 2.4a), the slab bends in one direction only, and the slab is called a one-way slab. The entire load

Fig. 2.3 Load transfer systems

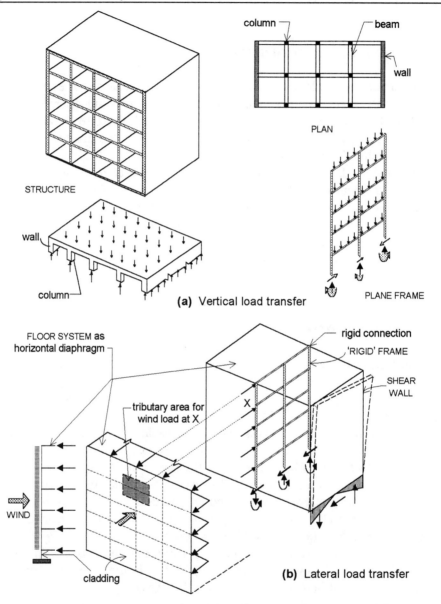

When the slab dimensions approach a square (Fig. 2.4c), the bending action along both spans becomes relatively significant. In this case, load is transmitted in both directions, and the slab is termed a two-way slab. In two-way slabs where corners are held down, there will be a zone of negative bending in the corner areas. In Fig. 2.4a, b, and c, the behaviour of the slab will be essentially unaltered if stiff beams (Fig. 2.4d) which deflect very little replace the walls. When the slab is fixed to or continuous over a

Fig. 2.4 One- and two-way slab action

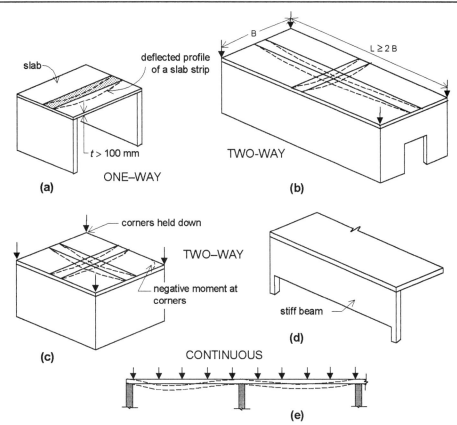

support (Fig. 2.4e), negative moments are developed in the slab over such supports.

Flat plate A flat plate floor is a reinforced concrete slab of uniform thickness supported directly on columns (Fig. 2.5). Edge (spandrel) beams are commonly added to stiffen exterior side frames that have heavy cladding loads. Flat plate floors are from 125 to 250 mm thick and are used with spans of 4.5 to 6 m. Flat plate construction is suitable for such buildings as apartments and hotels, where floor loads are small, long spans are not required, and where plate soffits can be used as ceilings. As the spans and/or loads increase, the shear and flexural stresses in the slab at the columns become very high. For moderate spans and loads, extra reinforcement may be used in this region, in the form of embedded steel beams or closely spaced bars. However, for longer spans and heavier loads, the effective depth of the slab at the support regions has to be increased, as is done in flat slab construction.

Flat slabs These are plates that are stiffened by drop panels and/or column capitals (Fig. 2.6). Longer spans and heavier floor loads are possible with flat slab

Fig. 2.5 *Flat plate*

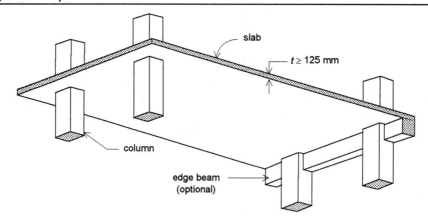

Fig. 2.6 *Flat slab*

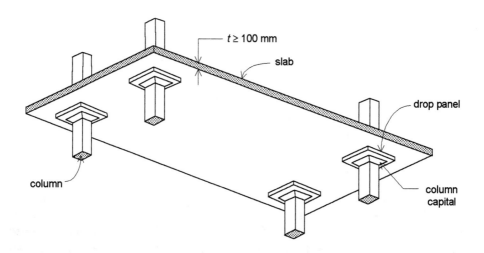

systems because of increased shear and negative moment capacity. Flat slabs are suitable for office construction when drop ceilings are provided and are used for spans up to about 10 m.

Concrete joist floors Also called ribbed slab, this consists of a thin slab (usually 50 to 100 mm thick) cast integrally with relatively narrow, closely spaced stems (ribs) arranged in a one or two-way pattern (Figs. 2.7 and 2.8). Ribs will have a thickness of 100 mm or more and a depth usually less than 3.5 times the minimum width. Two-way patterns are referred to as waffle slabs. Waffle slabs are built in 0.9 to 1.5 m modules, span from 9.0 to 12 m, and are about 300 to 600 mm deep overall. One-way joists use pans 500 to 800 mm wide to form the ribs, and have span-depth ratios similar to those of waffle slabs.

Fig. 2.7 One-way joist floor

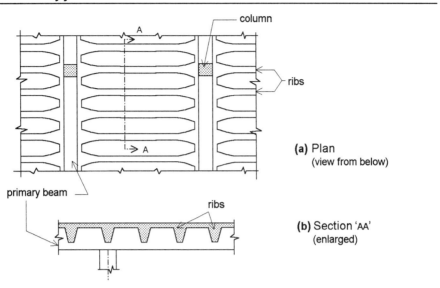

(a) Plan (view from below)

(b) Section 'AA' (enlarged)

Fig. 2.8 Two-way joist floor (waffle slab)

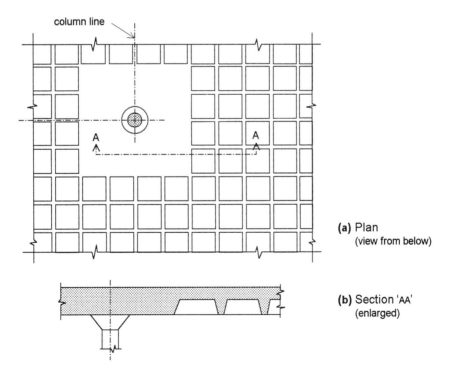

(a) Plan (view from below)

(b) Section 'AA' (enlarged)

Fig. 2.9 Beam and slab floor systems

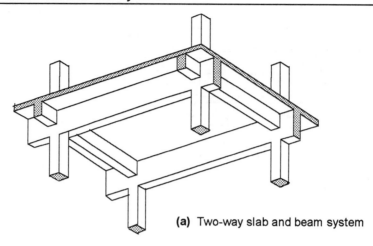

(a) Two-way slab and beam system

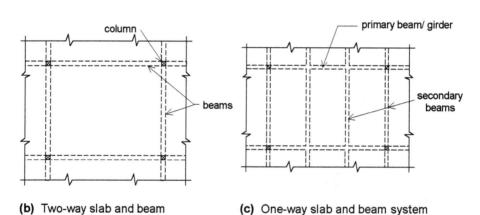

(b) Two-way slab and beam

(c) One-way slab and beam system

Beam and slab This type of construction consists of a slab built monolithically with beams (girders) forming a grid pattern, where the slab can be one- or two-way in nature depending on the dimensions of the slab panel (Figs. 2.9a, b, c). Slabs are commonly 100 to 150 mm thick and span 3 to 7.5 m. Beam and slab construction is a lightweight system which adapts easily to floor openings and can be used for a wide range of column spacings.

2.3.3 Vertical Framing Elements

The four types of vertical framing elements used in cast-in-place reinforced concrete construction include columns, walls, transfer girders, and hangers. Columns may be of any reasonable shape and vary in size depending upon the tributary floor area, number

CHAPTER 2 REINFORCED CONCRETE BUILDINGS

of storeys, and strength of concrete and reinforcing steel. A 300 by 300 mm square column is considered to be the minimum size for common column spacings. Frequently, high strength concrete (greater than 70 MPa) is used to minimise column size especially in the lower floors.

Bearing walls vary in thickness depending upon slab spans, the number of storeys, and the influence of lateral loads on their design. In medium-rise residential structures, walls are commonly 125 to 200 mm thick. Where walls form part of the major lateral load resisting system, such as core walls in office buildings, they may ratios similar to those of waffle slabs.

Transfer girders are occasionally necessary where the architect or client wishes to increase column spacing in the lower floors for convention halls, lobbies, or parking areas. The perimeter girder shown in Fig. 2.10 is 1.22 m by 3.05 m deep and is used to transfer the exterior loads at ground level to cast-in-place reinforced concrete columns, spaced at 15.2 m centres.

Fig. 2.10 Transfer girder

Courtesy SWR Engineering Limited

Steel hangers can be used to suspend the floors of a multi-storey structure from a central reinforced concrete core. Hanger loads can be transferred to the core by large cantilevered beams, cross-braced trusses, or Vierendeel trusses at the roof or at intermediate floors. Hangers take up very little of the floor space and provide an interesting effect at ground level where columns are eliminated.

2.3.4 Lateral Load Resisting Systems

Lateral load resisting structural systems are usually composed of one or more of the basic units, such as frames, shear walls and tubes (exterior wall frames). Frames are composed of columns and slabs or beams, and rely on the rigidity of the joints and the resistance to flexure of the individual elements to resist lateral loads (Fig. 2.3b, and 2.11a). They are used as the sole lateral load resisting system in buildings up to 25 storeys.

Shear walls are solid deep vertical elements in buildings such as core walls, transverse walls, or facade walls (Fig. 2.11b). Shear wall deformation under lateral load is similar to the bending of a solid vertical cantilever.

Frames are commonly interconnected to shear walls. This system combines the structural advantages of both frames and walls (Fig. 2.11c). In the lower storeys, the wall restrains frame deformations, and in the upper storeys the frame helps to reduce the bending deformation of the wall. Interacting frames and walls are considered economical for structures having to 40 to 60 storeys.

***Fig. 2.11** Lateral load resisting schemes*

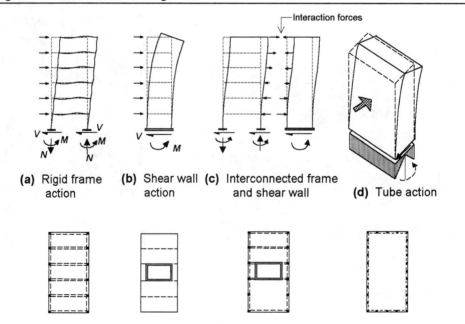

(a) Rigid frame action (b) Shear wall action (c) Interconnected frame and shear wall (d) Tube action

Tubular schemes (Fig. 2.11d), for building ranging from 40 to more than 100 storeys, are essentially peripheral moment resistant frame and wall systems. The National Life Assurance Company of Canada Building (Fig. 2.12) illustrates the tube concept with closely spaced exterior columns and relatively deep spandrel beams. The floor framing spans 12.2 m from perimeter to core providing column-free flexible office space.

Fig 2.12 *National Life Assurance Company of Canada Building, Toronto (Architects: Neish Owen Rowland & Roy, Toronto Structural Engineers: SWR Engineering Limited, Toronto)*

Courtesy SWR Engineering Limited

44 REINFORCED CONCRETE DESIGN

Fig. 2.13 Range of application of various framing systems

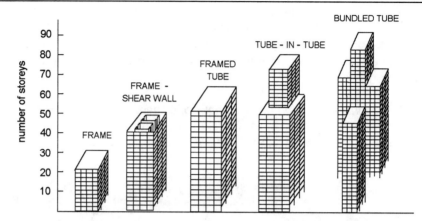

Tubes can also be combined with central shear cores in the tube-in-tube system. In this system, the stiff exterior frame (or tube) and the shear core mutually complement each other. Tube-in-tube schemes are efficient up to 80 storeys.

An extension of the tube system is the rigid or modular tube in which the exterior frame walls are linked by interior shear walls. These walls greatly increase the structural efficiency of the system by reducing the "shear lag" in the "flange column" of the tube. Shear lag in tube structures is the phenomenon in which all columns in the windward and leeward sides of the structure (flange columns) do not contribute equally in resisting lateral loads.

Figure 2.13 illustrates the basic lateral load resisting systems and their practical range of application for buildings. In summary, overall structural integrity is obtained by mutually interacting horizontal and vertical subsystems, which jointly resist both gravity loads and lateral loads.

2.4 STRUCTURAL ANALYSIS AND DESIGN

A structure is a three-dimensional system; however, it is generally constructed as an assemblage of two-dimensional or planar subsystems lying primarily in the horizontal and vertical planes, such as floors, roofs, walls, and plane frames. The planar subsystems themselves may be an assemblage of individual elements, such as slabs, beams, and columns. The purpose of the analysis for load effects is to obtain all the forces and moments acting on these elements, and the object of design is to provide adequate member sizes, reinforcement, and connection details to withstand these forces and moments satisfactorily. Thus, the basic design process involves the proportioning of individual members, such as slabs, beams, columns, walls, etc., knowing the forces acting thereon, and the detailing of their connections. The methods for the design of these basic elements are described in subsequent chapters. Methods of analysis of structures for member forces are detailed in books on structural analysis (Refs. 2.4 and 2.5). However, Codes often permit approximations and these are discussed in Chapter 8.

EXAMPLE 2.1

The framing details for an 8-storey office building to be located in Toronto, Ontario, are shown in Figs. 2.14a,b. Compute the lateral wind loads to be used in the design of an interior frame in the short direction.

Fig. 2.14 Example 2.1

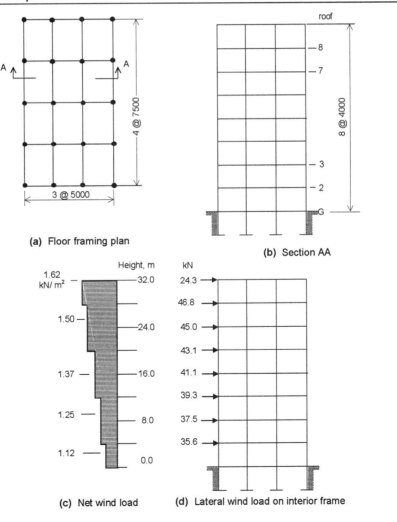

(a) Floor framing plan

(b) Section AA

(c) Net wind load

(d) Lateral wind load on interior frame

SOLUTION

The reference velocity pressure for a 30-year wind (to be used in the design of structural members) for Toronto (Ref. 2.1) is: $q = 0.48$ kN/m²

The exposure factor, C_e, for various heights specified in Ref. 2.2 is given in column 2 of Table 2.4 below. The gust effect factor, $C_g = 2.0$. Referring to Fig. B-14 in Commentary B of Ref. 2.2 (for height > width), external pressure coefficient, C_p, has values of 0.8 for the windward face and -0.5 (suction) for the leeward face.
For net wind load, $C_p = 0.8 - (-0.5) = 1.3$
The total net wind pressure is, $p = q\, C_e\, C_g\, C_p = 0.48\, C_e(2)(1.3) = 1.248\, C_e$ kN/m²

The variations of p with height are given in column 3 of Table 2.4. For convenience in computation, it will be conservatively assumed that the wind pressure intensity changes at the floor level as indicated in column 4 of Table 2.4 and in Fig. 2.14c. The tributary surface area for wind load per joint for wind along the short direction is $7.5 \times 4 = 30$ m² (except for the joint at roof level, for which area $= 7.5 \times 2 = 15$ m²). The wind loads are:

at level 2, $1.12 \times 15 + 1.25 \times 15 = 35.6$ kN
at level 3, $1.25 \times 30 = 37.5$ kN, and so on.

The lateral wind loads acting on the frame are shown in Fig. 2.14d.

Table 2.4   Example 2.1

Height (m)	C_e	$p=1.248\, C_e$ (kN/m²)	Use for Height (m)
0-6	0.9	1.12	0-4
6-12	1.0	1.25	4-12
12-20	1.1	1.37	12-20
20-30	1.2	1.50	20-28
30-44	1.3	1.62	28-32

REFERENCES

2.1   *National Building Code of Canada 1995*, National Research Council of Canada, Ottawa, 1995, 571 pp.
2.2   *User's Guide - NBC 1995 Structural Commentaries (Part 4)*, Canadian Commission on Building and Fire Codes, National Research Council of Canada, 1996, 135 pp.
2.3   *Canadian Highway Bridge Design Code*, Canadian Standards Association, Rexdale, Ontario, Canada, (in press).
2.4   Kassimali, A., *Matrix Analysis of Structures,* Brooks/Cole Publishing co., Pacific Grove, CA, 1999, 592 pp.
2.5   Wang, C.K., *Intermediate Structural Analysis*, McGraw-Hill, New York, 1982, 656 pp.

CHAPTER 3 Structural Safety

3.1 INTRODUCTION

Primary considerations in structural design are safety, serviceability, and economy. *Safety* requires that the structure function without damage under the normal expected loads (specified loads), and further that under abnormal but probable overloads, including earthquake or extreme winds, the likelihood of collapse (exceeding the load-carrying capacity, overturning, sliding, fracture and fatigue) be minimal. *Serviceability* requires that under the expected loads the structure performs satisfactorily with regard to its intended use, without discomfort to the user due to excessive deflection, permanent deformations, cracking, dynamic effects (vibration, acceleration), corrosion, or any other similar ill-effects. Furthermore, the overall structural system must have sufficient *structural integrity* to minimise the likelihood of a progressive type of collapse initiated by a local failure caused by an abnormal event or severe overload. Safety and serviceability can be improved by increasing the design margins of safety, but this increases the cost of the structure. In considering overall economy, any such increased cost associated with increased safety margins must be balanced against the potential losses that could result from any damage.

3.2 SAFETY CONSIDERATIONS

The major quantities considered in the design calculations, namely the loads, dimensions, and material properties, are subject to varying degrees of uncertainty and randomness. Furthermore, there are idealisations and simplifying assumptions used in the theories of structural analysis and design. There are also several other variable and often unforeseen factors that influence the prediction of ultimate strength and performance of a structure, such as construction methods, workmanship and quality control, probable service life of the structure, possible future change of use, frequency of loadings, etc. Thus, the problem facing the designer is to design economically on the basis of "prediction through imperfect mathematical theories of the performance of structural systems constructed by fallible humans from material with variable properties when these systems are subjected to an unpredictable natural environment" (Ref. 3.1). Any realistic and rational quantitative representation of safety must be based on statistical and probabilistic analysis.

Considering all the uncertainties and randomness, the load effects, S, on a structure and the resistance (or strength), R, of the structure may be represented by two

Fig. 3.1 Frequency distribution of load effects S and resistance R

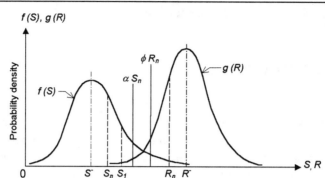

probability distribution curves, as shown schematically in Fig. 3.1. It is assumed here that R and S are independent, which is approximately true for the case of static loading. If $R > S$ the structure is safe and if $R < S$ the structure will fail. From Fig. 3.1, there is always a probability, however small, that failure may occur. For a load effect S_1, the probability that the resistance $R < S_1$ is given by:

$$\int_0^{S_1} g(R)dR$$

The probability that the load effect is S_1 is given by $f(S_1)$ and, accounting for all possible values of S_1 from 0 to 4, the probability of failure (that is, $R < S$) is obtained as:

$$P_F = \int_0^{\infty} f(S)[\int_0^S g(R)dR]dS \qquad (3.1)$$

Similarly, the probability of unserviceability, such as exceeding a deflection limit may be obtained (Fig. 3.2) as:

$$P_F = \int_{\Delta_{all}}^{\infty} h(\Delta)d\Delta \qquad (3.2)$$

where $h(\Delta)$ is the frequency distribution for the serviceability parameter (deflection, crack width, etc.) and Δ_{all} is the limiting allowable value. A rational solution to the design problem is to design the structure so as to limit its probability of failure (failure means collapse or unserviceability) to an "acceptable" low level. However, such a procedure is difficult, if not impossible, for standard structures for several reasons (Ref. 3.1). Since losses associated with failures are influenced by economic, social and moral considerations, which are difficult to quantify, generally acceptable failure probabilities are hard to define. Furthermore, frequency distributions, such as those shown in Figs. 3.1 and 3.2, are known only for a few simple cases. In addition much of the uncertainty associated with prediction of structural performance is of a non-statistical nature (e.g., relationship between laboratory and field strength, calculation and construction errors, relationship between assumed and actual load distributions) (Ref. 3.3).

Fig. 3.2 Probability of serviceability

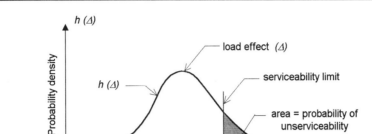

3.3 LIMIT STATES DESIGN

Because of the difficulties of a completely probabilistic design, simpler yet probabilistically sound design procedures have been developed (Refs. 3.1-3.5). One such procedure which has received increasing acceptance is the Limit States Design (LSD), also called Load and Resistance Factor Design (LRFD) (Refs. 3.1, 3.4, 3.5).

The concept of Limit States Design is simple. The probability of failure of a structure can be reduced by underestimating its resistance, R, and/or overestimating the load effects, S, and ensuring that $R \geq S$. Refining this concept on a probability basis, the LSD criteria (see Fig. 3.1) take the general form:

$$\phi R_n \geq \alpha S_n \tag{3.3}$$

where R is the *nominal resistance* computed on the basis of nominal material properties and dimensions; α is the *resistance factor* (also known as capacity reduction factor or performance factor), always less than unity, which reflects the uncertainties in determining R_n; S_n is the *nominal load effect* based on specified loads; and α is a *load factor*, usually greater than unity, which reflects potential overloads and uncertainties associated with the determination of S_n. The primary requirements of a structure are that it be serviceable and safe from collapse. The resistance, R_n, is computed as the value corresponding to a condition of a failure or *Limit state* being just reached, and Eq. 3.3 ensures that this limit state is avoided. *Limit states* define the various structural conditions representing collapse and unserviceability that are to be avoided. Those states concerning safety are called *ultimate limit states* (e.g., strength, buckling, overturning, sliding, fracture, fatigue, etc.), and those states that restrict the intended use and occupancy are called *serviceability limit states* (e.g., deflection, cracking, spalling, vibration, etc.). R_n is a generalised force (moment, axial force, shear force, etc.) for ultimate limit states and is an allowable limit of structural response (such as allowable deflection, crack width, stress, vibration, etc.) for serviceability limit states. An expression of the type in Eq. 3.3 must be satisfied for each limit state and for each set of load combinations applicable for the member or structure being designed.

Equation 3.3 is a simplified representation of Limit States Design criteria. In practice, the right hand side of this equation will be partitioned into the separate load effects (dead load, live load, wind load, etc.) with corresponding individual load factors so that:

$$\alpha S_n = \alpha_D S_{Dn} + \alpha_L S_{Ln} + \alpha_W S_{Wn} + ---- \qquad (3.4)$$

where S_{Dn}, S_{Ln}, S_{Wn}, are nominal load effects due to dead, live and wind loads, respectively, and α_D, α_L, α_W are corresponding load factors.

Individual load factors for each load type can better reflect the varying degrees of uncertainties associated with different types of load. For example, dead load, being predictable with better precision, would have a lower load factor than that for live loads. A load factor may also have different values depending on whether the effect of the load is to increase or decrease the total load effect. Thus, in considering failure by overturning, where the dead load helps prevent failure, the probability of failure is higher should the actual dead load be less than anticipated. In this case, the load factor for dead load will have to be less than unity. Equation 3.3 can be made more realistic by introducing additional factors or by adjusting the factors ϕ and α to reflect the lower probability of several loads acting concurrently at their full intensities (load combination factor), to account for the seriousness of failure of the member or structure under consideration (importance factor), to allow for the type of failure (sudden as opposed to gradual failure with warning), and to allow for uncertainties associated with structural analysis.

Acceptable values for all these factors (collectively termed partial safety factors) can be determined by statistical evaluation of data on strength, loads and other aspects related to the factor, and then calibrating the failure probability to those associated with current accepted practice (Section 3.4).

Equation 3.3 may be rearranged as:

$$\frac{\phi}{\alpha} R_n \geq S_n \qquad (3.5)$$

$$R_n \geq \frac{\alpha}{\phi} S_n \qquad (3.6)$$

These two forms are comparable with the allowable stress design procedure (Eq. 3.7) wherein a Factor of Safety (F.S.) is applied to the material strength (underestimating the resistance, R) and the plastic design procedure (Eq. 3.8) where all the working loads are multiplied by Load Factors (L.F.) (overestimating load effect, S), respectively.

$$\left(\frac{1}{F.S.}\right)R_n \geq S_n \qquad (3.7)$$

$$R_n \geq (L.F.)S_n \qquad (3.8)$$

Since the use of a single Factor of Safety as in Eq. 3.7 does not provide uniform reliability and economy, structural codes have been moving towards a method involving several partial safety factors (ACI, CSA, NBC). The separate consideration of loads, materials and performance by means of statistically determined partial safety factors in Eqs. 3.3 and 3.4 makes the design more responsive to the differences between types of loads, kinds of materials, types of structural behaviour, and degrees of seriousness of different modes of failure (failure here means unsatisfactory performance).

3.4   PARTIAL SAFETY FACTORS

A basis for the development of partial safety factors is presented in Ref. 3.4. Referring to Fig. 3.1, failure is characterised by $R < S$, or alternatively the probability of failure, P_F, is given by:

$$P_F = P\left[\ln\frac{R}{S} < 0\right] \qquad (3.9)$$

Figure 3.3 shows the frequency distribution for ln R/S, and the shaded area to the left of the origin gives the probability of failure. This area is a function of the number, β, of standard deviations, $\sigma_{\ln R/S}$, between the origin and the median, $(\ln R/S)_m$. The number β is a relative measure of the degree of safety and is defined as the *safety index*. The

Fig. 3.3 Definition of safety index

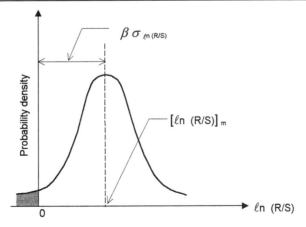

52 REINFORCED CONCRETE DESIGN

Fig. 3.4 Schematic representation of calibration of limit states design

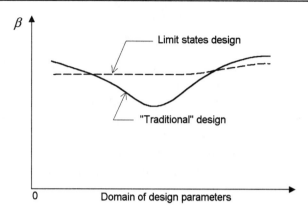

higher the value of β, the less the probability of failure and the higher the factor of safety. By adjusting β to be constant for all structural elements, a uniform safety could be achieved. If the actual distribution curves such as shown in Fig. 3.3 were known, the value of β could be evaluated and related to a desired probability of failure. In the absence of such data, acceptable values for β are established by a process of *calibrating* the probabilistic failure criteria to existing design standards (such as the allowable stress design method). On this basis, values of 3 to 4 have been used for β.

Reference 3.4 uses such a *calibration* technique to determine the required partial safety factors. Values for the resistance factor, ϕ, and the load factor, α, for different load combinations are assumed, and β is determined for standard structural elements and for a domain of relevant design parameters (such as dead load/live load ratios, dead load/wind load ratios). The variation of β for the same elements and over the same domains is also determined for the traditional design specification that on the basis of past experience represents a measure of acceptable safety factors. The values of the ϕ and α factors are then adjusted until the value of β is reasonably constant over the whole domain and is comparable with that of the traditional design practice over at least part of the domain (Fig. 3.4). This process ensures a safety index that is more consistent than, and comparable to, those inherent in the existing design practice.

3.5 LIMIT STATES DESIGN BY CANADIAN CODES

CSA Standard A23.3-94, *Design of Concrete Structures*, is based on NBC 1995 (Ref. 3.6), and these codes present a unified limit states design philosophy, with common *safety* and *serviceability* criteria for all materials and methods of construction. In this method, the design is aimed at avoiding the attainment of *Limit States* that define the onset of various types of collapse and unserviceability. Both ultimate limit states which concern safety (e.g., attainment of load carrying capacity, fracture, overturning, etc.) and serviceability limit states (e.g., limiting deflection, cracking, vibration, etc.) are considered in the design. The probability of attainment of any of the limit states (that is, the probability of failure) is kept to an acceptably low level with the use of load factors, α, load combinations factors, ψ, and importance factors, γ, applied to *specified loads*, and resistance factors, ϕ, applied to specified material properties or to the resistance of a member, connection, or structure, which for the limit state under consideration takes into account the variability of dimensions and material properties, workmanship, type of failure, and uncertainty in the prediction of resistance. The resistance of a cross-section, member, connection, or structure, including application of the appropriate resistance factors is termed the *factored resistance*. A structure and its components are designed such that:

$$\text{factored resistance} \geq \text{the effect of factored loads} \tag{3.10}$$

Here the *effect of factored loads* means the structural effect (such as bending moments, shear force, axial force, torsion) due to the factored load combinations obtained by multiplying the *specified loads* by appropriate load factors, α, a load combination factor, ψ, and an importance factor, γ. For load combinations not including earthquake, the factored load combination is taken as:

$$\alpha_D D + \gamma\psi[\alpha_L L + \alpha_W W + \alpha_T T] \tag{3.11}$$

and, correspondingly, Eq. 3.10 can be expressed as:

$$R_r \geq \text{effect of } [\alpha_D D + \gamma\psi(\alpha_L L + \alpha_W W + \alpha_T T)] \tag{3.11a}$$

where $R_r =$ factored resistance, including the application of resistance factors ϕ
and D, L, W, T are specified loads, with
- $D =$ dead loads
- $L =$ live loads due to intended use and occupancy (including vertical loads due to cranes); snow, ice and rain; earth and hydrostatic pressure; horizontal component of static or inertia forces
- $W =$ live loads due to wind
- $T =$ loads due to the cumulative effects of temperature, creep, shrinkage, and differential settlement
- $\alpha =$ load factors that take into account variability of the load and load patterns, and analysis of their effects
- $\psi =$ load combination factor which takes into account the reduced probability of simultaneous occurrence of several loads at their

54 REINFORCED CONCRETE DESIGN

factored values
$\gamma =$ importance factor to take into account the consequences of collapse
$\phi =$ resistance factor applied to specified material properties or to the resistance of a member to account for the variability of material properties and dimensions, workmanship, type of failure (brittle or ductile) and uncertainty in the prediction of resistance

For load combinations that include earthquake, the factored load combinations are taken as:

$$1.0D + \gamma(1.0E) \tag{3.11b}$$

and either:

$$1.0D + \gamma(1.0L + 1.0E) \tag{3.11c}$$

for storage and assembly occupancies, or

$$1.0D + \gamma(0.5L + 1.0E) \tag{3.11d}$$

for all other occupancies. There is no load combination factor, ψ, in Eqs. 3.11b, 3.11c, and 3.11d.

For serviceability limit states, the design criterion in Eq. 3.10 can be expressed in the form of Eq. 3.12, in which the left hand side, S_{lim}, represents the limiting value of a serviceability parameter, such as deflection, crack width, etc., and the right hand side is the same parameter computed for the combinations of specified loads shown (i.e., with load factors, $\alpha = 1.0$).

$$S_{lim} \geq \text{effect of } [D + \psi(L + W + T)] \tag{3.12}$$

No single structural theory for analysis and design (e.g., elastic theory, plastic theory, ultimate strength theory, etc.) may be universally applicable to all limit states. The appropriate theory to be used is generally indicated in the relevant standard (e.g., CSA A23.3-94 for concrete). In general, elastic theory is applicable to serviceability limit states and fatigue; and one of the ultimate strength theories (allowing for inelastic material behaviour) is applicable to the ultimate limit states. The factors applied to the specified loads (γ, ψ, α) are common to all materials and types of construction and are specified in Section 4.1 of NBC 1995 (Ref. 3.6). The resistance factors, ϕ, which are applied to the specified material properties or to the member resistance are specified in the respective structural material standard (e.g., CSA A23.3-94 for concrete building structures, CSA S16.1 for steel building structures, etc.). The specified loads and specified material properties used to compute member resistance are defined statistically on the basis of probability of occurrence (e.g., a return period of 10 to 100 years for wind, snow, and earthquake loads; and a 10 percent maximum probability of underrun for material properties (Ref. 3.7)). For live loads due to use and occupancy, for which statistical data is not available, nominal values are specified which represent upper limits of loads that could be expected normally. Specified loads and load combinations to be used in design of buildings are given in Section 4.1 of NBC 1995 (Ref. 3.6). Similar data for highway bridges are given in Ref. 3.8.

3.6 LOAD AND RESISTANCE FACTORS

The 1994 revision of CSA A23.3-94 uses the same load factors as those specified in the National Building Code of Canada and uses *material resistance factors* applied to the material strengths (although in a few cases the resistance factor is applied to the member resistance) as previously explained in Section 3.5.

The various factors to be used with Eqs. 3.10 to 3.12, as specified in CSA A23.3-94 are as follows:

(i) Load factor, α, to be used in Eq. 3.11:
 (a) $\alpha_D = $ 1.25, except that when considering overturning, uplift or reversal of load effect, where the effect of dead load is to resist these, α_D is taken as 0.85;
 (b) $\alpha_L = $ 1.5;
 (c) $\alpha_W = $ 1.5; and
 (d) $\alpha_T = $ 1.25.

(ii) Load combination factor, ψ:
 (a) $\psi = $ 1.0, when only one of the loads L, W, or T is included in Eq. 3.11 or 3.12;
 (b) $\psi = $ 0.70, when two of the loads L, W, or T are included in Eq. 3.11 or 3.12; and
 (c) $\psi = $ 0.60, when all three of the loads L, W, and T are included in Eq. 3.11 or 3.12.

Design is to be made for the most unfavourable effect considering L, W or T acting alone or in combinations with corresponding ψ values.

(iii) Importance factor, γ, to be used in Eqs. 3.11, 3.11a, 3.11b, 3.11c, and 3.11d, is not less than 1.0 for all buildings, except that for buildings where it can be shown that collapse is not likely to cause injury or other serious consequences, it is not less than 0.8.

(iv) Material resistance factors, ϕ:
 (a) $\phi_c = $ 0.60 for concrete strengths in checking ultimate limit states;
 (b) $\phi_p = $ 0.90 for prestressing tendons;
 (c) $\phi_s = $ 0.85 for reinforcing bars; and
 (d) $\phi_a = $ 0.90 for structural steel

Fig. 3.5 Factored material strength

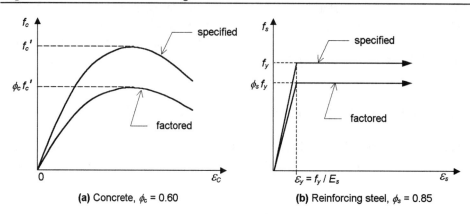

(a) Concrete, ϕ_c = 0.60
(b) Reinforcing steel, ϕ_s = 0.85

Thus, in calculation of resistance at ultimate limit state, the material strengths to be used are $\phi_c f_c'$ and $\phi_s f_s$, where f_c' is the specified concrete strength, and $f_s = E_s \varepsilon_s \leq f_y$ is the stress in steel as specified. The specified material strength and the factored material strength are shown in the stress-strain relationship in Fig. 3.5.

In addition to the material resistance factors given above, the Code also specifies member resistance factors of ϕ_m = 0.75 for use in certain slender column calculations and ϕ_m = 0.65 for certain brittle members in earthquake-resistant design.

3.7 NOTATION

The load effects, such as moment, shear force, etc., at vertical sections on members, as determined from the analysis of the structure under factored load combinations will be identified by a subscript f (e.g., factored moment, M_f, factored axial load, P_f, factored shear force, V_f, etc.). These factored load effects may also be designated as the *design moment*, *design shear*, etc., as these represent the minimum strengths for which the section/member must be designed. The subscript r will be used for the factored resistance, which is the resistance of a cross-section or member computed theoretically using factored material strengths (or, where applicable, a member resistance factor) and the design dimensions of the member. The design criteria for the ultimate limit state require that:

$$M_r \geq M_f, \quad V_r \geq V_f, \quad P_r \geq P_f$$

Therefore, the design basis is: $\quad M_r = M_f, \quad V_r = V_f, \quad P_r = P_f \quad$ (3.13)

3.8 TOLERANCES

Overall cross-sectional dimensions of reinforced concrete members such as slabs, beams, columns, and walls are normally specified in multiples of 10 mm, except that 5 mm increments may be used for thin slabs. Dimensional tolerances are specified as the maximum allowable variations (plus or minus) from the specified dimensions. Allowable variations for cross-sectional dimensions of beams, girders, and columns and for thickness of walls and slabs specified in CSA A23.3-94 are given in Table 3.1. Allowable tolerances for placing reinforcement are presented in Table 3.2.

Table 3.1 *Dimensional Tolerances for Cross-Sections of Beams, Girders, and Columns and for Thickness of Walls and Slabs Other Than Slabs on Grade (CSA A23.3-94)*

Dimension	Allowable variation
Less than 0.3 m	± 8 mm
Greater than 0.3 in but below 1 m	± 12 mm
Greater than 1 m	± 20 mm

Table 3.2 *Tolerances for Placing Reinforcement*

	Application	Tolerance
a)	Concrete cover	± 12 mm
b)	Where depth of a flexural member, thickness of a wall, or the smallest dimension of a column is:	
	(i) 200 mm or less	± 8 mm
	(ii) larger than 200 mm but less than 600 mm	± 12 mm
	(iii) 600 mm or larger	± 20 mm
c)	Lateral spacing of bars	± 30 mm
d)	Longitudinal location of bends and ends of bars	± 50 mm
e)	As item (d) at discontinuous ends of members	± 20 mm

Note: Regardless of these tolerances, the concrete cover shall in no case be reduced by more than one third of the specified cover.

REFERENCES

3.1 Cornell, C.A., *A Probability-Based Structural Code*, Journal of ACI, Vol. 66 (12), Dec. 1969, pp. 974-985.

3.2 Allen, D.E., *Limit State Design C A Unified Procedure for the Design of Structures*, Eng. Journal, Vol. 53 (2), Feb. 1970, pp. 18-29.

3.3 Ang, A.H.S., and Amin, M., *Safety Factors and Probability in Structural Design*, ASCE Journal of Struct. Div., Vol. 95 (7), July 1969, pp. 1389-1405.

3.4 Allen, D.E., *Limit States Design C A Probabilistic Study*, Can. Journal of Civ. Eng., Vol. 2, 1975, pp. 36-49.

3.5 Ravindra, M.K., and Galambos, T.V., *Load and Resistance Factor Design for Steel*, ASCE Journal of Struct. Div., Vol. 104, No. ST9, Sept. 1978, pp. 1337-1353.

3.6 *National Building Code of Canada 1995*, National Research Council of Canada, Ottawa, 1995, 571 pp.

3.7 *User's Guide - NBC 1995 Structural Commentaries (Part 4)*, Canadian Commission on Building and Fire Codes, National Research Council of Canada, 1996, 135 pp.

3.8 *Canadian Highway Bridge Design Code*, Canadian Standards Association, Rexdale, Ontario, Canada, (in press).

CHAPTER 4　Behaviour in Flexure

4.1　THEORY OF FLEXURE FOR HOMOGENEOUS MATERIALS

Sections of straight prismatic members loaded transversely through the shear centre are subjected to a bending moment, M, and a transverse shear force, V (Fig. 4.1). For equilibrium, the resultants of the internal stresses in the section must balance the moment and shear force due to external loads. For convenience, the influences of the bending moment and shearing forces are considered separately. The behaviour in shear will be considered in Chapter 6. Considering only the applied moment, M, for static equilibrium, the resulting internal normal stresses across the cross-section must reduce to a pure couple (Fig. 4.1c).

The fundamental assumption in flexural theory is that plane cross-sections taken normal to the beam axis before loading remain plane after the beam is subjected to bending. The assumption is generally valid for beams of usual proportions (Ref. 4.1). For initially straight members, this assumption means that the normal strains at points in the beam section are proportional to the distance of such points from the neutral axis (Figs. 4.2a, b). The flexural stress at any point such as A (Fig. 4.2b) in the cross-section depends on the strain, ε_A, at that level and is given by f_A, the stress corresponding to the strain ε_A, as determined from the stress-strain relationship for the material (Fig. 4.2c). The neutral axis location is such that the equilibrium condition $C = T$ is satisfied, and the moment of the internal resisting couple is $M = (C \text{ or } T)\,jd$, where jd is the lever arm of the couple. G_1 and G_2 represent the centres of compression and tension (locations through which C and T act), respectively.

Fig. 4.1 *Flexure of transversely loaded beam*

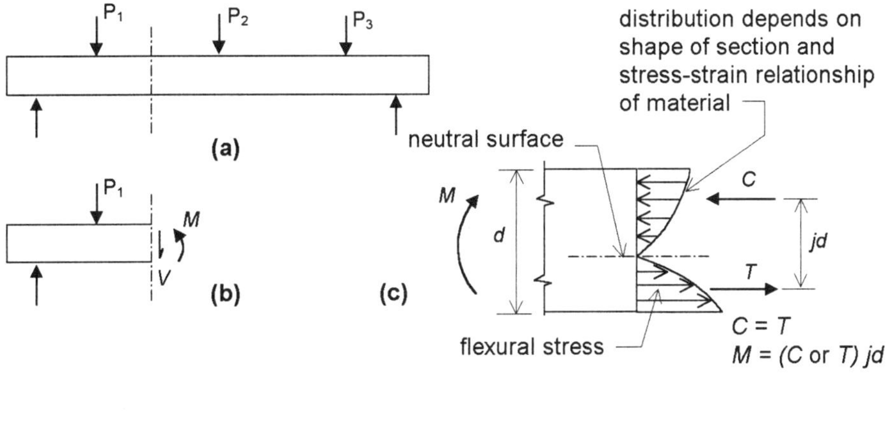

Fig. 4.2 Homogeneous section under flexure

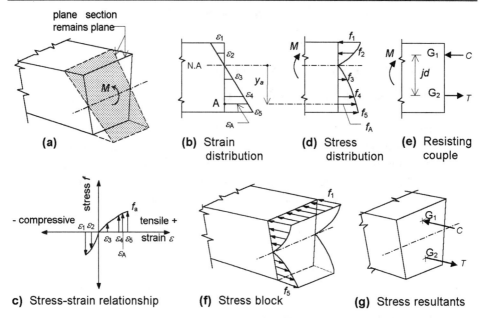

(a)
(b) Strain distribution
(c) Stress-strain relationship
(d) Stress distribution
(e) Resisting couple
(f) Stress block
(g) Stress resultants

If the material of the beam obeys Hooke's Law (that is, it is linearly elastic), the stress will be proportional to strain ($f = E\varepsilon$, where E is the Young's modulus of elasticity), and the stress distribution in the section will also be linear (Fig. 4.3). The material does not have to be *elastic* to satisfy these equations, and it is only necessary that stress be proportional to strain. However, traditionally a linear stress-strain relationship has been associated with elastic behaviour, and generally the latter is implied to mean the former as well. In this case (assuming that E has the same value in both tension and compression) it is easily shown (Ref. 4.2) that the neutral axis passes through the centroid of the section and the stress, f, at any point in the section at a distance y from the neutral axis, is given by Eq. 4.1, or alternatively the resisting moment in terms of the extreme fibre stress is given by Eq. 4.2.

$$f = \frac{My}{I} \quad (4.1)$$

$$M = \frac{f_{max} I}{y_{max}} \quad (4.2)$$

where f_{max} is the stress in the extreme fibre
 y_{max} is the distance of the extreme fibre from the neutral axis
 I is the moment of inertia of the section
 M is the bending moment at the section

Fig. 4.3 Linear elastic stress distribution in flexure

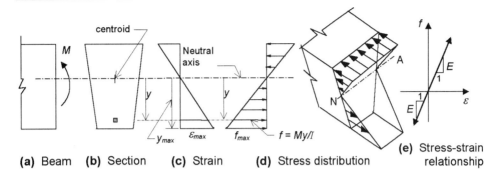

(a) Beam (b) Section (c) Strain (d) Stress distribution (e) Stress-strain relationship

4.2 ELASTIC BEHAVIOUR AND TRANSFORMED SECTIONS

The concept of a transformed section is used to calculate elastic stress distributions. In a composite beam made of two (or more) different elastic materials (Fig. 4.4a) in which the two parts are bonded together to act as an integral unit (with no slip between the two parts), the assumption that initially plane cross-sections remain plane while subject to bending is valid. Therefore, the strain variation in the section will be linear (Fig. 4.4b), and flexural stress at any point in the section is given by the product of the strain at that point and the elastic modulus of the material (Fig. 4.4c). At a depth y below the neutral axis, where the strain is ε, the stress in material 1 is $f_{1y} = E_1 \varepsilon_y$, and the stress in material 2 is $f_{2y} = E_2 \varepsilon_y = n f_{1y}$, where E_2/E_1 is termed the *modular ratio*. If the section is transformed into an equivalent homogeneous section all of one material, say material 1, the transformation of material 2 must be such that the forces across the section are unaltered. The total force in the element δA (Fig. 4.4a, c) of material 2 is

$$dF_2 = b_2 \, dy \, f_{2y} = b_2 \, dy (n f_{1y}) = (n b_2) dy \, f_{1y}$$

Thus $b_2 \, dy$ of material 2 is equivalent to $n b_2 \, dy$ of material 1. Therefore, material 2 may be transformed to material 1 by changing its dimension (b_2) parallel to the neutral axis by a factor equal to the modular ratio $n = E_2/E_1$. In the transformed homogeneous section (all of material 1), both strains and stresses vary linearly, and the neutral axis passes through its centroid. The stress distribution computed for the transformed section ($f_T = M y / I_T$) gives the actual stresses in the real section for the material of which the transformed section is made (here $f_1 = f_T$). For the second material, the actual stresses at any point will be n times the stress at that level computed for the transformed section ($f_2 = n f_T$).

Fig. 4.4 Transformed section

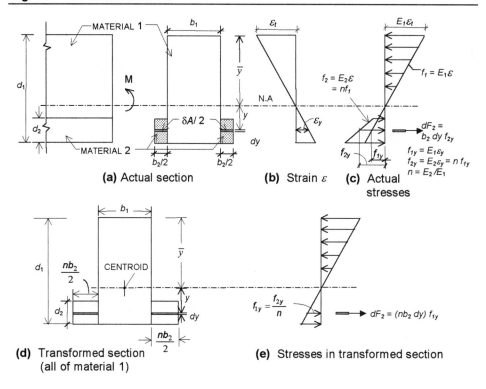

(a) Actual section
(b) Strain ε
(c) Actual stresses
(d) Transformed section (all of material 1)
(e) Stresses in transformed section

EXAMPLE 4.1

A reinforced concrete beam of rectangular section has the cross-sectional dimensions shown in Fig. 4.5a. The concrete, which is of normal density, has a compressive strength of 30 MPa and a modulus of rupture of 3.3 MPa. The yield strength of the steel is 400 MPa. Compute (*a*) the stresses due to an applied bending moment of 45 kN·m and (*b*) the bending moment at which cracking of concrete will be initiated.

SOLUTION

a) Initially it will be assumed that at the applied moment the concrete has not yet cracked. Prior to cracking the full section is effective, and, since both tensile and compressive stresses will be very low, the stress-strain relationship for concrete in both tension and compression may be assumed to be linear (Section 1.5) with the same value for elastic modulus E in both compression

Fig. 4.5 Example 4.1

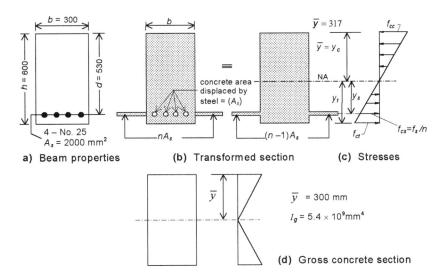

and tension. It is further assumed that there is perfect bond between the reinforcing bars and surrounding concrete.

From Eq. 1.1, for concrete, $E_c = 4500\sqrt{30}$ =24 648 MPa, and for steel E_s = 200 GPa. Reinforced concrete sections are usually transformed into equivalent concrete sections.

The modular ratio, $n = \dfrac{E_s}{E_c} = \dfrac{200\,000}{24\,648} = 8.1$

The transformed section is shown in Fig. 4.5b. Note that the transformed section consists of the *gross* concrete section plus n times the area of steel A_s minus the concrete area displaced by the embedded bars, A_s. The depth to neutral axis (centroid) $\bar{y}$ and the moment of inertia I_T of the transformed section may be computed as $\bar{y}$ = 317 mm and I_T = 6.10 × 10⁹ mm⁴. The compressive stress in the concrete at the top fibres of the section is:

$$f_{cc} = \dfrac{My_c}{I_T} = \dfrac{45 \times 10^6 \times 317}{6.10 \times 10^9} = 2.34 \text{ MPa}$$

and the tensile stress at the bottom fibres of concrete is:

$$f_{ct} = \dfrac{My_t}{I_T} = \dfrac{45 \times 10^6 \times 283}{6.10 \times 10^9} = 2.09 \text{ MPa} < f_r$$

The stress in steel is given by n times the stress in concrete at the steel level

$$f_s = n\left(\frac{f_{ct}}{y_t}y_s\right) = 8.1\frac{2.09 \times 213}{283} = 12.7 \text{ MPa}$$

Since the tensile stress in concrete is below the modulus of rupture, the assumption that the section is uncracked is correct. Both concrete and steel stresses fall within the initial linear ranges of the stress-strain curves, and, hence, the assumption of linear elastic behaviour is justified.

b) As the applied moment is increased, the stresses also increase, and cracking of concrete on the tension side will begin, as the maximum tensile stress, f_{ct}, reaches the tensile strength in flexure, i.e., the modulus of rupture, f_r. Equating the maximum tensile stress to f_r, the corresponding moment, M_{cr}, is given by

$$f_{cr} = \frac{M_{cr}y_t}{I_T}$$

$$M_{cr} = f_{cr}\frac{I_T}{y_t} = 3.3 \times \frac{6.10 \times 10^9}{283} \text{ N} \cdot \text{mm} = 71.1 \text{ kN} \cdot \text{m}$$

Notes: In the precracking stage, the influence of the reinforcement is usually negligible. In this example, if the reinforcing steel area is neglected and the section considered as having a gross concrete section 300 mm by 600 mm, (Fig. 4.5d), $\bar{y} = 300$ mm, and the moment of inertia $I_g = 5.4 \times 10^9 \text{ mm}^4$, the cracking moment is:

$$M_{cr} = \frac{3.3 \times 5.4 \times 10^9}{300} \times 10^{-6} = 59.4 \text{ kN} \cdot \text{m}$$

which is in error by 16.5 percent on the conservative side. For computing M_{cr} in deflection calculations, CSA A23.3-94(9.8.2.3) recommends the use of the gross concrete section properties instead of the more accurate full transformed section properties.

4.3 TRANSFORMED STEEL AREA

In reinforced concrete, the modular ratio, $n = E_s/E_c$, is usually taken as the nearest whole number, but not less than 6. The tensile reinforcement area, A_s, can be transformed to concrete by multiplying by the modular ratio, n, since the concrete in tension is usually neglected in strength computations. For compression reinforcement area, A_s', for elastic behaviour, the effective transformed area is $(n - 1)A_s'$, which allows for the concrete in compression displaced by the steel area A_s'.

The creep of concrete and the nonlinearity of the stress-strain relation for concrete at higher stresses lead to larger compressive strains in the compression steel and surrounding concrete than the elastic theory would indicate. To account for this, the Canadian Highway Bridge Design Code (Cl. 8.8) (Ref. 4.3) recommends the use of an effective modular ratio of $2E_s/E_c$ for transforming any compression reinforcement for stress computations in flexural members. The associated steel stress will then be $2nf_c$ where f_c is the compressive stress in the concrete at the steel level.

4.4   FLEXURE OF REINFORCED CONCRETE BEAMS

In the discussion that follows, explaining the general behaviour of a reinforced concrete section under flexure, the material resistance factors, ϕ_c and ϕ_s, are not taken into account. However, in design in accordance with the Code, the ϕ-values will have to be incorporated, as will be illustrated later on.

The computations in Example 4.1 show that concrete on the tension side of a beam begins to crack at very low stresses. A plain concrete beam will fail at a moment equal to the cracking moment M_{cr} due to the tensile failure of the concrete, while the maximum compressive stress in the concrete is still only a small fraction of the compressive strength. To utilise the higher compressive strength of concrete, the beam is reinforced with steel bars placed on its tension side. Thus, reinforced concrete is a nonhomogeneous material consisting of a judicious combination of concrete and reinforcing steel. It is assumed that there is perfect bond between the reinforcing bars and surrounding concrete (Chapter 9). Therefore the strain variation in the cross-section is linear (proportional to the distance from the neutral axis), with the strain in the reinforcement being the same as that in the surrounding concrete.

Figure 4.6 shows a beam with gradually increasing loading. In the initial phase when the applied moment is less than the cracking moment, M_{cr}, the maximum tensile stress in concrete is less than the tensile strength, the concrete is uncracked, the entire section is effective, and the strain and stress distribution is as shown in Fig. 4.6c (Example 4.1). As the applied moment is increased beyond the cracking moment, flexural tensile cracks develop in the bottom (tensile) fibres of the beam and propagate gradually towards the neutral axis (Fig. 4.6d). Since the cracked portion of the concrete is now ineffective in resisting tensile stresses, the effective concrete section is reduced by that much, and the neutral axis also shifts upwards. Cracks cannot be eliminated altogether from reinforced concrete flexural members under the normal range of applied loads. However, by proper design (Chapter 11) they can be controlled so that there will be several well-distributed fine hairline cracks rather than a few wide cracks. Minute hairline cracks do not seriously damage either the external appearance of the member or the corrosion protection of the steel bars, and, hence, are acceptable in normal situations.

66 REINFORCED CONCRETE DESIGN

Fig. 4.6 *Behaviour of reinforced concrete beam under increasing moment*

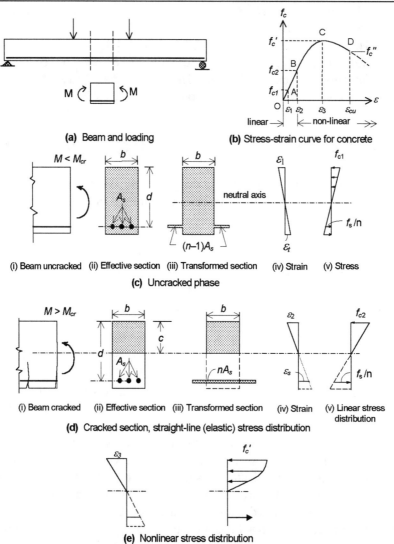

Because of the tensile cracking of concrete at very low stresses, it is generally assumed in flexural computations that concrete has no tensile resistance. This is equivalent to considering the concrete on the tension side of the neutral axis to be virtually nonexistent in the computations for flexural strength. The effective section of the member is as shown in Fig. 4.6d. However, the concrete on the tension side is necessary to hold the reinforcing bars in place, to resist shear and torsion, and to provide protection for the steel against corrosion and fire. During the first-time loading, a small part of the concrete on the tension side, and close enough to the neutral axis for

Fig. 4.6 Continued

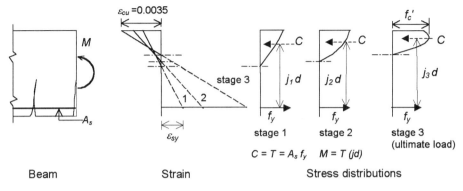

(f) Under-reinforced beam, tension failure

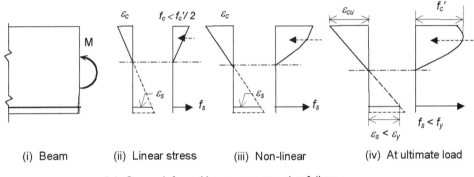

(g) Over-reinforced beam, compression failure

the tensile strain to be less than that corresponding to f_r, will still be uncracked and effective. However, the magnitude of the resulting tensile force and the internal moment due to it are negligibly small compared to the forces and moments due to the compressive stress in the concrete and tensile stress in the steel. Furthermore, on a previously loaded member, any prior overloading may have caused the tensile cracks to penetrate high enough into the beam to eliminate this small contribution from the tensile strength of concrete. Therefore, the assumption that concrete resists no flexural tensile stresses is satisfactory and realistic, and will be followed in subsequent discussions.

If the applied moment, M, in Fig. 4.6d is such that the maximum concrete stress, f_c, is less than about $0.4f_c'$, the stress will be nearly proportional to the strain (see part *OB* of curve in Fig. 4.6b, and Section 1.5). Then the distribution of both strain and stress in the section is linear, as shown in Fig. 4.6d, and is given by the linear elastic theory applied to the composite section (Section 4.2). The magnitude of stresses in usual structural members under the *specified loads* (service load conditions) is generally below about $0.4f_c'$, so that the service load stress distribution is closely approximated by the straight-line distribution shown in Fig. 4.6d. For this reason the straight-line distribution is generally used to check the serviceability criteria (crack control, see CSA A23.3-94(8.3, 9.8, and 10.6)) for which the applied loads are the actual service loads (specified loads with a load factor of 1.0). The linear elastic theory is also the basis of design in the *working stress design* method (also known as *allowable stress design* or *service load design*), where members are designed for the effects of the specified loads (service loads or working loads), with the resulting stresses limited to a fraction of the material strength to ensure a satisfactory margin of safety. Expressions for the stresses and moment of resistance for reinforced concrete sections using the linear elastic stress distribution and the concept of cracked-transformed sections are derived in Section 4.6.

As the applied loads and moments are further increased, the concrete strains and stresses enter into the nonlinear range *BCD* in Fig. 4.6b. For example, if the maximum compressive strain reaches a value of ε_3, which corresponds to a stress of f_c', the compressive stress distribution will have the shape of the curve *OBC* in Fig. 4.6b, as shown in Fig. 4.6e. The behaviour of the beam in the inelastic range prior to ultimate failure in bending is dependent on the amount of reinforcing steel and the two possible cases are discussed below.

The maximum usable strength of reinforcing bars is given by the yield strength f_y. Any increase in steel stress beyond f_y is accompanied with large strains and, hence, large curvatures, crack widths and deflections, which are normally unacceptable. Similarly, concrete is generally assumed to fail in compression upon reaching an ultimate compressive strain, ε_{cu}, values for which are recommended in codes (CSA A23.3-94(10.1.3) recommends $\varepsilon_{cu} = 0.0035$). A section in which these two limiting conditions, namely tension reinforcement yield strain, ε_y, and ultimate concrete strain, ε_{cu}, are attained simultaneously is termed a *balanced* section. A section containing reinforcement less than that of a balanced section will be termed an *under-reinforced* section. In under-reinforced sections, as the applied moment is increased, a stage will be reached when the tension reinforcement strain reaches ε_y, and the stress reaches its yield strength, f_y. At this stage the maximum concrete strain will be less than ε_{cu}, and the strain and stress distributions will be somewhat as shown by stage 1 in Fig. 4.6f, with $C = T = A_s f_y$ and $M = A_s f_y j_1 d$. A slight increase in load at this stage causes the steel to yield and stretch a large amount. The marked increase in tensile strains causes the neutral axis to shift upwards, thus reducing the area under compression. Since total tension, T, remains essentially constant at $A_s f_y$, to maintain equilibrium ($C = T$) with the reduced area in compression, the compressive stresses and strains increase. This situation is represented by stage 2 in Fig. 4.6f. The rise of the neutral axis causes a slight upward shift of the centre of compression, thus increasing the lever arm and moment of resistance ($M = A_s f_y j_2 d$) slightly. The relatively rapid increase in strain on the tension side also manifests in wider and deeper tensile cracks and increased beam deflection. This process continues until the maximum compressive strain reaches its ultimate value, and the reduced compressive area fails by crushing of concrete (stage 3, Fig. 4.6f). There is usually only a slight increase in the moment of resistance between stage 1 and stage 3, resulting from the marginal increase in the lever arm jd (and also due to any slight increase in steel stress above f_y when the steel has no sharply defined yield plateau, Fig. 1.5). However, there is a substantial increase in curvature, deflection, and width and spread of cracking between stage 1 and stage 3. In this type of beam, it is the yielding of tensile reinforcement that leads, as a secondary effect, ultimately to the crushing failure of concrete. This type of failure is termed a *tension failure*. A tension failure is gradual, with ample prior warning of impending failure in the form of increasing curvatures, deflections and extensive cracking, and, therefore, is preferred in design practice. A schematic presentation of the variation of moment with curvature for an under-reinforced beam is shown in Fig. 4.7a. The large increase in curvature (rotation per unit length) before failure at relatively constant load indicates a *ductile* type of failure. The strength of such sections can be taken as that at which the steel yields.

In an *over-reinforced* beam, containing reinforcement in excess of that required for a *balanced section*, the maximum compressive strain in the concrete reaches the failure strain, ε_{cu}, and the concrete falls in compression before the steel reaches its yield point. This type of failure is termed a *compression (or brittle) failure*. In this case the steel remains elastic to failure. Once the compressive strain in concrete enters the nonlinear range, compressive stresses do not increase proportionally with strain, whereas the tensile stresses do. Therefore, to keep the tensile and compressive forces equal ($C = T$), the area of concrete in compression has to increase causing a lowering of the neutral axis (Fig. 4.6g). As the concrete fails in compression, the strains in the member, and hence the curvature φ, the deflection, and the crack width all remain relatively low. Therefore, the failure occurs suddenly with little warning, and is often brittle and explosive. Because of this, CSA A23.3-94 does generally not permit over-reinforced flexural members. A typical moment-curvature relationship for an over-reinforced section is shown in Fig. 4.7b. Computation procedures for stresses and strength are described in subsequent Sections.

One other type of failure is possible, although very rare in normal practice. This is failure by fracture of the reinforcing steel, which can happen with extremely low amounts of reinforcements or under dynamic loading.

Under specified loads, a reinforced concrete flexural member will be in the cracked phase (Fig. 4.6d). Because the concrete is considered ineffective in tension, the *effective depth*, d, of the member is taken as the distance from the extreme compression fibre to the centroid of the tensile reinforcement. Since the moment of resistance of a beam is proportional to the lever arm, jd, the tensile reinforcing bars are placed as close to the extreme tension fibre as is practicable. However, a minimum amount of concrete cover for the bars is essential to protect the bars from damage due to corrosion and fire.

Fig. 4.7 Moment - curvature relationships

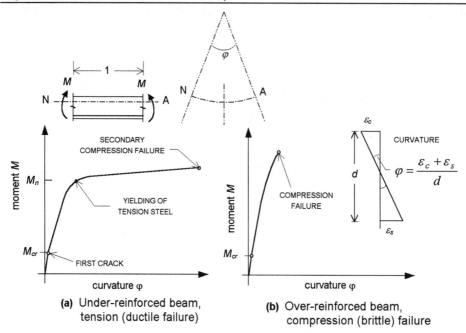

The requirements for placement of reinforcement are given in CSA A23.1-94(12), and are discussed in more detail in Section 5.2. To meet the minimum spacing requirements, reinforcing bars may have to be placed in two or more closely spaced layers. In such situations, it is usual to assume, for flexural computations, that all the steel area is concentrated at the centroid of the steel bars at the effective depth, d, and that all the bars sustain the same stress and strain.

4.5 ANALYSIS AND DESIGN

There are two kinds of problems encountered in reinforced concrete practice. In the first kind, termed *analysis* (or review) problems, the complete cross-sectional dimensions, including steel area and the material properties, of the member are known. It is desired to compute either (1) the stresses in the materials under a given load or (2) the loads (moments) that the member can resist. The second type of problem involves *design*. In this type of problem, the specified material properties and the load effects (moments, shears, etc.) are known, and it is required to compute the dimensions and reinforcement details of the member necessary to sustain satisfactorily the known load effects.

Fig. 4.8 Cracked section, elastic stress distribution

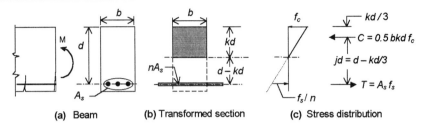

(a) Beam (b) Transformed section (c) Stress distribution

4.6 SERVICE LOAD STRESSES

4.6.1 Analysis for Stresses - Section Cracked and Elastic

Sections designed for strength (ultimate limit states) under factored loads must be checked for serviceability (deflection, crack width, etc.) under the specified loads. Under service loads, flexural members are generally in the cracked phase, with linear distribution of strains and stresses, as explained in Section 4.4 and Fig. 4.6d, and shown again in Fig. 4.8. The computation of stresses using the straight-line theory (elastic theory) for a cracked section is discussed below.

Figure 4.8a shows a beam of rectangular section, subjected to a specified load moment, M. For this beam, the corresponding transformed section, neglecting the concrete on the tension side of the neutral axis, is shown in Fig. 4.8b. The centroid of the transformed section locates the neutral axis. Expressing the depth to neutral axis as a fraction, k, of the effective depth, d, and equating the moments about the neutral axis of the compression and tension areas:

$$\frac{b(kd)^2}{2} = n A_s (d - kd) \qquad (4.3)$$

Solving this equation,
$$k = \frac{\sqrt{2B+1} - 1}{B} \qquad (4.4a)$$

where
$$B = \frac{bd}{nA_s} = \frac{1}{\rho n}$$

and
$$\rho = \frac{A_s}{bd} = \text{the reinforcement ratio}$$

Alternatively,
$$k = \sqrt{2\rho n + (\rho n)^2} - \rho n \qquad (4.4b)$$

72 REINFORCED CONCRETE DESIGN

The moment of inertia of the section, I_{cr}, is given by:

$$I_{cr} = \frac{b(kd)^3}{3} + nA_s(d-kd)^2 \qquad (4.5)$$

Knowing the neutral axis location and moment of inertia, the stresses in the concrete (and steel) in this composite section may be computed from the flexure formula $f = My/I$, as explained in Section 4.2 and illustrated in Example 4.1. The same results could be obtained more simply and directly considering the static equilibrium of internal forces and externally applied moment.

With the distribution of concrete and steel stresses shown in Fig. 4.8c, the internal resultant forces in concrete and steel forming the internal resisting couple are, respectively, C and T, and the lever arm of this couple is jd, where

$$C = b\,kd\,\frac{f_c}{2}, \quad T = A_s f_s \qquad (4.6)$$

$$jd = d - \frac{kd}{3}, \text{ or } j = \left(1 - \frac{k}{3}\right)$$

Equating the external applied moment M and the moment of resistance,

$$M = C\,jd, \text{ or } M = T\,jd$$

$$M = \frac{bkd\,f_c}{2} jd, \text{ or } M = A_s f_s jd \qquad (4.7)$$

from which

$$f_c = \frac{M}{\left(\frac{1}{2}kj\right)bd^2} \qquad (4.8)$$

and

$$f_s = \frac{M}{A_s jd} \qquad (4.9)$$

4.6.2 Beams of Other Shapes

Although the discussion above used a beam of rectangular section for the sake of simplicity, the procedure is identical for sections of other shapes, such as *Tee*, *Ell*, *I*, etc. Frequently beams are reinforced in compression, with steel bars placed near the compression edge of the cross-section. Such beams are termed *doubly reinforced*. The compression steel area is then transformed to concrete by multiplying the area by $n-1$, the -1 allowing for the concrete displaced by steel (see also Section 4.4 for inelastic effects). The stresses can thus be determined using the basic principles described above.

Fig. 4.9 Analysis for stress, straight line theory, Example 4.2

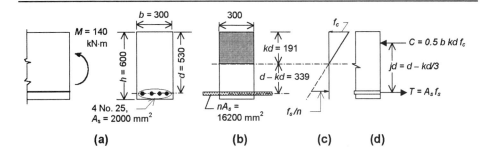

EXAMPLE 4.2

The beam shown in Fig. 4.9a is subjected to a specified load bending moment of 140 kN·m. The concrete strength f_c' is 30 MPa, and the yield strength of steel f_y is 400 MPa. Compute the stress in concrete and steel.

SOLUTION

Since concrete is assumed to take no flexural tension, the section will be treated as cracked. The modular ratio $n = E_s/E_c = 200\,000/(4500\sqrt{30}) = 8$, and the transformed section is shown in Fig. 4.9b. To compute the depth to the neutral axis, take moments of areas about the neutral axis (Eq. 4.3):

$$\frac{300\,(kd)^2}{2} = 8 \times 2000\,(530 - kd)$$

Solving, $kd = 191$ mm, $k = 0.360$, $j = 1 - \dfrac{0.360}{3} = 0.880$

Concrete stress (Eq. 4.8) is: $f_c = \dfrac{140 \times 10^6}{0.5 \times 0.360 \times 0.880 \times 300 \times 530^2} = 10.5$ MPa

Steel stress (Eq. 4.9) is: $f_s = \dfrac{140 \times 10^6}{2000 \times 0.880 \times 530} = 150$ MPa

Note that the same results are obtained by the flexure formula $f = \dfrac{My}{I}$

The moment of inertia of the cracked transformed section (Eq. 4.5) is calculated as $I_{cr} = 2.56 \times 10^9$ mm^4. The compressive stress in the top fibre of the concrete is:

$$f_c = \frac{My_c}{I_{cr}} = \frac{140 \times 10^6 \times 191}{2.56 \times 10^9} = 10.5 \text{ MPa}$$

74 REINFORCED CONCRETE DESIGN

The steel stress is n times the stress in the transformed section at the steel level so that

$$f_s = \frac{nM(d-kd)}{I_{cr}} = \frac{8 \times 140 \times 10^6 (530-191)}{2.56 \times 10^9} = 150 \text{ MPa} \text{ as before}$$

EXAMPLE 4.3

A T-beam having the cross-sectional dimensions shown in Fig. 4.10a is subjected to a specified load bending moment of 250 kN·m. Compute the steel and concrete stresses. The material properties are $f_c' = 40$ MPa and $f_y = 400$ MPa.

SOLUTION

Considerations about the width of flange that is effective in resisting flexural compression are described in Section 4.9.3. For the present it will be assumed that the full flange width is effective.

The modular ratio is: $n = \dfrac{E_s}{E_c} = 7$

The reinforcement area is transformed to concrete. The section is considered cracked, and the concrete area on the tension side of the neutral axis is ignored. For elastic stress distribution under specified load, the neutral axis passes through the centroid of the effective transformed section.

Fig. 4.10 Straight line theory, T - beams, Example 4.3

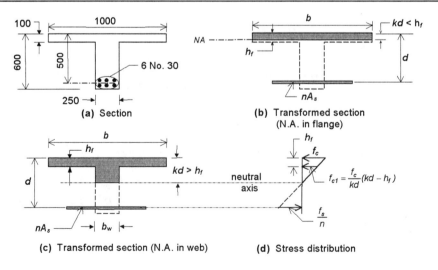

The neutral axis may lie in the flange, as shown in Fig. 4.10b (that is $kd < h_f$), or it may pass through the web, as in Fig. 4.10c, ($kd > h_f$).

To locate the neutral axis, first assuming that it lies in the flange, equating the moments of areas about neutral axis (Fig. 4.10b) gives the equation:

$$b\frac{(kd)^2}{2} = nA_s(d-kd) = 1000\frac{(kd)^2}{2} = 7 \times 4200(500-kd)$$

Solving, $kd = 145$ mm

Since this is greater than $h_f = 100$ mm, the assumption is not correct, and the neutral axis passes through the web. Taking moments of areas for the case in Fig. 4.10c:

$$b_w\frac{(kd)^2}{2} + (b-b_w)h_f\left(kd-\frac{h_f}{2}\right) = nA_s(d-kd)$$

Solving, $kd = 150$ mm $> h_f$ OK

The stresses may be found using the stress distribution or by using the flexure formula. Considering the stress distribution shown in Fig. 4.10d, the moment is given by the moment of the compressive forces about the centroid of steel. Taking the compressive area as the difference between two rectangles $b \times kd$ and $(b - b_w) \times (kd - h_f)$, the moment equation gives:

$$M = b \times kd \times \frac{f_c}{2}\left(d - \frac{kd}{3}\right) - (b-b_w)(kd-h_f)\frac{f_{c1}}{2}\left(d - h_f - \frac{kd-h_f}{3}\right)$$

Substituting for various terms and noting that

$$f_{c1} = \frac{f_c(kd-h_f)}{kd} = f_c\frac{(150-100)}{150} = 0.333 f_c,$$

$$250 \times 10^6 = 1000 \times 150\frac{f_c}{2}\left(500 - \frac{150}{3}\right) - (1000-250)(150-100)$$

$$\times \frac{0.333}{2}f_c\left(500 - 100 - \frac{50}{3}\right)$$

Solving $f_c = 7.97$ MPa

From the stress distribution diagram,

$$\frac{f_s/n}{f_c} = \frac{d-kd}{kd}$$

$$f_s = nf_c\frac{d-kd}{kd} = 130 \text{ MPa}$$

Alternatively, to use the flexure formula,

$$f = \frac{My}{I_{cr}}$$

$$I_{cr} = \frac{1000 \times 150^3}{3} - \frac{(1000-250) \times 50^3}{3} + 7 \times 4200(500-150)^2 = 4.70 \times 10^9 \text{ mm}^4$$

$$f_c = \frac{My_c}{I_{cr}} = \frac{250 \times 10^6 \times 150}{4.70 \times 10^9} = 7.97 \text{ MPa as before}$$

and $f_s = n\dfrac{My_s}{I_{cr}} = 7 \times \dfrac{250 \times 10^6 \times (500-150)}{4.70 \times 10^9} = 130$ MPa

EXAMPLE 4.4

The cross-section dimensions of a beam with compression reinforcement are shown in Fig. 4.11a. The beam is subjected to a specified load moment of 175 kN·m. Compute the specified load stresses in concrete and steel.
The material properties are $f_c' = 30$ MPa and $f_y = 400$ MPa.

SOLUTION

The specified load stresses are computed by the straight line theory applied to the cracked transformed section, shown in Fig. 4.11b. As explained in Section 4.3, in transforming the compression reinforcement (for stress computations here) the effective modular ratio is taken as $2n$. Allowing for the concrete displaced by A_s', the net transformed area for compression reinforcement is $(2n - 1)A_s'$. The modular ratio, $n = E_s/E_c = 200\,000/(4500\sqrt{30}) = 8$. To locate the neutral axis (centroid) of the transformed section, equating the moment of areas about the neutral axis (Fig. 4.11b),

Fig. 4.11 Example 4.4

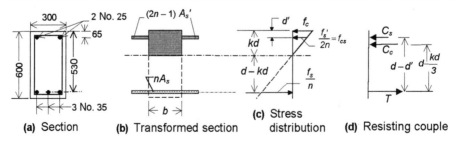

(a) Section (b) Transformed section (c) Stress distribution (d) Resisting couple

$$b\frac{(kd)^2}{2} + (2n-1)A_s'(kd-d') = nA_s(d-kd)$$

Substituting, $300\frac{(kd)^2}{2} + (16-1)1000(kd-65) = 8 \times 3000 \times (530-kd)$

Solving, kd = 199 mm

Referring to the stress distribution in Fig. 4.11c, the concrete stress at compression reinforcement level, f_{cs}, is:

$$f_{cs} = \frac{f_c}{kd}(kd-d') = \frac{f_c(199-65)}{199} = 0.673 f_c$$

Equating the applied moment to the moment of resisting forces (here taken as the moment of compressive forces C_c in concrete and C_s in steel, about the tension reinforcement):

$$M = b \times kd \times \frac{f_c}{2}\left(d - \frac{kd}{3}\right) + (2n-1) \times A_s' \times f_{cs} \times (d-d')$$

Substituting,

$$175 \times 10^6 = 300 \times 199 \times \frac{f_c}{2}\left(530 - \frac{199}{3}\right) + (2 \times 8 - 1) \times 1000 \times 0.673 f_c (530-65)$$

Solving, $f_c = 9.44$ MPa

$f_{cs} = 0.673 f_c = 6.35$ MPa

Compression steel stress $= 2n f_{cs} = 101$ MPa

Tension steel stress $= f_s = n f_c \frac{d-kd}{kd} = 8 \times 9.45 \frac{530-199}{199} = 126$ MPa

[Using flexure formula, $I_{cr} = 3.69 \times 10^9$ mm^4

and $f_c = \frac{My_c}{I_{cr}} = 9.44$ MPa as above]

4.6.3 Specified (Service) Load Moments

When it is desired to compute a specified or service load moment on a beam of known cross-section, such that the steel and concrete stresses are not to exceed specified stresses, the procedure is very similar to that given in Section 4.6.1 and Example 4.2. Referring to Fig. 4.8, the location of the actual neutral axis is given by Eq. 4.4. The moment capacity can be expressed in terms of the maximum specified steel and concrete stresses using Eq. 4.7.

Fig. 4.12

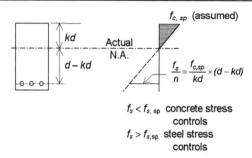

$f_s < f_{s,sp}$ concrete stress controls
$f_s > f_{s,sp}$ steel stress controls

Given a beam section, the specified stresses in both steel and concrete may not be reached simultaneously. Therefore, the lower of the two moments computed using either the specified steel stress or the specified concrete will give the specified moment and the corresponding stress (f_s or f_c) that will reach the specified maximum first.

Alternatively, from the known neutral axis location and the stress diagram, it is possible to determine first whether the steel or concrete stress controls. For example, referring to Fig. 4.12, if it is assumed that the concrete stress reaches the limit first, the associated steel stress is obtained as:

$$f_s = nf_c \frac{d-kd}{kd} \qquad (4.10)$$

If f_s so calculated is less than the specified steel stress, the concrete stress does in fact control, and the specified moment is calculated using the specified concrete stress. If, on the other hand, f_s from Eq. 4.10 exceeds the specified steel stress, it is the steel stress that reaches the limiting value first. Therefore, the specified moment is calculated using the specified steel stress.

EXAMPLE 4.5

For the beam in Example 4.2, compute the service load moment corresponding to a stress of $0.40f_c'$ for concrete in flexural compression and 165 MPa for tensile stresses in Grade 400 reinforcement.

SOLUTION

Specified $f_c = 0.4 \times 30 = 12$ MPa
Specified $f_s = 165$ MPa

Fig. 4.13

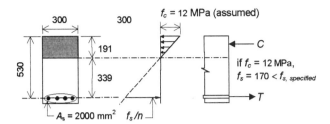

Depth to neutral axis (from Example 4.2), $kd = 191$ mm

Assuming the concrete stress reaches the specified value of 12 MPa, the steel stress will then be (Fig. 4.13):

$$f_s = 8 \times \frac{12}{191} \times 339 = 170 \text{ MPa} > 165 \text{ MPa specified}$$

Steel stress controls, and

$$M = A_s f_s jd = 2000 \times 165 \times 0.880 \times 530 \times 10^{-6} = 154 \text{ kN} \cdot \text{m}$$

4.7 STRENGTH IN FLEXURE

4.7.1 Using Stress-Strain Relationships

In Chapter 3, it was explained that in Limit States Design safety is ensured by designing members to have a factored resistance equal to or greater than the required strength, which is the load effects resulting from factored load combinations. The factored resistance, M_r, of a member is computed theoretically using factored material strengths and member dimensions. The evaluation of the *nominal resistance* in flexure, M_n, obtained using the specified material strengths without the ϕ-factors, and of the *factored resistance*, M_r, obtained using the specified material strengths with the ϕ-factors, is described in the following sections.

As described in Section 4.4, the ultimate failure in flexure of a reinforced concrete beam, whether under-reinforced or over-reinforced, takes place by crushing of concrete, with the maximum compressive strain in concrete reaching its failure value, ε_{cu} (Fig. 4.6f, g). At failure, the compressive strain in concrete is assumed to vary

80 REINFORCED CONCRETE DESIGN

linearly from zero at the neutral axis to ε_{cu} at the extreme fibre. If the material resistance factors, ϕ_c and ϕ_s are not included, the distribution of compressive stress in concrete (frequently referred to as the stress block) will be nearly identical to the stress-strain relationship for a concrete cylinder in compression, and will be as shown in Fig. 4.14c. (This assumes that the stress-strain relationship for a concrete cylinder in concentric compression is applicable to flexure.) The actual shape of the stress-strain relationship for concrete is not unique and depends on several factors, such as the cylinder strength, density, and rate and duration of loading. However, in order to compute the flexural strength corresponding to the stress pattern shown in Fig. 4.14c, it is not essential to know the exact shape of the stress distribution in concrete. It is sufficient to know only the location and magnitude of the resultant compressive force, C. Considering a beam of rectangular section (Fig. 4.14), the magnitude and location of the resultant compression, C, can be expressed in terms of two nondimensional parameters, namely, α, which represents the ratio of the average compressive stress in the compression zone to the cylinder strength, f_c', and β, which is the ratio of the depth of resultant force, C, to the depth of neutral axis, c (Fig. 4.14c). For the rectangular section, where the area under compression has a uniform width, α and β are basic properties of the stress-strain curve. Values of α and β can be obtained mathematically, if the actual shape of the stress-strain relationship is known or, more appropriately,

Fig. 4.14 *Conditions at ultimate load*

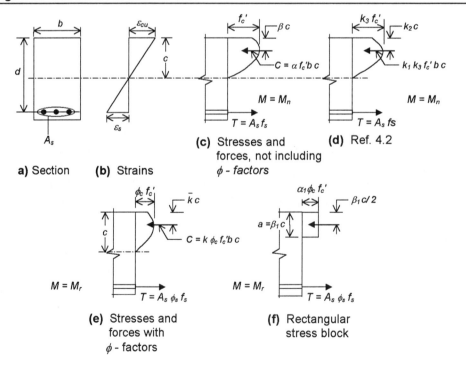

they may be determined experimentally (Refs. 4.1 and 4.4).

For the rectangular section in Fig. 4.14, knowing the values of ε_{cu}, α and β, considerations of equilibrium of forces and compatibility of strains are sufficient to compute the value of the nominal bending strength, M_n. Equating the compressive and tensile forces,

$$\alpha f_c' b c = A_s f_s \qquad (4.11)$$

Moment of the resisting couple $C = T$ is:

$$M_n = Cjd = \alpha f_c' b c (d - \beta c) \qquad (4.12a)$$

or
$$M_n = Tjd = A_s f_s (d - \beta c) \qquad (4.12b)$$

Equations 4.11 and 4.12 express equilibrium conditions and must always be satisfied. For a tension failure (under-reinforced beam), the steel yields prior to failure so that at the nominal strength, $f_s = f_y$, and Eq. 4.11 directly gives the value of c as:

$$c = \frac{A_s f_y}{\alpha f_c' b} \qquad (4.13)$$

Substituting for c in Eq. 4.12a or b, the ultimate strength is obtained as:

$$M_n = A_s f_y \left(d - \frac{\beta A_s f_y}{\alpha f_c' b} \right) \qquad (4.14)$$

Using the notations:

$$\rho = \frac{A_s}{bd} \qquad (4.15)$$

and
$$\omega = \frac{A_s f_y}{bd f_c'} \qquad (4.16)$$

where ρ is a reinforcement ratio expressing A_s as a fraction of effective concrete area, and ω is a tension reinforcement index. Equation 4.14 may also be expressed as:

$$M_n = \rho f_y \left(1 - \frac{\beta \rho f_y}{\alpha f_c'} \right) bd^2 \qquad (4.17)$$

$$= \omega \left(1 - \frac{\beta}{\alpha} \omega \right) f_c' bd^2 \qquad (4.18)$$

Equation 4.18 shows that, for a tension failure, the only property of the stress block needed to compute the flexural strength is the ratio, β/α.

For a compression failure (over-reinforced beam) the concrete crushes before the steel yields, so the steel stress at failure $f_s (< f_y)$ has to be computed before Eqs. 4.11 and 4.12 can be solved for M_n. For this, considering the strain distribution in

Fig. 4.12b, the steel strain, ε_s, and stress, f_s, can be computed as:

$$\varepsilon_s = \varepsilon_{cu} \frac{d-c}{c} \qquad (4.19)$$

and
$$f_s = E_s \varepsilon_s = E_s \varepsilon_{cu} \frac{d-c}{c} \qquad (4.20)$$

Substituting Eq. 4.20 for f_s in Eq. 4.11, one obtains:

$$\alpha f_c' bc = A_s E_s \varepsilon_{cu} \frac{d-c}{c} \qquad (4.21)$$

Equation 4.21 can be solved for c, which on substitution into Eq. 4.12 yields the flexural strength. In this case, it is necessary to know all three parameters, ε_{cu}, α, and β, related to the stress distribution.

Equations 4.11 to 4.21 can be applied to T-sections also, provided the neutral axis at the nominal strength is within the compression flange. When the area under compression is not a rectangle (as is the case with the neutral axis of a T-beam being below the flange), the entire shape of the stress block is required for an *exact* analysis.

4.7.1(a) Parameters ε_{cu}, α, β

Assuming a flexural stress distribution in the compression zone of a rectangular beam as shown in Fig. 4.14d, the parameters ε_{cu}, k_1, k_2, and k_3 defining the properties of the stress block have been evaluated experimentally (Ref. 4.1, 4.4). $k_1 k_3$ and k_2 correspond respectively to α and β in Fig. 4.14c. The experimental measurements show considerable scatter in the value of ε_{cu}. However, the generally conservative lower bound value of 0.0035 recommended by the Code is satisfactory. Over the normal range of concrete strengths ($f_c' = 20$ to 80 MPa), the average value of k_3 was found to range between 0.96 and 1.04 for normal density concrete and between 0.98 and 1.01 for low density concrete. This indicates that the peak compressive stress in flexure is in fact very nearly equal to the cylinder strength. These studies confirmed that the shape of the stress block is similar to the stress-strain curve in uniaxial compression. Lower bound values recommended for the Code constants α_1 and β_1 in Fig. 4.14f (Ref. 4.7) are given in Fig. 4.15.

4.7.1(b) Factored Moment Resistance, M_r

The factored moment resistance, M_r, is computed using the factored material strengths, $\phi_c f_c'$ and $\phi_s f_s$. The concrete stress distribution and the force in the reinforcement for this case are shown in Fig. 4.14e. Again, for a tension failure, when $\varepsilon_s \geq \varepsilon_y$, equating $C = T$ gives:

$$k\phi_c f'_c bc = A_s \phi_s f_y$$

$$c = \frac{A_s \phi_s f_y}{k \phi_c f'_c b} \quad (4.13a)$$

The moment of the couple, $C = T$ gives:

$$M_r = A_s \phi_s f_y (d - \bar{k}c)$$

$$= A_s \phi_s f_y \left(d - \frac{\bar{k} A_s \phi_s f_y}{k \phi_c f'_c b} \right) \quad (4.14a)$$

Note that the concrete stress distribution shown in Fig. 4.14e differs from that shown in Fig. 4.14c only by the constant factor ϕ_c. Therefore, the constants k and $\bar{k}$ (Fig. 4.14e) are the same as α and β (Fig. 4.14c), respectively. If the values for α_1 and β_1 given in Fig. 4.15 are used, Eq. 4.14a reduces to:

$$M_r = A_s \phi_s f_y \left(d - \frac{A_s \phi_s f_y}{2\alpha_1 f'_c b} \right) \quad (4.14b)$$

For compression failure, $\varepsilon_s < \varepsilon_y$ and the steel strain is given by Eq. 4.19. Equating C and T gives:

$$k \phi_c f'_c bc = A_s \phi_s E_s \varepsilon_{cu} \frac{d-c}{c} \quad (4.21a)$$

Equation 4.21a can be solved for c. The factored moment resistance is then obtained as the moment of the couple $C = T$:

$$M_r = A_s \phi_s f_s (d - \bar{k}c) \quad (4.21b)$$

4.7.1(c) Balanced Strain Condition

In Section 4.4 it was explained that a beam section is termed a balanced section when the tension reinforcement reaches its yield strain, ε_y, just as the concrete in compression reaches its ultimate strain, ε_{cu}. At the balanced strain condition, the ratio of neutral axis depth to effective depth, c/d, is obtained from Eq. 4.19 (Fig. 4.14b) as:

$$\left(\frac{c}{d} \right)_b = \frac{\varepsilon_{cu}}{\varepsilon_{cu} + \varepsilon_s} = \frac{\varepsilon_{cu}}{\varepsilon_{cu} + f_y/E_s} = \frac{E_s \varepsilon_{cu}}{E_s \varepsilon_{cu} + f_y} \quad (4.22)$$

With the values specified in CSA A23.3-94(8.5.4 and 10.1.4) of $E_s = 200\,000$ MPa and

84 REINFORCED CONCRETE DESIGN

$\varepsilon_{cu} = 0.0035$, the neutral axis depth ratio for balanced strain conditions is given by:

$$\left(\frac{c}{d}\right)_b = \frac{700}{700 + f_y} \quad (4.23)$$

To ensure a ductile failure, the Code limits the maximum reinforcement in flexural members, such that the actual neutral axis depth ratio, c/d, calculated using factored material strengths $\phi_c f_c'$ and $\phi_s f_y$ is less than that corresponding to the balanced strain conditions. That is, the reinforcement shall be such that:

$$\frac{c}{d} \leq \frac{700}{700 + f_y} \quad (4.24)$$

4.7.2 Equivalent Rectangular Stress Distribution

Referring to Fig. 4.14e, the strength is uniquely determined provided the magnitude and location of the resultant compressive force, C, in concrete is known, no matter what the shape of the stress block. Therefore, the stress block may be imagined to have any convenient shape, so long as the magnitude and location of C is unaltered. The simplest form of such a fictitious stress distribution is a rectangular stress block, shown in Fig. 4.14f, where a uniform stress of $\alpha_1 \phi_c f_c'$ acts over a depth $a = \beta_1 c$. In order to have the same location for the resultant force, C, as in Fig. 4.14e,

$$\frac{a}{2} = \frac{\beta_1 c}{2} = \bar{k}c$$

or the factor β_1 giving the depth of stress block in terms of c is

Fig. 4.15 Values of $\alpha_1\beta_1$ from tests of concrete prisms (Ref. 4.7)

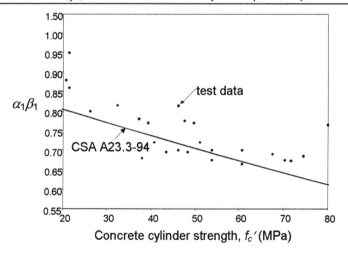

$$\beta_1 = 2\bar{k} \tag{4.25}$$

For the magnitude of C to remain unaltered,

$$C = k\phi_c f_c' bc = \alpha_1 \phi_c f_c' ba = \alpha_1 \phi_c f_c' b\beta_1 c$$

$$\alpha_1 = \frac{k}{\beta_1} = \frac{k}{2\bar{k}} \tag{4.26}$$

Experimental results in Ref. 4.4 show that, for concrete strengths in the range of 15 to 100 MPa, the mean value of $\bar{k}/k$ varies over a narrow range between 0.54 and 0.62 for normal density concrete (and between 0.57 and 0.70 for low density concrete). Therefore, the uniform stress in the equivalent rectangular stress block can be taken as $\alpha_1 \phi_c f_c'$. The depth of the stress block is given by the factor $a/c = \beta_1 = 2\bar{k}$, where the recommended value of $\bar{k}$ is given in Fig. 4.15. Values of $\alpha_1 \beta_1$, based on the recommended values k and $\bar{k}$ (Eqs. 4.25 and 4.26), are also plotted in Fig. 4.15.

4.8 CODE RECOMMENDATIONS

The assumptions forming the basis of flexural computations by CSA A23.3-94 are given in Clause 10.1. The strains in steel and concrete are assumed to vary linearly with distance from the neutral axis, with the maximum compressive strain in concrete limited to 0.0035. Steel stress is taken as $\phi_s f_s = \phi_s E_s \varepsilon_s \leq \phi_s f_y$. Tensile strength of concrete is neglected. In compression, the stress-strain relation for concrete may be assumed as trapezoidal, parabolic, or any other shape provided the strength so predicted is in agreement with test results. As one such acceptable stress distribution, the Code recommends an equivalent rectangular concrete stress distribution with a uniform stress of $\alpha_1 \phi_c f_c'$ over a depth $a = \beta_1 c$ (Fig. 4.16) where

$$\alpha_1 = 0.85 - 0.0015 f_c' \geq 0.67 \quad \text{and,} \quad \beta_1 = 0.97 - 0.0025 f_c' \geq 0.67$$

The uniform stress ($\alpha_1 \phi_c f_c'$) and the depth of stress block ($\beta_1 c$) recommended here are essentially the same as those determined experimentally (Ref. 4.4) and shown in Fig. 4.15.

4.9 MAXIMUM REINFORCEMENT RATIO

CSA Standard A23.3-94(10.1.4) limits the maximum tension reinforcement in flexural members, such that the ratio of neutral axis depth to effective depth, c/d, satisfies the condition:

86 REINFORCED CONCRETE DESIGN

$$\frac{c}{d} \leq \frac{700}{700+f_y} \tag{4.27}$$

For calculating the neutral axis depth, c, the factored concrete strength, $\phi_c f_c'$, and the factored reinforcement bar force, $A_s \phi_s f_s$, are used. The requirement of Eq. 4.27 ensures a tension (ductile) failure for the beam (also Section 4.7.1c).

In practice, reinforcement corresponding to the upper limit given by Eq. 4.27 will usually result in gross cross-sectional dimensions for the beam which are relatively small. Such slender beams may not have adequate stiffness to control deflections. Furthermore, it is difficult to place such a large amount of reinforcement in small cross-sections. Limitations of deflection, convenience in placement of reinforcement, and economy in design generally dictate larger overall beam dimensions, with correspondingly lower reinforcement ratios usually in the range of 30 to 40 percent of the maximum limit given by Eq. 4.27. These lower reinforcement ratios further improve the ductility of the beams.

4.9.1 Rectangular Section with Tension Reinforcement Only

The equivalent rectangular concrete stress distribution recommended in CSA A23.3-94 is shown in Fig. 4.16b. The right hand side of Eq. 4.27 corresponds to the balanced strain condition with $\varepsilon_{cu} = 0.0035$ and $\varepsilon_s = \varepsilon_y$. Therefore, at the upper limit of c/d given by Eq. 4.27, the tension reinforcement would have yielded. Equating $C = T$,

$$\alpha_1 \phi_c f_c' b (\beta_1 c) = A_s \phi_s f_y \tag{4.28}$$

$$c = \frac{A_s \phi_s f_y}{\alpha_1 \beta_1 \phi_c f_c' b}$$

$$\frac{c}{d} = \frac{A_s \phi_s f_y}{\alpha_1 \beta_1 \phi_c f_c' b d}$$

Equating this to the limiting value of c/d in Eq. 4.27 yields:

$$\frac{c_{max}}{d} = \left[\frac{\phi_s f_y}{\alpha_1 \beta_1 \phi_c f_c'} \frac{A_s}{bd} \right]_{max} = \left(\frac{700}{700+f_y} \right)$$

CHAPTER 4 BEHAVIOUR IN FLEXURE 87

Fig. 4.16 Rectangular stress block, CSA A23.3

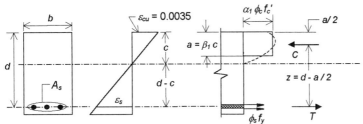

(a) CSA rectangular stress block

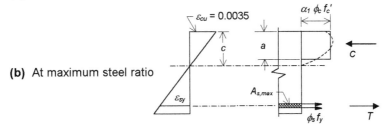

(b) At maximum steel ratio

Denoting $A_s/(bd)$ by ρ, the reinforcement ratio,

$$\rho_{max} = \overline{\rho} = \left(\frac{\alpha_1 \beta_1 \phi_c f_c'}{\phi_s f_y}\right)\left(\frac{700}{700 + f_y}\right) \qquad (4.29)$$

Equation 4.29 may also be obtained directly by substituting the maximum value of c from Eq. 4.27 into Eq. 4.28. Equation 4.29 gives the maximum reinforcement ratio, $\overline{\rho}$, permitted by the Code for a beam of rectangular section. As an example, for a specified concrete strength $f_c' = 30$ MPa and a steel yield strength $f_y = 400$ MPa, with $\alpha_1 = 0.805$, $\beta_1 = 0.895$ (Section 4.8), $\phi_c = 0.6$ and $\phi_s = 0.85$, substitution in Eq. 4.29 gives:

$$\rho_{max} = 0.0243 = 2.43\%$$

4.9.2 Rectangular Beam with Compression Reinforcement

There are situations where the overall dimensions of a beam section are restricted so that, even with the maximum permissible reinforcement provided, the strength of the section is inadequate to carry the loads. In such cases, additional tension and corresponding compression reinforcement is provided to enhance the strength of the section. There are also occasions when reinforcement may have to be provided near both top and bottom faces of the beam, such as at sections where there is a reversal of moment, at supports of continuous beams, for beams whose long time deflections are to be controlled, and for members detailed to resist seismic forces in a ductile manner. Such beams are called doubly reinforced beams.

Figure 4.17 shows the dimensions of a rectangular beam with compression reinforcement A'_s located at a depth d' below the compression face. The stress in the compression reinforcement, f'_s, depends on the strain, ε'_s, at its level, which may or may not have reached the yield level. From the strain triangle,

$$\varepsilon_s' = \frac{0.0035}{c}(c-d') = 0.0035\left(1-\frac{d'}{c}\right)$$

and
$$f_s' = E_s \varepsilon_s' = 700\left(1-\frac{d'}{c}\right) \leq f_y'$$

where f_y' is the specified yield strength of compression reinforcement. As the tension reinforcement reaches its maximum limit, the value of c is given by Eq. 4.27 as

$$c_{max} = \frac{700}{700+f_y}d$$

Substituting this value for c in the equation for f_s', the compression reinforcement stress corresponding to the maximum permissible amount of tension reinforcement is obtained as:

$$f_s' = 700\left(1 - \frac{700+f_y}{700}\frac{d'}{d}\right) \leq f_y' \qquad (4.30)$$

For convenience, the internal resisting couple may be split into two parts, the first consisting of the compressive force C_c, in the concrete of area, bc, and a corresponding tension steel area, A_{s1}, with force, T_1, and the second part consisting of

Fig. 4.17 Doubly reinforced section, maximum reinforcement ratio

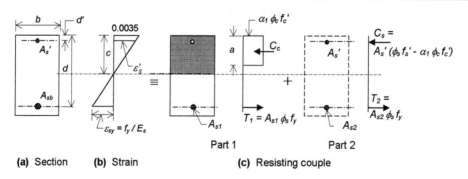

(a) Section   (b) Strain   (c) Resisting couple

the compressive force, C_s, in the compression reinforcement and an associated tensile force, T_2, on steel area, A_{s2}. This two-part approach is shown in Fig. 4.17c. The first part is identical to the rectangular beam with tension reinforcement only and the maximum reinforcement ratio for this part was derived as $\bar{\rho}$ in Section 4.9.1, so that:

$$\frac{A_{s1}}{bd} = \bar{\rho}$$

Equating the tensile and compressive forces for the second part,

$$T_2 = C_s$$
$$A_{s2}\phi_s f_y = A_s'\phi_s f_s'$$

In computing C_s, if allowance is made for the concrete area in compression actually displaced by A_s' but already included in C_c, the effective stress in A_s' may be more accurately taken as $(\phi_s f_s' - \alpha_1 \phi_c f_c')$ and the relation $T_2 = C_s$ becomes:

$$\phi_s A_{s2} f_y = A_s'(\phi_s f_s' - \alpha_1 \phi_c f_c')$$

$$\frac{A_{s2}}{bd} = \frac{A_s'}{bd} \times \frac{(\phi_s f_s' - \alpha_1 \phi_c f_c')}{\phi_s f_y} = \rho' \frac{(\phi_s f_s' - \alpha_1 \phi_c f_c')}{\phi_s f_y}$$

where $\rho' = A_s'/(bd)$ is the compression reinforcement ratio.

The total tension reinforcement ratio for limiting conditions is:

$$\rho_{max} = \frac{A_s}{bd} = \frac{A_{s1} + A_{s2}}{bd} = \bar{\rho} + \rho' \frac{(\phi_s f_s' - \alpha_1 \phi_c f_c')}{\phi_s f_y} \quad (4.31)$$

where $\bar{\rho}$ and f_s' are given by Eqs. 4.29 and 4.30, respectively. Note that a normal level of accuracy does not warrant the allowance for the compression concrete area displaced by A_s', so that the term $\alpha_1 \phi_c f_c'$ in Eq. 4.31 may be neglected. This is particularly so when the compression reinforcement stress, f_s', is high relative to f_c'. Equation 4.31 can then be simplified as:

$$\rho_{max} = \bar{\rho} + \rho' \frac{f_s'}{f_y}$$

When the compression reinforcement yields and has the same strength as the tension reinforcement, $\rho_{max} = \bar{\rho} + \rho'$.

As an illustrative example, for a beam of rectangular section with $f_c' = 30$ MPa, $f_y = 400$ MPa, $f_y' = 350$ MPa, $d'/d = 0.15$, and $\rho' = A_s'/bd = 0.012$, the maximum permissible tension reinforcement ratio, $\rho = A_s/bd$, can be worked as follows. Compression reinforcement stress at the maximum value of c/d is given by Eq. 4.30 as:

$$f_s' = 700\left(1 - \frac{700+400}{700} \times 0.15\right) = 535 > f_y'$$

so

$$f_s' = f_y' = 350 \text{ MPa}$$

The value ρ_{max} is given by Eq. 4.31. The maximum reinforcement ratio, $\bar{\rho}$, for a singly reinforced section for these material properties, was computed in the example at the end of Section 4.9.1 as $\bar{\rho} = 0.0243$. Thus Eq. 4.31 yields

$$\rho_{max} = 0.0243 + 0.012 \frac{0.85 \times 350 - 0.805 \times 0.6 \times 30}{0.85 \times 400} = 0.0343$$

If the simplified Eq. 4.31 is used, ρ_{max} is 0.0348.

4.9.3 T-Beams

Beams having T-shaped sections (Fig. 4.18a) are occasionally used in structures. Furthermore, floor slabs are usually cast monolithically with supporting beams, in which case a portion of the slab adjacent to the beam will effectively add to the latter's

Fig. 4.18 T - beams

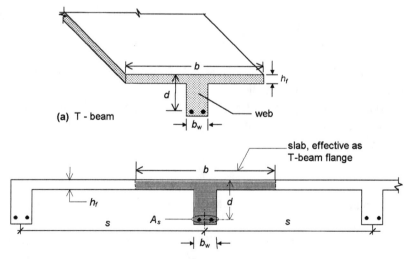

(a) T - beam

(b) T-beam action in beam and slab floor

strength, resulting in a beam of T-section (Fig. 4.18b). The slab portion of the T-section is termed the flange, and the beam portion below it the web or stem.

In a T-beam with a relatively wide flange, the flexural stress in the flange is not uniform over its width, but varies from a maximum over the web to progressively lower values at points further removed from the web. The theory of flexure used in reinforced concrete design assumes a uniform stress across the *width* of the section. Therefore for T-shaped sections, it is convenient to define a reduced effective flange width over which the stress distribution is considered to be uniform. CSA A23.3-94(10.3) defines the effective widths for flanged sections. To calculate the effective flange width, the overhang on either side of the web is to be not more than 1/5 of the span length for simply supported beams or 1/10 of the span length for continuous beams, 12 times the slab thickness, or 1/2 the clear distance to the next web. For simply supported beams, b = least of: 0.4 span + b_w, 24 h_f + b_w, or spacing s of beams.

The dimensions of a T-beam section considered fully effective are shown in Fig. 4.19. When the depth of neutral axis, c, falls within the flange thickness (Fig. 4.19a) (that is $c \leq h_f$), the analysis of the T-beam is identical to that of a rectangular beam of width b and depth d, since the difference in concrete area between these two shapes is entirely in the tension zone, an area which is neglected in the strength calculation. There is also the possibility where the neutral axis falls below the flange (Fig. 4.19b) ($c > h_f$), but the depth of stress block taken as $a = \beta_1 c$ is still within the flange thickness ($a \leq h_f$). Although the average stress $\alpha_1 f_c'$ and the β_1 values characterising the rectangular stress block are based on a compression concrete area of uniform width, extensive comparison with test results has shown (Ref. 4.5) that these values can be used with sufficient accuracy even when the concrete area under compression is not a rectangle. Thus, for all cases of $a \leq h_f$, the T-beam is usually analysed as a rectangular beam of width b.

For the case of the neutral axis passing through the web, with $a = \beta_1 c > h_f$, (Fig. 4.19c), the T-beam must be analysed to account for the different width of the compression areas in the flange and in the web. However, the rectangular stress block with uniform stress = $\alpha_1 f_c'$ is still a satisfactory assumption (Ref. 4.5). The maximum reinforcement condition with T-beam action is shown in Fig. 4.20. Once again, the resisting couple may be considered in two parts: (1) a web contribution consisting of the compressive force, C_w, in the concrete in the web portion and an equivalent steel area, A_{sw}, with tensile force, T_w; and (2) a flange contribution consisting of the compression, C_f, in the overhanging portions of the flanges and the corresponding tension, T_f, on steel area, A_{sf} (Fig. 4.20). For the first part, the maximum reinforcement ratio is as for a rectangular section, $b_w \times d$, so that

$$A_{sw} = \overline{\rho} b_w d$$

For the second part, equating $T_f = C_f$

92 REINFORCED CONCRETE DESIGN

$$A_{sf}\phi_s f_y = \alpha_1 \phi_c f'_c (b - b_w) h_f$$

$$A_{sf} = \frac{\alpha_1 \phi_c f'_c}{\phi_s f_y}(b - b_w) h_f$$

Designating $\qquad \rho_f = \dfrac{A_{sf}}{b_w d} = \dfrac{\alpha_1 \phi_c f'_c}{\phi_s f_y}\left(\dfrac{b - b_w}{b_w}\right)\dfrac{h_f}{d}$ (4.32)

the balanced reinforcement ratio in terms of the web area, $b_w d$ is:

$$\rho_{max} = \frac{A_s}{b_w d} = \frac{A_{sw} + A_{sf}}{b_w d} = \overline{\rho} + \rho_f \qquad (4.33)$$

In Eq. 4.33, $\overline{\rho}$ is the maximum reinforcement ratio for a rectangular beam, $b_w \times d$, with

Fig. 4.19 Possible neutral axis locations for T-sections

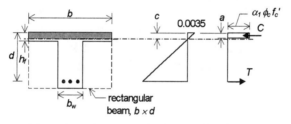

(a) $c \leq h_f$, as for rectangular beam, ($b \times d$)

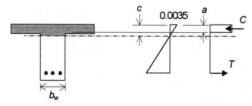

(b) $a < h_f < c$, may be treated as rectangular beam, ($b \times d$)

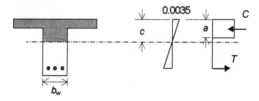

(c) $a > h_f$, treated as T- beam

Fig. 4.20 Balanced condition for T-beam

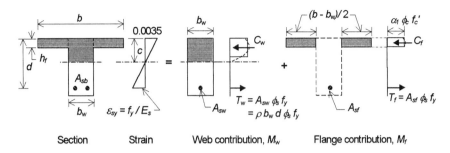

tension reinforcement only (Eq. 4.29).

In T-beams of usual proportions, because of the large concrete area in compression, the maximum reinforcement limitation is generally met unless extremely high (and costly) concentrations of reinforcement are provided in the narrower web.

4.10 MINIMUM REINFORCEMENT IN FLEXURAL MEMBERS

CSA A23.3-94(10.5) specifies minimum values for tension reinforcement in flexural members. For members other than slabs, $A_{s\,min}$ is given by Eq. 4.34 the factored moment resistance is at least 1/3 greater than the factored moment.

$$A_{s\,min} = \frac{0.2\sqrt{f'_c}}{f_y} b_t h \qquad (4.34)$$

Also, at every section of a flexural member, where tensile reinforcement is required by analysis, the minimum reinforcement provided will ensure that the factored moment resistance is at least equal to 1.2 times the value of the cracking moment, calculated using the modulus of rupture of concrete (Eq. 1.5). The requirements for minimum reinforcement in a flexural member ensure that brittle failure at first cracking

Fig. 4.21 Skin reinforcement for crack control

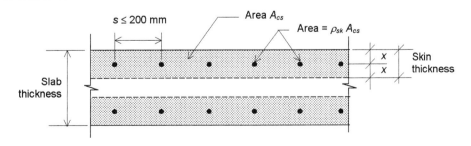

is avoided, and that the member has adequate post-cracking strength.

The minimum flexural reinforcement specified for slabs of uniform thickness is $0.002 A_g$ in each direction, where A_g is the gross area of section (CSA A23.3-94 (7.8)). This minimum reinforcement in slabs is meant to provide some control of the cracking due to shrinkage and temperature effects, and to tie the structure together after cracking. For exposure conditions where crack control is essential, reinforcement in excess of $0.002 A_g$ should be provided. In such cases, the concept of skin reinforcement given in CSA-A23.3-94(10.6.2) may be used for guidance. This is illustrated in Fig. 4.21. The area of skin reinforcement is equal to $\rho_{sk} A_{cs}$, where A_{cs} is the area of concrete in strips along the exposed face, shown shaded in Fig. 4.21, and $\rho_{sk} = 0.008$ for *interior exposure* and $\rho_{sk} = 0.01$ for *exterior exposure* condition.

4.11 ANALYSIS FOR STRENGTH BY CSA CODE

4.11.1 Rectangular Sections

The Code permits only under-reinforced beams with the maximum reinforcement ratio limited by Eq. 4.27. The value of ρ_{max} is computed by Eq. 4.29, and this condition is easily checked. If this requirement is satisfied (for over-reinforced beams, see Section 4.11.5), the member falls primarily in tension, which leads to a secondary compression failure, as explained in Section 4.4, and at the factored resistance, M_r, the steel stress equals $\phi_s f_y$ and the maximum concrete strain is equal to 0.0035. Using the rectangular stress block recommended in the Code, the internal stress resultants are shown in Fig. 4.22. Equating C and T,

$$\alpha_1 \phi_c f'_c b a = A_s \phi_s f_y$$

$$a = \frac{A_s \phi_s f_y}{\alpha_1 \phi_c f'_c b} = \frac{\rho \phi_s f_y}{\alpha_1 \phi_c f'_c} d \tag{4.35}$$

Calculating the moment of the resisting couple, the factored resistance M_r, is:

$$M_r = A_s \phi_s f_y \left(d - \frac{a}{2} \right) \tag{4.36}$$

$$= \rho \phi_s f_y \left(1 - \frac{\rho \phi_s f_y}{2 \alpha_1 \phi_c f'_c} \right) b d^2 \tag{4.36a}$$

$$= K_r b d^2 \tag{4.37}$$

where
$$K_r = \rho\phi_s f_y \left(1 - \frac{\rho\phi_s f_y}{2\alpha_1\phi_c f_c'}\right)$$

K_r has the unit of stress and is conveniently expressed in MPa (N/mm²). With b and d expressed in mm, the design strength is:

$$M_r = K_r b d^2 \, \text{N·mm}$$

$$= K_r b d^2 \times 10^{-6} \text{ kN·m} \qquad (4.38)$$

Values of K_r for different combinations of f_c' and f_y and for ρ ranging from ρ_{max} to ρ_{min} are tabulated in Table 4.1 as an aid to analysis. The table also gives the corresponding values of α_1, β_1, c_{max}/d, and ρ_{max}.

EXAMPLE 4.6

For the beam in Example 4.2, (Fig. 4.9a) compute the factored resistance by the Code.

SOLUTION

a) First the computation will be made using basic principles rather than the final formula. Referring to Fig. 4.22 for notation, the section properties are: b = 300 mm, d = 530 mm, h = 600 mm, A_s = 2000 mm², f_c' = 30 MPa, f_y = 400 MPa. For these values, ρ_{max} = 0.0243 (Eq. 4.29), and A_{smin} = 493 mm². The actual reinforcement ratio ρ = 2000/(300 × 530) = 0.0126 is less than ρ_{max} and A_s = 2000 mm² is greater than A_{smin}. The reinforcement is, therefore, within, the permissible limits. The stress block, and steel and concrete forces are as shown in Fig. 4.22.

Fig. 4.22

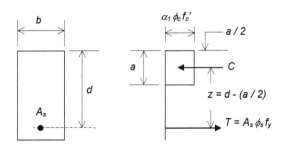

Table 4.1 Reinforcement Ratios ρ and Resistance Factors K_r for Rectangular Sections

$\rho = A_s/(bd)$; $\rho_{max} = \alpha_1 \beta_1 (700/(700+f_y))\phi_c f_c'/(\phi_s f_y)$; $(c_{max}/d) = 700/(700+f_y)$; $M_r = K_r bd^2 \times 10^{-6}$; $K_r = \rho \phi_s f_y [1 - \rho \phi_s f_y/(2\alpha_1 \phi_c f_c')]$; $\phi_s = 0.85$; $\phi_c = 0.6$; Units of f_c', f_y, K_r in MPa, M_r in kN·m, b and d in mm, A_s in mm^2.

f_y (c_{max}/d)	300 0.700				350 0.667				400 0.636			
f_c'	25	30	35	40	45	25	30	35	40	45	25	30	35	40	45
α_1	0.813	0.805	0.798	0.790	0.783	0.813	0.805	0.798	0.790	0.783	0.813	0.805	0.798	0.790	0.783
β_1	0.908	0.895	0.883	0.870	0.858	0.908	0.895	0.883	0.870	0.858	0.908	0.895	0.883	0.870	0.858
$\rho = \rho_{max}$	0.0304	0.0356	0.0406	0.0453	0.0497	0.0248	0.0291	0.0331	0.0370	0.0406	0.0207	0.0243	0.0277	0.0309	0.0339
K_r	5.28	6.23	7.15	8.03	8.88	5.14	6.07	6.95	7.81	8.63	5.01	5.90	6.76	7.59	8.38
$\rho = 0.95\rho_{max}$	0.0288	0.0338	0.0385	0.0430	0.0472	0.0235	0.0276	0.0315	0.0351	0.0386	0.0197	0.0231	0.0263	0.0293	0.0322
K_r	5.14	6.06	6.94	7.80	8.61	4.99	5.89	6.74	7.57	8.36	4.85	5.72	6.55	7.35	8.11
$\rho = 0.9\rho_{max}$	0.0273	0.0320	0.0365	0.0408	0.0448	0.0223	0.0262	0.0298	0.0333	0.0365	0.0186	0.0218	0.0249	0.0278	0.0305
K_r	4.98	5.87	6.72	7.54	8.33	4.83	5.69	6.52	7.31	8.07	4.69	5.52	6.33	7.09	7.83
$\rho = 0.85\rho_{max}$	0.0258	0.0303	0.0345	0.0385	0.0423	0.0211	0.0247	0.0282	0.0314	0.0345	0.0176	0.0206	0.0235	0.0262	0.0288
K_r	4.80	5.66	6.49	7.27	8.03	4.66	5.49	6.28	7.04	7.77	4.51	5.32	6.09	6.82	7.53
$\rho = 0.8\rho_{max}$	0.0243	0.0285	0.0325	0.0362	0.0398	0.0198	0.0232	0.0265	0.0296	0.0325	0.0166	0.0194	0.0221	0.0247	0.0271
K_r	4.62	5.44	6.23	6.99	7.71	4.47	5.27	6.03	6.76	7.45	4.33	5.10	5.83	6.54	7.21
$\rho = 0.75\rho_{max}$	0.0228	0.0267	0.0304	0.0340	0.0373	0.0186	0.0218	0.0248	0.0277	0.0304	0.0155	0.0182	0.0207	0.0232	0.0254
K_r	4.42	5.21	5.96	6.68	7.37	4.28	5.03	5.76	6.45	7.12	4.14	4.87	5.57	6.24	6.88
$\rho = 0.7\rho_{max}$	0.0213	0.0249	0.0284	0.0317	0.0348	0.0173	0.0203	0.0232	0.0259	0.0284	0.0145	0.0170	0.0194	0.0216	0.0237
K_r	4.21	4.96	5.68	6.36	7.01	4.07	4.79	5.48	6.14	6.76	3.93	4.63	5.29	5.92	6.53
$\rho = 0.65\rho_{max}$	0.0197	0.0231	0.0264	0.0294	0.0323	0.0161	0.0189	0.0215	0.0240	0.0264	0.0135	0.0158	0.0180	0.0201	0.0220
K_r	3.99	4.70	5.37	6.02	6.64	3.85	4.53	5.18	5.80	6.39	3.72	4.37	5.00	5.60	6.16

CHAPTER 4 BEHAVIOUR IN FLEXURE

f_y (c_{max}/d)	300 0.700				350 0.667				400 0.636			
f_c'	25	30	35	40	45	25	30	35	40	45	25	30	35	40	45
$\rho=0.6\rho_{max}$	0.0182	0.0214	0.0243	0.0272	0.0298	0.0149	0.0174	0.0199	0.0222	0.0244	0.0124	0.0146	0.0166	0.0185	0.0203
K_r	3.76	4.42	5.06	5.66	6.24	3.62	4.26	4.87	5.45	6.00	3.49	4.11	4.69	5.25	5.78
$\rho=0.55\rho_{max}$	0.0167	0.0196	0.0223	0.0249	0.0274	0.0136	0.0160	0.0182	0.0203	0.0223	0.0114	0.0133	0.0152	0.0170	0.0186
K_r	3.51	4.13	4.72	5.29	5.82	3.38	3.97	4.54	5.08	5.60	3.26	3.83	4.37	4.89	5.39
$\rho=0.5\rho_{max}$	0.0152	0.0178	0.0203	0.0226	0.0249	0.0124	0.0145	0.0166	0.0185	0.0203	0.0104	0.0121	0.0138	0.0154	0.0170
K_r	3.26	3.83	4.37	4.89	5.39	3.13	3.68	4.20	4.70	5.18	3.01	3.54	4.04	4.52	4.98
$\rho=0.45\rho_{max}$	0.0137	0.0160	0.0183	0.0204	0.0224	0.0112	0.0131	0.0149	0.0166	0.0183	0.00932	0.0109	0.0124	0.0139	0.0153
K_r	2.99	3.51	4.01	4.48	4.94	2.87	3.37	3.85	4.30	4.74	2.76	3.24	3.70	4.14	4.55
$\rho=0.4\rho_{max}$	0.0121	0.0142	0.0162	0.0181	0.0199	0.00991	0.0116	0.0132	0.0148	0.0162	0.00828	0.00971	0.0111	0.0123	0.0136
K_r	2.70	3.18	3.63	4.06	4.46	2.59	3.05	3.48	3.89	4.28	2.49	2.93	3.34	3.73	4.11
$\rho=0.35\rho_{max}$	0.0106	0.0125	0.0142	0.0158	0.0174	0.00867	0.0102	0.0116	0.0129	0.0142	0.00725	0.00850	0.00968	0.0108	0.0119
K_r	2.41	2.83	3.23	3.61	3.97	2.31	2.71	3.09	3.46	3.80	2.21	2.60	2.97	3.32	3.65
$\rho=0.3\rho_{max}$	0.00911	0.0107	0.0122	0.0136	0.0149	0.00744	0.00872	0.00994	0.0111	0.0122	0.00621	0.00728	0.00830	0.00926	0.0102
K_r	2.10	2.47	2.82	3.15	3.46	2.01	2.36	2.70	3.01	3.31	1.93	2.26	2.58	2.89	3.18
$\rho=0.25\rho_{max}$	0.00759	0.00890	0.0101	0.0113	0.0124	0.00620	0.00727	0.00828	0.00924	0.01015	0.00518	0.00607	0.00692	0.00772	0.00848
K_r	1.78	2.09	2.39	2.67	2.93	1.70	2.00	2.28	2.55	2.80	1.63	1.92	2.19	2.44	2.69
$\rho=0.2\rho_{max}$	0.00607	0.00712	0.00811	0.00906	0.00995	0.00496	0.00581	0.00662	0.00739	0.00812	0.00414	0.00485	0.00553	0.00617	0.00678
K_r	1.45	1.70	1.94	2.17	2.38	1.39	1.63	1.85	2.07	2.28	1.33	1.56	1.78	1.98	2.18
$\rho=0.15\rho_{max}$	0.00455	0.00534	0.00609	0.00679	0.00746	0.00372	0.00436	0.00497	0.00554	0.00609	0.00311	0.00364	0.00415	0.00463	0.00509
K_r	1.11	1.30	1.48	1.65	1.82	1.06	1.24	1.41	1.58	1.73	1.01	1.19	1.35	1.51	1.66
$A_{s\,min} \times b_t h$ mm^2	0.00333	0.00365	0.00394	0.00422	0.00447	0.00286	0.00313	0.00338	0.00361	0.00383	0.00250	0.00274	0.00296	0.00316	0.00335

For $f_c' = 30$ MPa, $\alpha_1 = 0.85 - 0.0015 \times 30 = 0.805$
Equating $C - T$,
$\alpha_1 \phi_c f_c' b a = A_s \phi_s f_y$
$0.805 \times 0.6 \times 30 \times 300 a = 2000 \times 0.85 \times 400$, $a = 156$ mm

Factored resistance, $M_r = A_s \phi_s f_y \left(d - \dfrac{a}{2} \right)$

$= 2000 \times 0.85 \times 400 \left(530 - \dfrac{156}{2} \right) \times 10^{-6} = 307$ kN·m

b) Using the design aid in Table 4.1
From table, $\rho_{max} = 0.0243$
Actual $\rho = 0.0126$, $< \rho_{max}$ OK

$\rho = 0.0126 = \dfrac{0.0126}{0.0243} \times \rho_{max} = 0.519 \rho_{max}$

From Table 4.1, $K_r = 3.54$ for $\rho = 0.5 \rho_{max}$
and $K_r = 3.83$ for $\rho = 0.55 \rho_{max}$
Interpolating, for $\rho = 0.519 \rho_{max}$, $K_r = 3.65$
$M_r = K_r b d^2 \times 10^{-6}$, $= 3.65 \times 300 \times 530^2 \times 10^{-6} = 307$ kN·m, as before

Fig. 4.23 *Analysis of slabs*

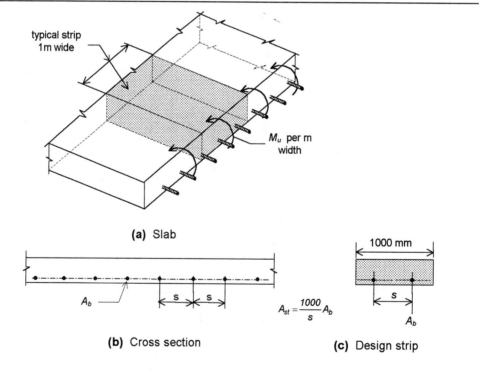

(a) Slab

(b) Cross section

(c) Design strip

4.11.2 Slabs as Rectangular Beams

Slabs of uniform thickness subjected to a moment distributed over its width (Fig. 4.23) may be considered as a wide shallow beam for purposes of analysis and design. In such slabs, reinforcing bars are usually spaced uniformly over the width of the slab. Therefore, for convenience, computations are generally based on a typical 1 m wide strip, considered as a beam (Fig. 4.23c). The area of reinforcement per metre wide strip is given by $A_s = 1000\, A_b/s$, where A_b is the area of one bar and s is the bar spacing in mm. The moment on the 1 m wide design strip is equal to M_f, the moment per metre width on the slab at the design location. The procedure is illustrated in Example 4.7.

EXAMPLE 4.7

A one-way slab (Fig. 4.24), having an overall thickness of 150 mm, is reinforced with No. 10 bars placed at an effective depth of 125 mm and at a spacing of 150 mm centre to centre. Use $f_c' = 30$ MPa and $f_y = 400$ MPa. Compute its factored resistance per metre width.

SOLUTION

Area of No. 10 bar = 100 mm^2

A_s per metre wide strip = $\dfrac{1000 \times 100}{150} = 667$ mm^2

ρ provided = $\dfrac{667}{1000 \times 125} = 0.005$

A_{smin} for slabs (CSA A23.3-94(7.8)) = $0.002 A_g = 0.002 \times 1000 \times 150 = 300$ mm^2
ρ_{max} (Table 4.1) = 0.0243
$\rho < \rho_{max}$ OK

Fig. 4.24 Example 4.7

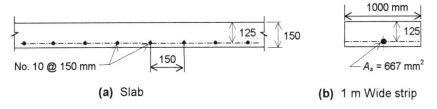

                    (a) Slab   **(b)** 1 m Wide strip

A_s per metre wide strip = 677 mm² > A_{smin} OK

For $f_c' = 30$ MPa,
From Table 4.1, $\alpha_1 = 0.805$ and $\beta_1 = 0.895$
Depth of stress block is obtained from $C = T$

$$\alpha_1 \phi_c f_c' b a = A_s \phi_s f_y$$

$0.805 \times 0.6 \times 30 \times 1000\, a = 667 \times 0.85 \times 400$, $a = 15.7$ mm

$$M_r = \phi_s A_s f_y \left(d - \frac{a}{2} \right)$$

$$= 0.85 \times 667 \times 400 \left(125 - \frac{15.7}{2} \right) \times 10^{-6} = 26.6 \text{ kN·m, per metre width of slab}$$

4.11.3 Rectangular Beam with Compression Reinforcement

It may be assumed first that the compression reinforcement also has yielded. This can be checked and calculations revised if it is found that the assumption is wrong. The factored strength may be considered as consisting of two parts (Fig. 4.25). The compressive force in the compression reinforcement, C_s, and an equal amount of tension, T_1, forms the first part, M_1. The remaining tensile reinforcement and the concrete compression, C_c, contribute the second part of the strength, M_2. Equating the forces, C_s and T_1, the tension reinforcement area, A_{s1}, contributing to M_1 is obtained as:

$$A_{s1} \phi_s f_y = A_s' (\phi_s f_y' - \alpha_1 \phi_c f_c')$$

$$A_{s1} = A_s' \left(\frac{\phi_s f_y' - \alpha_1 \phi_c f_c'}{\phi_s f_y} \right) \quad (4.39)$$

where f_y' is the yield strength of the compression reinforcement. The above equation allows for the concrete that is replaced by A_s' by reducing the effective stress in the compression reinforcement to $\phi_s f_y' - \alpha_1 \phi_c f_c'$. The area, A_{s2}, acting as tension reinforcement for the concrete part is:

$$A_{s2} = A_s - A_{s1} \quad (4.40)$$

Equating the forces in part 2, $(C_c = T_2)$

$$a = \frac{A_{s2} \phi_s f_y}{\alpha_1 \phi_c f_c' b} \quad (4.41)$$

Since the depth of neutral axis $c = a/\beta_1$, from the strain diagram, the strain in the

Fig. 4.25 Beam with compression reinforcement, all steel yielded

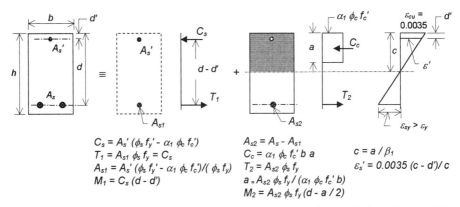

$C_s = A_s' (\phi_s f_y' - \alpha_1 \phi_c f_c')$
$T_1 = A_{s1} \phi_s f_y = C_s$
$A_{s1} = A_s' (\phi_s f_y' - \alpha_1 \phi_c f_c')/(\phi_s f_y)$
$M_1 = C_s (d - d')$

$A_{s2} = A_s - A_{s1}$
$C_c = \alpha_1 \phi_c f_c' b a$
$T_2 = A_{s2} \phi_s f_y$
$a = A_{s2} \phi_s f_y / (\alpha_1 \phi_c f_c' b)$
$M_2 = A_{s2} \phi_s f_y (d - a/2)$

$c = a/\beta_1$
$\varepsilon_s' = 0.0035 (c - d')/c$

compression reinforcement can now be computed to confirm whether that steel has in fact yielded.

$$\varepsilon_s' = \frac{0.0035}{c}(c - d')$$

If $\varepsilon_s' > \varepsilon_y' = f_y'/E_s$, the compression reinforcement has yielded, and the two parts of the factored strength are computed as:

$$M_1 = A_s'(\phi_s f_y' - \alpha_1 \phi_c f_c')(d - d')$$

$$M_2 = A_{s2} \phi_s f_y \left(d - \frac{a}{2}\right)$$

and $\qquad M_r = (M_1 + M_2)$ \hfill (4.42)

If it is found that $\varepsilon_s' < \varepsilon_y'$, the compression reinforcement stress $f_s' < f_y'$, and the

Fig. 4.26 Compression reinforcement not yielding

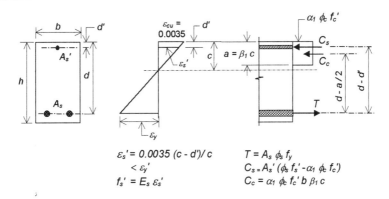

$\varepsilon_s' = 0.0035 (c - d')/c$
$< \varepsilon_y'$
$f_s' = E_s \varepsilon_s'$

$T = A_s \phi_s f_y$
$C_s = A_s'(\phi_s f_s' - \alpha_1 \phi_c f_c')$
$C_c = \alpha_1 \phi_c f_c' b \beta_1 c$

calculations have to be revised. One approach for this would be to use an iterative procedure, assuming to begin with, $f_s' = E_s\varepsilon_s'$, and then check and revise following the procedure above. Alternatively, an exact solution can be obtained as follows (Fig. 4.26).

Let c be the actual depth of neutral axis, which is as yet unknown. The compression reinforcement stress then is:

$$f_s' = E_s\varepsilon_s' = E_s \frac{0.0035}{c}(c-d')$$

Equating the total tension and compression across the section:

$$A_s\phi_s f_y = A_s'\left(\phi_s E_s \times 0.0035 \frac{c-d'}{c} - \alpha_1\phi_c f_c'\right) + \alpha_1\phi_c f_c' b\beta_1 c \qquad (4.43)$$

The only unknown, c, in this quadratic equation may be determined. Knowing c, the factored resistance may be computed as:

$$M_r = C_s(d-d') + C_c\left(d-\frac{a}{2}\right)$$

$$= A_s'\left(\phi_s E_s \times 0.0035\left(\frac{c-d'}{c}\right) - \alpha_1\phi_c f_c'\right)(d-d')$$

$$+ \alpha_1\phi_c f_c' b\beta_1 c\left(d-\frac{\beta_1 c}{2}\right) \qquad (4.44)$$

EXAMPLE 4.8

Analyse the beam shown in Fig. 4.27 and compute the factored moment that can be applied. Use $f_y' = 350$ MPa for the compression steel and $f_y = 400$ MPa for the tension steel.

SOLUTION

First check the steel ratio against allowable limits. For doubly reinforced beams, the maximum $f_c' = 30$ MPa steel ratio may be computed using Eq. 4.33.
Alternatively, starting from basics, the maximum depth of the neutral axis is:

$$c_{max} = \frac{700}{f_y + 700}d = \frac{700}{400+700} \times 530 = 337 \text{ mm}$$

For this depth, the compression steel strain is:

Fig. 4.27 Example 4.8

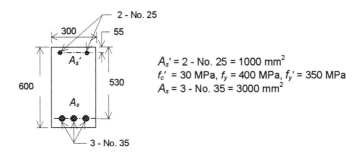

$$\varepsilon_s' = \frac{0.0035(c-d')}{c} = \frac{0.0035(337-55)}{337} = 0.00293$$

Therefore, the compression steel will be yielded under the limiting reinforcement condition, and $f_s' = 350$ MPa.

For $f_c' = 30$ MPa, from Table 4.1, $\alpha_1 = 0.805$ and $\beta_1 = 0.895$
Total compression at maximum tension reinforcement is
$C = C_s + C_c = A_s'(\phi_s f_s' - \alpha_1 \phi_c f_c') + \alpha_1 \phi_c f_c' b \beta_1 c$
$= 1000(0.85 \times 350 - 0.805 \times 0.6 \times 30) + 0.805 \times 0.6 \times 30 \times 300 \times 0.895 \times 337$
$= 1.59 \times 10^6$ N

$$A_s = \frac{C}{\phi_s f_y} = \frac{1.59 \times 10^6}{0.85 \times 400} = 4676 \text{ mm}^2$$

$$\rho_{max} = \frac{A_s}{bd} = \frac{4676}{300 \times 530} = 0.0294$$

$$A_{s\,min} = \frac{0.2\sqrt{f_c'}}{f_y} b_t h = \frac{0.2 \times \sqrt{30}}{400} \times 300 \times 600 = 493 \text{ mm}^2$$

Actual $\rho = \frac{3000}{300 \times 500} - 0.0189$

$\rho < \rho_{max}$ OK

$A_s > A_{s\,min}$ OK

The section is under-reinforced, and the tension reinforcement will yield as the factored strength is reached. To compute M_r (refer to Fig. 4.25) assume the compression reinforcement is also yielded (which is later confirmed).

104 REINFORCED CONCRETE DESIGN

$$A_{s1} = A_s' \frac{\phi_s f_y' - \alpha_1 \phi_c f_c'}{\phi_s f_y} = 1000 \frac{(0.85 \times 350 - 0.805 \times 0.6 \times 30)}{0.85 \times 400} = 832 \text{ mm}^2$$

$$A_{s2} = A_s - A_{s1} = 3000 - 832 = 2168 \text{ mm}^2$$

$$a = \frac{A_{s2} \phi_s f_y}{\alpha_1 \phi_c f_c' b} = \frac{2168 \times 0.85 \times 400}{0.805 \times 0.6 \times 30 \times 300} = 170$$

$$c = a/\beta_1 = 170/0.895 = 190 \text{ mm}$$

Check f_s':

$$\varepsilon_s' = \frac{0.0035(c-d')}{c} = \frac{0.0035(190-55)}{190} = 0.00249 > \varepsilon_y' = 0.00175$$

$f_s' = f_y' = 350$ MPa OK

Strength, $M_1 = A_s'(\phi_s f_y' - \alpha_1 \phi_c f_c')(d-d')$
$= 1000(0.85 \times 350 - 0.805 \times 0.6 \times 30)(530 - 55) \times 10^{-6} = 134$ kN·m

$$M_2 = A_{s2} \phi_s f_y \left(d - \frac{a}{2}\right) = 2170 \times 0.85 \times 400 \left(530 - \frac{170}{2}\right) \times 10^{-6} = 328 \text{ kN·m}$$

$M_r = M_1 + M_2 = 134 + 328 = 462$ kN·m

EXAMPLE 4.9

Compute the factored flexural strength of the beam shown in Fig. 4.28. Take $f_c' = 30$ MPa and $f_y = 400$ MPa.

Fig. 4.28 Example 4.9

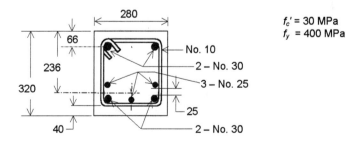

SOLUTION

Check the steel ratio first

$$c_{max} = \frac{700}{700+f_y}d$$

The distance to the centroid of tension and compression reinforcement may be computed from the given dimensions as $d = 316$ mm and $d' = 66$ mm, respectively.

$$c_{max} = \frac{700}{700+400} \times 236 = 150 \text{ mm}$$

$$f_s' = E_s \varepsilon_s' = 200\,000 \times \frac{0.0035(150-66)}{150} = 392 \text{ MPa} < f_y$$

Using Eq. 4.31,

$$\rho_{max} = \overline{\rho} + \rho' = \frac{(\phi_s f_s' - \alpha_1 \phi_c f_c')}{\phi_s f_y}$$

$$= 0.0243 + \frac{1400}{280 \times 236} \times \left(\frac{0.85 \times 392 - 0.805 \times 0.6 \times 30}{0.85 \times 400}\right) = 0.0442$$

Actual $\rho = \dfrac{2900}{280 \times 236} = 0.0439 < \rho_{max}$

If Eq. 4.31a is used, $\rho_{max} = \overline{\rho} + \rho'\dfrac{f_s'}{f_y} = 0.0243 + \dfrac{1400}{280 \times 236} \times \dfrac{392}{400} = 0.0451$

and still $\rho < \rho_{max}$ OK

$$A_{s\,min} = \frac{0.2\sqrt{f_c'}}{f_y}b_t h = \frac{0.2 \times \sqrt{30}}{400} \times 280 \times 320 = 245 \text{ mm}^2$$

$A_s = 2900 \text{ mm}^2 > A_{s\,min}$ OK

At factored strength, the compression reinforcement does not yield. (In the balanced section $f_s' < f_y$. The actual section being under-reinforced, its $c < c_{max}$ and $f_s' < (f_s')_{max} < f_y$). Assume that the actual depth to neutral axis is c. Equating $C_c + C_s = T$ yields (Eq. 4.43).

$$\alpha_1 \phi_c f_c' b \beta_1 c + A_s'\left[\phi_s E_s \times 0.0035 \frac{c-d'}{c} - \alpha_1 \phi_c f_c'\right] = A_s \phi_s f_y$$

Substituting for $f_c' = 30$ MPa, $b = 280$ mm, $\alpha_1 = 0.805$, $\beta_1 = 0.895$, $A_s' = 1400$ mm^2, $E_s = 200$ GPa, $d' = 66$ mm, $A_s = 2900$ mm^2 and $f_y = 400$ MPa, and solving the quadratic equation for c,

$c = 148$ mm

Fig. 4.29 Analysis of T-beam

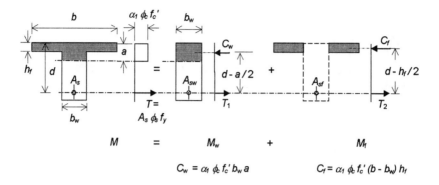

$$f_s' = E_s \times 0.0035 \frac{c-d'}{c} = 388 \text{ MPa} < f_y$$

The design strength is given by (Eq. 4.44).

$$M_r = C_c\left(d - \frac{\beta_1 c}{2}\right) + C_s(d - d')$$

$$M_r = [0.805 \times 0.6 \times 30 \times 280 \times 0.895 \times 148 \times \left(236 - \frac{0.895 \times 148}{2}\right)$$

$$+ 1400(0.85 \times 388 - 0.805 \times 0.6 \times 30)(236 - 66) \times 10^{-6}] = 166 \text{ kN} \cdot \text{m}$$

4.11.4 Analysis of T-Beams

When the depth of stress block, a, is less than or equal to the flange thickness, h_f, the analysis is identical to that for a rectangular beam. For $a > h_f$, the total compression area can be split up into a rectangular area, $b_w a$, corresponding to the web and an outstanding flange area, $(b - b_w) h_f$, as shown in Fig. 4.29. The average stress in compression is $\alpha_1 \phi_c f_c'$, and equating $C = T$ the actual value of a can be evaluated. The lever arm for the compression in the flange is $d - h_f/2$ and for the compression in the web portion is $d - a/2$. The relevant equations are

$$C_f + C_w = T$$

$$\alpha_1 \phi_c f_c'[(b - b_w)h_f + b_w a] = A_s \phi_s f_y$$

Fig. 4.30 Example 4.10

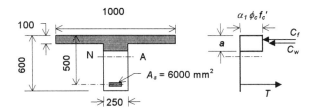

and hence:
$$a = \frac{A_s \phi_s f_y - \alpha_1 \phi_c f_c'(b - b_w) h_f}{\alpha_1 \phi_c f_c' b_w} \quad (4.45)$$

The design strength is:
$$M_r = C_f \left(d - \frac{h_f}{2}\right) + C_w \left(d - \frac{a}{2}\right) \quad (4.46)$$

EXAMPLE 4.10

A simply supported T-beam has a span of 8 m and the cross-section dimensions shown in Fig. 4.30. Compute its factored resistance in flexure.
Use $f_c' = 40$ MPa and $f_y = 400$ MPa.

SOLUTION

First check the effective flange width (CSA A23.3-94(10.3)). The overhang shall not exceed 1/5 × span = 1600 mm, nor 12 × h_f = 960 mm, nor one-half the clear distance to the next beam (not applicable here). Since the actual width is less than all these limitations, b = actual width = 1000 mm. Now check the reinforcement ratio:

From Eq. 4.34, $A_{smin} = \dfrac{0.2\sqrt{40}}{400} \times 250 \times 600 = 474$ mm^2

$A_s = 6000$ mm$^2 > A_{smin}$ OK

Actual $\rho = \dfrac{6000}{250 \times 500} = 0.048$

ρ_{max} may be computed using Eq. 4.33 or using the basic theory as follows:

Maximum neutral axis depth $c_{max} = \dfrac{700}{700 + f_y} d = 318$ mm

Here, $\alpha_1 = 0.790$ and $\beta_1 = 0.870$
$a_{max} = \beta_1 c_{max} = 0.870 \times 318 = 277$ mm $> h_f$

At maximum tension reinforcement conditions, the total concrete compression is:
$C_{max} = \alpha_1 \phi_c f'_c \times$ (concrete area over stress block depth)
$= 0.790 \times 0.6 \times 40 [250 \times 277 + (1000 - 250) \times 100] = 2.73 \times 10^6$ N

Since $A_s \phi_s f_y = T = C$

$$A_{s,max} = \frac{C}{\phi_s f_y} = \frac{2.73 \times 10^6}{0.85 \times 400} = 8044 \text{ mm}^2$$

Actual $A_s = 6000 \text{ mm}^2 < A_{s,max}$ OK

The section is under-reinforced, and the reinforcement will yield at the factored strength. The actual depth of stress block, a, is obtained by equating the tensile and compressive forces. Assuming $a > h_f$,

$T = C = C_w + C_f$

$A_s \phi_s f_y = \alpha_1 \phi_c f'_c b_w a + \alpha_1 \phi_c f'_c / (b - b_w) h_f$

$$a = \frac{A_s \phi_s f_y - \alpha_1 \phi_c f'_c (b - b_w) h_f}{\alpha_1 \phi_c f'_c b_w} \tag{4.45}$$

$$= \frac{6000 \times 0.85 \times 400 - 0.790 \times 0.6 \times 40(1000 - 250)100}{0.790 \times 0.6 \times 40 \times 250}$$

$= 130 \text{ mm} > h_f$, as assumed

The design strength is given by Eq. 4.46,

$$M_r = M_w + M_f = \left[C_w \left(d - \frac{a}{2}\right) + C_f \left(d - \frac{h_f}{2}\right) \right]$$

$$= \left[0.790 \times 0.6 \times 40 \times 250 \times 130 \left(500 - \frac{130}{2}\right) + 0.790 \times 0.6 \times 40(1000 - 250) \right.$$

$$\left. \times 100 \left(500 - \frac{100}{2}\right) \right] 10^{-6} = 908 \text{ kN·m}$$

4.11.5 Tension Reinforcement Not Yielded (Over-Reinforced Beams)

All designs conforming to CSA A23.3-94 have to be under-reinforced (Section 4.9), so that the tension reinforcement will have yielded before the ultimate strength is reached. However, there may be special occasions when the nominal strength of an over-reinforced beam has to be evaluated.

In all beams at nominal ultimate strength, the compressive strain at the extreme fibre in concrete will be $\varepsilon_{cu} = 0.0035$. If the depth to the neutral axis, c, is also known, the steel strains and stresses can be evaluated. Knowing steel stresses and assuming the standard rectangular stress block for concrete in compression, the internal forces and the flexural strength can be evaluated from statics. Thus, the problem reduces to finding the correct value of c. The value of c must be such that the equilibrium condition that the total compressive force is equal to the total tensile force is satisfied. Therefore, this condition can be used to evaluate c, either by a trial and error approach, or by setting up an algebraic equation in terms of c and solving it. After the correct value of c is determined, the strength can be evaluated using statics. Example 4.11 illustrates this procedure.

EXAMPLE 4.11

For the doubly reinforced section shown in Fig. 4.31 compute the design strength. Use $f_c' = 30$ MPa, $f_y = f_y' = 400$ MPa.

SOLUTION

First check whether the section is under-reinforced. The maximum reinforcement ratio can be found using Eq. 4.30 and 4.31. Eq. 4.30 gives, for limiting reinforcement ratio:

$$f_s' = 700 \times \left(1 - \frac{400 + 700}{700} \times \frac{55}{500}\right) = 490 \text{ MPa} > f_y$$

So $f'_{s,\max} = 400$ MPa. From Table 4.1, for a rectangular section with tension reinforcement only, $\overline{\rho} = 0.0243$. The maximum reinforcement ratio (Eq. 4.31) is:

$$\rho_{\max} = \overline{\rho} + \rho' \frac{\phi_s f_s' - \alpha_1 \phi_c f_c'}{\phi_s f_y}$$

Fig. 4.31 Example 4.11

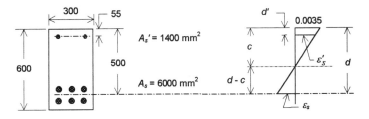

$$= 0.0243 + \frac{1400}{300 \times 500}\left(\frac{0.85 \times 400 - 0.805 \times 0.6 \times 30}{0.85 \times 400}\right) = 0.0332$$

actual $\rho = \frac{6000}{300 \times 500} = 0.04 > \rho\,\text{max}$

Therefore, the section is over-reinforced.

To compute M_r by the trial and error method, assume the tension reinforcement stress f_s as, say,

$$f_s = 400 \times \frac{0.0332}{0.04} = 332\,\text{MPa}$$

Note that assuming f_s is similar to assuming a value for c.

$$\varepsilon_s = f_s = \frac{332}{200\,000} = 0.00166, \quad c = \frac{d \times 0.0035}{\varepsilon_s + 0.0035} = 339\,\text{mm}$$

$$\varepsilon_s' = \frac{0.0035(c-d')}{c} = 0.00293 > \varepsilon_y$$

so, $f_s' = f_y$

$a = \beta_1 c = 303\,\text{mm}$

$C = 0.805 \times 0.6 \times 30 \times 300 \times 303 + 1400(0.85 \times 400 - 0.805 \times 0.6 \times 30) = 1.77 \times 10^6\,\text{N}$

$T = 6000 \times 0.85 \times 332 = 1.69 \times 10^6\,\text{N}$

With the first trial $C > T$. Therefore, revise with a higher value for f_s. Try 348 MPa which gives a $T = 6000 \times 0.85 \times 348 = 1.77 \times 10^6$ N, which is slightly higher than the average of C and T obtained in the first trial. With $f_s = 348$ MPa, $\varepsilon_s = 0.00174$, $c = 334$ mm, $\varepsilon_s' > \varepsilon_y'$, $f_s' = f_y = 400$ MPa, $a = 299$ mm, $C = 1.76 \times 10^6$ N. Since $C \approx T$, this trial value is sufficiently accurate. The factored strength is:

$$M_r = \left[\alpha_1 \phi_c f_c' b a \left(d - \frac{a}{2}\right) + A_s'(\phi_s f_y - \alpha_1 \phi_c f_c')(d - d')\right]$$

$$= [0.805 \times 0.6 \times 30 \times 300 \times 299\left(500 - \frac{299}{2}\right)$$

$$+ 1400(0.85 \times 400 - 0.805 \times 0.6 \times 30)(500 - 55) \times 10^{-6}] = 658\,\text{kN}\cdot\text{m}$$

To solve the above example directly, assume the depth of neutral axis as c. Since the compression reinforcement is close to the top edge where the strain is 0.0035, it may be assumed to have yielded so that $f_s' = f_y$ (and later confirmed). From the strain distribution,

$$f_s = E\varepsilon_s = 200\,000 \times \frac{0.0035(d-c)}{c} = 700\frac{(500-c)}{c}$$

Equating $C = T$,

$$= 0.805 \times 0.6 \times 30 \times 300(0.895 \times c) + 1400(0.85 \times 400 - 0.805 \times 0.6 \times 30)$$
$$= 6000 \times 0.85 \times \frac{700(500 - c)}{c}$$

Solving this quadratic equation, $c = 335$ mm, $a = 300$ mm, $f_s = 345$ MPa, $\varepsilon_s' = 0.00292 > \varepsilon_y, f_s' = f_y$ (confirmed), and the factored strength is:

$$M_r = \left[0.805 \times 0.6 \times 30 \times 300(300)\left(500 - \frac{300}{2}\right) \right.$$
$$\left. + 1400(0.85 \times 400 - 0.805 \times 0.6 \times 30) \times (500 - 55) \right] \times 10^{-6} = 659 \text{ kN} \cdot \text{m}$$

PROBLEMS

1. The rectangular beam shown in Fig. 4.32 has $f_c' = 30$ MPa and $f_y = 350$ MPa.
   a) Compute the stresses developed in concrete and steel under a service load moment of 175 kN·m. Check the computations using the flexure formula $f = Mc/I$.
   (b) If the service load stresses are to be limited to $0.4 f_c'$ in concrete and 150 MPa in steel, compute the service load moment for the section. Indicate whether the steel stress or the concrete stress controls.

2. (a) For the rectangular beam shown in Fig. 4.33, $f_c' = 40$ MPa and $f_y = 400$ MPa. The service load stresses are not to exceed $0.4 f_c'$ in concrete and 165 MPa in steel. Compute the allowable service load moment for the section.
   (b) Repeat the problem at (a) with the flexural reinforcement taken as 3 - No. 30 bars ($A_s = 2100$ mm^2).

3. A beam simply supported on a span of 6 m is subjected to the service loads (excluding self weight of the beam) shown in Fig. 4.34a. The cross-section dimensions of the beam are given in Fig. 4.34b. Compute the service load stresses in concrete and steel. Take $f_c' = 30$ MPa and $f_y = 300$ MPa.

4. For each of the beams in problems 1, 2, and 3 above, using basic principles, check whether the beams are under-reinforced or over-reinforced based on their behaviour at the ultimate limit state. Check whether the beams satisfy CSA A23.3-94 requirements for flexure. Also compute the factored resistance of each section.

Fig. 4.32 Prob. 1

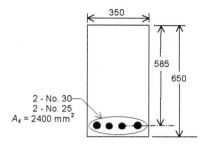

Fig. 4.33 Prob.2

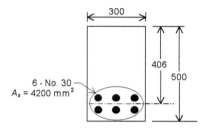

Fig. 4.34 Prob.3

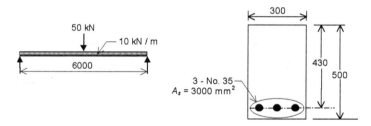

5. The cross-section dimensions of a doubly reinforced beam are shown in Fig. 4.35. The beam is subjected to a service load moment of 70 kN·m.
   (a) Compute the stresses in concrete and steel under the specified loads. Take $f_c' = 30$ MPa and $f_y = 300$ MPa.
   (b) Compute the service load moment if the concrete and steel stresses are not to exceed $0.4 f_c'$ and 140 MPa, respectively.

6. Compute the flexural strength of the beam in Problem 5 based on CSA A23.3-94 requirements.

7. For the beam shown in Fig. 4.36, compute the factored resistance, M_r. Use $f_c' = 25$ MPa, $f_y = 400$ MPa and $f_y' = 300$ MPa (for compression reinforcement).

8. The cross-sectional dimensions of a T-beam are given in Fig. 4.37. Use $f_c' = 40$ MPa and $f_y = 400$ MPa.
   (a) Compute the steel and concrete stresses under a service load moment of 300 kN·m.
   (b) Compute the factored flexural resistance of the beam.

9. A simply supported T-beam with a span of 6 m has the cross-section dimensions shown in Fig. 4.38. Compute the flexural strength of the beam. Use $f_c' = 30$ MPa, $f_y = 350$ MPa.

10. A one-way slab having an overall thickness of 100 mm is reinforced with No. 10 bars placed at an effective depth of 75 mm and at a spacing of 125 mm c/c (Fig. 4.39). Material properties are $f_c' = 25$ MPa and $f_y = 400$ MPa. Compute the factored flexural resistance of the slab per metre width.

Fig. 4.35 Prob. 5

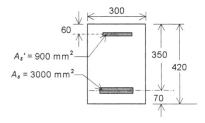

Fig. 4.36 Prob. 7

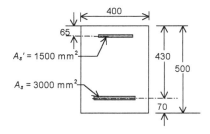

Fig. 4.37 Prob. 8

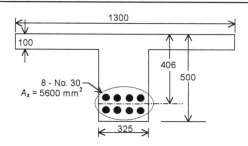

Fig. 4.38 Prob. 9

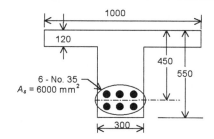

Fig. 4.39 Prob. 10

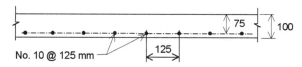

REFERENCES

4.1 Hognestad, E., Hanson, N.W., and McHenry, D., *Concrete Stress Distribution in Ultimate Strength Design*, J. of ACI, Vol. 52, Dec. 1955, pp. 455-479.

4.2 Popov, E.P., *Engineering Mechanics of Solids*, Prentice-Hall, Inc., Englewood Cliffs, New Jersey, 1999, 864 pp.

4.3 *Canadian Highway Bridge Design Code*, Canadian Standards Association, Rexdale, Ontario, Canada, (in press).

4.4 Kaar, P.H., Hanson, N.W., and Capell, H.T., *Stress-Strain Characteristics of High-Strength Concrete*, Proc., Douglas McHenry Int. Symp., ACI Special Publication SP-55, 1978, pp. 161-185.

4.5 Mattock, A.H., Kriz, L.B., and Hognestad, E., *Rectangular Concrete Stress Distribution in Ultimate Strength Design*. J. of ACI, Vol. 57, Feb. 1961, pp. 875-928.

CHAPTER 5 Design of Beams and One-Way Slabs for Flexure

5.1 INTRODUCTION

In Chapter 4, behaviour of reinforced concrete beams in flexure was explained, and procedures were given for the analysis of sections. Analysis may involve calculations of stresses under known service loads, of service load moments, and/or of factored flexural strength. From the latter two values, the service loads and the factored loads in flexure may also be computed. In these computations, since the complete cross-sectional dimensions and the material properties are fixed (given) the results are unique, being dictated solely by the conditions of equilibrium of forces and comparability of strains (although simplifying assumptions may be used regarding material response).

The design problem is somewhat the reverse of the above. The external loads, material properties, and the skeletal dimensions (centreline dimensions such as span, height, etc.) are given, and it is required to arrive at the cross-sectional dimensions, including area of reinforcement, necessary to carry all the load effects (including dead loads) satisfactorily. The dead load depends on the cross-sectional dimensions, as yet unknown, and will have to be initially assumed, and later verified and revised if needed. The flexural strength is dependent on the width and depth of sections, and on the area of reinforcement. Therefore, there are several combinations of these that could satisfactorily carry a given moment. Thus, unless all but one of the variables affecting the strength are fixed, there is no unique solution to a design problem. Often one or more of these variables (or their ratios) may be limited to narrow ranges by consideration of aesthetics, economy, deflection control, crack control, ductility requirements, and other structural constraints (for example, b/d ratios, headroom limitations, reinforcement ratio limits, etc.).

In addition, to determine the required overall cross-sectional dimensions and the area and disposition of flexural reinforcement at all sections of peak moments (which essentially follow in reverse the analysis procedures discussed in Chapter 4), a complete design of a beam involves the following considerations:
1. the detailing of longitudinal reinforcement in the section, with adequate concrete protection, without congestion, with proper bonding, and in a manner avoiding wide cracks;
2. detailing bar cut-off (or bend points) along the span, as the applied moment decreases and some of the bars become no longer necessary to carry the reduced moment;
3. designing for effects of transverse shear force and torsion; and

4. checking that serviceability requirements (deflection, crack width, etc.) are met.

Details of the design for shear and considerations of bond (development length) will be presented in Chapters 6, 8 and 9. Serviceability requirements and computations are detailed in Chapter 11. Design for torsion is presented in Chapters 7 and 8.

Methods for calculating the load effects on members, such as factored bending moments, shear forces, axial forces and torsion, are given in standard books on structural analysis (Refs. 5.1, 5.2). However, some of the approximate methods and simplified procedures permitted by the Code are presented in Chapter 10. For the purpose of design examples in this Chapter, it will be assumed that the factored moments have already been computed and are known, or they will be calculated using standard formulas.

General requirements for placing of flexural reinforcement are reviewed in the next Section, as these somewhat influence the choice of beam sizes. This is followed by design for flexure and bar cut-off procedures.

5.2 CONCRETE COVER AND BAR SPACING REQUIREMENTS

It is convenient in the construction of form work for the overall sizes of beams and thickness of slabs to be selected in round figures, such as in multiples of 5 mm for slabs, 10 mm for beams, and 20 mm for deeper girders. The effective depth of the member used in flexural computations depends on the reinforcing bar details and the amount of clear concrete cover. To protect the reinforcing bars from corrosion and fire, codes specify minimum concrete cover for all reinforcement, the amount of cover depending on the type of member and the exposure conditions. The minimum clear covers specified by CSA A23.1-94 are given in Table 5.1.

Codes also specify minimum and maximum limits for the spacing between parallel reinforcing bars in a layer. The minimum limits are necessary to ensure that the concrete flows readily in between and around bars to facilitate proper placement. These limits can be met generally without difficulty in slabs, because of the large width available and the relatively low area of reinforcement required for flexure. However, in selecting the width of beams, the minimum spacing between parallel bars is a major consideration, because of the limited width within which bars can be placed, as stirrups have to be accommodated, and cover requirements have to be met at the two sides (Fig. 5.1). If the beam width is inadequate to accommodate all the reinforcement in one layer with the necessary clearance between bars, the alternatives are: (1) to increase the beam width; (2) to place the reinforcement in two or more layers properly separated; and (3) to bundle groups of parallel bars in contact. In beams and girders, bundling of bars up to four in a bundle may be made, provided the bars are no larger than No. 35,

Table 5.1 Minimum Specified Clear Concrete Cover for Reinforcement in Cast-in-Place Concrete, CSA A23.1-94(12.6)

Exposure to earth or weather	Exposed	Not exposed
(a) When cast against and permanently exposed to earth	75 mm	
(b) For		
(i) Beams, girders, columns and piles;		
Principal reinforcement, No. 35 and smaller	50 mm	40 mm
Ties, stirrups, and spirals	40 mm	30 mm
(ii) Slabs, walls, and joists, shells and folded plates:		
No. 20 and smaller	30 mm	20 mm
(c) For bars with a diameter d_b larger than listed above, the cover shall be at least but need not be more than 60 mm	$1.5 d_b$	$1.0 d_b$
(d) The ratio of the cover to the nominal maximum aggregate size shall be at least	1.5	1.0
(e) The cover for a bundle of bars shall be the same as that for a single bar with an equivalent area.		

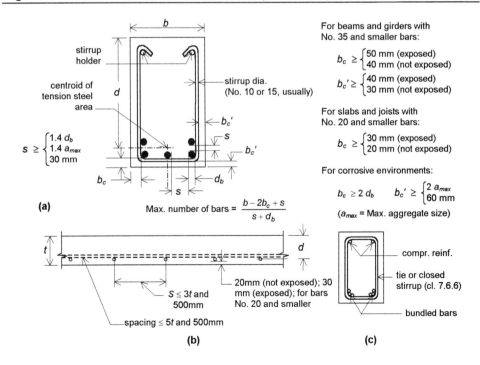

Fig. 5.1 Limitations for flexural reinforcement placement (CSA A23.1, Cl. 12)

and the bundle is enclosed by stirrups or ties. The complete requirements for fabrication and placement of reinforcement for building structures are given in CSA A23.1-94(12), and A23.3-94(7). Those requirements that influence the flexural design are summarised below. The other requirements (such as for ties, splices, anchorage, etc.) will be considered later when discussing associated design details. Reinforcement placement requirements for bridge structures are somewhat different and are given in the Canadian Highway Bridge Design Code (Cl. 8.14).

CSA A23.1-94 specifies a clear distance between parallel bars in a layer of not less than 1.4 times the diameter of bars, 1.4 times the maximum aggregate size, and 30 mm. When the reinforcement is placed in two or more layers, the bars in the upper layers must be placed directly above those in the bottom layer, with the same clear distance between layers as for parallel bars in a layer. These clear distance limitations also apply to the clear distance between the bars in a contact lap splice and adjacent parallel bars. Three and four bar bundles are arranged in triangular, L-shaped or square patterns. The bars in a bundle must be tied, wired or otherwise fastened together. For spacing and cover limitations, a unit of bundled bars is treated as a single bar of an equivalent diameter which would give the same total area. Cut off of individual bars in a bundle, within the span of flexural members should be staggered by at least 40 bar diameters. In beams and girders, compression reinforcement must be enclosed by ties or closed stirrups at least 6 mm in diameter, or welded wire fabric of equivalent area.

In slabs, the spacing of principal reinforcement should not exceed 3 times the slab thickness, 1.5 times the slab thickness for negative moment reinforcement at critical sections of two-way slabs, or 500 mm. The spacing of shrinkage and temperature reinforcement in slabs should not exceed either 5 times the slab thickness, or 500 mm. The clearance requirements for beams and slabs are shown in Fig. 5.1.

In addition to these, the CSA A23.3-94(10.6) provides for crack control by controlling the distribution of reinforcement in the zone of concrete tension. For a given area of reinforcement, providing several bars well-distributed over the zone of maximum concrete tension at a moderate spacing is more effective in controlling cracks and in improving bond than providing fewer bars of larger size. The two aspects related to detailing that affect crack width are the thickness of concrete cover and the area of concrete in the zone of maximum tension surrounding each individual bar. The Code requirement in Clause 10.6.1 translates into a maximum spacing limitation for one-way slabs and into a minimum required number of bars for a given width of beam stem. Here again, a bar bundle is treated as an equivalent single round bar giving the same area. While the basis for crack control is discussed in Chapter 11, the detailing requirements resulting from Clause 10.6.1 are illustrated in Fig. 5.2.

The present cover and spacing requirements in the Code are primarily based on considerations of protection of reinforcement and efficient placement of concrete. However, the clear cover and bar spacing are important factors that influence the bond resistance of the bar (Chapter 9). Research indicates that if the clear cover on the main

Fig. 5.2 Crack control (CSA A23.3-94, Cl. 10.6.1)

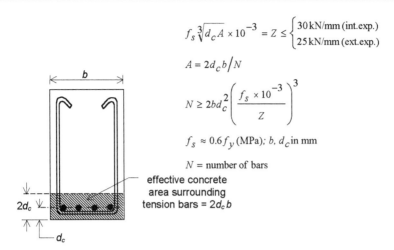

$$f_s \sqrt[3]{d_c A} \times 10^{-3} = Z \le \begin{cases} 30 \text{ kN/mm (int.exp.)} \\ 25 \text{ kN/mm (ext.exp.)} \end{cases}$$

$$A = 2d_c b / N$$

$$N \ge 2bd_c^2 \left(\frac{f_s \times 10^{-3}}{Z} \right)^3$$

$f_s \approx 0.6 f_y$ (MPa); b, d_c in mm

N = number of bars

reinforcement is much less than about 50 mm, or the clear spacing between bars is less than about 100 mm, the present Code requirements for development length may be inadequate (unconservative), particularly for higher grade steels (Section 9.9, and Ref. 5.3).

In relatively deep flexural members, a substantial portion of the web will be in tension. Properly distributed tension reinforcement will control the crack width at the level of the flexural tension reinforcement; however, higher up in the web wider cracks may develop. In order to control such cracks, CSA A23.3-94(10.6.2) requires auxiliary longitudinal skin reinforcement to be placed near the web faces in the tension zone of members with web depth in excess of 750 mm.

5.3 SELECTION OF MEMBER SIZES

Further to the requirements for placement of reinforcement discussed in the previous Section, an additional major consideration that influences the overall cross-sectional dimensions is the need to control deflections. The effective flexural stiffness of a reinforced concrete section is dependent to a larger extent on the gross section dimensions than on the area of tension reinforcement (Chapter 11). Table 5.2 gives minimum thickness to span ratios specified by CSA A23.3-94(9.8.2.1) for beams and one-way slabs. When these minimum thicknesses are provided, under normal situations in building structures, deflections will be satisfactory, and the design will satisfy Code requirements. However, under unusual conditions, there is no guarantee that providing these minimum thicknesses will ensure satisfactory deflection behaviour. More elaborate recommendations for minimum thickness of beams and one-way slabs are given in Ref. 5.4

The deflection may also be controlled indirectly by using a relatively small tension reinforcement ratio ($0.25\rho_b$ to $0.4\rho_b$) in the flexural design where ρ_b is the steel ratio obtained using Eq. 4.29 without the ϕ-factors. The low area of reinforcement leads to a deeper member, resulting in a higher flexural stiffness. Tension reinforcement ratios recommended for deflection control (Ref. 5.4) are presented in Table 5.3.

Table 5.2 *Thickness Below Which Deflections Must Be Computed for Non-prestressed Beams or One-Way Slabs Not Supporting or Attached to Partitions or Other Construction Likely to Be Damaged by Large Deflections, CSA A23.3-94(9.8.2.1).*

	Minimum Thickness, h*			
	Simply Supported	One End Continuous	Both Ends Continuous	Cantilever
Solid one-way slabs	$l_n/20$	$l_n/24$	$l_n/28$	$l_n/10$
Beams or ribbed one way slabs	$l_n/16$	$l_n/18.5$	$l_n/21$	$l_n/8$

* The span length *l* and the thickness *h* are both in mm.
+ The values given in this Table shall be used directly for non-prestressed reinforced concrete members made with *normal density concrete* ($\gamma_c > 2150$ kg/m^3) with reinforcement having a yield strength of 400 MPa. For other conditions, the values shall be modified as follows:
(a)   For structural low density concrete and structural semi-low density concrete the values in the Table shall be multiplied by $1.65-0.0003\gamma_c$, but not less than 1.00 where γ_c, is the mass density in kg/m^3.
(b)   For reinforcement having yield strengths other than 400 MPa the values in the Table shall be multiplied by $(0.4 + f_y/670)$.
Note: With the permission of CSA International, this material is reproduced from CSA Standard A23.3-94, *Design of Concrete Structures*, which is copyrighted by the CSA International, 178 Rexdale Boulevard, Rexdale, Ontario, M9W 1R3.

For a given reinforcement ratio, the resistance varies in direct proportion to the width, *b*, and to the square of the effective depth, *d* (Eq. 4.36). Therefore, an increase in depth is effective to one order higher than an increase in width. However, deeper beams result in loss of headroom or overall increase in the building height. For economy in rectangular sections, the proportion of overall depth to width is generally in the range of 1.5 to 2 for beams and may be higher (up to 3 or even more) for girders carrying heavy loads. These proportions also apply for the web portion of T-beams. The width and depth are also governed by the shear force on the section (shear stress limitation, Chapter 6). The beam depth must also be sufficient to provide effective anchorage for the shear reinforcement (Chapter 6). Often in T-beams the flange may be part of a slab, in which case the flange dimensions are known.

The design sequence generally follows the progression of load transfer from the structure to the foundation. For example, in a slab-beam-girder floor system supported on columns, the design order will be slab, beam, girder, column, and foundation, as the member sizes of and loading on each of these elements are needed for the design of the subsequent element.

Table 5.3 Recommended Tension Reinforcement for Non-prestressed One-Way Construction for Deflections to Be Within Acceptable Limits* (Ref. 5.4)

Members	Cross Section	Normal Density Concrete	Low Density Concrete
Not supporting or not attached to nonstructural elements, likely to be damaged by large deflections	Rectangular	$\rho \leq 35\% \rho_b$	$\rho \leq 30\% \rho_b$
	"T" or Box	$\rho_w \leq 40\% \rho_b$	$\rho_w \leq 35\% \rho_b$
Supporting or attached to nonstructural elements, likely to be damaged by large deflections	Rectangular	$\rho_w \leq 25\% \rho_b$	$\rho_w \leq 20\% \rho_b$
	"T" or Box	$\rho_w \leq 30\% \rho_b$	$\rho_w \leq 25\% \rho_b$

* For continuous members, the positive region steel ratios only may be used.
ρ_b refers to the balanced steel ratio corresponding to balanced strain conditions, computed without using ϕ factors (i.e., $\phi_c = \phi_s = 1.0$) in Eq. 4.29 and is approximately 1.4 times ρ_{max}. $\rho_w = A_s/b_w d$

5.4 DESIGN OF RECTANGULAR SECTIONS FOR BENDING – TENSION REINFORCEMENT ONLY

The problem generally is to determine the required section properties, b, d (hence h) and A_s, to carry a known factored moment, M_f. The material properties, f_c' and f_y, are generally specified for the job, and, if not, can be selected on the basis of availability and economy. For normal applications, the strengths used are 20 to 60 MPa for concrete and 400 MPa for reinforcement. For under-reinforced sections, the influence of f_c' on the factored flexural resistance is relatively small.

The Code (10.1.4) requires that all sections be designed as under-reinforced. For under-reinforced sections, the factored flexural resistance, M_r, is given by Eq. 4.36, (Section 4.11). The design equation for flexure thus is (Eq. 3.13):

$$M_f = M_r = \rho \phi_s f_y \left(1 - \frac{\rho \phi_s f_y}{2\alpha_1 \phi_c f_c'}\right) \quad (5.1)$$

Obviously there are several combinations of ρ, b, and d that will carry a given M_f and, as a result, many different and satisfactory designs are possible. A general procedure for design, when all section properties (b, d, h, A_s) are to be determined, will be to select an appropriate reinforcement ratio ρ, and then compute the other values (b, d, h) accordingly. This procedure is explained in Section 5.4.1.

There are also occasions where b and h (and hence d) are known, and it is only necessary to compute the required area of reinforcement to carry the given moment. For example, the beam section of a continuous beam is first designed for the absolute maximum bending moment in the beam, which normally occurs at one of the support sections where the beam is continuous. The gross concrete section dimensions, $b \times h$, so determined may be provided for the entire length of the beam, since a prismatic member is often desired and may be more economical. Therefore, in the design of the beam at other sections of peak moments, such as at midspan and at other support sections, the only unknown is the required area of reinforcement. The design in this case is simpler and may be made as explained in Section 5.4.2.

5.4.1 Complete Design To Determine b, d, h, A_s

Equation 5.1 can be rearranged as:

$$\frac{M_r}{bd^2} = K_r = \rho \phi_s f_y \left(1 - \frac{\rho \phi_s f_y}{2\alpha_1 \phi_c f'_c}\right) \tag{5.2}$$

Step 1
Select a suitable ρ, less than the upper limit of ρ_{max} and resulting in an A_s greater than A_{smin}. These limits are given in Sections 4.9 and 4.10. For economy and control of deflections, ρ selected should preferably be in the range of 0.35 ρ_{max} to 0.50 ρ_{max}. For the selected ρ, the right hand side of Eq. 5.2 gives the strength factor K_r.

$$K_{r,trial} = \rho_{trial} \phi_s f_y \left(1 - \frac{\rho_{trial} \phi_s f_y}{2\alpha_1 \phi_c f'_c}\right)$$

Step 2
Compute required bd^2 from

$$bd^2 = M_r / K_{r,trial}$$

Now a suitable combination of b and d can be selected to yield the required bd^2. The value of b is taken as a convenient round figure, and d is selected such that the resulting h/b ratio is in the economic range (1.5 to 2 for beams and up to 3 for girders). For slabs, since design is made for a typical 1 m wide strip, $b = 1000$ mm and d can be directly calculated. The choice of d may also be guided by the minimum thickness limitations in Table 5.2 (deflection control).

Step 3
Select actual b, h, d. The overall depth h is selected as a round figure, and such that, with allowances made for clear cover, stirrup, size and bar size, an effective depth slightly in excess of the required value computed in Step 2 is obtained.

Step 4
Compute actual reinforcement area required. Since the actual d provided in Step 3 may differ slightly from the value computed in Step 2, a revised value for ρ is now computed. Using the dimensions b and d selected,

$$\text{actual } K_r = \frac{M_r}{bd^2}$$

ρ required is given by:

$$K_r = \rho \phi_s f_y \left(1 - \frac{\rho \phi_s f_y}{2\alpha_1 \phi_c f'_c}\right)$$

or

$$\rho = \frac{\alpha_1 \phi_c f'_c}{\phi_s f_y}\left(1 - \sqrt{1 - \frac{2K_r}{\alpha_1 \phi_c f'_c}}\right) \qquad (5.3)$$

Required $A_s = \rho b d$

Step 5
Check the design. Select size and number of bars required, and locate the bars in one or more layers to meet the clearance requirements. Check whether the actual effective depth (distance from the compression face to the centroid of reinforcement area) is satisfactory. Note that even a slight reduction in d may be satisfactory if the bar selection gives an area correspondingly greater than the A_s computed in Step 4 (provided $\rho < \rho_{max}$). The best way to check the design is to use the final dimensions selected to compute the factored resistance, M_r, of the section using Eq. 5.1 and, thereby, ensure that this is greater than the factored moment. If the design is not satisfactory, it is revised once again.

Frequently, during the analysis of the structure to compute the factored moments, the self-weight of the member itself is initially based on an assumed member size. In such cases, at Step 2, it can be checked whether the assumptions regarding self-weight are satisfactory, and, if not, the moment calculation can be revised taking the value of b and h obtained in Step 2 as a guide.

EXAMPLE 5.1

A rectangular beam is simply supported on two walls, 240 mm thick and 6 m apart (centre to centre). The beam has to carry, in addition to its own weight, a distributed live load of 15 kN/m and a dead load of 8 kN/m. Design the beam section for maximum moment at midspan. Use $f_c' = 30$ MPa, and $f_y = 400$ MPa.

SOLUTION

For $f_c' = 30$ MPa, $\alpha_1 = 0.805$ and $\beta_1 = 0.895$. The effective span length (CSA A23.3-94(9.2.2)) will in this case be taken as the distance between supports = 6 m. Assuming a trial cross-section of 250 mm by 400 mm, the self-weight of the beam per metre length is:

$$0.25 \times 0.4 \times 2400 \times \frac{9.81}{1000} = 2.35 \text{ kN}$$

With the load factors specified in Section 3.6, the factored load is:
$$w_f = \alpha_D D + \gamma \psi \alpha_L L = 1.25(8+2.35) + 1.5 \times 15 = 35.4 \text{ kN/m}$$

$$M_f = \frac{w_f l^2}{8} = \frac{35.4 \times 6^2}{8} = 159 \text{ kN} \cdot \text{m}$$

The section will be designed first with the maximum allowable reinforcement ratio. The maximum allowable reinforcement ratio is (Eq. 4.29):

$$\rho_{max} = \frac{700}{700+400} \times \frac{0.805 \times 0.895 \times 0.6 \times 30}{0.85 \times 400} = 0.0243$$

For $\rho_{max} = 0.0243$,

$$\frac{M_r}{bd^2} = K_r = \rho \phi_s f_y \left(1 - \frac{\rho \phi_s f_y}{2\alpha_1 \phi_c f_c'}\right)$$

$$= 0.0243 \times 0.85 \times 400 \left(1 - \frac{0.0243 \times 0.85 \times 400}{2 \times 0.805 \times 0.6 \times 30}\right) = 5.91 \text{ MPa}$$

Required $bd^2 = M_r/K_r = \dfrac{159 \times 10^6}{5.91} = 26.9 \times 10^6 \text{ mm}^3$

Possible combinations of b and d are:

b	d
200 mm	367 mm
220 mm	350 mm
250 mm	328 mm

Fig. 5.3 Example 5.1

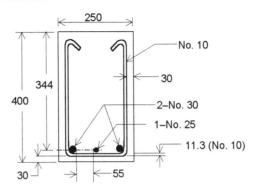

Select $b = 250$ mm and $d = 328$ mm. The minimum overall depth required allowing for 30 mm clear cover (not exposed), No. 10 stirrups (diameter = 11.3 mm), and No. 30 bars (diameter = 29.9 mm) is $h_{min} = 328 + 30 + 11.3 + 29.9/2 = 384$ mm. Since the ρ selected is the maximum, the depth cannot be decreased. Therefore, select an overall member size of 250×400 mm. Actual effective depth then will be $d = 400 - (30 + 11.3 + 29.9/2) = 344$ mm.

Actual $K_r = \dfrac{M_r}{bd^2} = \dfrac{159 \times 10^6}{250 \times 344^2} = 5.37$ MPa $= \rho \phi_s f_y \left(1 - \dfrac{\rho \phi_s f_y}{2\alpha_1 \phi_c f_c'}\right)$

Solving for ρ, $\rho = 0.0209 < \rho_{max}$; area $A_s = \rho bd = 0.0209 \times 250 \times 344 = 1797$ mm^2

Two No. 30 bars and one No. 25 bar (Fig. 5.3) give an area of 1900 mm^2. The clear spacing between the bars is $1/2$ $(250 - 2 \times 30 - 2 \times 11.3 - 2 \times 29.9 - 25.2) = 54$ mm. The minimum clearance is the greater of 1.4 $d_b = 42$ mm, 1.4 times maximum aggregate size = 28 mm (assuming 20 mm size aggregate), or 30 mm. The clearances are adequate. The cross-sectional dimensions designed are shown in Fig. 5.3. For this section, to check the actual M_r,

$a = A_s \phi_s f_y /(\alpha_1 \phi_c f_c' b) = 1900 \times 0.85 \times 400 / (0.805 \times 0.6 \times 30 \times 250) = 178$ mm

$M_r = A_s \phi_s f_y \left(d - \dfrac{a}{2}\right) = 1900 \times 0.85 \times 400 \,(344 - 178/2) \times 10^{-6}$

$\qquad = 165$ kN·m > 159 kN·m   OK

Check crack control (Fig. 5.2),
$Z = 0.6\,(400)\,(2 \times 56 \times 2 \times 56 \times 250 / 3)^{1/3} \times 10^{-3}$
$\quad = 24.4$ kN/mm < 25 kN/mm (assuming exterior exposure)   OK

EXAMPLE 5.2

The beam in Example 5.1 will now be redesigned with a moderate reinforcement ratio.

SOLUTION

To improve the ductility and economy, and to reduce deflection, select a reinforcement ratio $\rho = 0.5\rho_{max} = 0.5 \times 0.0243 = 0.0122$. This will result in an A_s greater than A_{smin}. From Eq. 5.2, the value of K_r for $\rho = 0.0122$ is:

$$K_r = 0.0122 \times 0.85 \times 400 \left(1 - \frac{0.0122 \times 0.85 \times 400}{2 \times 0.805 \times 0.6 \times 30}\right) = 3.55 \text{ MPa} = M_r/bd^2$$

required $bd^2 = M_r/K_r = 159 \times 10^6 / 3.55 = 44.8 \times 10^6 \text{ mm}^3$

Trial sizes are:

	b		d	
	= 240 mm		=	432 mm
	250 mm			423 mm
	260 mm			415 mm
	280 mm			400 mm

The trial sizes make it clear that the self-weight has been underestimated. It appears that an overall size of 250 × 480 mm is adequate. This would give an effective depth close to the 423 mm indicated above. The larger moment due to higher self-weight will result in a reinforcement ratio slightly higher than the value of $0.5\rho_{max}$, which is acceptable.

Corrected self-weight:
$0.25 \times 0.48 \times 24 = 2.88$ kN/m (unit of concrete $\approx 2400 \times 9.81/1000 \approx 24$ kN/m^3)
$w_f = 1.25(8 + 2.88) + 1.5 \times 15 = 36.1$ kN/m
$M_f = 36.1 \times 6^2 / 8 = 162$ kN·m
with $h = 480$ mm, $d = 480 - (30 + 11.3 + 29.9/2) = 424$ mm
Actual $M_r/bd^2 = K_r = 162 \times 10^6 /(250 \times 424^2) = 3.60$ MPa

For $K_r = 3.60$ MPa, using Eq. 5.3, reinforcement ratio required is:

$$\rho = \frac{0.805 \times 0.6 \times 30}{0.85 \times 400}\left(1 - \sqrt{1 - \frac{2 \times 3.60}{0.805 \times 0.6 \times 30}}\right) = 0.0124$$

Required $A_s = 0.0124 \times 250 \times 424 = 1314$ mm^2

Three No. 25 bars give $A_s = 1500$ mm^2, with adequate clearances

To check M_r of the section,
$$a = \frac{A_s \phi_s f_y}{\alpha_1 \phi_c f'_c b} = \frac{1500 \times 0.85 \times 400}{0.805 \times 0.6 \times 30 \times 250} = 141 \text{ mm}$$
$$M_r = A_s \phi_s f_y \left(d - \frac{a}{2}\right) = 1500 \times 0.85 \times 400(424 - 141/2) \times 10^6$$
$$= 180 \text{ kN·m} > 162 \text{ kN·m} \qquad \text{OK}$$

Check crack control (Fig. 5.2),
$Z = 0.6 (400) (2 \times 56 \times 2 \times 56 \times 250 / 3)^{1/3} \times 10^{-3}$
  $= 24.4$ kN/mm < 25 kN/mm (assuming exterior exposure) OK

5.4.2 Design with Known Concrete Section Dimensions

In this case b and h are known, and, from the latter, d can be estimated. Therefore, it is only necessary to determine the area of steel A_s. The design involves only the Steps 4 and 5 of the complete design process detailed in Section 5.4.1. The reinforcement obtained in Step 5 has to be checked against the maximum and minimum limits.

EXAMPLE 5.3

Design the beam in Example 5.1 if the beam size is fixed as 250 mm × 450 mm from architectural considerations.

SOLUTION

Self-weight $= 0.25 \times 0.45 \times 2400 \times \dfrac{9.81}{1000} = 2.65$ kN/m

$w_f = 1.25(8 + 2.65) + 1.5 \times 15 = 35.8$ kN/m

$M_f = 35.8 \times 6^2 / 8 = 161$ kN·m

Assuming No. 10 stirrups, No. 30 bars, and interior exposure:
$d = 450 - (30 + 11.3 + 29.9/2) = 394$ mm

$$\frac{M_r}{bd^2} = K_r = \frac{161 \times 10^6}{250 \times 394^2} = 4.15 \text{ MPa}$$

$$K_r = \rho \phi_s f_y \left(1 - \frac{\rho \phi_s f_y}{2\alpha_1 \phi_c f_c'}\right)$$

Solving, $\rho = 0.0148$, $A_s = \rho bd = 1458 \text{ mm}^2$

Three No. 25 bars give $A_s = 1500 \text{ mm}^2$. Clear distance between bars
= 1/2(250 - 2 × 30 - 2 × 11.3 - 3 × 25.2) = 45.9 mm, which is adequate.

With No. 25 bars, $d = 450 - (30 + 11.3 + 25.2/2) = 396$ mm

$$\rho = \frac{1500}{250 \times 396} = 0.0152, \quad \rho_{max} = 0.0243, \quad \rho < \rho_{max} \quad \text{OK}$$

$$A_{s\,min} = 0.2\sqrt{30} \times 250 \times 450 / 400 = 308 \text{ mm}^2$$

$$A_s = 1500 \text{ mm}^2 > A_{s\,min} \quad \text{OK}$$

To check ultimate strength:

$$a = \frac{A_s \phi_s f_y}{\alpha_1 \phi_c f_c' b} = \frac{1500 \times 0.85 \times 400}{0.805 \times 0.6 \times 30 \times 250} = 141 \text{ mm}$$

$$M_r = A_s \phi_s f_y \left(d - \frac{a}{2}\right) = 1500 \times 0.85 \times 400(396 - 141/2) \times 10^{-6}$$

$$= 166 \text{ kN} \cdot \text{m} > 161 \text{ kN} \cdot \text{m} \quad \text{OK}$$

Check crack control (Fig. 5.2),
$Z = 0.6\ (400)\ (2 \times 54 \times 2 \times 54 \times 250/3)^{1/3} \times 10^{-3}$
$= 23.8$ kN/mm < 25 kN/mm (assuming exterior exposure)   OK

5.5 DESIGN AIDS FOR RECTANGULAR BEAMS

The design procedure explained in Section 5.4 and the previous examples show that the design essentially involved the solution of Eq. 5.2, either to get a required $bd^2 = M_r/K_r$ for a selected reinforcement ratio, or to get a required ρ for a selected bd^2 from $K_r = M_r/(bd^2)$. The solution can be simplified by expressing the relation between K_r and ρ, either graphically or in a tabular form, for commonly used values of f_c' and f_y. Table 5.4 presents such a relation between K_r in MPa and 100ρ. Table 5.4 is similar to Table 4.1, except that it is rearranged in terms of selected K_r values at closer intervals. This table also indicates the maximum limits for ρ. For design using the table, a suitable ρ is first selected, and the resulting K_r read off from the table using the column for the specified values of f_c' and f_y. Knowing K_r, the required $bd^2 = M_r \times 10^6/K_r$, where b and d are in mm and M_r in kN·m. Conversely, when b and d are known or appropriately selected, the value $K_r = M_r \times 10^6/(bd^2)$ is computed first. For this K_r, the required value of ρ is read off from Table 5.4. For values in between those given in Table 5.4 linear interpolation may be used.

EXAMPLE 5.4

Design the beam in Example 5.1 using Table 5.4

SOLUTION

As in Example 5.1, required $M_r = M_f = 159$ kN·m.
Selecting $\rho = \rho_{max} = 0.0243$, from Table 5.4, $K_r = 5.90$ MPa.

Required $bd^2 = 159 \times 10^6/5.90 = 26.9 \times 10^6$ mm^3
Select $b = 250$ mm,

Required $d = \sqrt{\dfrac{26.9 \times 10^6}{250}} = 328$ mm ; select $h = 400$ mm.

Allowing for cover, actual $d = 400 - (30 + 11.3 + 29.9/2) = 344$ mm

With $b = 250$ mm and $d = 344$ mm

Actual $K_r = \dfrac{M_r}{bd^2} = \dfrac{19 \times 10^6}{250 \times 344^2} = 5.37$ MPa

From Table 5.4,   for $K_r = 5.2$ MPa, $100\,\rho = 2.00$
                  and $K_r = 5.4$ MPa, $100\,\rho = 2.11$

Interpolating, for $K_r = 5.37$ MPa, $100\,\rho = 2.09$

$A_s = \rho bd = 0.0209 \times 250 \times 344 = 1797$ mm^2 as before

The rest of the design follows that in Example 5.1.

Table 5.4 Reinforcement Percentage, 100ρ, for Resistance Factor, K_r

$$M_r = K_r b d^2 \times 10^6; \quad K_r = \rho \phi_s f_y [1 - \rho \phi_s f_y / (2\alpha_1 \phi_c f_c')]; \quad \rho = A_s/bd = \alpha_1 \phi_c f_c' [1 - \sqrt{1 - 2K_r/(\alpha_1 \phi_c f_c')}]/(\phi_s f_y); \quad \phi_c = 0.6; \phi_s = 0.85$$

Units of M_r in kN·m; b and d in mm; K_r in MPa

f_y (MPa)	300					350					400				
f_c' (MPa)	25	30	35	40	45	25	30	35	40	45	25	30	35	40	45
α_1	0.813	0.805	0.798	0.790	0.783	0.813	0.805	0.798	0.790	0.783	0.813	0.805	0.798	0.790	0.783
β_1	0.908	0.895	0.883	0.870	0.858	0.908	0.895	0.883	0.870	0.858	0.908	0.895	0.883	0.870	0.858
K_r 100ρ															
0.20	0.079	0.079	0.079	0.079	0.079	0.068	0.068	0.068	0.068	0.068	0.059	0.059	0.059	0.059	0.059
0.40	0.160	0.159	0.159	0.159	0.158	0.137	0.136	0.136	0.136	0.136	0.120	0.119	0.119	0.119	0.119
0.60	0.241	0.240	0.240	0.239	0.239	0.207	0.206	0.205	0.205	0.205	0.181	0.180	0.180	0.179	0.179
0.80	0.325	0.323	0.322	0.321	0.320	0.278	0.277	0.276	0.275	0.274	0.244	0.242	0.241	0.240	0.240
1.00	0.410	0.407	0.405	0.403	0.402	0.351	0.349	0.347	0.345	0.344	0.307	0.305	0.303	0.302	0.301
1.20	0.496	0.492	0.489	0.487	0.485	0.425	0.422	0.419	0.417	0.416	0.372	0.369	0.367	0.365	0.364
1.40	0.585	0.578	0.574	0.571	0.569	0.501	0.496	0.492	0.489	0.487	0.439	0.434	0.431	0.428	0.426
1.60	0.675	0.667	0.661	0.656	0.653	0.579	0.571	0.566	0.563	0.560	0.506	0.500	0.496	0.492	0.490
1.80	0.768	0.756	0.749	0.743	0.739	0.658	0.648	0.642	0.637	0.633	0.576	0.567	0.561	0.557	0.554
2.00	0.862	0.848	0.838	0.831	0.825	0.739	0.726	0.718	0.712	0.708	0.647	0.636	0.628	0.623	0.619
2.20	0.959	0.941	0.928	0.920	0.913	0.822	0.806	0.796	0.788	0.783	0.719	0.705	0.696	0.690	0.685
2.40	1.06	1.04	1.02	1.01	1.00	0.907	0.888	0.875	0.865	0.859	0.794	0.777	0.765	0.757	0.751
2.60	1.16	1.13	1.11	1.10	1.09	0.995	0.971	0.955	0.944	0.936	0.870	0.849	0.836	0.826	0.819
2.80	1.27	1.23	1.21	1.19	1.18	1.08	1.06	1.04	1.02	1.01	0.949	0.924	0.907	0.895	0.887
3.00	1.37	1.33	1.31	1.29	1.27	1.18	1.14	1.12	1.10	1.09	1.03	1.00	0.980	0.966	0.956
3.20	1.49	1.44	1.41	1.38	1.37	1.27	1.23	1.20	1.19	1.17	1.11	1.08	1.05	1.04	1.03
3.40	1.60	1.54	1.51	1.48	1.46	1.37	1.32	1.29	1.27	1.25	1.20	1.16	1.13	1.11	1.10
3.60	1.72	1.65	1.61	1.58	1.56	1.48	1.42	1.38	1.35	1.34	1.29	1.24	1.21	1.18	1.17
3.80	1.85	1.76	1.71	1.68	1.66	1.58	1.51	1.47	1.44	1.42	1.39	1.32	1.29	1.26	1.24

f_y (MPa)	300					350					400				
f_c' (MPa)	25	30	35	40	45	25	30	35	40	45	25	30	35	40	45
α_1	0.813	0.805	0.798	0.790	0.783	0.813	0.805	0.798	0.790	0.783	0.813	0.805	0.798	0.790	0.783
β_1	0.908	0.895	0.883	0.870	0.858	0.908	0.895	0.883	0.870	0.858	0.908	0.895	0.883	0.870	0.858
K_r, 100ρ															
4.00	1.98	1.88	1.82	1.78	1.75	1.70	1.61	1.56	1.53	1.50	1.48	1.41	1.37	1.34	1.32
4.20	2.12	2.00	1.93	1.89	1.85	1.81	1.71	1.66	1.62	1.59	1.59	1.50	1.45	1.41	1.39
4.40	2.26	2.12	2.04	1.99	1.96	1.94	1.82	1.75	1.71	1.68	1.69	1.59	1.53	1.49	1.47
4.60	2.41	2.25	2.16	2.10	2.06	2.07	1.93	1.85	1.80	1.77	1.81	1.69	1.62	1.58	1.55
4.80	2.58	2.38	2.28	2.21	2.17	2.21	2.04	1.95	1.90	1.86	1.93	1.79	1.71	1.66	1.62
5.00	2.75	2.52	2.40	2.32	2.27	2.36	2.16	2.06	1.99	1.95		1.89	1.80	1.74	1.70
5.20	2.95	2.66	2.52	2.44	2.38		2.28	2.16	2.09	2.04		2.00	1.89	1.83	1.79
5.40		2.81	2.65	2.56	2.49		2.41	2.27	2.19	2.14		2.11	1.99	1.92	1.87
5.60		2.97	2.79	2.68	2.61		2.55	2.39	2.30	2.23		2.23	2.09	2.01	1.95
5.80		3.14	2.93	2.80	2.72		2.70	2.51	2.40	2.33		2.36	2.19	2.10	2.04
6.00		3.33	3.07	2.93	2.84		2.85	2.63	2.51	2.43			2.30	2.20	2.13
6.20		3.52	3.22	3.06	2.96			2.76	2.62	2.54			2.42	2.30	2.22
6.40			3.38	3.20	3.08			2.90	2.74	2.64			2.53	2.40	2.31
6.60			3.54	3.34	3.21			3.04	2.86	2.75			2.66	2.50	2.41
6.80			3.72	3.48	3.34			3.19	2.98	2.86				2.61	2.50
7.00			3.91	3.63	3.47				3.11	2.98				2.72	2.60
7.20				3.79	3.61				3.25	3.09				2.84	2.71
7.40				3.95	3.75				3.39	3.22				2.96	2.81
7.60				4.12	3.90				3.54	3.34					2.92
$100\rho_{max}$	3.04	3.56	4.06	4.53	4.97	2.48	2.91	3.31	3.70	4.06	2.07	2.43	2.77	3.09	3.39
$K_{r\,max}$	5.28	6.23	7.15	8.03	8.88	5.14	6.07	6.95	7.81	8.63	5.01	5.90	6.76	7.59	8.38

5.6 DESIGN OF ONE-WAY SLABS

Slabs are designed for a design strip of one metre width, so that $b = 1000$ mm. Otherwise, the design of a slab is identical in procedure to that of a beam of rectangular section. Slabs in general are not reinforced for shear, as shear stresses are kept in check by providing adequate depth. Therefore, no allowance for stirrups need be made in the design of one-way slabs. Since the one metre wide strip is only a typical design strip, it is not necessary to specify a certain number of reinforcing bars for the design section of one metre width. Instead, what is required is the selection of a suitable size of bar and its spacing necessary to provide the calculated steel area. If A_s is the calculated reinforcement area required per one metre wide strip, and A_b is the area of one bar of the selected size, the required spacing s in mm is found from:

$$A_s = \frac{1000}{s} A_b, \qquad s = \frac{A_b}{A_s} \times 1000$$

To facilitate the selection of bar size and spacing for slabs, Table 5.5 gives the area, A_s, per metre width obtained by providing different size bars at different spacings. The principal reinforcement in slabs must be no less than that required for shrinkage and temperature stresses (CSA-A23.3-94(7.8)). When the principal reinforcement extends only in one direction, as in one-way slabs, reinforcement required for shrinkage and temperature stresses is provided perpendicular to the principal reinforcement.

EXAMPLE 5.5

The plan of a beam and slab floor is shown in Fig. 5.4a. The specified floor loading consists of a live load of 6.5 kN/m^2 and a dead load (due to floor finish, partitions, etc.) of 1 kN/m^2, in addition to the self-weight. Use $f_c' = 35$ MPa and $f_y = 400$ MPa. Design the slab thickness and the reinforcement area required at all critical sections.

SOLUTION

The slab is one-way continuous and integral with the beam supports spaced at 3 m clear. A one metre wide design strip can be treated as a 4 span continuous beam.
The design moments at all critical sections (at midspan for positive moments and at supports for negative moments) can be obtained by the analysis of this continuous beam bearing in mind that the dead loads act on all spans, whereas the live loads have to be placed so as to cause the worst effect at the section considered. Such a detailed analysis will be discussed and illustrated in Chapter 10. In lieu of such an analysis the CSA A23.3-94(9.3) permits the use of moment coefficients when the structure and loading meet the limitations in Clause 9.3.3. This approximate procedure will be used

in this example for computing the design moments.

To compute the self-weight of the slab, an estimate for the slab thickness may be taken as the minimum thickness specified in Table 5.2. (CSA A23.3-94(9.8.2.1)). Considering the end span, which is more critical as one end is discontinuous (note that since the slab is monolithic with the spandrel beam at the end, there is some fixity at this end as well, but this will be discounted in estimating h_{min}), $h_{min} = l_n/24$.

The span l_n is the clear span and is defined in CSA A23.3-94(9.0) as the

Fig. 5.4 Example 5.5

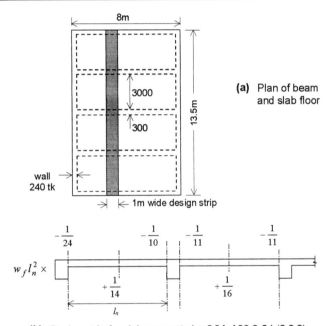

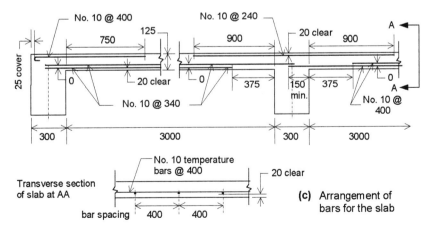

Table 5.5 *Area of Reinforcing Bars in Slabs in mm² Per Metre Width*

Bar Size →	10	15	20	25	30	35
Nominal Diameter mm →	11.3	16.0	19.5	25.2	29.9	35.7
Area of 1 Bar mm² →	100	200	300	500	700	1000
Bar Spacing ↓ mm						
60	16667	3333	5000	8333	11667	16667
80	1250	2500	3750	6250	8750	12500
100	1000	2000	3000	5000	7000	10000
120	833	1667	2500	4167	5833	8333
140	714	1429	2143	3571	5000	7143
150	667	1333	2000	3333	4667	6667
160	625	1250	1875	3125	4375	6250
180	556	1111	1667	2778	3889	5556
200	500	1000	1500	2500	3500	5000
220	455	909	1364	2273	3182	4545
240	417	833	1250	2083	2917	4167
250	400	800	1200	2000	2800	4000
260	385	769	1154	1923	2692	3846
280	357	714	1071	1786	2500	3571
300	333	667	1000	1667	2333	3333
320	313	625	938	1563	2188	3125
340	294	588	882	1471	2059	2941
350	286	571	857	1429	2000	2857
360	278	556	833	1389	1944	2778
380	263	526	789	1316	1842	2632
400	250	500	750	1250	1750	2500
420	238	476	714	1190	1667	2381
440	227	455	682	1136	1591	2273
450	222	444	667	1111	1556	2222
460	217	435	652	1087	1522	2174
480	208	417	625	1042	1458	2083
500	200	400	600	1000	1400	2000

length of the span from the face of one support to the face of the opposite support.

Clear span $l_n = 3$ m

$$h_{min} = \frac{3000}{24} = 125 \text{ mm}$$

Self-weight of slab $= 1 \times 1 \times 0.125 \times 2400 \times \dfrac{9.81}{1000} = 2.94 \text{ kN/m}^2$

Total dead load = 1 + 2.94 = 3.94 kN/m^2,
live load = 6.5 kN/m^2

Since there are four equal spans, and the factored live load is less than two times the factored dead load, the requirements of CSA A23.3-94(9.3.3) are met, and the moment coefficients may be used.

Total factored load per metre length for the one metre wide strip is:

$$w_f = 1.25 \times 3.94 + 1.5 \times 6.5 = 14.7 \text{ kN/m}$$

The design moments at critical sections are (Fig. 5.4b):

(i) positive moment in end span (discontinuous end integral with supporting beam):

$$\frac{1}{14} w_f l_n^2 = \frac{14.7 \times 3^2}{14} = 9.45 \text{ kN} \cdot \text{m}$$

(ii) positive moment in interior spans:

$$\frac{1}{16} w_f l_n^2 = 8.27 \text{ kN} \cdot \text{m}$$

(iii) negative moment at interior face of exterior support (integral beam support):

$$\frac{1}{24} w_f l_n^2 = 5.51 \text{ kN} \cdot \text{m}$$

(iv) negative moment at exterior face of first interior support (more than two spans):

$$\frac{1}{10} w_f l_n^2 = 13.2 \text{ kN} \cdot \text{m}$$

(v) negative moment at other faces of interior supports:

$$\frac{1}{11} w_f l_n^2 = 12.0 \text{ kN} \cdot \text{m}$$

The slab depth is selected for the largest moment which is the negative moment at supports = 13.2 kN·m.

Selecting a steel ratio of 0.4 ρ_{max}, from Table 4.1,

$$\rho = 0.4\, \rho_{max} = 0.01108, \text{ for which } K_r = 3.34 \text{ MPa}$$

Required $bd^2 = M_r/K_r = 13.2 \times 10^6 / 3.34 = 3.95 \times 10^6 \text{ mm}^2$

$$d = \sqrt{3.95 \times 10^6 / 1000} = 62.8 \text{ mm}$$

Assuming No. 10 bars and 20 mm clear cover, the required overall thickness is $62.8 + 11.3/2 + 20 = 88.5$ mm

Select an overall thickness of 125 mm for the slab throughout. The initial self-weight computation is satisfactory. The actual effective depth is:

$$d = 125 - 20 - 11.3/2 = 99.4 \text{ mm}$$

The area of reinforcement and bar spacing are computed for the interior support sections below. Similar computations will give the bar details required at all critical sections. At interior supports negative $M_r = 13.2$ kN·m

$$\text{Actual } K_r = \frac{M_r}{bd^2} = \frac{13.2 \times 10^6}{1000 \times 99.4^2} = 1.34 \text{ MPa}$$

For $K_r = 1.34$, $\rho = 0.00411$ (Eq. 5.3. or Table 5.4)

$$A_s = 0.00411 \times 1000 \times 99.4 = 409 \text{ mm}^2$$

$$\text{Spacing} = \frac{1000 A_b}{A_s} = \frac{1000 \times 100}{409} = 244 \text{ mm}$$

No. 10 bars at 240 mm is adequate. The bar spacing may also be selected from Table 5.5, which shows that No. 10 bars at 240 mm gives $A_s = 417$ mm² compared to the required 409 mm².

The calculations for other critical sections are set forth in Table 5.6. It is necessary also to check that the ρ provided is not less than the specified minimum for slabs, and that the spacing of bars does not exceed the maximum specified (CSA A23.3-94(7.4.1.2)).

In a one-way slab such as this with principal reinforcement designed for flexure only in one direction (bars placed along the short span), secondary reinforcement must be provided at right angles to the main reinforcement to take care of temperature and shrinkage stresses (CSA A23.3-94(7.8)). The minimum temperature and shrinkage reinforcement in this example = $0.002 \times$ gross area = $0.002 \times 1000 \times 125 = 250$ mm² per metre width. No. 10 bars at 400 mm give an area of 250 mm² per metre, as required. The maximum spacing for this is the lesser of 500 mm and $5h = 625$ mm (CSA A23.3-94(7.8.3)). The spacing, 400 mm, is satisfactory. The actual arrangement of slab reinforcement is discussed in Section 5.7. Crack control, as per Fig. 5.2, must be checked.

Table 5.6 *Computation for Flexural Reinforcement at All Critical Sections, Example 5.5*

Location of Critical Section	End Span			Interior Span			Code A23.3 Ref. Clause
	End Support	Span	First Int. Support	Int. Support	Int. Span	Int. Support	
Moment Sign and Location of Reinforcement (top/bottom)	Negative	Positive	Negative	Negative	Positive	Negative	9.3
Moment coefficient	-1/24	+1/14	-1/10	-1/11	+1/16	-1/11	
Design moment M_r=Coeff $\times 14.7 \times 3^2$; kN·m	5.51	9.45	13.2	12.0	8.27	12.0	
Actual K_r $=\dfrac{M_r}{bd^2}=\dfrac{M_r \times 10^6}{1000 \times 99.4^2}$; MPa	0.558	0.956	1.34	1.21	0.837	1.21	
Required ρ (Table 5.4 or Eq. 5.3)	0.00167	0.00290	0.00411	0.00370	0.00253	0.00370	7.8.1
ρ_{min} on gross area				0.002			
Required area $A_s = \rho \times 1000 \times 99.4$, mm^2	166	288	409	368	251	368	
$A_{s,min} = 0.002 \times$ gross area	$0.002 \times 1000 \times 125 = 250$ mm^2, $A_s > A_{s,min}$						
Required spacing of No. 10 bars, $\dfrac{1000 \times 100}{A_s}$, mm	250	347	244	272	398	272	
For principal reinforcement, maximum allowable spacing, lesser of $3h$ or 500, mm			375				7.4.1.2
For minimum reinforcement, lesser of $5h$ or 500, mm			500				7.8.3
Select No. 10 bars at (mm)	400	340	240	260	400	260	
Area provided, mm^2	250	294	417	385	250	385	

138 REINFORCED CONCRETE DESIGN

5.7 ARRANGEMENT OF REINFORCEMENT IN CONTINUOUS ONE-WAY SLABS

The positive and negative moment reinforcement required in the span and over the support regions may be provided in one of two ways, shown in Fig. 5.5a and b. The straight bar arrangement of Fig. 5.5a is used almost exclusively, and separate reinforcement is detailed for the positive moments and the negative moments. Prior to 1965, the negative reinforcement over a support region was commonly provided in part by bending over to the top some of the positive moment reinforcement from the spans on either side, as the bottom reinforcement becomes no longer needed in the support region (Fig. 5.5b). Detailing of bar cut-off, bending and extensions are discussed in Section 5.9. These details are dependent on the moment envelope, giving the upper and lower limits of the moment for the entire span. Determination of moment envelopes for continuous spans is explained in Chapter 10. However, for uniformly loaded continuous slabs of nearly equal spans, such detailed calculations are often unnecessary, and cut-off (and bending points) may be approximately related to span dimensions. Typical bar details for one-way solid slabs recommended in Ref. 5.5 are shown in Fig. 5.6. An arrangement of straight bars for the slab in Example 5.5 is given in Fig. 5.4c.

5.8 DESIGN OF RECTANGULAR BEAMS WITH COMPRESSION REINFORCEMENT

When the cross-sectional dimensions of a beam are limited so that even with the maximum allowable reinforcement ratio of ρ_{max} the factored resistance of the section is inadequate to carry the moment, M_f, due to the factored loads, the section is designed

Fig. 5.5 *Arrangements of main reinforcement for continuous slabs*

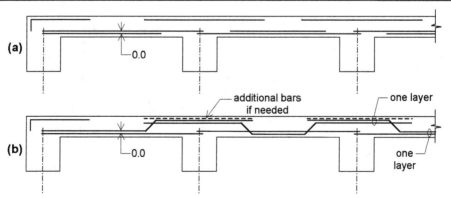

with compression reinforcement and additional tension reinforcement to carry the extra bending moment. All compression reinforcement must be enclosed by ties or closed stirrups conforming to CSA A23.3-94(7.6.6), to prevent their buckling. The design of doubly reinforced rectangular beams is illustrated with Example 5.6 below.

EXAMPLE 5.6

If the beam in Example 5.1 is limited in size to $b = 250$ mm and $h = 400$ mm, and it has to carry, in addition to the loads already mentioned, a concentrated live load of 30 kN, design the reinforcement.

Fig. 5.6 *Recommended typical bending details (Ref. 5.5)*

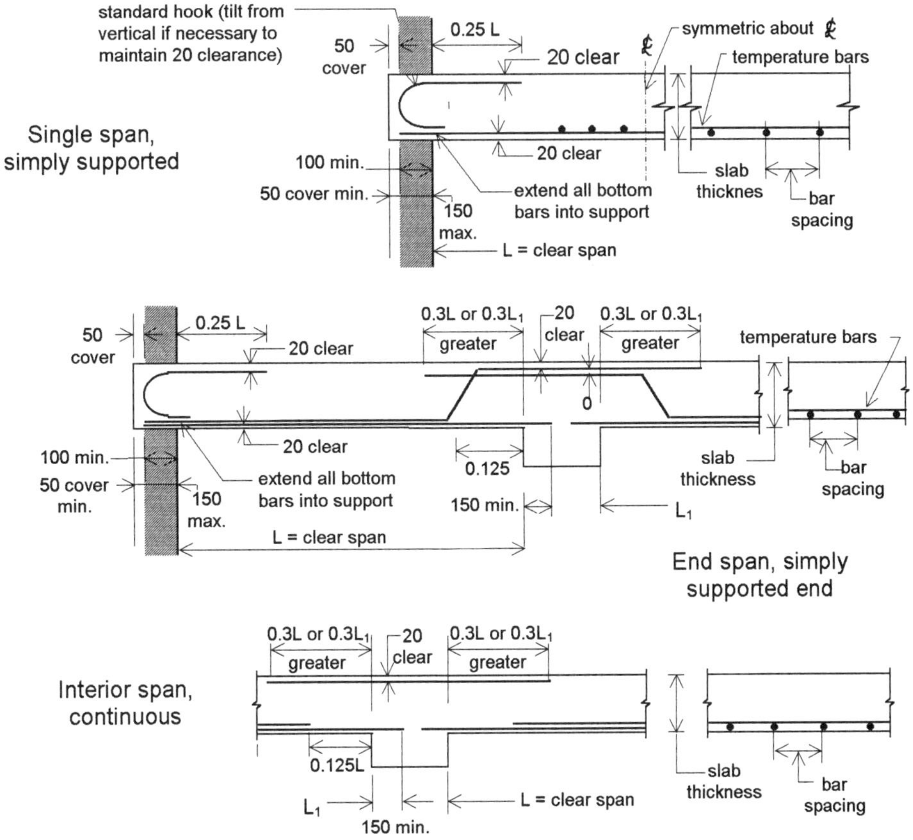

Note: Ref. 5.5 is the *Reinforcing Steel Manual of Standard Practice*, copies of which can be purchased from the Reinforcing Steel Institute of Canada, 70 Leek Crescent, Richmond Hill, Ontario, L4B 1H1.

SOLUTION

Self-weight = $0.25 \times 0.4 \times 2400 \times 9.81 \times 10^{-3}$ = 2.35 kN/m
Factored distributed load = $1.25(8 + 2.35) + 1.5 \times 15$ = 35.4 kN/m

Moment due to distributed load $= \dfrac{w_f l^2}{8} = \dfrac{35.4 \times 6^2}{8} = 159 \text{ kN} \cdot \text{m}$

The concentrated load is to be placed at midspan for maximum moment.

This moment is:
$$\dfrac{P_f l}{4} = (1.5 \times 30) \times \dfrac{6}{4} = 67.5 \text{ kN} \cdot \text{m}$$

Total design moment, M_f = 227 kN·m
The maximum strength of the section with tension reinforcement only, M_1, is checked first. As computed in Example 5.1, or from Table 4.1

$\rho_{max} = \overline{\rho} = 0.0243$, for which K_r = 5.91 MPa

Anticipating that tension reinforcement will be provided in two layers of No. 30 bars,

$d \approx 400 - (30 + 11.3 + 29.9 + 1.4 \times 29.9/2) = 308$ mm
$A_{s1} = 0.0243 \times 250 \times 308 = 1871 \text{ mm}^2$
$M_1 = K_r b d^2 = 5.91 \times 250 \times 308^2 \times 10^{-6} = 140$ kN·m

Since $M_1 < M_f$, the section has to be additionally reinforced in compression and tension to carry the excess moment $M_2 = M_f - M_1 = 227 - 140 = 87$ kN·m. The compression reinforcement will be placed at a depth,

$d' = 30 + 11.3 + 29.9/2 \approx 56$ mm.

The internal resisting couples are shown in Fig. 5.7. The additional tension reinforcement area A_{s2} required is given by:

$$M_2 = A_{s2} \phi_s f_y (d - d')$$
$$A_{s2} = \dfrac{87 \times 10^6}{0.85 \times 400(308 - 56)} = 1015 \text{ mm}^2$$

CHAPTER 5 DESIGN OF BEAMS AND ONE-WAY SLABS

Fig. 5.7 Example 5.6

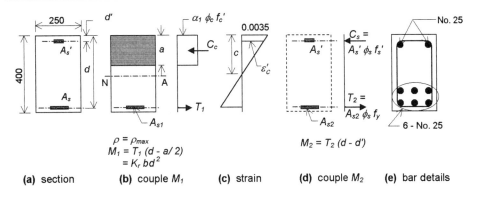

(a) section (b) couple M_1 (c) strain (d) couple M_2 (e) bar details

Total tension reinforcement required = 1871 + 1015 = 2886 mm²

If the compression reinforcement yields, $A'_s = A_{s2}$. To check this, the strain at the compression reinforcement level is computed using the strain distribution shown in Fig. 5.7.

For the moment M_1, $\rho = \rho_{max}$, $c/d = c_{max}/d = 0.636$ as given in Table 4.1, and $c = 0.636 \times 308 = 196$ mm. Alternatively, equating $C_c = T_1$, $\alpha_1 \phi_c f'_c b a = A_{s1} \phi_s f_y$

$$a = \frac{A_{s1} \phi_s f_y}{\alpha_1 \phi_c f'_c b} = \frac{1871 \times 0.85 \times 400}{0.805 \times 0.6 \times 30 \times 250} = 176 \text{ mm}$$

$$c = \frac{a}{\beta_1} = \frac{176}{0.895} = 197 \text{ mm}$$

$$\varepsilon'_s = \frac{0.0035(c-d')}{c} = \frac{0.0035(197-56)}{197} = 0.00251$$

$$\varepsilon'_s > \varepsilon_y = \frac{400}{200000} = 0.002$$

The compression steel yields, and the stress in it is $f'_s = 400$ MPa.

$$A'_s = A_{s2} = 1015 \text{ mm}^2$$

$$\rho' = \frac{A'_s}{bd} = \frac{1015}{250 \times 308} = 0.0132$$

Note that if allowance is made for the concrete area in compression which is displaced by A_s', the net effective steel stress will be only $(\phi_s f_y' - \alpha_1 \phi_c f_c')$, and A_s' will be obtained as:

$$A_s' = \frac{A_s^2 \phi_s f_y}{(\phi_s f_y' - \alpha_1 \phi_c f_c')} = 1060 \text{ mm}^2$$

The tension reinforcement ratio in this example can be expressed as:

$$\rho = \frac{A_{s1}}{bd} + \frac{A_{s2}}{bd} = \overline{\rho} + \rho' = 0.0375$$

The required reinforcement areas can be provided by six No. 25 placed in two layers on the tension side ($A_s = 3000$ mm^2), and two No. 25 bars on the compression side ($A_s = 1000$ mm^2) as shown in Fig. 5.7. This arrangement meets the minimum clearance requirements. Crack control, as per Fig. 5.2, must also be checked.

An exact evaluation of the actual ultimate strength of this section can be made by the procedure outlined in Section 4.11.3. Such an analysis (Example 4.9) shows that the actual strength is 238 kN·m > M_f = 227 kN·m. (The actual strength is slightly higher, because with the reinforcement arrangement used $d = 316$ mm and $d' = 54$ mm, rather than the 308 mm and 56 mm assumed, respectively, in the design.)

5.9 DESIGN OF T-BEAMS

The complete design of an isolated T-beam (such as may be used for a footbridge) involves determining of the flange and web dimensions and the area of reinforcement. In such a case, the overall flange width is normally specified, and the flange thickness and flange reinforcement (transverse) required may be computed considering the transverse bending of the overhanging flange as a cantilever slab. However, the most common use of T-beams is in the monolithic slab-beam floor system, where the slab dimensions are already known from the results of the slab design, which precedes the T-beam design. Thus, the effective flange width and the flange thickness are generally known, and the T-beam design involves determining web dimensions and the area of reinforcement. (design provisions for the effective flange width of T- and L-shapes are given in CSA A23.3-94(10.3)).

In a continuous T-beam, the section over the support is under negative moment, with the flange of the T being on the tension side of the section. Here, the beam acts as a rectangular section, with compression in the bottom fibres of the web. In such cases, the rectangular section dimensions determined for the maximum negative moment, which generally exceeds the maximum positive moment, also control the web size in the positive moment region, and only the area of reinforcement in the T-beam remains to be determined.

In simply supported T-beams, however, the web dimensions must also be designed. The width of web b_w of a T-beam must be sufficient to provide the tension reinforcement in one or more convenient layers, and the depth, d, must be adequate to provide the area $b_w \times d$ to resist shearing forces (Chapter 6). The depth must also be adequate to control the deflection, and the minimum thicknesses specified in Table 5.2 may be used as guidance, although with caution, as the loads on T-beams vary considerably depending on the type of structure, whether this be a building, bridge, or other structure.

In T-beam design procedure, the web dimensions are first assumed, if not already known, and the area of reinforcement required is computed. If the area so determined is beyond allowable limits, or cannot be placed conveniently, the design is revised with modified web dimensions. Because of the very large compressive concrete area contributed by the flange in T-beams of usual proportions, the depth of stress block a will be less than the flange thickness h_f, so that the T-beam effectively acts as a rectangular beam of width b. Actual T-beam action ($a > h_f$) occurs only when the flange dimensions are small while the beam is very deep and very heavily reinforced, as often occurs with closely spaced bridge girders in a T-beam bridge.

Flanges of T-beams must have adequate transverse reinforcement provided near the top of the flange. Such reinforcement is usually present in the form of the negative moment reinforcement for the slab, which forms the flange of the T-beam, and generally spans across the T-beams. When this is not the case, the required transverse reinforcement must be designed by considering the over-hanging portion of the flange as a cantilever slab, carrying the design loads applied on that portion of the slab. The spacing of this reinforcement is limited to the lesser of 5 times the flange thickness or 500 mm (CSA A23.3-94(7.8.3)).

At negative moment regions in a T-beam, the flange is in tension, and the beam is designed as a rectangular section having the width of the web. However, placing all the tension reinforcement in the web will leave the integral flange unprotected against wide tensile cracking. Therefore, CSA A23.3-94(10.5.3) specifies that a part of such tension reinforcement be distributed over a width of an overhang equal to the lesser of 1/20 of the beam span, or the width defined in Clause 10.3 (Section 4.9.3). The area of this reinforcement must be at least 0.004 times the gross area of the overhanging flange. Where necessary, the outer portions of the flange must be protected by some additional longitudinal reinforcement.

EXAMPLE 5.7

Design the interior beam in the floor system in Example 5.5.

SOLUTION

The slab is one-way, spanning the beams. Each interior beam carries a load from a width of 3.3 m of the slab (1.65 m on either side).

Dead load from slab including slab weight

$$= 3.3 \text{ m} \times \left(1 + 0.125 \times 2400 \times \frac{9.81}{1000}\right) = 13 \text{ kN/m}$$

The overall depth will be assumed as $l_n/16$ for deflection control,

$$= \frac{8000 - 2(240)}{16} = 470 \text{ mm, choose 500 mm}$$

Self-weight of rib = $0.3 \times (0.500 - 0.125) \times 2400 \times \frac{9.81}{1000} = 2.65$ kN/m

Live load = $3.3 \text{ m} \times 6.5 \text{ kN/m}^2 = 21.5$ kN/m

Total factored load = $1.25(13 + 2.65) + 1.5(21.5) = 51.8$ kN/m

Design moment = $51.8 \times \frac{(8 - 0.24)^2}{8} = 390$ kN·m

The effective width of flange is the least of
 (1) $0.4\,l + b_w = 0.4 \times (8000 - 240) + 300 = 3404$ mm
 (2) $24\,h_f + b_w = 24 \times 125 + 300 = 3300$ mm   (controls)
 (3) spacing = 3300 mm

Assuming that reinforcement will be provided in two layers, take the effective depth $d \approx 400$ mm.

The steel area limitations will be computed first. The maximum allowable reinforcement ratio is given either by Eq. 4.31 or may be computed from first principles as follows:

$$c_{max} = \frac{700}{700 + f_y}\,d = 0.636 \times 400 = 255 \text{ mm}$$

$a_{max} = 0.883 \times 255 = 225 \text{ mm} > h_f$

$C_{max} = [125(3300 - 300) + 300 \times 225] \times 0.798 \times 0.6 \times 35 = 7.42 \times 10^6$ N

$$A_{s\,max} = \frac{7.42 \times 10^6}{400} = 18539 \text{ mm}^2$$

$A_{s\,min} = 0.2 \times \sqrt{35} \times 300 \times \frac{500}{400} = 444 \text{ mm}^2$

The procedure for calculation of reinforcement area will depend on whether the depth of stress block for the given moment is less than h_f or not (whether rectangular beam action or T-beam action is to be considered). This may be checked by computing the factored moment resistance for $a = h_f$ and comparing this with the applied moment M_f. If $a = h_f$,

$$M_r = bh_f \times \alpha_1 \phi_c f'_c \left(d - \frac{h_f}{2}\right)$$

$$= 3300 \times 125 \times 0.798 \times 0.6 \times 35 \left(400 - \frac{125}{2}\right) \times 10^{-6}$$

$$= 2333 \text{ kN} \cdot \text{m} \gg M_f = 390 \text{ kN} \cdot \text{m}$$

Therefore, the depth $a < h_f$ and the beam acts like a rectangular beam of width $b = 3300$ mm and $d = 400$ mm. Following the design procedure for rectangular beams:

$$K_r = \frac{M_r}{bd^2} = \frac{390 \times 10^6}{3300 \times 400^2} = 0.739 \text{ MPa}$$

$$= \rho \phi_s f_y \left(1 - \frac{\rho \phi_s f_y}{2\alpha_1 \phi_c f'_c}\right)$$

$$\rho = 0.00222$$

$$A_s = 0.00222 \times 3300 \times 400 = 2930 \text{ mm}^2$$

This is within the allowable limits. Provide six No. 25 bars in two layers, giving an area of 3000 mm^2 as shown in Fig. 5.8. The actual strength of this section may be computed as $M_r = 415$ kN·m > 390 kN·m. (See Example 5.10 for computations.)

Alternative Procedure

The method of solution illustrated in Example 5.7 above gives the exact value of A_s required directly. However, the solution can be simplified by a trial-and-revision procedure, as explained and illustrated below.

For T-beams of normal proportions, the depth of stress block, a, is normally within the flange. The lever arm, z, between the resultant compression in concrete, C, and tension in reinforcement, T, can be approximately estimated initially as the larger of $d - h_f/2$ and $0.9d$. With this value of z, the area of reinforcement necessary to carry a design moment M_r may be estimated as:

$$A_s = \frac{M_r}{\phi_s f_y z}$$

146 REINFORCED CONCRETE DESIGN

Fig. 5.8 Example 5.7

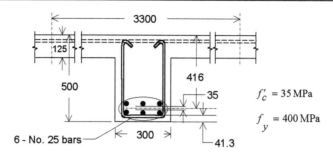

Using this approximate value of A_s, the depth of stress block, a, may be computed by equating $C = T$. Usually $a < h_f$, and even if $a > h_f$, in normal situations it is admissible to ignore the compressive concrete area below the flange and assume $a = h_f$. Thus, equating $C = T$ yields,

$$\alpha_1 \phi_c f'_c b \times (a \le h_f) = A_s \phi_s f_y$$

$$a = \frac{A_s \phi_s f_y}{\alpha_1 \phi_c f'_c b} \le h_f$$

and revised
$$z = d - \frac{a}{2}$$

Revised
$$A_s = \frac{M_r}{\phi_s f_y z}$$

Usually this revised value of A_s is very close to the exact value obtained by the detailed procedure. To illustrate the procedure, the problem in Example 5.7 will be solved by this method.

$$\text{Approximate } z = d - \frac{h_f}{2} \text{ or } 0.9 \times 400 = 360 \text{ mm}$$

$$\text{Approximate } A_s = \frac{390 \times 10^6}{0.85 \times 400 \times 360} = 3186 \text{ mm}^2$$

$$a = \frac{3186 \times 0.85 \times 400}{0.798 \times 0.6 \times 35 \times 3300} = 19.6 \text{ m}$$

$$\text{Revised } z = d - \frac{a}{2} = 390 \text{ mm}$$

$$\text{Revised } A_s = \frac{390 \times 10^6}{0.85 \times 400 \times 390} = 2941 = \text{mm}^2$$

compared to the value of 2930 mm^2 obtained by the detailed procedure.

In continuous T-beams, the negative moments at supports are usually greater than the positive moments in the span. Under negative moments, the flange is on the tension side, and the beam acts as a rectangular section. In such cases, the overall dimensions of the T-beam will be established first by the design of the beam section at the support. When this section is designed as an under-reinforced section (with tension reinforcement only), the T-section under the lesser positive moment in the span will naturally be under-reinforced, and there is no need to check this.

EXAMPLE 5.8

A continuous T-beam has the cross-sectional dimensions shown in Fig. 5.9. The web dimensions have been determined from considerations of negative moment at supports and the shear strength requirements. The spans are 10 m, and the design moment at midspan under factored loads is 700 kN·m. Determine the flexural reinforcement requirement at midspan. Take $f_c' = 30$ MPa and $f_y = 400$ MPa.

SOLUTION

The effective flange width is the least of: $24 h_f + b_w = 2700$ mm, actual width = 1000 m (controls), or $0.20 \times$ span $+ b_w = 2300$ mm.

Assuming that two layers of reinforcement are required, the effective depth is estimated as $d \approx 600 - (40 + 29.9 + 1.4 \times 29.9/2) = 510$ mm. (Here, with the clear cover of 30 mm, the nominal bar diameter of 10 mm is used, instead of its true diameter, this being acceptable for calculations.) First it will be checked whether the depth of stress block $a < h_f$ or not.

If $a = h_f = 100$ mm, the moment capacity will be given by:

Fig. 5.9 Example 5.8

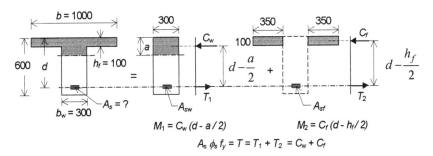

$$M = \alpha_1 \phi_c f_c' b h_f \left(d - \frac{h_f}{2} \right)$$

$$= 0.805 \times 0.6 \times 30 \times 1000 \times 100 \left(510 - \frac{100}{2} \right) \times 10^{-6} = 667 \text{ kN} \cdot \text{m}$$

Since design moment $M_r = 700$ kN·m is larger than this, a must be greater than h_f. To find the correct value of a, equate M_r to the moment of the compressive forces in the web, C_w, and in the overhanging flange, C_f (Fig. 5.9).

$$M_r = C_w \left(d - \frac{a}{2} \right) + C_f \left(d - \frac{h_f}{2} \right)$$

$$= \alpha_1 \phi_c f_c' b_w a \left(d - \frac{a}{2} \right) + \alpha_1 \phi_c f_c' (b - b_w) h_f \left(d - \frac{h_f}{2} \right)$$

$$= 700 \times 10^6 = 0.805 \times 0.6 \times 30 \left[300 \times a \left(510 - \frac{a}{2} \right) + (1000 - 300)100 \left(510 - \frac{100}{2} \right) \right]$$

Solving this quadratic equation, $a = 129$ mm.
The steel area can now be obtained from $T = A_s \phi_s f_y = C_w + C_f$

$$A_s = \frac{C_w + C_f}{\phi_s f_y} = \frac{0.805 \times 0.6 \times 30 [300 \times 129 + (1000 - 300) \times 100]}{0.85 \times 400} = 4633 \text{ mm}^2$$

Now check whether this area is within allowable limits.

$$A_{s\,min} = 0.2 \times \sqrt{30} \times 300 \times \frac{600}{400} = 493 \text{ mm}^2$$

Maximum reinforcement such that $\dfrac{c}{d} \leq \dfrac{700}{700 + f_y} = 0.636$

Actual $c/d = a/(\beta d) = 129/(0.895 \times 510) = 0.28 << 0.636$ OK
The area designed, $A_s = 4633$ mm^2, is within the allowable limits.

5.10 CUT-OFF OF FLEXURAL REINFORCEMENT

5.10.1 Introduction

In simply supported beams, the maximum bending moment is at or near the midspan, and the flexural design of the cross-section is first worked out for this section. Similarly, in continuous spans the maximum negative moment is at a support section and the maximum positive moment near midspan. Therefore, the required concrete section dimensions and the respective flexural reinforcement areas are determined by designing these sections. (For negative moment, CSA A23.3-94(9.2.2.2) permits the use of the moment at the face of the support for design of beams.) At sections away from these critical design sections, the moment progressively decreases; however, the same overall section dimensions are usually provided throughout the length of the beam. Such a prismatic member is usually more economical than a member of varying cross-section, because of the greater fabrication costs for the latter. Thus, away from the midspan section the positive moment decreases, and, as a result, the flexural reinforcement required also decreases. Therefore, some of the bars provided at the midspan section can be terminated as they become no longer necessary to resist bending moment or, alternatively, in continuous beams, bent over to the top side of the beam and continued to the support for the negative moment requirement over the supports. In the same manner, away from the face of the support towards midspan in a continuous beam, some of the bars provided for the peak negative moment at the support section may be cut off or bent down. Typically, the continuous beams of older structures have bent bars, but bending the reinforcement to support positive or negative moments is no longer common practice in Canada.

5.10.2 Theoretical Cut-Off Points Due to Moment Only

In a prismatic beam, as the bending moment decreases along the length of the member, the required flexural tension reinforcement decreases approximately in direct proportion to the moment. If M is the moment and A_s is the tension reinforcement area required at the critical section (Fig. 5.10b, c),

$$A_s = \frac{M}{\phi_s f_y z_1}$$

At a section where the moment decreases to M_1, the required reinforcement area is:

$$A_{s1} = \frac{M_1}{\phi_s f_y z_1}$$

and

$$\frac{A_{s1}}{A_s} = \frac{M_1 z}{M z_1} \approx \frac{M_1}{M}, \text{ since } \frac{z}{z_1} \approx 1$$

As the moment decreases, the lever arm z increases slightly, as is evident from Fig. 5.10c; however, it is sufficiently accurate (and slightly conservative) to assume $z/z_1 = 1$. Therefore, the steel area requirement varies in proportion to the variation of the moment. Thus, at a section where the moment is, say, 60 percent of M, the reinforcement area required is only 60 percent of the designed area, A_s, and the remaining 40 percent may be cut off as far as flexural requirement is concerned.

Fig. 5.10 *Theoretical bar cut-off (or bend) points*

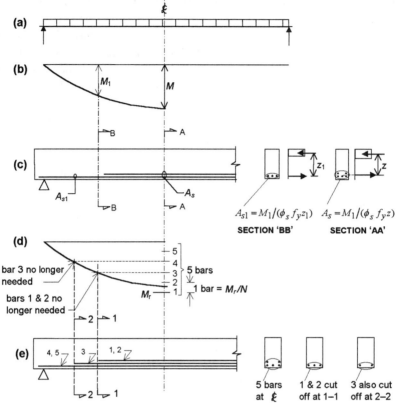

In practice, it may be that the actual reinforcement area, A_s, provided at the critical design section is in excess of the exact design requirement for M, as bars are available only in discrete sizes and can be provided only in full numbers. Furthermore, bars to be cut off are selected by numbers of bars, rather than percentage of areas. Therefore, it would be more appropriate to use the factored resistance, M_r, of the critical section, and the strength per bar M_r/N (where N is the total number of bars at the design section, assuming all of them to be of the same size) to detail bar cut-off. For example, the first two bars are no longer required for flexure where the moment reduces to $M = (N-2) \times M_r/N$; a third bar also becomes no longer needed where $M = (N-3) \times M_r/N$, and so on, as illustrated in Fig. 5.10d. In so determining the theoretical bar cut-off locations, the bending moment diagram must represent the possible maximum at each section (moment envelope). In continuous spans, the loading patterns for the maximum negative moment at supports and for the maximum positive moment in the span are different (Chapter 10). Therefore, the points of inflection for the negative and the positive moment diagrams are usually different (Fig. 5.11). (Also, the bending moment diagram will be plotted on the side on which it causes tension).

Fig. 5.11 Moment envelope for continuous spans

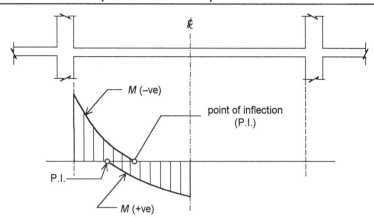

5.10.3 Actual Point of Cut-Off

Although the theoretical location where a reinforcing bar is no longer required for flexure can be determined as above, the actual point of cut-off is selected more conservatively to allow for other factors, such as possible diagonal tension cracks, unexpected shift in maximum moments, development length requirements, and possible deterioration of shear strength.

When a bar is terminated in the tension zone at the theoretical point of cut-off, the discontinuity effects tend to open up flexural cracks, which could develop into diagonal tension cracks (Chapter 6). Such a diagonal tension crack in a flexural member without shear reinforcement is shown in Fig. 5.12. Neglecting any dowel action in the reinforcing bars and any aggregate interlocking forces, the equilibrium of forces indicated on the freebody in Fig. 5.12c shows that the reinforcement force at section *b-b* depends on the moment at section *a-a*. Thus, the reinforcement area necessary to provide for the moment M_1 at section *a-a* is needed up to section *b-b*.

A similar situation exists near points of inflection in a continuous beam, (Fig. 5.13). The development of flexural cracks and inclined flexure-shear cracks transforms the load transfer mechanism in a beam loaded on its top face and supported on the bottom into one approaching tied-arch action; with an inclined compression strut formed by the concrete between the load and reaction points, (Fig. 5.13b). As a result, the zone of tension in the longitudinal reinforcement may extend beyond the point of inflection (Fig. 5.13c), although the bending moment diagram would indicate zero tensile stress in the longitudinal bars at the inflection point.

Also, the theoretical bending moment diagrams are only idealisations or best estimates, and the actual bending moments may differ, because of the approximations

152 REINFORCED CONCRETE DESIGN

involved in the calculation of loads and load effects, possible changes in loading, yielding of supports, lateral load effects, etc. In order to allow for these factors, CSA A23.3-94(12.10.3) requires that the flexural reinforcement bars be extended beyond the theoretical point of cut-off for a distance equal to the larger of the effective depth d or 12 times the bar diameter d_b. Individual bars in a bundle, if cut off, must be terminated at different points at least 40 bar diameters apart. Bars for cut-off are preferably selected such that the continuing bars form a symmetrical arrangement with respect to the vertical centreline of the beam cross-section. When bars are placed in two or more layers, the curtailment starts with bars in the innermost layer.

Instead of cutting off tension reinforcement in the zone of tension, it is frequently anchored into the compression zone by bending it across the web and making it continuous with the reinforcement on the opposite face of the member, or anchoring it there (CSA A23.3-94(12.10.1)). Such bent bars restrain the spread of diagonal tension cracks, adding to the shear resistance of the section (Chapter 6). The discontinuity effects discussed in this Section (Fig. 5.12), and the following one, are less severe for bars that are bent compared to bars cut off. Therefore, a requirement that bend points be extended beyond the theoretical points of cut off by the larger distance of d or $12d_b$ may be too conservative. However, for bar extension, the Code does not distinguish between cut off bars and bent bars.

Fig. 5.12 Influence of diagonal tension cracks on reinforcement stress

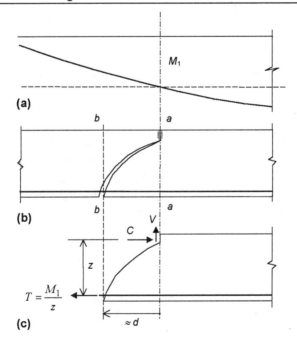

Fig. 5.13 Reinforcement stresses at inflection points in cracked member

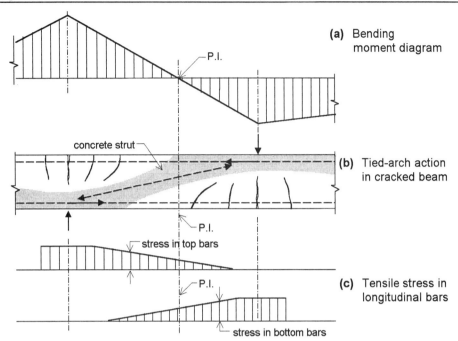

5.10.4 CSA A23.3-94 Requirements on Bar Extension

The requirement of bar extension beyond the theoretical cut-off point by a distance equal to the larger of d or $12d_b$, (CSA A23.3-94(12.10.3)) was explained in the foregoing. Figure 5.12 shows that if a bar that is no longer required for moment at section a-a is cut off at section b-b, the remaining bars are stressed to their yield stress at section b-b. To develop this yield stress, the continuing bars must extend beyond section a-a for an embedment length of at least one full development length l_d, plus the effective depth of the member or $12d_b$, whichever is greater (CSA A23.3-94(12.10.4)). Development length is discussed in detail in Chapter 9, but simply stated it is the minimum length of embedment required beyond the point of peak bar stress (critical section) so that the bar force ($A_s f_y$) at the critical section can be fully transferred to the surrounding concrete through bond stress along the interface. This minimum extension by the length l_d (or equivalent) beyond a critical section where the reinforcement stress is f_y is required for *all* bars (Section 9.12). Requirements of Clauses 12.10.3, 12.10.4 and 12.10.5, (Section 5.10.5) do not apply if both the stirrups and the longitudinal reinforcement are designed using the *General Method* for shear given in CSA A23.3-94(11.4), as the latter accounts for these effects directly.

154 REINFORCED CONCRETE DESIGN

Fig. 5.14 Code requirements for flexural reinforcement cut-off

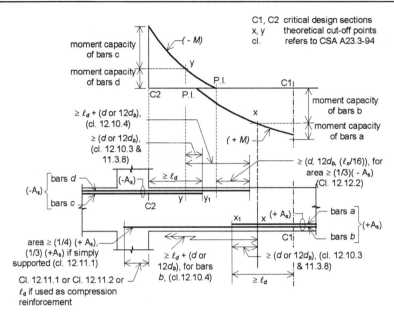

To further provide for the possible shifting of the moment diagram, the Code (Clauses 12.11 and 12.12) specifies that at least one-third of the positive moment reinforcement in simply supported members, and one-fourth in continuous members, be continued straight and into the supports for a length of at least 150 mm. However, if the member is part of a lateral load resisting system, to provide some strength and ductility under load reversals, the anchorage of this minimum positive moment reinforcement continued into the support must be adequate to develop the full yield stress f_y. (This does not apply for reinforcement continued into the support in excess of the minimum requirements.) Similarly at least one-third of the negative moment reinforcement required over a support must be continued beyond the point of inflection of the negative moment diagram for a distance equal to the largest of the effective depth, d, 12 times the bar diameter, or 1/16 of the clear span. These Code requirements concerning curtailment of flexural reinforcement are summarised in Fig. 5.14. For completeness, Fig. 5.14 also shows the development length requirement beyond all critical sections (minimum embedment, l_d, beyond critical

The need for an embedment equal to a development length of l_d (l_d is a function of bar diameter d_b as described in Chapter 9) beyond a critical section at which the reinforcement stress is f_y has already been discussed. For positive moment reinforcement continued beyond the section of zero moment (centre of simple supports and point of inflection for continuous spans), the portion l_d (Eq. 5.4) of the actual embedment beyond the point of zero moment, that is considered to effectively

Fig. 5.15 *Positive reinforcement bar-size limitation at zero moment location (CSA A23.3-94(12.11.3))*

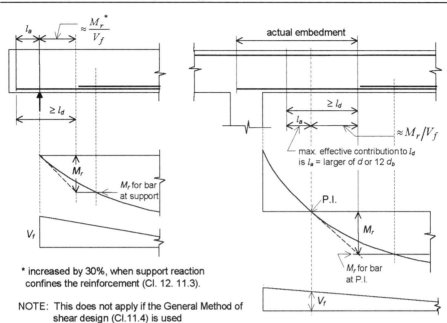

contribute to l_d is limited to:
1. at simple supports, the embedment length beyond the centre of support, and
2. at a point of inflection, the larger of the effective depth of $12d_b$, but not exceeding the actual bar extension beyond the point of inflection.

However, the detailed consideration of bond and development and the methods for computing l_d are given in Chapter 9.

CSA A23.3-94(12.11.3) specifies that (except for reinforcement terminating in a standard hook extending l_{dh} beyond the centreline of simple supports or a mechanical anchorage at least equivalent to a standard hook) the total effective embedment length thus available, taken as $l_a + M_t/V_f$, must satisfy (Section 9.12 for explanation):

$$l_d \leq l_a + M_r/V_f \tag{5.4}$$

Here, M_r is the factored moment resistance of the section for the reinforcement that is continued, V_f is the factored shear force at the point of zero moment, and l_a is the embedment length beyond the centre of the support (Fig. 5.15), The development length of standard hook in tension is l_{dh}. If Eq. 5.4 is not satisfied, either smaller size bars have to be used, so that the required l_d is reduced, or the area of reinforcement at the section has to be increased to increase M_r.

5.10.5 Effect of Bar Cut-Off on Shear Strength

Cutting off bars in the tension zones substantially lowers the shear strength and ductility of beams (Ref. 5.6). This results from the premature flexural cracks initiated by the discontinuity effects near the cut bar, which may develop into diagonal tension cracks if the shear stress at this section is also relatively high. To safeguard against this, CSA A23.3-94(12.10.5) permits termination of flexural reinforcement in a tension zone only if one of the following three conditions is satisfied:
1. the shear at the section does not exceed 2/3 the permitted value; and
2. excess stirrups, having an area

$$A_v > \frac{b_w s}{3 f_y}$$

are provided over a distance $0.75d$ from the cut-off point, at a spacing, s, not exceeding

$$\frac{d}{8\beta_b}$$

where β_b is the ratio of area of bars cut off to the total area of bars at the section.

Clause 12.10.5 does not apply if shear reinforcement is designed by the *General Method* for shear design (Cl. 11.4). Examples of bar cut-off calculations given below consider only the moment requirement. The consideration of shear strength at bar cut-off points is illustrated in Examples 6.1 and 6.2 in Chapter 6.

EXAMPLE 5.9

In the beam designed in Example 5.3, locate the point where one of the three longitudinal reinforcement bars can be terminated.

SOLUTION

The bending moment diagram is parabolic with a central ordinate of 161 kN·m, as shown in Fig. 5.16b.

The flexural capacity with all three bars is = 166 kN·m. If one bar is terminated, the flexural capacity with the remaining two bars is $\approx$ 2/3 × 166 = 111 kN·m. In the beam, the location x where the bending moment is this value is obtained as:

$$M_x = w_f \frac{l}{2} x - w_f \frac{x}{2} = 35.8 \frac{6}{2} x - 35.8 \frac{x^2}{2} = 111 \text{kN} \cdot \text{m}$$

Fig. 5.16 Example 5.9

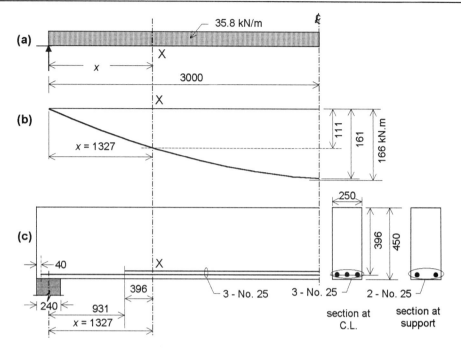

Solving $x = 1.33$ m (or 4.67 m)

The middle bar is no longer required for moment at 1327 mm from the support. However, the bar should be continued beyond this point a distance equal to the larger of $d = 396$ mm (controls) or $12d_b = 12 \times 25.2 = 302$ mm.

Actual point of termination = 1327 - 396 = 931 mm

The middle bar may be terminated at 930 mm from the centre of the support. Since the bar is terminated in a tension zone, the requirements in CSA A23.3-94(12.10.5) concerning the shear strength at this section must be satisfied. This is investigated in Example 6.1, Chapter 6.

Check bar size limitation at support (CSA A23.3-94(12.11.3)).

$$\left(1.3 \frac{M_r}{V_f} + l_a\right) \text{ must exceed } l_d$$

For No. 25 bars, with $f_y = 400$ MPa and $f_c' = 30$ MPa, (using Table 9.1, Chapter 9), $l_d = 829$ mm

$$V_f = 35.8 \times \frac{6}{2} = 107 \text{ kN}$$

158 REINFORCED CONCRETE DESIGN

For two bars continued, $M_r = 111$ kN·m and $l_a = 120 - 40$ (cover) = 80 mm (Fig. 5.16)

$$1.3 \frac{M_r}{V_f} + l_a = \frac{1.3 \times 111 \times 10^6}{107 \times 10^3} + 80 = 1740 \text{ mm} > l_d = 829 \text{ mm} \qquad \text{OK}$$

Check development length:
1. The cut off bar must have a length $\geq l_d$ beyond midspan (critical section).
   Length provided = 3000 - 930 = 2070 mm > l_d = 829 mm   OK
2. The continuing bar must have an embedment length of at least $l_d + d$ (or $l_d + 12d_b$) beyond the theoretical point of cut-off.
   Length provided = 1327 + 80 = 1407 mm
   $l_d + d$ = 829 + 396 = 1225 mm < 1407 mm   OK
3. The longitudinal reinforcement at the support must be capable of resisting a tensile force of $V_f - 0.5V_s$, where V_s is calculated from the stirrup design (Chapter 6) at that location (CSA A23.3-94(11.3.8.2)). Since stirrup design is not given, this check cannot be undertaken.

EXAMPLE 5.10

Work out the bar cut-off details for the beam designed in Example 5.7, considering bending moments only. Fifty percent of the bars are to be terminated, in two stages.

SOLUTION

From Example 5.7, $A_s = 3000$ mm^2, $d = 416$ mm, $f_c' = 35$ MPa, $f_y = 400$ MPa
Total factored load $w_f = 51.8$ kN/m, and $M_{\max} = 51.8 \times (8-0.24)^2/8 = 390$ kN·m
The flexural strength of the section is first computed.

$$a = \frac{A_s \phi_s f_y}{b_f \times \alpha_1 \phi_c f_c'} = \frac{3000 \times 0.85 \times 400}{3300 \times 0.798 \times 0.6 \times 35} = 18.4 \text{ mm}, \; h_f = 125 \text{ mm}$$

$$M_r = A_s \phi_s f_y \left(d - \frac{a}{2} \right) = 3000 \times 0.85 \times 400 \left(416 - \frac{18.4}{2} \right) \times 10^{-6}$$

$$= 415 \text{ kN·m} > 390 \text{ kN·m} \qquad \text{OK}$$

The bending moment diagram is shown in Fig. 5.17b. The bars will be terminated in two stages, first the middle bar in the top layer at section 1-1 and then the two remaining bars in the top layer at location 2-2 (Fig. 5.17c). The flexural strength with

Fig. 5.17 Example 5.10

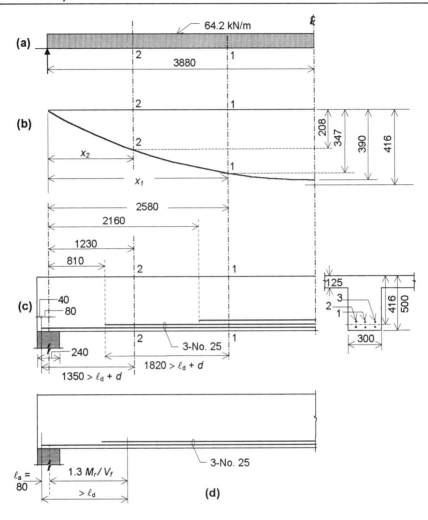

the continuing bars at sections 1-1 and 2-2 can be taken approximately (and conservatively) as:

$$M_1 = \frac{5}{6} \times 415 = 346 \text{ kN} \cdot \text{m} \text{ and } M_2 = \frac{3}{6} \times 415 = 208 \text{ kN} \cdot \text{m}$$

The location of sections 1-1 and 2-2 where the bending moments have values of $M_1 = 346$ kN·m and $M_2 = 208$ kN·m may be determined graphically, or by computation as follows:

At section 1-1, distance x_1 from the centre of the left support,

160 REINFORCED CONCRETE DESIGN

$$M_1 = \frac{w_f l}{2} x_1 - \frac{w_f x_1^2}{2}$$

$$346 = \frac{51.8 \times 7.76 \times (x_1)}{2} - \frac{51.8 \times (x_1^2)}{2}$$

Solving, $x_1 = 2.58$ m
Similarly for section 2-2, $x_2 = 1.23$ m

The bars must be continued beyond the above theoretical points for a distance equal to the larger of: effective depth = 416 mm (controls) or $12 \times d_b = 12 \times 25.2 = 302$ mm

Actual points of termination are (Fig. 5.17d):
for bar 1, 2580 - 416 = 2164 ≈ 2160 mm from centre of support
and for bars 2 and 3, 1230 - 416 = 814 ≈ 810 mm from centre of support.

Checks
1. The bars continuing must have an embedment length of at least $l_d + d$ beyond theoretical point of cut-off (Clause 12.10.4).
   For No. 30 bars, $f_y = 400$ MPa, $f_c' = 35$ MPa, $l_d = 768$ mm (using, Table 9.1).

   From Fig. 5.17c showing bar details,
   Minimum provided at section 1-1 = 2580 - 810 = 1770 mm
   $l_d + d = 768 + 416 = 1184$ mm < 1770 mm   OK
   Minimum provided at section 2-2 = 1230 mm
   $l_d + d = 1184$ mm < 1230 mm   OK

2. The bar that is cut off shall have an embedment length of at least l_d beyond midspan section.
   Minimum length given (bar 1) = 3880-2160=1720 mm > l_d OK

3. Check maximum size of bar at support (Clause 12.11.3). At support, flexural strength of bars continued to support (three bars), $M_r = 208$ kN·m
   Shear at support, $V_f = \frac{64.2 \times 8}{2} = 257$ kN

   The bar diameter must be small enough so that the development length

   $$l_d \leq \left(1.3 \frac{M_r}{V_f} + l_a\right) \quad \text{(Fig. 5.17d)}$$

$$1.3\frac{M_r}{V_f} + l_a = 1.3 \times \frac{208 \times 10^6}{257 \times 10^3} + 80 = 1132 \text{ mm} > l_d = 768 \text{ mm} \qquad \text{OK}$$

4. The shear strength at the bar cut-off points should be checked to conform with Clause 12.10.5. This is done in Chapter 6, Example 6.2.

5. Check the anchorage at the support (Clause 11.3.8.2). Since stirrup design is not known, this check cannot be undertaken.

PROBLEMS

1. A rectangular beam of span 8 m (centre to centre) is to carry specified load moments of 120 kN·m due to live load and 100 kN·m due to dead load (in addition to its self-weight). Using $f_c' = 25$ MPa and $f_y = 350$ MPa, design the beam with (a) maximum permissible reinforcement and (b) with a moderate reinforcement ratio of $\rho = 0.4\ \rho_{max}$. Design using first principles and check using Table 5.3.

2. Design the beam in Example 5.1 if the overall beam size is restricted to $b = 350$ mm and $h = 500$ mm.

3. The floor plan of a building is shown in Fig. 5.18. The specified floor loading consists of a live load of 10 kN/m^2 (including floor finish, partitions, etc.), in addition to the self-weight. The material properties are $f_c' = 40$ MPa and

Fig. 5.18 Problems 5.3 – 5.5

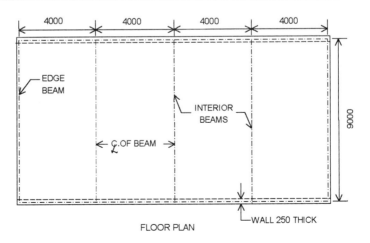

$f_y = 400$ MPa. Design the slab thickness and the reinforcement area required at all critical sections.

4. Design the interior beam of the floor in Problem 5.3, considering the beam as simply supported.

5. Design the edge beam (L-beam) of the floor system in Example 5.3 with the restriction on stem width of $b_w = 250$ mm and considering the beam as simply supported. Neglect the torsional effect.

6. A simply supported, 9 m span T-beam is subjected to a dead load of 20 kN/m (including self-weight) and a live load of 30 kN/m. The overall size of the beam is shown in Fig. 5.19. Design the beam for tensile reinforcement and detail the bar cut-off. Assume that 50 percent of bars are to be cut off. Use $f_c' = 30$ MPa and $f_y = 400$ MPa and width of support as 300 mm.

Fig. 5.19 Problem 5.6

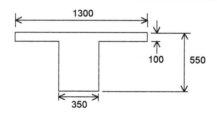

Fig. 5.20

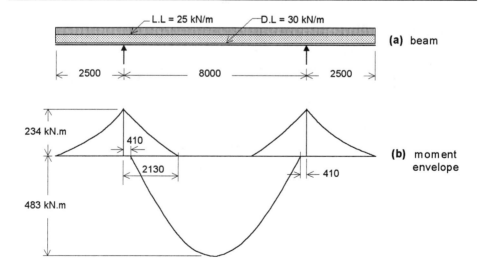

7. The beam of rectangular section with overhang at both ends, (Fig. 5.20a) is subjected to a dead load, including estimated self-weight, of 30 kN/m and a live load of 25 kN/m. The factored moment envelope is determined to be as shown in Fig. 5.20b. Design the beam and detail the bar cut-off. Assume that 50 percent of bars are to be cut off for both positive and negative moment reinforcement. Take $f_c' = 30$ MPa, $f_y = 350$ MPa and width of support as 300 mm.

REFERENCES

5.1 Wang, C.K., *Statically Indeterminate Structures*, McGraw-Hill, New York, 1983, 650 pp.

5.2 Sennett, R.E., *Matrix Analysis of Structures*, Prentice Hall Inc, New Jersey, 1993, 228 pp.

5.3 Ferguson, Phil M., *Small Bar Spacing or Cover - A Bond Problem for the Designer*, J. of ACI, Vol. 74, Sept. 1977, pp. 435-439.

5.4 ACI Committee 435, *Proposed Revisions by Committee 435 to ACI Building Code and Commentary Provisions on Deflections*, J. of ACI, Vol. 75, No. 6, June 1978, pp. 229-238.

5.5 *Reinforcing Steel - Manual of Standard Practice*, Reinforcing Steel Institute of Canada, Richmond Hill, Ontario, 1992. (3rd edition, 1996).

5.6 ASCE-ACI Task Committee 426, *The Shear Strength of Reinforced Concrete Members*, ASCE, Journal of Struct. Div., Vol. 99, No. ST6, June 1973, pp. 1091-1187.

CHAPTER 6 Design for Shear

6.1 INTRODUCTION

The ultimate limit state considered in Chapters 4 and 5 dealt with failure in bending. Bending moments are generally accompanied by shear forces, and sometimes also by axial loads and torsion. Premature failure in shear can reduce the strength of members below their capacity in flexure and reduce the ductility of the elements considerably.

Failure in shear of reinforced concrete takes place under combined stresses resulting from an applied shear force, bending moments and, where applicable, axial loads and torsion. Because of the non-homogeneity of material, nonuniformity and nonlinearity in material response, presence of cracks, presence of reinforcement, combined load effects, etc., the behaviour of reinforced concrete in shear is very complex, and the current understanding of and design procedures for shear effects are, to a large measure, based on analysis of results of extensive tests and simplifying assumptions, rather than on an exact universally acceptable theory.

Prior to its 1984 revision, CSA Standard A23.3 recommended a method for Shear and Torsion design based on the traditional method adopted by the ACI Code (Ref. 6.1). The background for this procedure, the basic concepts, and a review of extensive test results and bibliographic references on shear strength of reinforced concrete beams are presented in Ref. 6.2 and 6.3. In 1984 the Canadian Standard A23.3, while retaining a shortened version of the traditional method followed in previous codes (under the designation *Simplified Method*), introduced an alternative method termed *General Method* for shear and torsion design, based on the so-called *Compression Field Theory*.

The 1994 revision of the Canadian Standard (CSA A23.3-94) recommends various design procedures for shear and torsion, which are essentially updated and/or modified versions of the methods recommended in the 1984 version of the Code. The recommended methods (CSA A23.3-94) and their applicability can be summarised as in the Table 6.1.

The most widely used method for typical flexural members is the *Simplified Method* (CSA A23.3-94(11.3)), which is an updated version of the Simplified Method contained in the 1984 Code. However, the method is restricted to the shear design of beams, columns, or walls (or such portions of the member as are) designed by the conventional theory of flexure, in which the assumption that "plane sections remain plane" is reasonably valid, and in which the member is *not subjected to significant axial tension*. In the Simplified Method, the transverse reinforcement is designed for the shear, while the longitudinal reinforcement is designed for the combined effects of flexure and axial load (compression only). The effects of shear on the longitudinal reinforcement (Section 5.10.3) are taken care of by bar detailing requirements. Transverse and longitudinal reinforcement required for coexisting torsion, if any, are designed separately and added on to that already determined for flexure, shear, and axial loads. The design by the Simplified Method is dealt with in this Chapter for shear and in Chapter 7 for torsion. Accordingly, much of the information on shear strength that follows in this Chapter is based on Refs. 6.1-6.3 and gives the basis for the Simplified Method of CSA A23.3-94(11.3).The Simplified Method is one that may be used, in lieu of the General Method, for flexural members not subjected to significant axial tension. Otherwise, flexural regions, where the assumption of plane sections remaining plane is valid are to be designed for shear by any of the other two methods (the General Method or the Strut-and-Tie Model).

Table 6.1 Design Procedure for Shear and Torsion in CSA A23.3-94

No	Method	Applicability
1.	The "General Method" OR the "Strut-and-Tie Model", (Cl. 11.4 or 11.5):	For "flexural regions" of members (i.e. where the traditional beam theory which assumes that 'plane sections remain plane' is reasonably applicable).
2.	The "Simplified Method", (Cl. 11.3):	In lieu of Cl. 11.4, for "flexural regions" of members, but not subjected to significant axial tension.
3.	The "Strut-and-Tie Model", (Cl. 11.5):	For regions near discontinuities where the 'plane section remaining plane' assumption is not applicable; including effects of axial force, if any.
4.	The "Shear Friction" concept, (Cl. 11.6):	For interface shear transfer, in situations involving the possibility of shear failure in the form of sliding along a plane of weakness.
5.	The "Two-way Shear" or "Punching Shear" concept, (Cl. 13.4 & 13.5):	For thin slabs and footings with two-way action and subjected to concentrated loads, with the possibility of a 'punch-through' type of failure.

The General Method for Shear and Torsion design (CSA A23.3-94 (11.4)) is presented in Chapter 8. In this method, the member is designed for the combined effects of flexure, shear, axial load, and torsion. The method based on the "Strut-and-Tie Model" (CSA A23.3-94(11.5)) is also presented in Chapter 8. The Code requires that regions near discontinuities, where the assumption that plane sections remain plane is not applicable, should be designed for shear and torsion using the Strut-and-Tie Model. Such regions include deep beams, parts of members with a deep shear span, brackets and corbels, and regions with abrupt changes in cross-section, such as regions of web openings in beams. The shear-friction procedure is described in this Chapter (Section 6.18) and two-way (slab) shear in Chapter 14.

To gain an insight into the causes of shear failure in reinforced concrete, the stress distribution in a homogeneous elastic beam of rectangular section will be

166 REINFORCED CONCRETE DESIGN

reviewed briefly. In such a beam, loaded as shown in Fig. 6.1a, a transverse section, X–X, is subjected, in general, to a bending moment M and a shear force V.

As described in texts on mechanics of materials (Ref. 6.4), the flexural stress, f_x, and the shear stress, v, at a point in the section distant y from the neutral axis are given by:

$$f_x = \frac{My}{I} \text{ and } v = \frac{VQ}{Ib} \tag{6.1}$$

***Fig. 6.1** Stress distribution in rectangular beams*

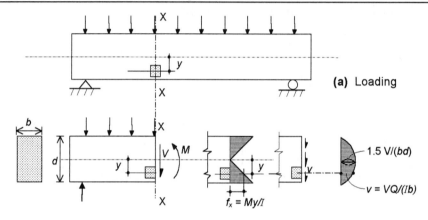

(a) Loading

(b) Flexural and shear stresses

(c) Principal stresses

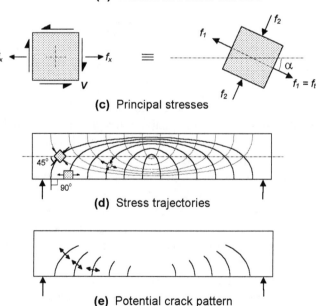

(d) Stress trajectories

(e) Potential crack pattern

where Q is the first moment of the area of section above the depth y, about the neutral axis, and I is the moment of inertia of the section. The distribution of these stresses is as shown in Fig. 6.1b. Considering an element at depth y (Fig. 6.1c), the flexural and shear stresses can be resolved into equivalent principal stresses, f_1, and f_2, acting on orthogonal planes inclined at an angle α, where

$$f_{1,2} = \frac{1}{2} f_x \pm \sqrt{\left(\frac{1}{2} f_x\right)^2 + v^2} \qquad (6.2)$$

and
$$\tan 2\alpha = \frac{2v}{f_x} \qquad (6.2a)$$

In general, neglecting the influence of any vertical normal stresses due to loads, the stress f_1 is tensile (say $= f_t$) and f_2 is compressive. The relative magnitudes of f_1 and f_2 and their directions depend on the relative values of f_x and v. In particular, $f_t = f_x, f_2 = 0$, and $\alpha = 0$ at points along the bottom face; $f_t = v_{max}, f_2 = -v_{max}$ and $\alpha = 45°$ at points along the neutral axis. Considering the variation of stresses f_x and v over the depth at a section and the variations of M and V (and hence of f_x and v) along the span, the principal stress trajectories indicating the directions of principal stresses will follow the pattern shown in Fig. 6.1d. (Stress trajectories are a set of orthogonal curves, whose tangent at any point is in the direction of the principal stress at that point.) The broken lines indicate the directions of the principal tensile stresses. In a material that is weak in tension, tensile cracks would occur at right angles to the tensile stresses, and hence the compressive stress trajectories indicate *potential* crack patterns, depending on the magnitude of the tensile stress as shown in Fig. 6.1e. (Note that if a crack is developed, the stress distributions assumed here are no longer valid in that region, and redistribution of the internal stresses takes place.) The location of the absolute maximum principal tensile stress will depend on the variation of f_x and v, which in turn depends on the shape of the cross section and on the span and loading. For instance, in a relatively deep beam with a thin web (such as an I-section) and short spans subjected to high shear force and low bending moment, the peak shear stress, v_{max}, may far exceed the peak flexural stress $f_{x,max}$, and hence the maximum principal tensile stress, $f_{t,max}$, may be at the neutral axis level at an inclination of $\alpha = 45°$. In contrast, for a shallow rectangular beam with high moment and low shear the maximum principal tensile stress may be the flexural stress $f_{x,max}$ in the outer fibre in the peak moment regions.

The general influence of shear is to induce tensile stresses on an inclined or diagonal plane. Concrete is weak in tension. Furthermore, as shown in Fig. 1.8, a biaxial state of combined tension and compression stresses, as exists in the beam shown in Fig. 6.1, significantly reduces both the tensile and compressive strengths of concrete. Hence, the tensile strength of the concrete in a reinforced concrete beam subjected to combined bending and shear is likely to be even less than the uniaxial tensile strength of concrete. Thus, one can naturally expect some inclined tension cracks to develop in reinforced concrete beams subjected to any appreciable shear force. Failure of concrete beams in shear is triggered by the development of these inclined cracks (referred to as *diagonal tension* cracks) under combined stresses. To avoid a failure of the concrete in compression, it is also necessary to ensure that the principal compressive stress, f_2, is less than the compressive strength of concrete under the biaxial state of stress.

Although several theories of failure have been proposed for concrete under combined stresses, for the traditional method of shear and torsion design, the principal tensile stress theory has been followed.

6.2 NOMINAL SHEAR STRESS

The maximum shear stress at the neutral axis level in the homogeneous elastic beam of rectangular section is given in Fig. 6.1b as:

$$v_{max} = 1.5\frac{V}{bd} = kv_{av} \tag{6.3}$$

where the average shear stress, $v_{av} = \dfrac{V}{area}$

The value of the constant, k, in Eq. 6.3 depends on the cross-sectional shape. Equations 6.1 and 6.3 are not applicable for a composite, cracked and inelastic material such as reinforced concrete, and the determination of the distribution and the maximum value of the shear stress in a reinforced concrete member is complex. Hence, as a means of comparing and correlating with test results, and as a parameter to aid design and control shear stress, the average shear stress $v = V/(bd)$, is generally used as a *measure* of the shear stresses resulting from the shear force, V.

6.3 EFFECT OF SHEAR ON BEAM BEHAVIOUR

If a reinforced concrete beam is subjected to bending only, as in Fig. 6.2a, the principal tensile stresses will be parallel to the axis of the member, and hence tension cracks will develop from the outer fibres, where the tensile stress is maximum, and spread inwards, in a direction at right angles to the beam axis. The addition of shear stresses alters the magnitude and direction of the principal tensile stresses, so that inclined tension cracks, called *diagonal tension cracks,* may develop.

Fig. 6.2 Shear cracks

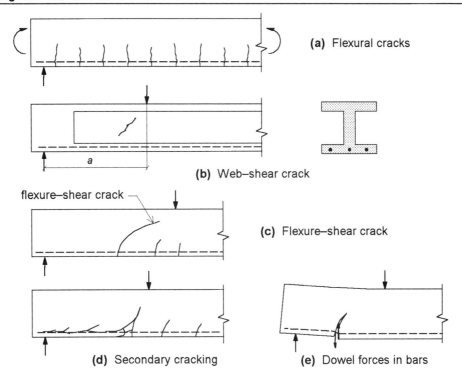

6.3.1 Modes of Inclined Cracking

Inclined cracks have been classified as *web-shear cracks* and *flexure-shear cracks* (Fig. 6.2). When the shear stress, v, is large in relation to the flexural stress, f_x, the maximum principal tensile stress is near the neutral axis level at an inclination of nearly 45° to the beam axis. When this stress reaches the limiting value, cracks may initiate in the web, independent of any flexural cracks. This type of crack is termed *web-shear crack*, and may occur in relatively deep beams with thin webs and low M/V ratios (Fig. 6.2b) or near inflection points of continuous beams. In general, web-shear cracks are extremely rare in reinforced concrete beams of usual proportions and are likely only in prestressed beams with thin webs. The web-shear cracking load may be computed with reasonable accuracy by equating the principal tensile stress in the web at or near the neutral axis to the tensile strength of the concrete.

In reinforced concrete beams of normal proportions, inclined cracks usually form as an extension of the flexural cracks. The flexural cracks generally form first, but these are controlled, and the tensile stresses are carried by the tensile reinforcement.

However, these cracks lead to increased shear stresses and, hence, diagonal tensile stresses at the head of the crack, resulting in the extension of a flexure crack into a *diagonal tension* crack. This type of inclined crack is called a flexure-shear crack (Fig. 6.2c). Often, combined with flexure-shear cracks, secondary cracks (sometimes termed splitting cracks) may develop, extending from the inclined crack and along the tension reinforcement toward the region of lower moment (Fig. 6.2d). These cracks may be due to the splitting forces generated by the wedging action of the tension bar deformations and the vertical downward forces exerted on the cover concrete by the tension bars in transferring shear across the crack by dowel action (Fig. 6.2e).

6.3.2 Shear Transfer Mechanisms in Reinforced Concrete

There are several mechanisms by which shear is transmitted between two planes in a concrete member. The prominent amongst these are identified in Fig. 6.3, which shows the freebody of one of the segments of a reinforced concrete beam separated by an inclined crack. The applied bending moment is resisted by the couple formed by flexural forces, C and T. The major components contributing to the shear resistance are: (1) the shear strength, V_{cz}, of the uncracked concrete; (2) the vertical component, V_{ay}, of the *interface shear*, V_a; (3) the dowel force, V_d, in the longitudinal reinforcement; and (4) the shear, V_s, carried by the shear reinforcement, if present. The interface shear, V_a, (also called aggregate interlock), is a tangential force transmitted along the plane of the crack, resulting from the resistance to relative movement (slip) between the two rough interlocking surfaces of the crack, much like frictional resistance. So long as the crack is not too wide, the force V_a may be significant. The dowel force in the longitudinal tension reinforcement is the transverse force developed in these bars, functioning as a dowel across the crack, resisting relative transverse displacements between the two segments of the beam. In deep beams, part of the applied load may be transmitted to the reaction points by tied-arch action, thereby reducing the effective shear force at a section. The equilibrium of vertical forces in Fig. 6.3 results in the

Fig. 6.3 *Internal forces acting at inclined crack (adapted from Ref. 6.3)*

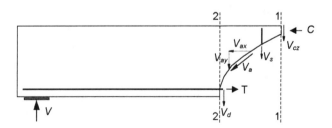

relationship
$$V_{ext} = V_{cz} + V_{ay} + V_d + V_s \tag{6.4}$$

Different mechanisms dominate in members of different types, and the relative magnitudes of the terms in Eq. 6.4 also depend on the loading stage. Thus, prior to flexural cracking, the applied shear is resisted almost entirely by the uncracked concrete. At flexural cracking, there is a redistribution of internal stresses and some interface shear and dowel action develops. Prior to diagonal cracking, however, there is little tensile stress developed in the shear reinforcement. At diagonal cracking, the shear reinforcement intercepted by the crack receives a sudden increase in tensile stress, and all the components in Eq. 6.4 now effectively contribute to the shear strength. The post-inclined cracking behaviour, the failure mode, and the ultimate strength all depend on the relative values of these individual components and on to what extent these forces can be redistributed successfully.

In beams without shear reinforcement, the breakdown of any of the shear transfer mechanisms may soon cause failure. In such beams there are no stirrups enclosing the longitudinal bars and restraining them against splitting failure, and the value of V_d is usually small. The component V_{ay} also decreases progressively due to the unrestrained opening up of the crack. The spreading of the crack into the compression zone decreases the area of uncracked concrete section contributing to V_{cz}. However, in relatively deep beams, tied-arch action may develop following inclined cracking, which in turn could transfer part of the load to the supports, thereby reducing the effective shear force at the section. Because of the uncertainties in all these effects, it is difficult to predict precisely the behaviour and strength beyond diagonal cracking for beams without shear reinforcement.

In members with moderate amounts of shear reinforcement, the shear resistance continues to increase even after inclined cracking, until the shear reinforcement yields, and V_s can increase no more. Any further increase in applied shear force leads to increases in V_{cz}, V_d, and V_{ay}. With progressively widening crackwidth (which is now accelerated by the yielding of the shear reinforcement), V_{ay} begins to decrease, forcing V_{cz} and V_d to increase at a faster rate until either a splitting (dowel) failure occurs, or the concrete in the compression zone fails under the combined shear and compression forces. Thus, in general, the failure of shear reinforced members is more gradual (ductile).

The opening up of the diagonal tension crack, as shown in Fig. 6.3, also indicates, qualitatively, the influence of shear on the longitudinal reinforcement requirement. Thus, the flexural steel requirement at section 2-2 is dependant on the bending moment at section 1-1, and the axial component V_{ax} of V_a has to be compensated by an increase in the tensile force T in the flexural steel.

6.3.3 Shear Failure Modes

The magnitude and direction of the maximum principal tensile stress (Eq. 6.2) and the development and growth of inclined cracks are affected by the relative magnitudes of the flexural stress, f_x, and the shear stress, v. As an approximation, stresses f_x and v can be considered, respectively, proportional to $M/(bd^2)$ and $V/(bd)$, where M is the applied bending moment, V the shear force at the section, b is the width of section, and d is the effective depth. The stress ratio f_x/v can be expressed as:

$$\frac{f_x}{v} = \frac{F_1 M/(bd^2)}{F_2 V/(bd)} = F_3 \frac{M}{Vd} \tag{6.5}$$

where F_1, F_2, F_3 are constants of proportionality. Equation 6.5 shows that the dimensionless parameter $M/(Vd)$ strongly influences the development and propagation of inclined cracks, as well as the shear strength at inclined cracking. For beams subjected to concentrated loads, as in Fig. 6.4a, the ratio M/V is the so-called *shear span*, a, and $M/(Vd) = a/d$ is the shear span to depth ratio.

Typical shear failure modes of reinforced concrete beams, and the influence of the a/d ratio, are illustrated in Fig. 6.4 with reference to a simply supported rectangular beam subjected to symmetrical two-point loading (Refs. 6.3 and 6.5).

In very deep beams ($a/d < 1$) without web reinforcement, inclined cracking transforms the beam into a tied-arch (Fig. 6.4b). The tied-arch can fail by either a breakdown of its tension element, namely the longitudinal reinforcement (by yielding, fracture or failure of anchorage), or by a breakdown of the concrete compression chord by crushing (Fig. 6.4b).

In relatively short beams, with a/d in the range of 1 to 2.5, the failure is initiated by an inclined crack, usually a flexure-shear crack. The actual failure may take place by (1) crushing of the reduced concrete section above the head of the crack under combined shear and compression (termed *shear-compression* failure, Fig. 6.4c), or (2) secondary cracking along the tension reinforcement, resulting in loss of bond and anchorage of the tension reinforcement (*shear-tension* failure, Fig. 6.4c). This type of failure usually occurs before the flexural strength of the section is attained.

In beams with relatively short shear spans (a/d less than about 2.5), and with loads and reactions applied on the top (compressive) and bottom surfaces of the beam, respectively, the effect of the vertical compressive stresses between the load and reaction may be significant and result in substantial increases in shear strength above the inclined cracking strength. Such beams are treated as *deep beams* (Section 8.5).

Normal beams have a/d ratios, in excess of approximately 2.5. Such beams may fail in shear or in flexure. The limiting a/d ratio above which flexural failure occurs depends on the tension reinforcement ratio, yield strength of reinforcement, and

Fig. 6.4 *Shear failure modes (adapted from Ref. 6.3)*

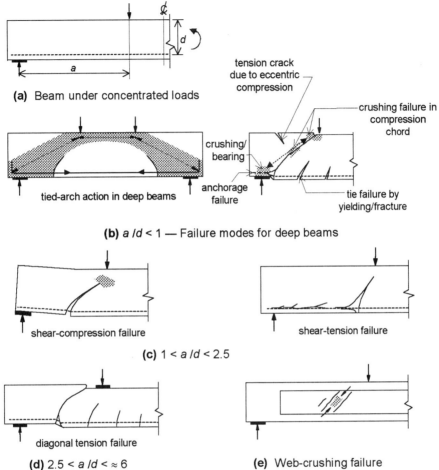

concrete strength, and is in the neighbourhood of 6. For beams with a/d ratios in the range of 2.5 to 6, flexural tension cracks develop early on; however, before the ultimate flexural strength is reached, the beam may fail in shear by the development of inclined flexure-shear cracks, which, in the absence of web reinforcement, rapidly extend through the beam, as shown in Fig. 6.4d. This type of failure is usually sudden, without warning, and is termed *diagonal-tension* failure. Addition of web reinforcement in such beams leads to either a shear-compression failure or a flexural failure.

In addition to these different modes, thin webbed members, such as I-beams with web reinforcement, may fail by crushing of the concrete in the web portion between inclined cracks under the diagonal compression forces (Fig. 6.4e).

6.4 SHEAR STRENGTH WITHOUT SHEAR REINFORCEMENT

All shear failures result from or are affected by inclined cracks that precede them. In beams without shear reinforcement, the shear failure load may equal or exceed the load at which inclined cracks develop, depending on several variables such as ratio $M/(Vd)$, thickness of web, influence of vertical normal stresses, concrete cover, and resistance to splitting (dowel) failure, etc. Furthermore, the margin of strength beyond diagonal cracking is subject to considerable fluctuation. Hence, as a design criterion for beams of normal proportions ($M/(Vd) >$ about 2.5), the shear force, V_{cr}, causing formation of the first inclined crack is generally considered as the usable ultimate strength for beams without shear reinforcement (Ref. 6.2).

6.5 SHEAR STRENGTH AT INCLINED CRACKING

(a) Web-Shear Cracking Load

Inclined cracks in concrete beams are of two types and they develop in different ways. *Web-shear* cracks are initiated in the web portion before any flexural cracks develop in its vicinity. Hence, assuming the principal tensile stress theory of failure, the web-shear cracking load may be estimated by relating the principal tensile stress in the uncracked web to the tensile strength of concrete. In the zone of potential web-shear cracking, which is close to the neutral axis level, the flexural stress, f_x, is very small, and the maximum principal tensile stress is primarily dependent on the shear stress, v. This concept, together with correlation to extensive tests results (Ref. 6.2), has resulted in an equation for the average shear stress at web-shear cracking, v_{cw}, as:

$$v_{cw} = \frac{V_{cw}}{bd} = 0.3\sqrt{f'_c} \text{ MPa} \qquad (6.6)$$

where V_{cw} is the shear force at web-shear cracking.

(b) Flexure-Shear Cracking Load

The prediction of the *flexure-shear* cracking load is complicated by the presence of prior flexural cracks. A semi-rational expression developed in Ref. 6.2 is the basis for the current ACI Code (Ref. 6.1) and the simplified method in CSA A23.3-94(11.3), and the derivation of this equation is briefly described below.

If it is assumed that, at the point of inclined cracking, the principal tensile stress is equal to the tensile strength of concrete, f'_t, using Eq. 6.2:

$$f'_t = \frac{1}{2}f_x + \sqrt{\left(\frac{1}{2}f_x\right)^2 + v^2} \qquad (6.7)$$

If the flexural tensile stress f_x is assumed proportional to the steel stress f_s, and the shear stress v is assumed proportional to the average shear stress,

$$f_x \propto \frac{f_s}{n} = F_1 \frac{M}{n\rho bd^2}$$

and
$$v = F_2 \frac{V}{bd}$$

where F_1 and F_2 are constants of proportionality, $n = E_s/E_c$, and ρ is the tension reinforcement ratio. Substituting for f_x and v, and recognising that both the tensile strength f_t and modulus E_c are generally considered proportional to $\sqrt{f'_c}$, Eq. 6.7 can be expressed as a relation between two dimensionless parameters:

$$X = \frac{\rho V d}{\sqrt{f'_c}M} \text{ and } Y = \frac{V}{bd\sqrt{f'_c}}$$

The importance of the three major variables, $M/(Vd)$, ρ, and $\sqrt{f'_c}$, has also been observed in tests. Having established, in this manner, the relevance of the two parameters X and Y, they are correlated to extensive test data on an X–Y plot, yielding a valid relationship between X and Y. A simple expression so developed in Ref. 6.2, giving a lower bound estimate of the shear strength, V_{cr}, at flexure-shear cracking for beams without shear reinforcement,

$$\frac{V_{cr}}{bd\sqrt{f'_c}} = 0.16 + (17)\frac{\rho}{\sqrt{f'_c}}\frac{Vd}{M} \qquad (6.8)$$

where the constant 17 has the unit of MPa.

Combining Eq. 6.6 and 6.8, the nominal shear strength of reinforced concrete beams without shear reinforcement is considered equal to the strength corresponding to inclined cracking and is given by

$$V_{cn} = \left(0.16\sqrt{f'_c} + 17\rho\frac{Vd}{M}\right)bd \leq 0.3\sqrt{f'_c}bd \qquad (6.9)$$

where the term within the brackets has the unit of MPa.

The equivalent of Eq. 6.9, in imperial units, with rounding off of the numerical factors is:

$$V_{cn} = \left(1.9\sqrt{f_c'} + 2500\rho\frac{Vd}{M}\right)bd \leq 3.5\sqrt{f_c'}bd \qquad (6.10)$$

where V_{cn} is in pounds, f_c' in psi, b and d in inches, and ratios ρ and Vd/M are dimensionless.

Equation 6.10 is the expression recommended by the ACI Code (Ref. 6.1) for the *nominal shear strength* carried by concrete. Since the value of the moment M is zero at points of inflection, and to limit V_{cn} near such points, the value of Vd/M in Eq. 6.10 is limited to a maximum of 1.0. As a convenient simplification useful for most designs, the ACI Code also permits the use of an approximate value for the second term within brackets of Eq. 6.10 of $0.1\sqrt{f_c'}$, so that the shear strength provided by concrete may be computed as:

$$V_{cn} = 2\sqrt{f_c'}bd \qquad (6.10a)$$

where the various terms have the same units as in Eq. 6.10.

The 1977 revision of CSA A23.3 had recommended Eq. 6.9 and, as a simplification, the SI units equivalent of Eq. 6.10a, namely $V_{cn} = 0.17\sqrt{f_c'}\,b_w d$ for the nominal shear strength carried by the concrete. The V_{cn} so calculated was multiplied by a ϕ-factor of 0.85 to get the factored resistance. With a reduced material resistance factor $\phi_c = 0.6$ for concrete strength introduced in the 1984 revision of CSA A23.3, the coefficient 0.17 above was increased to 0.2 to partially offset the decrease in computed resistance on account of the reduced ϕ_c value. Introducing the material resistance factor ϕ_c and the factor λ to account for low density concrete (Section 1.6), CSA A23.3-94 (11.2.8.2 and 11.3.5.1) recommends Eq. 6.11 for computing the *factored shear resistance attributed to concrete*, V_c, for flexural members not subjected to significant axial tension.

$$V_c = 0.2\lambda\phi_c\sqrt{f_c'}b_w d \qquad (6.11)$$

where V_c is in Newtons, f_c' in MPa, and the web width b_w and effective depth d are in mm. Equation 6.11 may overestimate the shear strength of large size lightly reinforced members without stirrups. Hence, Eq. 6.11 is restricted to members with at least the minimum specified amount of transverse reinforcement, or with an effective depth not exceeding 300 mm.

For members with effective depth greater than 300 mm and with no stirrups (Cl. 11.3.5.2) or with transverse reinforcement less than the minimum specified), the Code recommends Eq. 6.11a for computing V_c:

$$V_c = \left[\frac{260}{1000+d}\right]\lambda\phi_c\sqrt{f_c'}b_w d \geq 0.10\lambda\phi_c\sqrt{f_c'}b_w d \qquad (6.11a)$$

6.6 BEHAVIOUR OF BEAMS WITH SHEAR REINFORCEMENT

Shear reinforcement, also known as web reinforcement and transverse reinforcement in reinforced concrete members may consist of: (a) stirrups or ties perpendicular to the member axis; (b) inclined stirrups making an angle of 45° or more with the longitudinal tension bars in order to intercept potential diagonal cracks; (c) welded wire fabric with transverse wires located at right angles to the member axis, provided that the transverse wires can undergo a minimum elongation of 4 percent measured over a gauge length of at least 100 mm, which includes at least one cross-wire; (d) longitudinal bars (No. 35 or smaller) bent at an angle of 30° or more with the longitudinal tensile bars to cross potential diagonal cracks, with the centre three-fourths of the inclined portion being considered effective; (e) spirals; or (f) a combination of stirrups and bent longitudinal bars (Fig. 6.5). By far the most common type of shear reinforcement for beams is the U-shaped stirrup placed perpendicular to the axis of the member. However, for the Simplified Method of design for shear (Cl. 11.3), the Code permits only transverse reinforcement *perpendicular* to the longitudinal axis of the member.

Fig. 6.5 Types of shear reinforcement

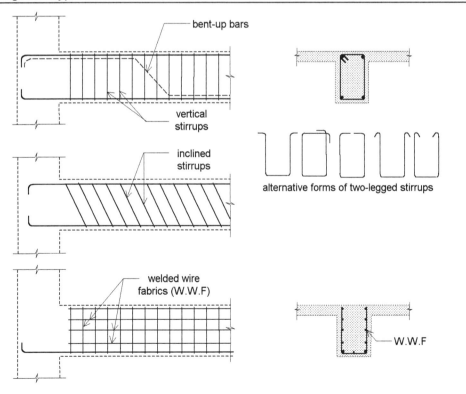

Fig. 6.6 *Classical truss analogy for action of web reinforcement*

(a) truss analogy

(b) analogous truss

(c) effect of shear on tensile force

$$C = \frac{M}{jd} - \frac{0.5V}{\tan\theta}$$

$$T = \frac{M}{jd} + \frac{0.5V}{\tan\theta}$$

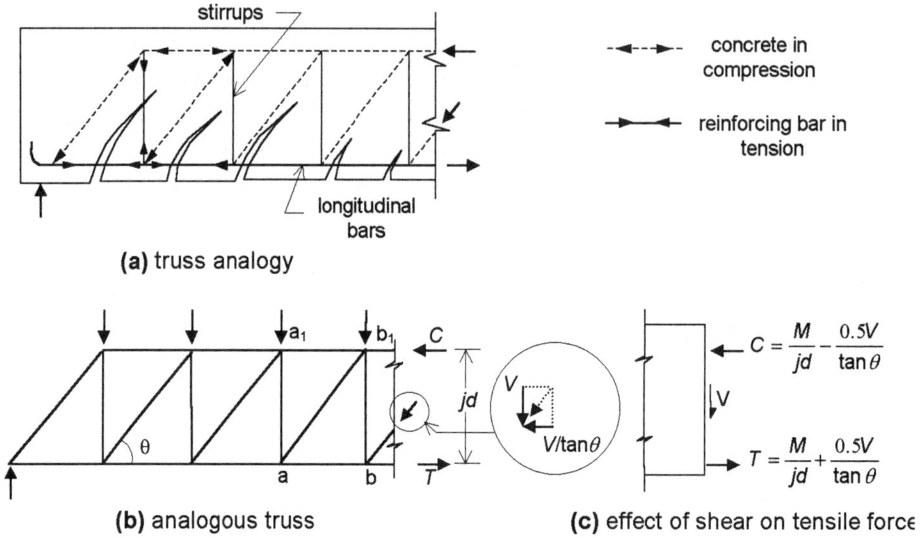

Traditionally, the action of web reinforcement in reinforced concrete beams has been explained with the aid of the *truss analogy*, the simplest form of which is shown in Fig. 6.6. In this model, a reinforced concrete beam with inclined cracks is replaced with a pin-jointed truss, of which the compression chord represents the concrete compression zone at the top, the bottom tension chord represents the tension reinforcement, the web members in tension represent the stirrups, and the diagonal web members in compression represent the concrete in compression between the inclined cracks. More complex forms of the truss, as well as other analogies (such as arch analogy, frame analogy) have also been proposed to better represent the complex stress field in a real reinforced concrete beam (Ref. 6.5). Figure 6.6 shows discrete compression diagonals at a spacing of $jd/\tan\theta$ along the length of the member. In reality, the compression diagonals are distributed throughout, with a continuous field of compression parallel to the diagonal member. The diagonal truss member may be considered to represent the aggregate diagonal compression over the length $jd/\tan\theta$ contributory to it. Similarly, it is not intended that the transverse reinforcement (stirrups), represented by the vertical members of the truss, be spaced at $jd/\tan\theta$ as in Fig. 6.6(b). The actual spacing may be different, and the vertical member of the truss may be considered to represent all stirrups tributary to a length of $jd/\tan\theta$. The truss model is a helpful tool to visualise the nature of stresses in the stirrups and in the concrete, and to form a basis for simplified design concepts and methods. It may also be used to derive equations for the design of shear reinforcement. However, it does not recognise fully the true action of web reinforcement and its effect on the various types of shear transfer mechanisms identified in Fig. 6.3.

The necessity for extension of flexural tension reinforcement beyond the theoretical point of cut-off (Section 5.10.3 and Fig. 5.12) can also be explained with the aid of the truss analogy. For the truss shown isolated in Fig. 6.6b, the tensile force in any bottom chord such as ab is dependent on the bending moment at section $b_1 b$ whereas the compression in the top chord $a_1 b_1$ is dependent on the moment at section $a_1 a$. Therefore, the tension reinforcement designed for section $b_1 b$ is in fact needed up to section $a_1 a$. Thus flexural tension reinforcement must extend for a length $\approx jd/\tan\theta$ beyond the theoretical point of cut-off. Alternatively, from the equilibrium of forces at a section, due to the shear force V, the tension T in a panel of the truss (or at any section of the reinforced concrete beam) exceeds the compressive force C by an amount equal to $V/\tan\theta$. Thus, the truss analogy also illustrates that to resist shear, a beam requires both stirrups and additional longitudinal reinforcement at a section.

6.7 ACTION OF WEB REINFORCEMENT

Measurements of strain in web reinforcement have shown that, prior to the formation of inclined tension cracks, there is little or no tensile stress developed in them. Hence, web reinforcement does not appreciably influence the location of inclined cracking or the inclined cracking load V_{cr}, and the latter can be computed with reasonable accuracy by the methods described in Section 6.5 for members without web reinforcement.

Following the development of inclined cracking, a portion of the shear is resisted by the web reinforcement intercepted by the crack. In this phase all the major shear transfer mechanisms are effective, and the overall shear resistance can be expressed by Eq. 6.4. The web reinforcement contributes significantly to the overall shear strength by increasing or maintaining all the individual components of Eq. 6.4. Thus, first, there is the direct contribution V_s to the strength resulting from the tensile stress induced in the web reinforcement. Secondly, web reinforcement crossing the inclined crack restricts the widening of the crack, and thereby helps maintain the aggregate interlock and the associated strength contribution V_{ay}. Also, by containing the crack and reducing its penetration deep into the compression zone, the shear resistance of the uncracked portion of the concrete, V_{cz}, is enhanced. Thirdly, the web reinforcement in the form of stirrups (which wrap around the longitudinal tension bars) restrains the longitudinal tension bars against vertical (dowel) displacement, and thereby improves their contribution by dowel action, V_d. Furthermore, when sufficient web reinforcement is present, the shear failure is gradual, with adequate warning in the form of large increases in the width of the inclined cracks as the web reinforcement reaches its yield point. Other favourable influences attributed to web reinforcement include: (1) improvement in concrete compressive strength and ductility due to the

Fig. 6.7 Secondary effects of web reinforcement

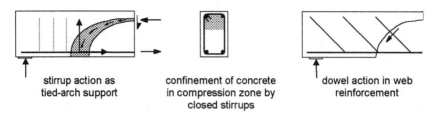

lateral confinement offered by the stirrups; (2) transfer of some force across the crack by developing dowel action in the web reinforcement crossing the crack; and (3) provision vertical support for the concrete between inclined cracks facilitating arch action (Fig. 6.7).

As described in Section 6.3.2, the ultimate shear strength of a web reinforced section is reached soon after the yielding of this reinforcement. As stirrups yield, V_s remains constant, and, on further loading, the accelerated widening and spreading of the crack which follows soon leads to the breakdown of one or more of the remaining components of shear transfer (V_{ay}, V_d and V_{cz}) and to failure. Thus, the shear strength and mode of shear failure of beams with web reinforcement are influenced by the amount, distribution and yield strength of web reinforcement, in addition to the variables f_c', ρ and $M/(Vd)$, which were shown to influence the cracking load and failure mode of beams without web steel. If the web reinforcement is so small that the sudden increase in stress in it at the formation of the inclined crack is sufficient to cause its yielding, the strength and behaviour of the beam may be little different from those of a similar beam without web reinforcement. For instance, a sudden diagonal tension failure may occur in beams of low $\rho\,(Vd)/M$ ratios. This suggests the need for a minimum amount of web steel in members where the sudden formation of inclined cracking may lead to distress. On the other hand, if a member is heavily reinforced with web steel, particularly in short deep members, the compression zone of the concrete may fail in shear-compression before the web steel reaches yield point. (Shear-compression failure before yielding of web steel may also occur under dynamic loading.) Such failure may occur without warning, and hence an upper limit on the shear strength with web reinforcement is also desirable. A shear-compression failure before the yielding of web steel could also occur if the web steel has a very high yield strength. Within the two extremes considered above, for moderate amounts of web reinforcement, there is an increase in shear strength beyond diagonal cracking due to the web reinforcement effects. Furthermore, the shear failure of such beams follows the yielding of the web reinforcement and is more gradual, with sufficient warning of impending failure in the form of widening cracks.

6.8 SHEAR STRENGTH OF BEAMS WITH WEB REINFORCEMENT

The overall shear strength of beams with web reinforcement was expressed by Eq. 6.4. The magnitudes of the strength components in this equation due to individual shear transfer mechanisms vary widely in members of different types, and they are also interdependent. For simplicity, the ultimate nominal shear strength, V_n, is generally expressed as

$$V_n = V_{cn} + V_{sn} \qquad (6.12)$$

where V_{cn} is referred to as the *shear carried by the concrete* at ultimate and includes all the components V_{cz}, V_{ay}, and V_d of Eq. 6.4; and V_{sn} is the shear carried by the web reinforcement. Although the relative magnitudes of the components of V_{cn} vary, depending on the stage of loading and the state of cracking, Ref. 6.3 recommends that their aggregate value represented by V_{cn} can be assumed to be constant and equal to the shear strength at diagonal cracking, V_{cr}. Such an approximation was also found to result in good correlation with test results. V_{cr}, the shear strength at diagonal cracking, is independent of the web reinforcement and, hence, may be computed using Eq. 6.9 or its simplified version – the SI equivalent of Eq. 6.10a. An expression for the shear, V_{sn}, carried by web reinforcement is derived below on the basis of some simplifying assumptions.

Figure 6.8a shows one segment of a beam separated by a diagonal tension crack. This is an idealisation of Fig. 6.3 by assuming that the diagonal crack is straight and at an angle θ to the beam axis, and that it extends over the full depth of the beam. In Fig. 6.8a, at the section, only the forces the web reinforcement that constitute the strength V_{sn} are shown. The web reinforcement is assumed to be placed at an angle α with the beam axis and spaced s apart along the beam axis. If A_v is the total cross-section area of one stirrup (considering all legs) and f_v is the stress in it, the vertical component of the tensile force in one stirrup is $A_v f_v \sin \alpha$. The total number, N, of stirrups intercepted by the diagonal crack is seen from Fig. 6.8a to be:

$$N = \frac{d(\cot\theta + \cot\alpha)}{s}$$

The total vertical component of the stirrup forces across the crack, which make up the shear strength contribution, V_{sn}, of web steel is:

$$\begin{aligned} V_{sn} &= N A_v f_v \sin\alpha \\ &= \frac{A_v f_v d}{s} \sin\alpha (\cot\theta + \cot\alpha) \end{aligned} \qquad (6.13)$$

In particular, for vertical stirrups and a crack inclination of θ,

$$V_{sn} = \frac{A_v f_v d}{s} \cot \theta \qquad (6.13a)$$

Two further simplifying assumptions are usually made. First, the diagonal crack is assumed to occur at 45° to the beam axis; and secondly, the web reinforcement is assumed to yield as the *ultimate strength* is attained. The latter can be ensured by limiting the amount of web reinforcement (or alternatively by limiting the ultimate strength including the web reinforcement) and by specifying an upper limit for the yield strength of web steel. With these two assumptions, at ultimate strength Eq. 6.13 reduces to:

$$V_{sn} = \frac{A_v f_v d}{s} (\sin \alpha + \cos \alpha) \qquad (6.14)$$

Equation 6.14 gives an expression for the nominal shear strength for web reinforcement placed at an angle α.

When vertical stirrups are used as web reinforcement, as in Fig. 6.8b, $\alpha = 90°$, and the shear strength is given by:

$$V_{sn} = \frac{A_v f_v d}{s} \qquad (6.15)$$

An equation similar to Eq. 6.15 could also be derived from the truss analogy. Because of this, the design procedure using Eq. 6.12 and 6.14 (which is the basis for both ACI 318-R92 and CSA A23.3-94(11.3)) is often referred to as the *Modified Truss Analogy* procedure.

Web reinforcement is also provided in the form of bent-up longitudinal bars. When there is a series of parallel bent up bars at different distances from the support, the shear strength can be computed, as for inclined stirrups, using Eq. 6.14. However, when a single bar or a group of parallel bars are all bent-up at the same location at an angle α, the vertical component of the tension in these bars gives:

$$V_{sn} = A_v f_y \sin \alpha \qquad (6.16)$$

Fig. 6.8 *Shear strength due to web reinforcement*

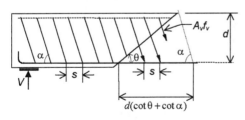
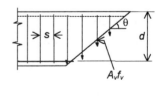

(a) Inclined stirrups (b) Vertical stirrups

For bent-up bars, only the centre three-fourths of the inclined portion is considered effective as shear reinforcement.

Note that Eqs. 6.14 to 6.16 give the *nominal ultimate shear strength* provided by shear reinforcement. The corresponding *factored shear resistance* provided by shear reinforcement, V_s, is obtained by multiplying f_y in these equations, with ϕ_s, the resistance factor for reinforcing bars. The three equations then respectively yield:

(i) for inclined stirrups,

$$V_s = A_v \phi_s f_y \frac{d}{s}(\sin\alpha + \cos\alpha) \tag{6.17}$$

(ii) for shear reinforcement perpendicular to the axis of the member,

$$V_s = A_v \phi_s f_y d/s \tag{6.18}$$

(iii) for a single bar or a group of parallel bars, all bent up at the same location,

$$V_s = A_v \phi_s f_y \sin\alpha \tag{6.19}$$

6.9 SHEAR DESIGN OF BEAMS BY SIMPLIFIED METHOD

(a) Design Basis

The strength criterion in Eq. 3.13 applied to shear design leads to the requirement:

Factored shear force at section ≤ Factored shear resistance of the section

$$V_f \leq V_r \tag{6.20}$$

where V_f is the applied shear force under factored load combinations, and V_r is the factored shear resistance.

The simplified design procedure for shear given in CSA A23.3-94(11.3) is applicable for flexural members/regions (in which the assumption that plane sections remain plane is reasonable) when the member is not subjected to significant axial tension. The procedure is based on the assumptions that (1) the factored shear resistance of a reinforced concrete member can be taken as the sum of the factored shear resistance, V_c, provided by the concrete and the factored shear resistance, V_s, provided by the shear reinforcement; (2) the strength, V_c, provided by the concrete is equal to the shear resistance at inclined cracking (see Section 6.5, Eqs. 6.11 and 6.11a); and (3) V_s is computed by Eq. 6.18 developed in Section 6.8 (only shear reinforcement placed perpendicular to the longitudinal axis of the member is permitted in this method of design).

On this basis, the Code (Clause 11.3) recommends Eqs. 6.20 to 6.28 for the shear design of flexural members not subjected to significant axial tension. (Torsion design is considered in Chapter 7.) The factored shear resistance, V_r, is given by:

$$V_r = V_c + V_s \qquad (6.21)$$

The factored shear resistance provided by concrete, V_c, is given by:
(i)   for sections having either at least the minimum of transverse reinforcement given in Eq. 6.28; or an effective depth not exceeding 300 mm,

$$V_c = 0.2\lambda\phi_c \sqrt{f_c'} b_w d \qquad (6.22)$$

(ii)   for sections with effective depth exceeding 300 mm and with no transverse reinforcement (or transverse reinforcement below the minimum specified),

$$V_c = \left[\frac{260}{1000+d}\right]\lambda\phi_c\sqrt{f_c'}b_w d \geq 0.10\lambda\phi_c\sqrt{f_c'}b_w d \qquad (6.23)$$

where V_c is in Newtons, f_c' in MPa, and the web width, b_w, and effective depth, d, are in mm. The factored shear resistance provided by shear reinforcement, V_s, for shear reinforcement perpendicular to the axis of the member (vertical stirrups) is given by:

$$V_s = \frac{\phi_s A_v f_y d}{s} \qquad (6.24)$$

Combining Eqs. 6.20, 6.21 and 6.24,

$$V_f \leq V_c + \frac{\phi_s A_v f_y d}{s}$$

Denoting the factored shear stress, $V_f/(b_w d)$, by v_f, the factored shear stress resistance provided by the concrete, $V_c/(b_w d)$, by v_c, and rearranging the above equation, Eq. 6.25 may be obtained:

$$s \leq \frac{\phi_s A_v f_y}{b_w(v_f - v_c)} \qquad (6.25)$$

In Eqs. 6.23 and 6.24,
     $A_v =$ cross-sectional area of shear reinforcement perpendicular to the axis of the member within a distance of s
     $b_w =$ web width or diameter of circular section
     $d =$ distance from the extreme compression fibre to the centroid of the longitudinal tension reinforcement
     $s =$ spacing of shear reinforcement measured parallel to the member axis

The area A_v represents the total area of shear reinforcement within the spacing s. Therefore for a U-shaped stirrup with 2 legs, A_v is equal to twice the cross-sectional area for the bar size used.

In an "analysis" problem, Eq. 6.21 can be applied directly to evaluate the factored shear strength of a given section (Example 6.1). In a "design" problem (to design the required web reinforcement for a given V_f) the factored resistance provided by the concrete, V_c, is first computed using Eq. 6.22 or 6.23 and the minimum required shear reinforcement contribution V_s can be obtained from Eq. 6.21. Knowing V_s, Eq. 6.24 can be used to determine the amount and distribution of shear reinforcement (Example 6.2). Alternatively, Eq. 6.25 can be used directly.

(b) Maximum shear stress or minimum size of cross-section
As mentioned in Section 6.7, members heavily reinforced with web steel may fail in shear-compression (diagonal compression) before the web steel yields. Such failures are usually sudden and brittle. Hence, a limitation is placed on the shear strength provided by shear reinforcement. Accordingly, CSA23.3-94(11.3.4) limits V_s to:

$$V_s \leq 0.8\lambda\phi_c\sqrt{f'_c}b_w d \tag{6.26}$$

When V_s exceeds the value given by Eq. 6.26, it is necessary to provide a larger beam cross section. The limitation on V_s given in Eq. 6.26 also serves as a serviceability limit for the control of width of inclined cracks. The latter is a function, among other things, of the stirrup strain, which in turn is a function of V_s.

In thin webbed members, failure may occur by crushing of the web prior to yielding of stirrups. To avoid such failures, the diagonal compression stress in the web and, hence, the maximum shear stress $v_f = V_f/(b_w d)$ must be limited. Such limits for v_f have been suggested in the range of $0.2 f_c'$ (Ref. 6.2) without the inclusion of the ϕ-factors. CSA A23.3-94(11.3.9.8) limits $V_f/(b_w d)$ to a maximum of $0.25\phi_c\sqrt{f'_c}$. Equations 6.22 and 6.26 together give the upper limit for $V_f/(b_w d)$ as $1.0\lambda\phi_c\sqrt{f'_c}$, which is more conservative than $v_f < 0.25\phi_c f_c'$. Hence, Eq. 6.26 also covers the requirement to prevent crushing of the web. The minimum size of cross-section that is required from considerations of shear is in effect determined by Eq. 6.26.

(c) Longitudinal Reinforcement
The influence of shear and diagonal cracking on longitudinal reinforcement requirements was explained in Sections 5.10.3, 6.3.2, and 6.6. To provide for this, in the Simplified Method for shear design, the Code (Cl. 11.3.8) requires that the flexural tension reinforcement be extended for a distance of d or $12d_b$, whichever is greater, beyond the location required for flexure alone. Here d_b is the nominal diameter of the longitudinal bar concerned, and the provision is applicable for locations other than at the support of simple spans, exterior support of continuous spans and at free ends of cantilevers with concentrated loads. This provision is equivalent to the outward shifting of the design moment diagram by a distance of d or $12d_b$ (Fig. 6.9a).

Fig. 6.9 Effect of shear on longitudinal reinforcement

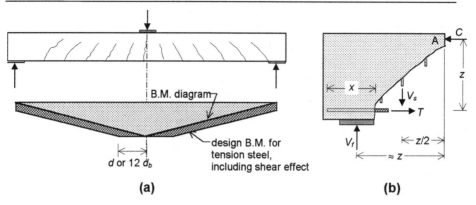

At simple supports and near free ends of cantilevers, the flexural tension reinforcement should be capable of resisting a tensile force of $V_f - 0.5 V_s$ at the inside edge of the bearing area, where V_s is the factored shear resistance provided by the shear reinforcement in this location. The derivation of this expression can be made from equilibrium considerations of the forces in the free body diagram of the support region separated by a diagonal crack, (Fig. 6.9b). Taking moments about A and neglecting small quantities of second order,

$$V_f z \approx T z + V_s z/2$$

$$T = V_f - 0.5 V_s \qquad (6.27)$$

If the actual straight embedment length available at the support, x, is less than the development length, l_d, the stress that can be developed in the bar at the critical section at the inside edge of the bearing area (Fig. 6.9b) may be taken as:

$$f_s = \phi_s f_y (x/l_d)$$

Alternatively, the embedment length required to develop the stress f_s in the bar can be computed as:

$$x = \frac{f_s}{\phi_s f_y} l_d$$

(d) Limitations Set in the Code
   1. Yield Strength f_y The Code limits the specified yield strength, f_y, of reinforcement to a maximum of 500 MPa. The shear strength computations are based on the assumption that web reinforcement yields. The width of the inclined crack is a function of the stirrup strain. The yield strength limitation ensures that the diagonal crack widths are not excessive. Furthermore, higher strength reinforcement may be

brittle at sharp bends, which often occur in stirrups.

For welded wire fabric used as shear reinforcement, the Code (Cl. 11.2.2.1b) also specifies a minimum elongation requirement of 4 percent over a gauge length of at least 100 mm. To mobilise the shear resistance, V_s, of the reinforcement, the transverse wires have to undergo substantial strains. The elongation requirement is to guard against premature failure of the transverse wires at or between the cross-wire intersection.

2. Minimum Shear Reinforcement When the shear force, V_f, does not exceed $V_c/2$ (that is, the nominal shear stress $V_f/(b_w d) \leq 0.1\lambda\phi_c\sqrt{f_c'}$), beams need not be reinforced for shear. As explained in earlier sections, in some members, formation of an inclined crack may lead to sudden failure. To restrain the spread of inclined cracking and, thereby, provide increased ductility, with a gradual failure and adequate warning of distress, the Code specifies a minimum area of shear reinforcement in reinforced concrete flexural members whenever the factored shear force, V_f, exceeds $0.5V_c$, (where V_c, in lieu of more detailed calculations, may be computed by Eq. 6.22), except in slabs and footings, concrete joist floors and certain classes of relatively wide and shallow beams. The minimum shear reinforcement is:

$$A_{v,\min} = 0.06\sqrt{f_c'}\frac{b_w s}{f_y} \tag{6.28}$$

In satisfying the requirement for $A_{v,\min}$ in Eq. 6.28, inclined reinforcement and transverse reinforcement used to resist torsion may be included.

The factored shear resistance of shear reinforcement corresponding to the minimum A_v required in Eq. 6.28 is obtained by substituting $A_v = A_{v,\min}$ in Eq. 6.24 as:

$$V_s = 0.06\phi_s\sqrt{f_c'}b_w d$$

(Note that this corresponds approximately to 50% of V_c given by Eq. 6.22) Hence, for the case of stirrups placed perpendicular to the beam axis, the minimum reinforcement controls as long as $V_s \leq 0.06\phi_s\sqrt{f_c'}\, b_w d$. In summary, the shear reinforcement area is controlled by the following requirements:

Zone	Condition	Requirement
Zone I	$V_f \leq V_c/2$	No reinforcement required
Zone II	$V_f > V_c/2$, and $V_s = 0.06\phi_s\sqrt{f_c'}b_w d$	Provide minimum reinforcement area, Eq. 6.28
Zone III	$V_s > 0.06\phi_s\sqrt{f_c'}b_w d$, and $V_f < 0.1\lambda\phi_c f_c' b_w d$	Provide reinforcement for $V_s = V_f - V_c$, Eq. 6.24, 6.25; Spacing limitations (see below)

Zone IV $V_s < 0.8\lambda\phi_c\sqrt{f_c'}b_w d$
$V_f \geq 0.1\lambda\phi_c f_c' b_w d$
[and $V_f < 1.0\lambda\phi_c\sqrt{f_c'}b_w d$]

Cl. 11.2.11, Provide reinforcement for V_s, Eq. 6.24 and 6.25. Maximum spacing limitations to be one-half of previous case (Cl. 11.2.11, see below)

Zone V $V_s > 0.8\lambda\phi_c\sqrt{f_c'}b_w d$
[and $V_f > 1.0\lambda\phi_c\sqrt{f_c'}b_w d$]

Provide larger beam cross section (Cl. 11.3.4)

3. Spacing limits for shear reinforcement Web reinforcement becomes effective when it is crossed by inclined cracks. Furthermore, dowel action will depend on the longitudinal bars being restrained by closely spaced stirrups. The Code (Cl. 11.2.11) limits the spacing of transverse shear reinforcement, s, measured parallel to the longitudinal axis of the member, to a maximum of:

600mm or $0.7d$, for $V_f/(b_w d) < 0.1\lambda\phi_c f_c'$, and

300mm or $0.35d$, for $V_f/(b_w d) \geq 0.1\lambda\phi_c f_c'$ (6.29)

The controlling limits for shear reinforcement design given above are presented diagrammatically in Fig. 6.10.

Fig. 6.10 Controlling limits for stirrups in simplified method of design

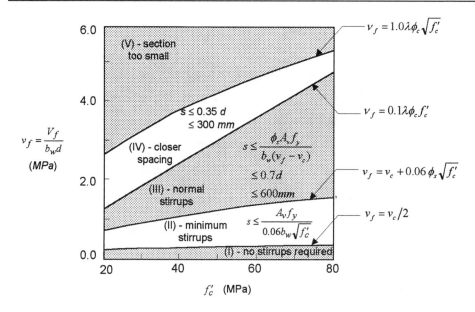

6.10 SHEAR DESIGN NEAR SUPPORTS

Usually the maximum shear force in a member occurs at the face of supports and near concentrated loads. When a support reaction introduces transverse compressive stresses in the end region of the member, the shear strength of this region is enhanced, and inclined cracks do not generally form for a distance from the face of the support equal to the effective depth, d. Accordingly, for such cases, the Code permits sections located less than a distance d from the face of the support to be designed for the same factored shear force, V_f, as that computed for the section at a distance d. Critical sections to be used in computing V_f for shear design for a few typical support conditions are shown in Fig. 6.11.

When a load is applied to the side of a member, through brackets, ledges or cross beams (Fig. 6.12), the detailing of the reinforcement in the connection region may be based on the strut-and-tie model (Chapter 8). Alternatively, the procedure recommended in the 1984 version of the Code may be adopted. According to this, additional full depth transverse reinforcement is necessary in both members in the vicinity of the interface, for "hanging up" a portion $(1 - h_b/h_1)$ of the interface shear. The transverse reinforcement is designed to transfer a tensile force of $V_f(1 - h_b/h_1)$. (Fig. 6.12 gives notation.) The effective region in each member within which this additional reinforcement should be placed in order to be effective is also shown in Fig. 6.12. For the supporting member, this is a zone within a distance of h_b from the shear interface, and for the supported member it is a zone within a distance of $d/4$ on each side of the shear interface. The value of h_b need not be taken as less than 75 mm.

Fig. 6.11 Critical sections for shear at support

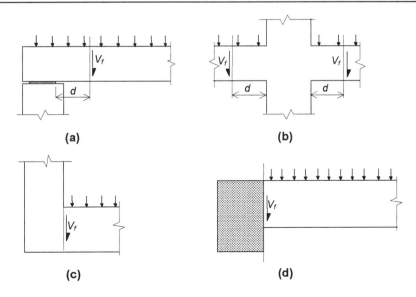

190 REINFORCED CONCRETE DESIGN

Fig. 6.12 Hanging up bars for indirect support

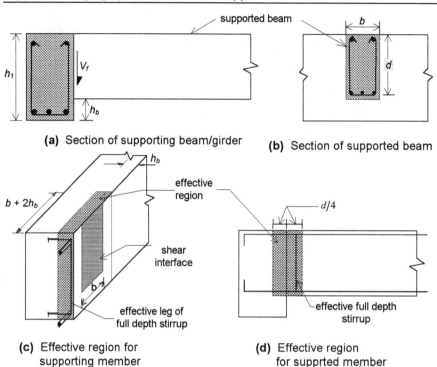

(a) Section of supporting beam/girder

(b) Section of supported beam

(c) Effective region for supporting member

(d) Effective region for supprted member

6.11 ANCHORAGE OF SHEAR REINFORCEMENT

Web reinforcement is designed on the assumption that the yield stress ($f_y \leq 500$ MPa) is developed at ultimate strength. Hence, in order to be fully effective, such reinforcement must be adequately anchored at both ends so as to develop the yield strength. Web reinforcement is carried as close to the compression and tension surfaces as practical, consistent with minimum cover requirements.

Minimum requirements for effective anchorage are given in CSA A23.3-94(12.13). For the most common type of web reinforcement, the U-stirrup, the Code requirements are shown in Fig. 6.13. The maximum stress in the stirrup, equal to design f_y, is assumed to be developed at a depth $d/2$ from the compression face. In Fig. 6.13, l_d represents the development length (see Chapter 9). When the standard hook (Fig. 6.13c) is used, to satisfy the requirement in Fig. 6.13a, a reduction in stirrup bar size and spacing, or alternatively an increase in the effective depth, d, may become necessary. Mechanical anchorages capable of developing the yield strength of the bar may also be used.

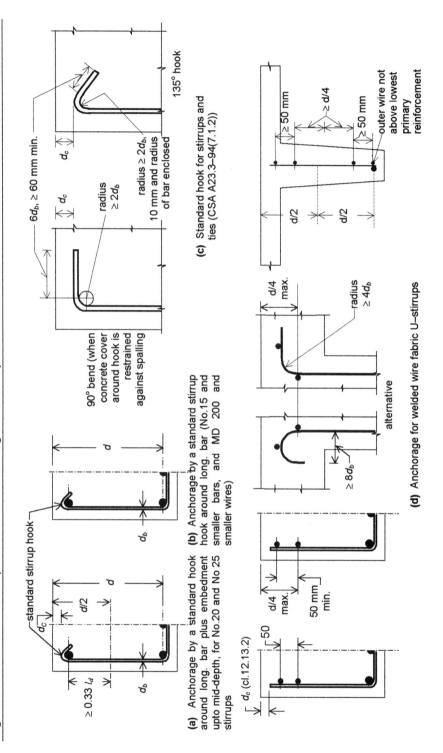

Fig. 6.13 CSA A23.3-94 requirements for anchorage of U-stirrups

6.12 ADDITIONAL COMMENTS ON SHEAR REINFORCEMENT DESIGN

In designing for shear, the effects of axial tension due to creep, shrinkage, and temperature effects in restrained members should be considered where applicable. The Simplified Method is for flexural members *not subjected to significant axial tension*. Here, axial tension causing an increase in reinforcement stress at crack locations in excess of 50 MPa may be considered as *significant axial tension* (Ref. 6.6).

Although the Code does not require shear reinforcement in portions of beams where $V_f < V_c/2$, it is good practice to provide the minimum web reinforcement in this region to provide ductility and restrain cracks, in the event of accidental overloading. In beams subjected to reversed bending to the full design capacities, stirrups may be provided for the full factored shear force, neglecting the shear strength provided by concrete. Such stirrups should be in the form of closed ties.

Shear reinforcement is designed usually by selecting a suitable stirrup bar size (which is kept the same for the entire span) and then computing the required spacings. Smaller diameter, closely spaced stirrups (for example, welded wire fabric stirrups) give better crack control than stirrups of larger size bars, placed farther apart. In general, the required spacing will vary continuously along the length of the beam due to the variation in shear force, V_f. However, stirrups are usually arranged with the spacing kept uniform over portions of the span, the spacing varying between these portions as necessitated by strength and limiting spacing considerations. The first stirrup is placed at not more than one-half space from the face of support. Between the anchored ends, each bend in the continuous portion of a stirrup must enclose a longitudinal bar.

Termination of flexural reinforcement in the tension zone may lower the shear strength of beams (Section 5.10.5). Therefore, such sections may also be critical and have to be checked for shear and, if necessary, provided with additional shear reinforcement to satisfy the requirement in CSA A23.3-94 (12.10.5) (Example 6.1).

6.13 SHEAR DESIGN EXAMPLES

EXAMPLE 6.1

The simply supported beam shown in Fig. 6.14a is provided with web reinforcement consisting of 6 mm plain bar U-stirrups at a uniform spacing of 190 mm. Check the adequacy of the shear design. If necessary revise the design. Given: Dead load including self-weight = 10.65 kN/m; L.L = 15 kN/m; $f_c' = 30$ MPa; $f_y = 400$ MPa.

Fig. 6.14 Example 6.1

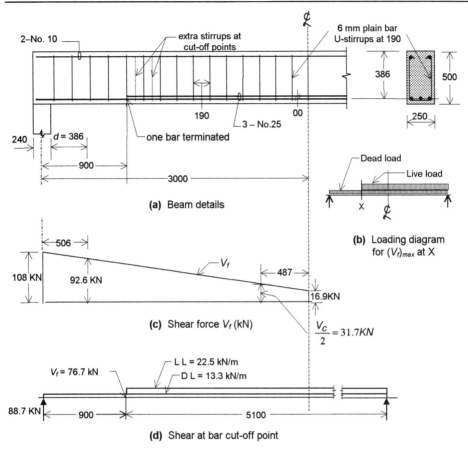

(a) Beam details
(b) Loading diagram for $(V_f)_{max}$ at X
(c) Shear force V_f (kN)
(d) Shear at bar cut-off point

SOLUTION

1. The factored loads are:
   Dead load = 1.25 (10.65) = 13.3 kN/m; Live load 1.5 × 15 = 22.5 kN/m

2. The shear force envelope giving the maximum factored shear force at each location along the span is first drawn for one half of the span. The placement of live load to give the maximum shear force at a section, X, on the left half of the span is as shown in Fig. 6.14b. Dead load generally acts on the entire span. The actual shape of the shear envelope is curvilinear. However, it is sufficiently accurate and conservative to consider the shear force due to live load to vary as a straight line between the maximum values computed for the

support and for the midspan. In this problem,

$$V_{f,\text{support}} = \frac{(13.3+22.5)6}{2} = 108 \text{ kN}$$

$$V_{f,\text{centreline}} = 0 + \frac{22.5 \times 3 \times 1.5}{6} = 16.9 \text{ kN}$$

The critical section for shear (Cl. 11.3.2) is at a distance $d = 386$ mm from face of support, or $120 + 386 = 506$ mm from centre of support.
Factored shear force at critical section:

$$V_f = 16.9 + \frac{(108-16.9)(3000-506)}{3000} = 92.6 \text{ kN}$$

The factored shear force diagram is shown in Fig. 6.14c.

3. The factored shear resistance provided by concrete, V_c, using Eq. 6.22 is:

$$V_c = 0.2\lambda\phi_c\sqrt{f_c'}b_w d = 0.2 \times 1.0 \times 0.6 \times \sqrt{30} \times 250 \times 386 \times 10^{-3} = 63.4 \text{ kN}$$

4. Check adequacy of section
Maximum permitted $V_s = 0.8\lambda\phi_c\sqrt{f_c'}b_w d = 4 \times 63.4 = 254$ kN
Actual required $V_s = 92.6 - 63.4 = 29.2$ kN < maximum permitted.
Section size is adequate.

5. Check adequacy of shear reinforcement
Since $V_f > V_c/2$, shear reinforcement is required.
Minimum shear reinforcement required is:

$$A_{v,\text{min}} = 0.06\sqrt{f_c'}\frac{b_w s}{f_y} = \frac{0.06 \times \sqrt{30} \times 250 \times 190}{400} = 39.0 \text{ mm}^2$$

A_v provided $= 2 \times 28.3 \times 56.6 >$ minimum   OK
Required spacing for the V_s is obtained from Eq. 6.24:

$$\text{Required } s = = \frac{\phi_s A_v f_y d}{V_s} = \frac{0.85 \times 56.6 \times 400 \times 386}{29.2 \times 10^3} = 254 \text{ mm}$$

Provided $s = 190$ mm < required   OK
Alternatively, the factored resistance of the section, V_r, is:

$$V_r = V_c + V_s = 63.4 + \frac{\phi_s A_v f_y d}{s}$$

$$= 63.4 + \frac{0.85 \times 56.6 \times 400 \times 386}{190} \times 10^{-3} = 63.4 + 39.1$$

$= 103$ kN $> V_f = 92.6$ kN   OK

6. Check spacings
   To check condition in Code Clause 11.2.11
   $0.1\lambda\phi_c f'_c b_w d = 0.1 \times 1 \times 0.6 \times 30 \times 250 \times 386 \times 10^{-3} = 173.7$ kN
   Since $V_f = 92.6$ kN $< 0.1\lambda\phi_c f'_c b_w d$, the maximum spacing is as in Clause 11.2.11(a). Maximum spacing limit is the least of:
   $0.7d = 0.7 \times 386 = 270$ mm (controls) or 600 mm
   $s_{max} = 270$ mm
   $s_{provided} = 190$ mm $< s_{max}$ OK

7. Check anchorage
   Since the stirrups are of 6mm bars (i.e., smaller than No. 15), the standard 135° stirrups around longitudinal reinforcement is adequate (Cl. 12.13.2 (a)).

8. Check shear strength at bar cut-off point
   One longitudinal tension reinforcing bar is terminated in the tension zone, at distance of 900 mm from support. Hence, this section should be checked for shear to conform to requirements in Clause 12.10.5. Loading condition for maximum shear at this section is shown in Fig. 6.14d.
   Factored shear at cut-off point, $V_f = 76.7$ kN
   Shear resistance of this section: $V_r = V_c + V_s = 63.4 + 39.1 = 103$ kN
   $\frac{2}{3}$ shear resistance = $\frac{2}{3} \times 103 = 68.3$ kN
   Since $V_f > \frac{2}{3}$ that permitted = 68.3 kN, Cl. 12.10.5a is not satisfied, and additional stirrups must be provided. Here, adding two additional stirrups along the last portion of the cut-off bar, as shown by broken lines in Fig. 6.14a, will reduce the spacing to 190/2 = 95 mm. Then the shear strength at the cut-off point is: $V_c + V_s = 63.4 + 2 \times 39.1 = 142$ kN
   $V_f < \frac{2}{3}$ permitted = (2/3) × 142 = 94.4 kN OK
   The requirement in Cl. 12.10.5a is now satisfied. Alternatively, the additional stirrups can be shown to satisfy the requirements in Cl. 12.10.5b.
   The reinforcement provided corresponds to the minimum spacing. Since the design is safe at the critical section for shear, it is safe at other points along the beam. According to the Code, shear reinforcement is not required where
   $V_f < V_c/2$ or $V_f < 63.4/2 = 31.7$ kN
   Referring to Fig. 6.14c, $V_f < V_c/2$, for a distance of 487 mm from centre of span, and the stirrups in this region (five in all for the entire span) are not essential.

9. Check longitudinal reinforcement at support
Tension in bars at inside edge of bearing area is (Eq. 7.27):
$T = V_f - 0.5 V_s = 92.6 - 0.5 \times 39.1 = 73$ kN
Stress in bars = $73 \times 10^3 / 1000 = 73$ MPa $< \phi_s f_y$,   OK
Development length of No. 25 bars is (Code Cl. 12.2.3):

$$l_d = 0.45 k_1 k_2 k_3 k_4 \frac{f_y d_b}{\sqrt{f_c'}} = 0.45 \times 1 \times 1 \times 1 \times 1 \times \frac{400}{\sqrt{30}} \times 25.2 = 828 \text{ mm}$$

Embedment length required to develop in the bar the required stress of 88.5 MPa:

$$X = l_d \left(f_s / \phi_s f_y \right) = 828 \times \frac{73}{0.85 \times 400} = 178 \text{ mm}$$

Straight length available, with a cover of 25 mm is
$240 - 25 = 215$ mm, adequate   OK

EXAMPLE 6.2

Design the shear reinforcement for the T-beam shown in Fig 6.15a, given the following data: $b = 3180$ mm, $b_w = 300$ mm, $h = 500$ mm, $h_f = 120$ mm, d at support = 434 mm, d at midspan = 406 mm, A_s at support (3 No. 30 bars) = 2100 mm^2, A_s at midspan (6 No. 30 bars) = 4200 mm^2, Span centre to centre of support = 8 m, Width of bearing = 240 mm, Dead load including self weight = 15.7 kN/m, Live load =29.7 kN/m, $f_c' = 20$ MPa, $f_y = 400$ MPa.

SOLUTION

1. Factored loads are:
   Dead load = $1.25 \times 15.7 = 19.6$ kN/m; Live load = $1.5 \times 29.7 = 44.6$ kN/m
   Total = 64.2 kN/m

2. Compute shear force envelope
   Maximum shear at support = $\frac{62.2 \times 8}{2} = 257$ kN

   Maximum shear at midspan (live load only on one half of span
   $= 0 + \frac{1}{4} \frac{44.6 \times 8}{2} = 44.6$ kN
   Shear force diagram may be assumed conservatively as a straight line shown

Fig. 6.15 Example 6.2

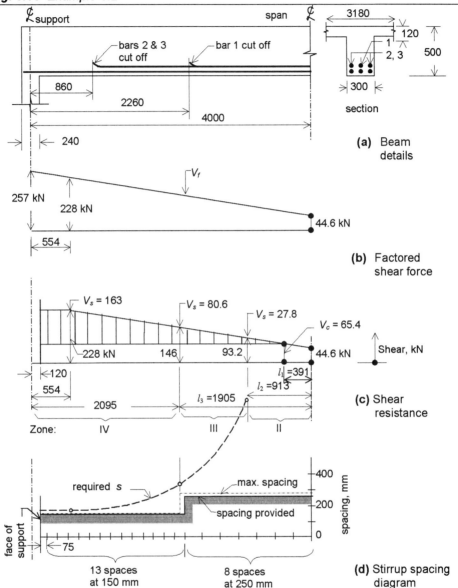

in Fig. 6.15b. Critical section at support is at a distance d away from the face of support, or $120 + 434 = 554$ mm from the centre of support.

Shear at critical section $= 44.6 + \dfrac{(257 - 44.6)(4 - 0.554)}{4} = 228$ kN

3. Factored shear resistance of concrete, V_c:
Assuming the section will have more than the minimum amount of transverse reinforcement (Cl. 11.3.5.1),
$$V_c = 0.2\lambda\phi_c\sqrt{f'_c}b_w d$$
At support section,
$V_c = 0.2 \times 1.0 \times 0.6 \times \sqrt{20} \times 300 \times 434 \times 10^{-3} = 69.9$ kN
At midspan section, $d = 406$ mm and
$V_c = 0.2 \times 1.0 \times 0.6 \times \sqrt{20} \times 300 \times 406 \times 10^{-3} = 65.4$ kN
Variations of V_f, V_c and V_s are shown in Fig. 6.15c
Note that the effective depth, d, varies along the span as tension reinforcement bars are terminated. Here, the lower value of $V_c = 65.4$ kN computed for midspan section will be considered as the factored shear resistance contributed by concrete for the entire length of the beam. Normal levels of accuracy desired in practice do not generally warrant consideration of variation of d along span.
The excess shear to be carried by shear reinforcement, V_s, is represented by the shaded area in Fig. 6.15c. The location of the section where $V_f = V_c$ is calculated from similar triangles, in Fig. 6.15c as:
$$\frac{l_c}{3446} = \frac{65.4 - 44.6}{228 - 44.6} \;;\; l_c = 391 \text{ mm}$$

4. Check adequacy of overall depth
Maximum permitted $V_s = 0.8\lambda\phi_c\sqrt{f'_c}b_w d = 4 \times 65.4 = 262$ kN
Actual maximum $V_s = 228 - 65.4 = 163$ kN < 262 kN   OK

5. The zones with different shear reinforcement requirements are calculated below and identified in Fig. 6.15c
Zone I: $V_f < V_c/2 = 32.7$ kN
No shear reinforcement required
In this example $V_{f,min} = 44.6$ kN $> V_c/2$. Hence, no such zone in this beam.
Zone II: $V_f > \dfrac{V_c}{2}$ and $V_s = 0.06\lambda\phi_s\sqrt{f'_c}b_w d$
$0.06 \times \sqrt{20} \times 0.85 \times 300 \times 406 \times 10^{-3} = 27.8$ kN
or $V_f \le 65.4 + 27.8 = 93.2$ kN

Minimum reinforcement requirement controls. Location for this is:
$$l_2 = \frac{(93.2 - 44.6)}{228 - 44.6} \times 3446 = 913 \text{ mm}$$
Zone II is from midspan to 913 mm.

Zone III: For 93.2kN $< V_f < 0.1\lambda\, \phi_c f_c' b_w d$
$0.1 \times 1 \times 0.6 \times 20 \times 300 \times 406 \times 10^{-3} = 146$ kN
Reinforcement designed for excess shear V_s, and maximum spacing limitations as in CSA A23.3-94(11.2.11a). The upper limit to this is at $V_f = 146$ kN, and its location is given by:

$$l_3 = \frac{(146 - 44.6)}{228 - 44.6} \times 3446 = 1905 \text{ mm from midspan}$$

Zone IV:
$0.1\lambda\phi_c f_c' b_w d < V_f < 1.0\lambda\phi_c \sqrt{f_c'} b_w d = 5 \times 65.4 = 327$ kN
Reinforcement designed for excess shear V_s, but maximum spacing limitations reduced to one-half of that for Zone III. The region from $l_3 = 1905$ mm to face of support falls in this category.

Note that the zones have been identified here for illustration purposes only. For designing shear reinforcement, such a detailed zoning is not necessary and it is sufficient to locate the region where shear reinforcement is necessary ($V_f > V_c/2$) and the section ($V_s = 0.1\lambda\phi_c f_c' b_w d$) beyond which the closer spacing requirement controls.

6. Design of shear reinforcement
   Try No. 10 bar U-stirrups placed vertically.
   $A_v = 2 \times 100$ (for 2 legs) $= 200$ mm^2.
   Check adequacy of beam depth for anchorage of stirrups (Cl.11.2.5).
   For No. 10 bar stirrups, 135° standard stirrup hooks around longitudinal reinforcement provides adequate anchorage. There is no special embedment length requirement. Hence OK.
   The spacing requirement is given by (Eq. 6.24),

   $$s = \frac{\phi_s A_v f_y d}{V_s} = \frac{0.85 \times 200 \times 400 \times 406}{V_s} = \frac{27600 \times 10^3}{V_s}$$

   The required stirrup arrangement can be determined by plotting a spacing curve that gives the variation of required s as computed by the above equation wherein the spacing limitations are also superimposed. Having done this, convenient bar spacings are selected which are on the safe side of the requirement. The calculations for the spacing curve for this example are set up in tabular form below.

Distance of section from support, mm		V_s kN	Required $s = 27\,600/V_s$ mm
Critical section	554	163	170
For maximum spacing	2095	80.6	342
For minimum reinforcement	3087	27.8	993

The maximum allowable spacing in Zones I to III is the least of:
$0.7d = 0.7 \times 406 = 284$ mm (controls) or 600 mm; use 250 mm
Maximum allowable spacing in Zone IV is:
$(1/2)(0.7d) = (1/2)(0.7 \times 434) = 151.9$ mm; use 150 mm
(Note: Zone IV where top layer tension reinforcement is cut off, $d = 434$ mm.)
The curve for required spacing is shown in Fig. 6.15d, wherein the maximum spacing limitations are also indicated. A stirrup layout meeting these requirements is detailed along the base of Fig. 6.15d, and the shaded area shows the corresponding spacing diagram. In this layout, the first stirrup is placed at $s/2 = 75$ mm from face of support and this is followed in turn by 13 spaces (stirrups) at 150 mm and 8 at 250 mm making up one-half of the span. This spacing can be selected as final if no flexural reinforcement bar is terminated in the tension zone. In this Example, since the main flexural reinforcement is terminated in the tension zone, the shear at these locations must be checked to satisfy CSA A23.3-94(12.10.5)

7. Check shear strength at bar cut-off points

   The cut-off point closest to support, which is more critical, will be checked first. The section is 0.86 m from centre of support. Since the shear here may be critical and in order to avoid too much conservatism, the actual loading condition for maximum shear (shown in Fig. 6.16a) and the effective depth at support will be used to check the strength of this section.
   From Fig. 6.16a, at the cut-off point $V_f = 204$ kN
   Shear carried by concrete for $d = 434$ mm is: $V_c = 69.9$ kN
   Shear carried by shear reinforcement (at spacing of 150 mm provided at this location) is:
   $$V_s = \frac{\phi_s A_v f_y d}{s} = \frac{0.85 \times 200 \times 400 \times 434 \times 10^{-3}}{150} = 197 \text{ kN}$$
   Admissible shear without additional stirrups (CSA A23.3-94(12.10.5)) is:
   $(2/3)(V_c + V_s) = (2/3) \times (69.9 + 197) = 178$ kN
   $V_f = 204 >$ admissible,   not OK
   Shear strength at cut-off point is not adequate for the arrangement of stirrups worked out at step 6 above. Excess stirrup should be provided as per Cl. 12.10.5(b).
   Revise stirrup layout with first stirrup placed at 75 mm from face of support, followed in turn by two at 150 mm, four at 100 mm, nine at 150 mm and seven at 250 mm, making up one-half of the span (total 46 stirrups for the whole span) as shown in Fig. 6.16 (b). With this layout, the spacing at bar cut-off point at 860 mm from the support is 100 mm. Corresponding $V_s = 197 \times 150 / 100 = 295$ kN.
   $(2/3)(V_c + V_s) = 2/3 (69.9 + 295) = 243$ kN $> V_f = 204$,   OK.

Fig. 6.16 Example 6.2

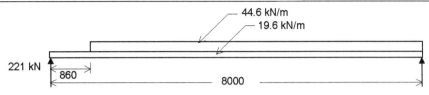

(a) Loading for maximum shear at cut-off point

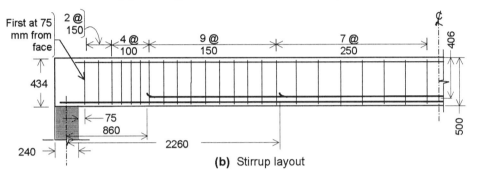

(b) Stirrup layout

A similar evaluation at the cut-off point at 2260 mm from the support (where the stirrup spacing is 150 mm) will show that the factored shear force here is less than 2/3 of the shear strength permitted for the spacing of 150 mm [(2/3)×(69.9 + 197) = 178 kN] and hence shear strength at this location is adequate with the stirrup spacing designed in step 6. Hence a final satisfactory stirrup layout is as shown in Fig. 6.16b.

8. Check adequacy of longitudinal reinforcement at support
Tensile force to be resisted by flexural tension reinforcement at the critical section at the inside edge of bearing area is (Eq. 6.27):
$T = V_f - 0.5 V_s = 228 - 0.5 \times 197 = 129.5$ kN.
Stress in reinforcement bar $= f_s = 129 \times 10^3/2100 = 61.6$ MPa << 400MPa.
The reinforcement area is adequate.
The development length for No. 30 bar is given by (Code Cl. 12.2.3).

$$l_d = 0.45 k_1 k_2 k_3 k_4 \left(f_y / \sqrt{f'_c}\right) d_b = 0.45 \times \frac{400}{\sqrt{20}} \times 29.9 = 1203 \text{ mm}$$

Embedment length required to develop stress f_s is

$$X = l_d (f_s / \phi_s f_y) = 1203 \times \frac{61.6}{0.85 \times 400} = 218 \text{ mm}$$

The straight embedment length available = 240 – 25 mm cover = 215 mm is just adequate. Hence the bars may be extended to the far side of the support and anchored with a standard 90° hook, which provides adequate development length (minimum development length of standard hook (Cl. 12.5.1) is = $8d_b$ = 240 mm, which is more than the shortage of 218 − 215 = 3 mm).

6.14 SHEAR STRENGTH OF FLEXURAL MEMBERS SUBJECTED TO AXIAL LOADS

Frequently, flexural members in frames are subjected to axial loads in addition to bending moment and shear. Axial tension may also result from restraint of deformations due to temperature changes, creep, and shrinkage. In determining shear resistance, effects of axial forces, in particular of tensile forces, must be considered wherever applicable.

Inclined cracking is related to the attainment of a limiting principal tensile stress in the concrete. The addition of an axial force alters the stress distribution and, hence, the shear strength at inclined cracking. In particular, axial compression decreases the normal tensile stress, f_x. Furthermore, it decreases the shearing stress, v, by decreasing the height of flexural cracks and thereby increasing the effective concrete area resisting shear. Both these effects reduce the principal tensile stress and, hence, tend to increase the shear strength at diagonal cracking. In contrast, axial tension will have just the opposite effects. For this reason, the Simplified Method for shear design is restricted to members that are not subjected to significant axial tension. For flexural members subjected to substantial axial loads, particularly tensile force, shear design should be based on the General method or the Strut-and-Tie model (Chapter 8).

6.15 STRUCTURAL LOW DENSITY CONCRETE

A study of diagonal tension test results of structural low density concrete beams has shown (Ref. 6.7) that the general principles of shear strength discussed in Sections 6.3-6.5 for normal density concrete apply equally well to low density concrete members. Once again, the major variables that affect the inclined cracking strength of such beams are the ratio $M/(Vd)$, ρ, and tensile strength of concrete. Furthermore, these variables can be grouped into the same parameters as used in the development of Eq. 6.10. However, the tensile strength of low density concrete is highly dependent on the type of aggregate, and is somewhat less than that of normal density concrete of the same specified compressive strength, f_c'. As a simplification, it can be assumed that for a given compressive strength of concrete, the tensile strength of lightweight concrete is a fixed proportion of the tensile strength of normal density concrete (Ref. 6.7). Thus, CSA A23.3-94(8.6.5) has introduced a modification factor, λ, which appears in equations for tensile strength of concrete and for shear carried by concrete, to account for the density of concrete. The factor λ has values of unity for normal density concrete, 0.85 for structural semi-low density concrete in which all the fine aggregate is natural sand, and 0.75 for structural low density concrete in which none of the fine aggregate is natural sand.

6.16 MEMBERS OF VARYING DEPTH

In flexural members of varying depth, as shown in Fig. 6.17, the vertical component of the flexural tensile force may affect the shear force for which the section is to be designed. Assuming the horizontal component of the tension in the flexural reinforcement as $T = M_{fv}/z \approx M_{fv}/d$, it can be seen from Fig. 6.16 that the net shear force for which the section should be designed is:

$$V_{net} = V_f \pm \frac{M_{fv}}{d}\tan\beta \qquad (6.30)$$

Fig. 6.17 Design shear force in members of variable depth

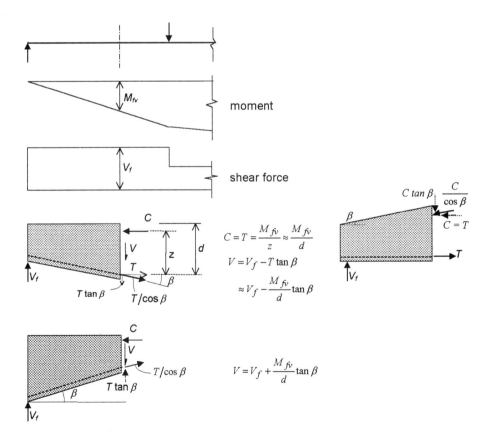

where V_f and M_{fv}, are the applied shear force and moment at the section, and β is the inclination of the tension reinforcement to the beam axis. The negative sign applies where M_{fv} increases in the same direction as depth, d, and the positive sign applies where M_{fv} decreases in this direction.

A similar adjustment to the shear, V_f, is necessary when the flexural compression is inclined to the beam axis. CSA A23.3-94 (11.2.2) specifies that when the effect of the vertical component is unfavourable (that is, additive to V_f), it should be taken into account in designing for shear. The effect may also be considered when favourable.

6.17 INTERFACE SHEAR AND SHEAR FRICTION

6.17.1 Shear-Friction

There are situations where shear has to be transferred across a defined plane of weakness, nearly parallel to the shear force and along which slip could occur. Examples are planes of existing or potential cracks, interface between dissimilar materials, interfaces between elements such as webs and flanges, and interface between concrete placed at different times (Fig. 6.18). In such cases, possible failure involves sliding along the plane of weakness rather than diagonal tension. Therefore it would be appropriate to consider shear resistance developed along such planes in the form of resistance to the tendency to slip. The *shear-friction* concept is a method to do this.

Fig. 6.18 *Typical cases where shear friction is applicable (adapted from Ref. 6.8)*

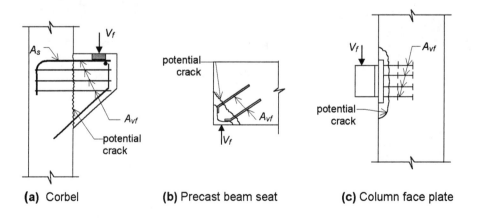

(a) Corbel   (b) Precast beam seat   (c) Column face plate

Fig. 6.19 Shear-friction analogy

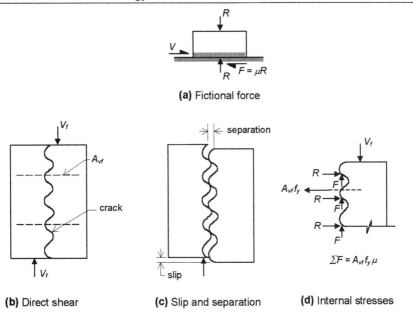

When two bodies are in contact with a normal reaction, R, across the surface of contact, the frictional resistance, F, acting tangential to this surface and resisting relative slip is known to be $F = \mu R$, where μ is the coefficient of friction (Fig. 6.19a). Figure 6.19b shows a cracked concrete specimen loaded in shear. In such a specimen, a clamping force between the two faces of the crack can be induced by providing reinforcement perpendicular to the crack surface (shear-friction reinforcement, A_{vf}). Any slip between the two faces of the rough irregular crack causes the faces to ride upon each other, which opens up the crack. This in turn induces tensile forces in the reinforcement, which ultimately yields (Fig. 6.19c, d). If the area of reinforcement is A_{vf} and yield stress f_y, at ultimate, the clamping force between the two faces is $R = A_{vf} f_y$, and the frictional resistance is $A_{vf} f_y \mu$.

In reality, the actual resistance to shear, V_r, is composed of this frictional force, the resistance to shearing off of the protrusions on the irregular surface of the crack, the dowel force developed in the transverse reinforcement, and when there are no cracks developed yet, the cohesion between the two parts as well. The nominal shear resistance, V_{sn}, due to the friction between the crack faces, is given by Eq. 6.31. Other less simple methods of calculation have been proposed (Refs. 6.9, 6.10) which result in predictions of shear transfer resistance in substantial agreement with comprehensive test results.

For shear-friction reinforcement placed perpendicular to the shear plane,

$$V_{sn} = A_{vf} f_y \mu \tag{6.31}$$

Fig. 6.20 Inclined shear-friction reinforcement

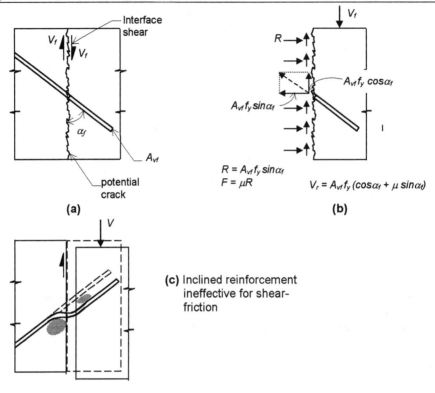

Where, V_{sn} = nominal shear resistance due to the assumed friction part alone contributed by reinforcement stress
A_{vf} = area of shear-friction reinforcement, placed normal to the plane of possible slip
μ = coefficient of friction.

Shear-friction reinforcement may also be placed at an angle α_f to the shear plane, such that the shear force produces tension in the shear-friction reinforcement, as shown in Fig. 6.20a, b (i.e., $\alpha_f \leq 90°$). As the shear-friction reinforcement yields, the tensile force in the reinforcement is $A_{vf}f_y$, which has a component parallel to the shear plane of $A_{vf}f_y \cos\alpha_f$, and a component normal to the plane equal to $A_{vf}f_y \sin\alpha_f$. The latter produces the clamping force. The total force resisting shear is then obtained as $A_{vf}f_y \cos\alpha_f + \mu.A_{vf}f_y \sin\alpha_f$, and the nominal shear resistance is given by:

$$V_{sn} = A_{vf}f_y (\cos\alpha_f + \mu \sin\alpha_f) \tag{6.32}$$

If the area of concrete section at the interface resisting shear transfer is A_{cv}, the nominal shear resistance per unit area can be expressed (from Eq. 6.32) as:

$$v_{sn} = \rho_v f_y \cos\alpha_f + \mu(\rho_v f_y \sin\alpha_f) \tag{6.32a}$$

where $\rho_v = A_{vf}/A_{cv}$, and v_{sn} = nominal shear resistance due to the transverse reinforcement. If there is a load, N, normal to the interface, this will increase or decrease the effective normal pressure across the interface,

$$R = A_{vf} f_y \sin\alpha_f$$

and correspondingly the shear resistance associated with shear-friction, depending on whether it is compressive or tensile.

Reinforcement inclined at an angle $\alpha_f > 90°$ is ineffective in resisting interface shear, because, as relative slip between the two parts occurs and the reinforcement deforms, the effect is to separate the two parts farther rather than to introduce any clamping forces (Fig. 6.20c). Hence reinforcement with $\alpha_f \leq 90°$ (i.e., placed such that shear force produces tension in the bar) only is effective as shear-friction reinforcement.

If allowance is made for the shear strength contribution due to the cohesion between the two parts across the interface, the nominal shear resistance (for the general case of inclined shear-friction reinforcement) can be obtained as:

$$v_{rn} = c + \mu(\rho_v f_y \sin\alpha_f + N/A_g) + \rho_v f_y \cos\alpha_f \tag{6.33}$$

where c = stress due to cohesion
  N = load across shear plane (positive if compressive and negative if tensile)

6.17.2 Interface Shear Transfer by Code

The CSA A23.3(11.6) recommends that the factored interface shear resistance based on the shear-friction concept, v_r, be calculated using Eq. 6.34

$$v_r = \lambda\phi_c(c + \mu\sigma) + \phi_s \rho_v f_y \cos\alpha_f \tag{6.34}$$

where, $\sigma = \rho_v f_y \sin\alpha_f + N/A_g$ (6.35)
  A_{cv} = area of concrete section resisting shear
  A_g = gross area of section transferring N
  A_{vf} = area of shear-friction reinforcement
  c = resistance due to cohesion
  f_y = yield stress of shear-friction reinforcement

and

N = unfactored permanent compressive load perpendicular to the shear plane
v_r = V_r/A_{cv} = factored shear stress resistance
α_f = inclination of shear-friction reinforcement with shear plane
ρ_v = A_{vf}/A_{cv} = ratio of shear-friction reinforcement
μ = coefficient of friction
ϕ_c, ϕ_s = material resistance factors for concrete and steel reinforcement
λ = factor to account for low density concrete

Eq. 6.34 is identical to Eq. 6.33 except for the introduction of material resistance and density factors. The Code recommends the following values for c and μ:

Table 6.2 Values of c and μ to be used with Eq. 6.34

Case	Concrete placed against:	c (MPa)	μ
1	Hardened concrete	0.25	0.60
2	Hardened concrete, clean and intentionally roughened	0.50	1.00
3	Monolithically	1.00	1.40
4	As-rolled structural steel and anchored by headed studs or reinforcing bars	0.00	0.60

An upper limit on the first term of Eq. 6.34 is specified, equal to $0.25\phi_c f_c' \leq 7.0\ \phi_c$ MPa, to avoid failure of concrete by crushing.

Any direct tension, N_f, across the shear plane must be provided for by additional reinforcement having an area equal to $N_f/(\phi_s f_y)$. Such tensile forces may be caused by restraint of deformations due to temperature change, creep and shrinkage, etc. Although there is a beneficial effect of a permanently occurring net compressive force across the shear plane that reduces the amount of shear-friction reinforcement required, it is prudent to ignore this effect. When there is a bending moment acting on the shear plane, the flexural tensile and compressive forces balance each other, and the ultimate compressive force across the plane (which induces the frictional resistance) is equal to $A_s f_y$. Hence, the flexural reinforcement area, A_s, can be included in the area A_{vf} for computing V_r. When there is no bending moment acting on the shear plane, the shear-friction reinforcement is best distributed uniformly along the shear plane in order to minimise crack widths. When a bending moment also exists, most of the shear-friction reinforcement is placed closer to the tension face to provide the required effective depth. Since it is assumed that the shear-friction reinforcement yields at the ultimate strength, it must be anchored on both sides of the shear plane so as to develop the specified yield strength in tension.

PROBLEMS

1. A simply supported beam of 6 m span, shown in Fig. 6.21, is to carry a uniform dead load of 25 kN/m (including beam weight) and a uniform live load of 40 kN/m. The width of the supporting wall is 250 mm.

   a) Determine the adequacy of the No. 10 U-stirrups as shear reinforcement.

   b) If two of the tension reinforcement bars are terminated at 300 mm from the centre of the support, check the shear strength at the bar cut-off point.
   Use $f_c' = 30$ MPa and $f_y = 350$ MPa.

Fig. 6.21 Problem 6.1

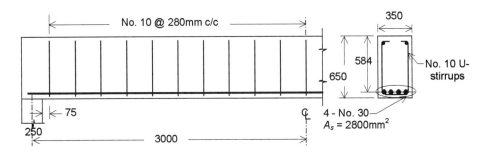

Fig. 6.22 Problem 6.2

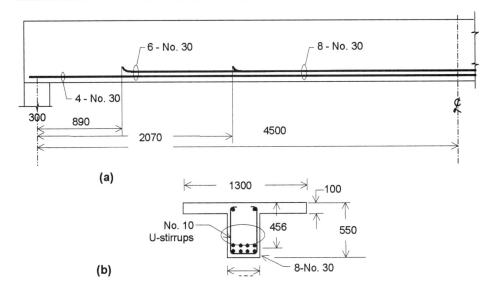

2. A simply supported T-beam of 9 m span is subjected to a dead load (including self weight) of 20 kN/m and a live load of 30 kN/m. Details of the section and bar cut-offs are shown in Fig. 6.22a and b. Design and detail the shear reinforcement using vertical stirrups. Use $f_c' = 30$ MPa and $f_y = 400$ MPa.

3. A simply supported beam with overhangs at both ends shown in Fig. 6.23 is subjected to a dead load (including self weight) of 30 kN/m and a live load of 25 kN/m. Design and detail the shear reinforcement using vertical stirrups. Use $f_c' = 30$ MPa and $f_y = 350$ MPa. The beam is symmetric about the centreline.

4. Figure 6.24 shows a uniformly loaded cantilever beam. The dead load including self-weight of the beam is 20 kN/m and the live load is 50 kN/m. Design the shear reinforcement using vertical stirrups. The bar cut-off details are as shown. Assume $f_c' = 30$ MPa and $f_y = 400$ MPa.

Fig. 6.23 Problem 6.3

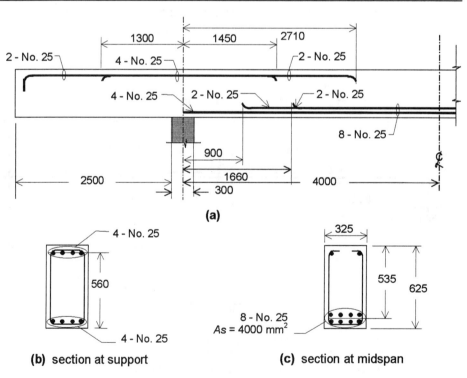

(a)

(b) section at support

(c) section at midspan

Fig. 6.24 Problem 6.4

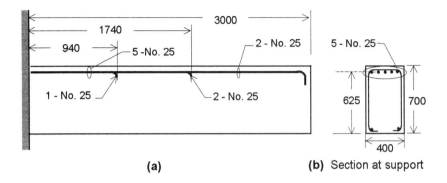

(a) (b) Section at support

REFERENCES

6.1 ACI Standard 318-95, *Building Code Requirements for Structural Concrete*, (ACI 318R-95), American Concrete Institute, Detroit, Michigan, 1995, 369 pp.

6.2 ACI-ASCE Committee 326 Report, *Shear and Diagonal Tension*, J. of ACI, Vol. 59, Jan., Feb., and Mar., 1962, pp. 1-30, 277-334 and 352-396.

6.3 ASCE-ACI Committee 426, *The Shear Strength of Reinforced Concrete Members*, ASCE Journal, Struc. Div., Vol. 99, No. ST 6, June 1973, pp. 1091-1187.

6.4 Popov, E.P., *Engineering Mechanics of Solids*, Prentice Hall Inc., Englewood Cliffs, New Jersey, 1998, 864 pp.

6.5 Bresler, B., and MacGregor, J.G., *Review of Concrete Beams Failing in Shear*, ASCE Journal, Struc. Div., Vol. 93, No. ST 1, Feb. 1967, pp. 343-372.

6.6 Explanatory Notes on CSA Standard A23.3-94, *Concrete Design Handbook*, Canadian Portland Cement Association, Ottawa, Canada, 1995, pp 1-220.

6.7 Hanson, J.M., *Tensile Strength and Diagonal Tension Resistance of Structural Lightweight Concrete*, J. of ACI, Vol. 58, No. 1, July 1961, pp. 1-37.

6.8 *Commentary on Building Code Requirements for Structural* Concrete (ACI 318R-95), American Concrete Institute, Detroit, Michigan, 1995.

6.9 Mattock, A.H., *Shear Transfer in Concrete Having Reinforcement at an Angle to the Shear Plane, Shear in Reinforced Concrete*, SP-42, American Concrete Institute, Detroit, 1974, pp. 17-42.

6.10 Mattock, A.H., Li W.K., Wang, T.C., *Shear Transfer in Lightweight Reinforced Concrete*, J. PCI, Vol. 21, No. 1, Jan.-Feb. 1976, pp. 20-39.

CHAPTER 7   Design for Torsion

7.1   GENERAL

In this chapter, the basic concepts of the behaviour of concrete members under torsion are briefly discussed. The basis and the methods for design for torsion by the Simplified Method of CSA A23.3-94(11.3) are also presented.

Torsional moments are developed in elements of reinforced concrete structures in two distinct forms as follows:

1. Statically Determinate Torsion In this case, the torsional moment (torque) is developed to maintain equilibrium, and is not dependent on the magnitude of the stiffness of the members. The torque can be determined from statics alone. Therefore the member is designed for the full torsion, as no redistribution to other members is possible. Two examples of this type of torsion are shown in Fig. 7.1. In usual structures, this type of torsion is relatively rare.

2. Statically Indeterminate Torsion Most of the torsional moments which arise in reinforced concrete structural members are of this type. Here, the torsion is induced in the member because it is restrained, and has to undergo a twist in order to maintain compatibility of deformations. The amount of torsional moment in the member depends on its torsional stiffness relative to the stiffness of the connecting members. An example of statically indeterminate torsion is the spandrel beam *ABC* in Fig. 7.2. The flexure of the beam *BD* introduces a rotation, θ, at the end *B*. Since both *ABC* and *BD* are monolithically connected at *B*, this introduces a twist equal to θ in the beam *ABC*, and hence torsion. Here, changes in relative stiffnesses due to cracking and yielding can lead to redistribution and reduction of the torsional moment. At torsional

Fig. 7.1 Statically determinate torsion

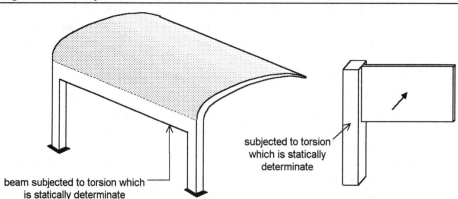

Fig. 7.2 Statically indeterminate torsion

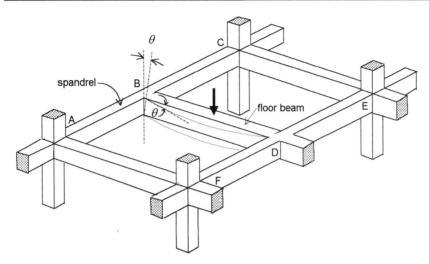

cracking, reinforced concrete members undergo a fairly large twist. Furthermore, with only moderate amounts of torsional reinforcement, the post-cracking torsional stiffness is only a small fraction of the pre-cracking value. Both these factors lead to a redistribution of forces and moments, resulting usually in a reduction of the twisting moment in the torsional member. Although heavy torsional reinforcement can increase the torsional strength, such a high strength can be attained only with very large twisting and accompanying deformations of the structure. Thus, in most practical situations, the maximum torsional moment in a reinforced concrete member under this type of torsion is likely to be the value corresponding to torsional cracking of the member.

The preceding discussion indicates that torsion is not a serious problem in usual concrete structures. Statically determinate type of torsion occurs rarely. Statically indeterminate type of torsion, when it occurs, is usually limited to the cracking torque. The latter can be predicted with reasonable accuracy, and has a relatively low value.

7.2 TORSION FORMULAS

Details of the theory of torsion (St. Venant torsion) of prismatic members having circular, noncircular, and thin-walled tubular cross-sections are given in standard tests on mechanics of materials (Refs. 7.1, 7.2). For rectangular sections under elastic behaviour, the distribution of shear stress over the cross-section is as shown in Fig. 7.3. The maximum shear stress and angle of twist are given by (Ref. 7.1):

$$v_{t,\max} = \frac{T}{\alpha x^2 y} \qquad (7.1)$$

Fig. 7.3 Torsional shear stresses in a beam of rectangular section

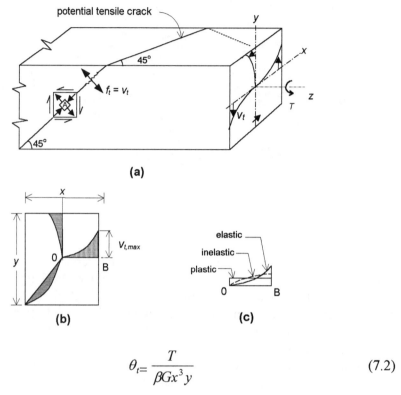

$$\theta_t = \frac{T}{\beta G x^3 y} \quad (7.2)$$

where
- G = modulus of rigidity of the material
- T = applied torsional moment
- $v_{t,max}$ = maximum shear stress, which occurs at the middle of the wider face
- x = length of shorter side of the cross section
- y = length of longer side of the section
- α, β = constants, depending on the ratio, y/x (α varies from 0.208 to 1/3, and β from 0.141 to 1/3 (Refs. 7.1, 7.2))
- θ_t = angle of twist per unit length of the member

The torsional stiffness of the member, represented by GC, is the torque per unit twist, T/θ_t, and is obtained from Eq. 7.2 as:

$$GC = G\beta x^3 y \quad (7.3)$$

where C is a property of the section having the same relationship to torsional stiffness as does the polar moment of inertia for a circular section. An approximate expression for C has been obtained as (Ref. 7.2):

$$C = (1 - 0.63 \, x/y) \, x^3 \, y/3 \tag{7.4}$$

For sections composed of rectangular elements, such as T-, L-, and E-shapes, C may be computed by summing up the individual values for each of the component rectangles, as given by Eq. 7.5. The value of C so obtained will always be less than the correct value (as shown by the membrane analogy, Ref. 7.1), so that the subdivision of the section into component rectangles may be made such as to yield the highest possible value for C.

For a section composed of several rectangular elements, x by y,

$$C = \sum (1 - 0.63 \, x/y) \, x^3 \, y/3 \tag{7.5}$$

Equation 7.5 is used in the computation of C values for torsional members in Chapter 14.

For thin-walled tubular sections of any shape, as shown in Fig. 7.4, the torsion formulas are (Ref. 7.1):

$$q = v_t t = \frac{T}{2 A_o} \tag{7.6}$$

$$v_t = \frac{T}{2 A_o t} \tag{7.7}$$

$$\theta_t = \frac{T}{G 4 A_o^2} \oint \frac{ds}{t} \tag{7.8}$$

$$GC = \frac{T}{\theta_t} = G \frac{4 A_o^2}{\oint \frac{ds}{t}}$$

where A_o = area enclosed by the centreline of the wall
q = the constant shear flow across the wall thickness
t = thickness of the wall, at element ds along the perimeter
v_t = torsional shear stress at the wall section having thickness t
and $\oint$ = indicates integration around the perimeter of the tube

For a tube of uniform wall thickness,

$$\oint \frac{ds}{t} = p_o / t$$

where p_o is the perimeter of the tube measured along the wall centreline (Fig. 7.4). With this substitution, Eq. 7.8 for torsional stiffness reduces to:

$$GC = G \, 4 \, A_o^2 \, t / p_o \tag{7.9}$$

Fig. 7.4 Torsion of thin-walled tubular section

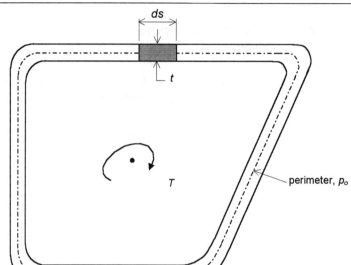

7.3 STRENGTH AND BEHAVIOUR OF CONCRETE MEMBERS IN TORSION

7.3.1 General Behaviour

For simplicity, the behaviour of concrete members subjected to pure torsion is discussed with reference to a member of rectangular cross section. A plain concrete member under pure torsion develops shear stresses in the cross section with a distribution in the early stages of loading close to that given by the elastic theory, (Fig. 7.3). The maximum shear stress occurs at the middle of the wider face. The state of pure shear develops direct tensile and compressive stresses along the diagonal directions as shown in the element at A in Fig. 7.3a. The principal tensile and compressive stress trajectories spiral around the beam in opposite directions at 45° to the beam axis. One such line, on which at all points the principal tensile stress is at right angles (and compressive stress tangential) is shown in Fig. 7.3a. This is a potential line of crack. In a concrete beam, such a crack would develop when the diagonal tensile stress reaches the tensile strength of concrete. The material in the outer portions of the cross section contributes the bulk of the torsional strength of the section. (The stresses in the outer fibres are the largest and have the greatest lever arm.) Therefore, in a plain concrete member, the diagonal torsional cracking in the outer fibres would immediately lead to a sudden failure of the entire section.

The distribution of shear stress in the cross section may not be elastic, particularly as the failure strength is reached. An inelastic stress distribution will result in somewhat larger stresses in the interior fibres, and a fully plastic behaviour will give uniform shear stress distribution over the entire cross-section (Fig. 7.3c).

A typical torque-twist relation for a plain concrete section is shown by curve a in Fig. 7.5 (Ref. 7.3). Torque and twist increase nearly linearly up to failure, which is sudden and brittle, and occurs immediately after the formation of the first torsional crack. The failure strength of a plain concrete member, T_{cr}, is generally computed by equating a theoretical nominal maximum shear stress, $v_{t,max}$ (which is a measure of the diagonal tension), to the tensile strength of concrete. Expressions for T_{cr}, have been derived based on (a) elastic theory, (b) plastic theory, (c) skew bending theory, and (d) the equivalent tube analogy (Refs. 7.4, 7.5). Each of these theories gives slightly different expressions for T_{cr}, and hence has to be correlated, experimentally, with an appropriate measure of the tensile strength of concrete to be used with it. CSA A23.3-94 requirements are based on the equivalent thin-walled tube approach (Ref. 7.5). The derivation of an expression for T_{cr} on this basis is presented later.

The torsional cracking due to the diagonal tensile stresses causes the failure of a plain concrete member in torsion. Therefore, the ideal form of reinforcing for torsion is in the form of a spiral along the direction of principal tensile stresses. However, this is often impractical, and the usual form of torsional reinforcement consists of a combination of (a) longitudinal reinforcement distributed around the cross section, and (b) closed stirrups placed perpendicular to the member axis. A combination of both types is needed to resist the inclined tensile stresses. Furthermore, these tensile stresses develop on all faces of the beam. Therefore, the stirrups must be closed. Alternatively, a closed cage of welded wire fabric with the transverse wires located perpendicular to the member axis can be used.

Members with torsional reinforcement behave similarly to the plain section until the formation of the first torsional crack, at a torque, identified as the cracking torque, T_{cr}, as shown in Fig. 7.5 (Ref. 7.6). The value of T_{cr} is insensitive to the amount of torsional reinforcement, and is very nearly the same as the failure strength of an identical plain section. When cracking occurs, there is a large increase in twist under nearly constant torque (Fig. 7.5). Beyond this, however, the strength and behaviour depend on the amount of torsional reinforcement.

For very small amounts of torsional reinforcement, the failure occurs soon after the first crack in a brittle manner. Increasing the torsional reinforcement will increase the ultimate torsional strength, but this is accompanied by larger twist. The torsional stiffness after cracking is primarily dependent on the amount of torsional reinforcement, and is usually only a small fraction (0 to 10 percent) of its value before cracking.

Fig. 7.5 *Typical torque-twist curves for concrete members in pure torsion*

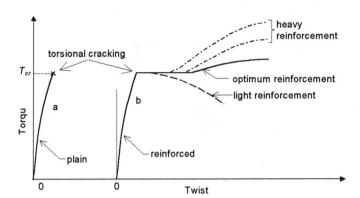

7.3.2 Strength in Torsion

(a) Cracking Torque Using Equivalent Tube Concept

As already indicated, the failure strength of a plain concrete member and the strength at torsional cracking, T_{cr}, of a reinforced member are essentially the same. The latter is also essentially independent of the amount of reinforcement. Although several methods have been used to compute T_{cr}, only the equivalent thin-walled tube approach (Ref. 7.5, 7.10), which forms the basis of the CSA A23.3-94 requirements, is briefly described below.

In this approach, an equivalent thin-walled tube having the same external dimensions as the actual section (Fig. 7.6) replaces the actual cross section. The uniform wall thickness, t_c, of the equivalent tube is taken as:

$$t_c = 0.75\, A_c/p_c \tag{7.10}$$

where p_c = the external perimeter of the actual section, and A_c = the area enclosed within perimeter p_c. For actual hollow sections, t_c must not be taken greater than the actual wall thickness.

Using Eq. 7.6 for thin-walled tubular sections, the torque, T, in the equivalent tube is:

$$T = 2 A_{oc} t_c v_t \tag{7.11}$$

where T = the torque, and A_{oc} = area enclosed within the centreline of the wall thickness, and v_t is the torsional shear stress.

Assuming that diagonal cracking occurs when the torsional shear stress, v_t, reaches $0.4\sqrt{f'_c}$, at cracking, Eq. 7.11 reduces to:

$$T_{cr} = 2 A_{oc} t_c \times 0.4\sqrt{f'_c} \tag{7.12a}$$

Fig. 7.6 *Equivalent tube, before cracking*

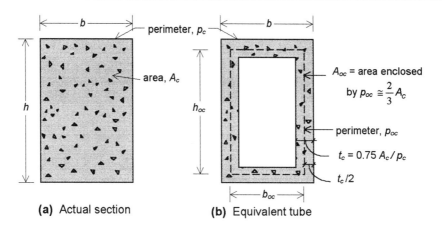

(a) Actual section (b) Equivalent tube

For design purposes, for rectangular sections of usual proportions the approximation $A_{oc} \approx 2/3\, A_c$ may be used. With this, and substituting Eq. 7.10 for t_c,

$$T_{cr} = A_c^2 / p_c \times 0.4 \sqrt{f_c'} \qquad (7.12b)$$

Now introducing the material resistance factor ϕ_c and the λ-factor to allow for low density concrete, the factored torsional cracking resistance in pure torsion is:

$$T_{cr} = (A_c^2 / p_c)\, 0.4 \lambda \phi_c \sqrt{f_c'} \qquad (7.12c)$$

CSA A23.3-94(11.2.4) recommends Eq. 7.12c for computing T_{cr}.

The Code Clause 11.2.9.1 allows torsion to be neglected if the design torque, determined by analysis using stiffnesses based on uncracked sections, does not exceed $0.25\, T_{cr}$. This corresponds to a nominal shear stress of $0.1 \lambda \phi_c \sqrt{f_c'}$. Torsion of such magnitude does not cause any significant reduction in the strength in either flexure or shear (Ref. 7.7).

The torsional stiffness before cracking, GC_{gross}, may be computed by the elastic theory (Eq. 7.4). For this, the modulus of rigidity, G, for concrete may be taken as (assuming the Poisson ratio, $\mu = 0$):

$$G = 0.5 E_c$$

where E_c is the elastic modulus of concrete.

Alternatively, the uncracked torsional stiffness may be computed using the tubular section formula (Eq. 7.9) applied to the equivalent tube in Fig. 7.6. On this basis,

$$GC_{gross} = G\frac{4A_{oc}^2 t_c}{p_{oc}} \qquad (7.13)$$

where p_{oc} is the perimeter of the tube measured along the wall centreline.

(b) Torsional Strength of Cracked, Reinforced Concrete Members

Several theories have been proposed for the computation of the torsional strength of reinforced concrete members, notable amongst them being (1) the ultimate equilibrium method, (2) the skew bending theory, and (3) the space truss analogy (Refs. 7.8, 7.9, 7.10).

CSA A23.3-94(11.3) requirements for design for torsion (ie, the Simplified Method) are based on the space truss analogy and its further developments (Refs. 7.5, 7.10, 7.11). In this approach, after torsional cracking, the longitudinal and transverse torsional reinforcement and surrounding layer of concrete in the shape of a thin-walled tube (Fig. 7.7) may be considered as a space truss. The corner longitudinal bars act as stringers, the closed stirrup legs as transverse ties, and the concrete between diagonal cracks as compression diagonals. Also, the tube wall is assumed to have a thickness t_o, as shown in Fig. 7.7. However, at higher loads the concrete cover will spall off, and the CSA A23.3-94 equations are based on the assumption that the effective outer dimensions of the tube coincide with the centreline of the exterior closed transverse torsion reinforcement (Fig. 7.7d).

Considering a transverse section of the tube, the inclined diagonal compressive force, C_d, in the concrete has a tangential component providing the shear stress, v_t, and the shear flow, $q = v_t t_o$, necessary to resist the applied torque, T. The normal component of the diagonal compression is balanced by the tensile forces in the longitudinal bars. Considering a transverse section, the face AB of the element in Fig. 7.7b, the normal component of the diagonal compression is balanced by the stirrup tension, and the tangential component provides the complementary shear v_t. In the thin-walled tube, the area enclosed by the shear flow path is A_o, and the perimeter of area A_o is p_o, as shown in Fig. 7.7c. (Compare with Fig. 7.4.)

Assuming that at failure the longitudinal and stirrup reinforcements yield, the equilibrium of the axial and tangential forces on the cross section in Fig. 7.7a. and of the forces normal to section AB in Fig. 7.7b gives:

$$A_l f_y = \frac{q}{\tan\alpha} p_o \qquad (7.14)$$

$$T = 2A_o q \qquad (7.15)$$

$$A_t f_y = s\, q \tan\alpha \qquad (7.16)$$

Fig. 7.7 The space truss model

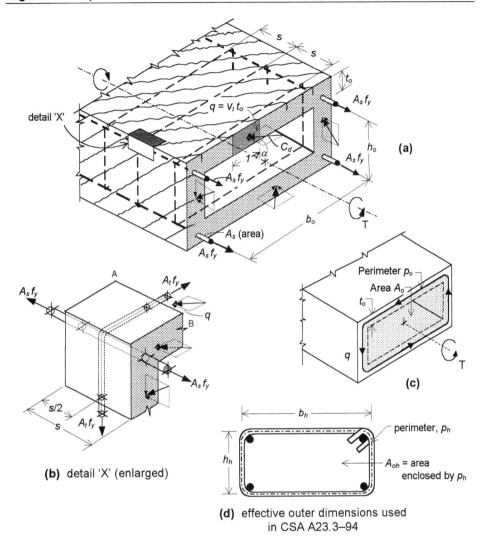

(a)

(b) detail 'X' (enlarged)

(c)

(d) effective outer dimensions used in CSA A23.3–94

and the shear stress, v_t, is

$$v_t = q/t_o = T/(2 A_o t_o)$$

where A_l = total area of symmetrically placed longitudinal reinforcement
        A_t = area of one leg of the closed stirrup
        A_o = area enclosed by the shear flow path
        p_o = the perimeter of area A_o
        q = shear flow across thickness of the tube

s = spacing of stirrups
T = torsional resistance
v_t = torsional shear stress

Eliminating q (and α) in Eqs. 7.14 and 7.16, the relation between A_l and A_R and the torsional resistance can be obtained as:

$$\frac{A_t}{s \tan \alpha} = \frac{A_l \tan \alpha}{p_o} \qquad (7.17)$$

$$T = \frac{2 A_o A_t f_y}{s \tan \alpha} \qquad (7.18a)$$

$$= 2 A_o f_y \sqrt{\frac{A_t}{s} \frac{A_l}{p_o}} \qquad (7.18b)$$

Note that the same expression for T as Eq. 7.18b can be obtained by using the thin-walled tube concept (Eq. 7.6a), with an equivalent steel tube having wall thickness, t_s, given by:

$$t_s = \sqrt{\frac{A_t}{s} \frac{A_l}{p_o}} \qquad (7.19)$$

It is generally assumed, and this indeed is the case in pure torsion, that $\alpha = 45°$. In that case, Eq. 7.17 simplifies to Eq. 7.20 giving the condition that the volume of longitudinal torsional reinforcement be equal to the volume of closed transverse stirrups, and the torsional resistance is given by Eq. 7.21:

$$\frac{A_t}{s} = \frac{A_l}{p_o} \qquad (7.20)$$

$$T = \frac{2 A_o A_t f_y}{s} \qquad (7.21)$$

Introducing the material resistance factor, ϕ_s, the factored torsional resistance, T_r, is obtained as:

$$T_t = \frac{2 A_o A_t \phi_s f_y}{s} \qquad (7.22)$$

and the corresponding total area, A_l, of symmetrically distributed longitudinal reinforcement required (from Eq. 7.20) is

$$A_l = A_t p_o / s \qquad (7.23a)$$

In adopting Eqs. 7.22 and 7.23a, CSA A23.3-M77 had taken A_o as the area enclosed by a line connecting the centres of longitudinal bars at the corners of the stirrups (i.e., $A_o = b_o \times h_o$, in Fig. 7.7a). The CEB Code (Ref. 7.18) also uses a similar definition for the area A_o.

CSA A23.3-94(11.3) specifies Eqs. 7.22 and 7.23b, below, for the design for torsion.

$$A_l = A_t\, p_h/s \tag{7.23b}$$

where p_h = perimeter of the centreline of closed transverse torsion reinforcement, and the area A_o in Eq. 7.22 may be taken as $0.85\,A_{oh}$, A_{oh} being the area enclosed by perimeter p_h.

If a member is not to fail suddenly in a brittle manner after the development of torsional cracks (Fig. 7.5), the torsional strength of the cracked section must at least equal the cracking torque T_{cr} given by Eq. 7.12c. For this condition to hold,

$$T_r = \frac{2A_o A_t \phi_s f_y}{s} \geq \frac{A_c^2\, 0.4\lambda\phi_c\sqrt{f_c'}}{p_c}$$

$$\frac{A_t}{s} = \frac{A_l}{p_o} \geq \frac{A_c^2}{A_o p_c}\frac{0.2\lambda\phi_c\sqrt{f_c'}}{\phi_s f_y}$$

The torsional stiffness of a cracked reinforced concrete member, GC_{cr}, is only a small fraction (usually less than 10 percent) of its value before cracking. Expressions for the postcracking stiffness have been derived on the basis of the space truss model (Ref. 7.12). However, a much simpler relation is obtained by using the equivalent thin-walled tube concept (Ref. 7.5). On this basis, assuming a steel tube of wall thickness, t_s, given by Eq. 7.19, and substituting in Eq. 7.9,

$$\begin{aligned}GC_{cr} &= G_s\, 4A_o^2 t_s/p_o \\ &\approx \frac{E_s}{2}\frac{4A_o^2 t_s}{p_o}\end{aligned} \tag{7.24}$$

7.4 COMBINED LOADINGS

In practice, torsion exists in reinforced concrete members mostly in combination with bending moments and transverse shear. Several forms of moment-torque interaction curves and moment-shear-torque interaction surfaces have been developed, and a summary of the early theories is presented in Ref. 7.13. For the most part, these theories are quite complex and awkward for design applications.

For design, CSA A23.3-94 (11.3) adopts the simple and conservative procedure of merely adding the reinforcement required for torsion to that required for flexural, axial, and shear loads, that act simultaneously. This procedure is simple and has been found to give satisfactory designs (Refs. 7.10, 7.14). However, the general method in Clause 11.4 takes an interactive approach for the design of longitudinal reinforcement required for shear and torsion coexisting with flexure and axial force.

7.5 CSA CODE APPROACH TO ANALYSIS AND DESIGN FOR TORSION

7.5.1 General

The Simplified Method in CSA A23.3-94(11.3) is based on the 45° space truss model. When the factored torsion, T_f, (determined by analysis using stiffnesses based on uncracked sections) exceeds 25 percent of the pure torsional cracking resistance, T_{cr}, given by Eq. 7.12c, the torsional reinforcement, consisting of closed stirrups perpendicular to the member axis and longitudinal bars distributed symmetrically around the section, is designed to resist the full torsion, T_f, using Eq. 7.22 and 7.23b. This reinforcement is provided in addition to the reinforcement required for shear, flexure, and axial loads that act in combination with the torsion. However, in the flexural compression zone, the area of longitudinal torsion steel required may be reduced by an amount equal to $M_f/(0.9\ d\ f_y)$, where M_f is the factored moment at the section acting simultaneously with the torsion (Cl.11.3.9.6). This is because the flexural compression counteracts the required tension due to torsion. The Code also requires that the cross-sectional dimensions of the member be such that,

$$\frac{V_f}{b_w d} + \frac{T_f p_h}{A_{oh}^2} \le 0.25 \phi_c f_c' \qquad (7.25)$$

This limitation is to ensure that the concrete in the web does not fail by crushing due to the diagonal compression prior to the development of full torsional resistance, T_r, with yielding of reinforcement. The maximum diagonal compression is dependent on the maximum nominal web shear stress due to shear and torsion combined, represented by the left hand side of Eq. 7.25.

The Code requires a minimum area of transverse reinforcement whenever $T_f > 0.25\ T_{cr}$, except in slabs and footings, concrete joist floors, and certain classes of relatively wide and shallow beams (Cl. 11.2.5.1). The minimum transverse reinforcement, as in the case of shear (Eq. 6.26), is given by:

$$A_v = 0.06 \sqrt{f_c'} b_w s / f_y \qquad (7.26)$$

For hollow sections, the distance measured from the centreline of the transverse torsion reinforcement to the inside face of the wall should be greater than or equal to 0.5 A_{oh}/p_h. Sections located less than a distance d from the face of supports may be designed for the same torsion, T_f, as that at a distance d, provided that the support reaction introduces compression into the end region of the member.

7.5.2 Statically Indeterminate Torsion

This type of torsion arises out of a requirement to twist, and hence its magnitude depends on the torsional stiffness of the member. The design considerations are, therefore, directly related to the torque-twist relationship of the reinforced concrete member such as shown in Fig. 7.5. (Although members may be subjected to combined loadings, the torsional stiffness characteristics are not seriously altered.) The major factors related to this curve which affect the analysis and design procedures are:

1. The uncracked stiffness, GC_{gross}, is effective prior to torsional cracking, and the maximum torque that can be reached with this stiffness is T_{cr}.
2. At cracking, a large twist occurs at nearly constant torque, resulting in redistribution of force in the structure (Refs. 7.14, 7.15, 7.16). Furthermore, torsional cracking drastically reduces the torsional stiffness in relation to the flexural stiffness. For practical ranges of twist, a previously cracked member, on subsequent loading, will develop torque far less than that at the initial loading.
3. Although increasing the torsional reinforcement will increase the torsional capacity of the section, often it may not be possible to utilise the increased strength, as it can be attained only with very large twist and associated cracking.
4. For the torsional member to behave in a ductile manner after cracking, torsional reinforcement must be provided sufficient to develop a strength equal to T_{cr}.

Recognising these factors, in situations where redistribution of internal forces can occur, the Code 11.2.9.2 permits the reduction of the maximum factored torsion, T_f, to 0.67 T_{cr}, provided corresponding adjustments are made to the bending moments, shear forces and torques in the member and in the members framing into it.

In many practical situations, the torsion on members in indeterminate concrete frames may be extremely low, and there is no possibility of torsional cracking. In such cases, the minimum torsional reinforcement requirement may be omitted. One example of such a case is the negligible torsion introduced in an interior beam, such as beam EF in Fig. 7.2, due to the bending of the column at its end. The torque, below which no torsional reinforcement need be provided, is $0.25T_{cr}$. However, in this range, the effective torsional stiffness is GC_{gross}, and must be used in the analysis for torsion.

7.5.3 Detailing of Reinforcement

In members having significant torque, the torsional reinforcement should not be less than that given in Eq. 7.26. For satisfying this requirement, the closed stirrups provided to resist transverse shear may be included. Torsional reinforcement should consist of both longitudinal and transverse (closed stirrups, closed cage of WWF or spiral) reinforcements.

The Code requirements for detailing torsion reinforcement are given in CSA A23.3-94(11.2 & 11.3) and shown in Fig. 7.8. These requirements are meant to avoid premature brittle failures and to control crack widths. The functions of the reinforcement are evident from the equilibrium conditions of the internal forces in the space truss shown in Fig. 7.7; their relations to detailing requirements are explained in Ref. 7.17.

Because stirrup legs are needed as ties on all sides of the member for torsion, they must be provided in the closed form. To ensure proper anchorage, even after the spalling of the concrete cover, the free ends of the closed stirrups must be bent into the concrete within the stirrups with 135° hooks. In regions where the concrete surrounding the anchorage is restrained against spalling, standard 90° hooks, as in Fig. 6.13c may be used. The stirrup spacing limits are as in the case of transverse reinforcement for shear (Section 6.9). The longitudinal bars are distributed around the perimeter of the stirrups, at a spacing not exceeding 300 mm (Cl.11.3.9.5). The spacing limitations are intended to control the crack widths. Inside each corner of the closed stirrup at least one longitudinal bar, having a diameter not less than $s/16$, must be provided and anchored to provide full development in tension (Cl. 11.2.7). This is necessary to distribute the resultant stirrup force at the corners where the stirrup changes its direction. These stirrup spacing and corner bar size requirements have been found to be adequate to prevent the outward buckling of the corner bar between the stirrups, due to the outward thrust exerted by the compression diagonal of the truss.

Fig. 7.8 *Details of torsion reinforcement*

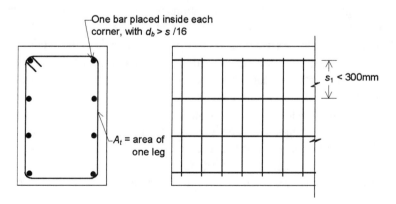

EXAMPLE 7.1

A reinforced concrete member, subjected to pure torsion, has the cross-sectional dimensions shown in Fig. 7.9. $\sqrt{f'_c}$ = 30 MPa and f_y = 400 MPa. Compute (1) the torsional cracking resistance, (2) the torsional stiffness prior to cracking, (3) the factored torsional resistance of the section, and (4) the torsional stiffness after cracking.

SOLUTION

1. Torsional cracking resistance
   Area, A_c = 300 × 500 = 150 × 10³ mm²
   Perimeter = p_c = 2 (300 + 500) = 1600 mm
   Using Eq. 7.12c:
   $$T_{cr} = \frac{A_c^2}{p_c} 0.4\lambda\phi_c \sqrt{f'_c}$$
   $$= \frac{(150 \times 10^3)^2}{1600} \times 0.4 \times 1.0 \times \sqrt{30} \times 10^{-6} = 18.5 \text{ kN} \cdot \text{m}$$

2. Precracking stiffness
   If the elastic theory is applied to the gross concrete section, the property C (Eq. 7.4) is:
   $$C = \left(1 - 0.63 \times \frac{300}{500}\right) \frac{300^3 \times 500}{3} = 2.80 \times 10^9 \text{ mm}^4$$
   $G = E_c/2 = 4500\sqrt{30}/2 = 12.3 \times 10^3$ MPa
   $GC_{gross} = 12.3 \times 10^3 \times 2.80 \times 10^9 = 34.4 \times 10^{12}$ N·mm²

   Alternatively, using the equivalent tube formula, Eq. 7.13, thickness of equivalent tube, before cracking, is,
   $$t_c = \frac{0.75 A_c}{p_c} = \frac{0.75 \times 150 \times 10^3}{1600} = 70.3 \text{ mm}$$
   b_{oc} = 300 - 70.3 = 229.7 mm, h_{oc} = 500 - 70.3 = 429.7 mm
   A_{oc} = 229.7 × 429.7 = 98.7 × 10³ mm²
   p_{oc} = 2(229.7 + 429.7) = 1319 mm
   $$GC_{gross} = G \frac{4 A_{oc}^2 t_c}{p_{oc}}$$
   $$= 12.3 \times 10^3 \times \frac{4(98.7 \times 10^3)^2 \times 70.3}{1319} = 25.5 \times 10^{12} \text{ N} \cdot \text{mm}^2$$

Fig. 7.9 Example 7.1

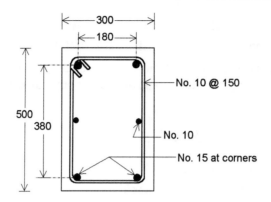

3. Factored torsional resistance, T_r
   Width between stirrup centrelines, $b_h = 180 + 16 + 11.3 = 207.3$ mm
   Depth between stirrup centrelines, $h_h = 380 + 16 + 11.3 = 407.3$ mm
   Area enclosed by stirrup centrelines, $A_{oh} = 207.3 \times 407.3 = 84.4 \times 10^3$ mm^2
   Perimeter of A_{oh} is $p_h = 2(207.3 + 407.3) = 1229$ mm
   Area enclosed by shear flow path, $A_o \approx 0.85 A_{oh} = 71.7 \times 10^3$ mm^2
   Area of transverse reinforcement, $A_t = 100$ mm^2

   Using Eq. 7.22,
   $$T_r = \frac{2A_o A_t \phi_s f_y}{s} = \frac{2 \times 71.7 \times 10^3 \times 100 \times 0.85 \times 400}{150} \times 10^{-6} = 32.5 \text{ kN} \cdot \text{m}$$

   For this to be valid, check:
   (i) longitudinal steel area required,
   $$A_l = A_t p_h / s = 100 \times 1229/150 = 819 \text{ mm}^2$$

   Area provided = $4 \times 200 + 2 \times 100 = 1000$ mm^2 OK

   (ii) minimum area of transverse reinforcement required (Eq. 7.26),
   $$A_{v,min} = \frac{0.06\sqrt{30} \times 300 \times 150}{400} = 37.0 \text{ mm}^2$$
   A_v provided = $2 \times 100 = 200$ mm^2 OK

   (iii) adequacy of section dimensions (Eq. 7.25),
   $$0.25 \phi_c f'_c = 0.25 \times 0.6 \times 30 = 4.5 \text{ MPa}$$

$$\frac{T_r p_h}{A_{oh}^2} = \frac{32.5 \times 10^6 \times 1229}{(84.4 \times 10^3)^2} = 5.61 \text{ MPa} > 4.5 \text{ MPa} \qquad \textbf{not OK}$$

The condition in Eq. 7.25 is not satisfied, and the nominal shear stress is excessive. Therefore the nominal shear stress (diagonal compression failure) in the concrete controls. In order to satisfy Eq. 7.25,

$$T_r \leq 0.25 \phi_c f_c' \left(\frac{A_{oh}^2}{p_h} \right)$$

$$\leq 4.5 \times \frac{(84.4 \times 10^3)^2}{1229} \times 10^{-6} = 26.1 \text{ kN} \cdot \text{m}$$

Factored torsional resistance, $T_r = 26.1$ kN·m

4. Stiffness after cracking
   Wall thickness, t_s, of equivalent steel tube, using Eq. 7.19, is:

   $$t_s = \sqrt{\frac{A_t}{s} \frac{A_l}{p_o}} = 0.772 \text{ mm}$$

   $$G_s \approx E_s/2 = 100 \text{ GPa}$$

   Using Eq. 7.24:

   $$GC_{cr} = G_s \frac{4 A_o^2 t_s}{p_o}$$

   $$= 100 \times 10^3 \times \frac{4(68.4 \times 10^3)^2 \times 0.772}{1120} = 1.29 \times 10^{12} \text{ N} \cdot \text{mm}^2$$

   Note: The member in this example has torsional reinforcement only slightly in excess of the minimum required to resist the cracking torque. It is seen that the post-cracking stiffness is negligible, compared to the precracking value. This member, if used to resist torsion in an indeterminate frame, will develop very little torque after it has cracked. To compare, the unit twist, θ_t, when $T = T_{cr}$ during the first time loading (uncracked stiffness) is:

   $$\theta_{t,gross} = \frac{T_{cr}}{GC_{gross}} = \frac{18.5 \times 10^6}{25.5 \times 10^{12}} = 0.725 \times 10^{-6} \text{ rad/mm}$$

   This kind of twist is in the probable range. If the member is cracked, $GC = GC_{cr}$, and in order to develop a torque equal to T_{cr}, the unit twist has to be,

   $$\theta_t = \frac{T_{cr}}{GC_{cr}} = \frac{18.5 \times 10^6}{1.29 \times 10^{12}} = 14.3 \times 10^{-6} \text{ rad/mm}$$

230 REINFORCED CONCRETE DESIGN

This is 20 times the precracking value and is unusually large. Therefore, the procedure of analysing the structure assuming $GC = 0$ is a satisfactory one and is realistic, once the torsional member is cracked. However, the member may be subjected to $T = T_{cr}$ before cracking has occurred, and the minimum reinforcement to resist $T = T_{cr}$ is essential to avoid a sudden brittle failure at the formation of the first crack. Though the failure strength for this member is 26.1 kN·m, it can seldom be realized if the torsion is of the statically indeterminate type.

EXAMPLE 7.2

Figure 7.10 shows a continuous beam supporting a monolithically cast cantilever slab. The specified live load on the slab is 2.4 kN/m². In addition, the beam supports a dead load of 12 kN/m, applied directly over the beam width; $f'_c = 30$ MPa and $f_y = 400$ MPa. Design the reinforcement for the beam at a critical section near the first interior support.

Fig. 7.10 Example 7.2

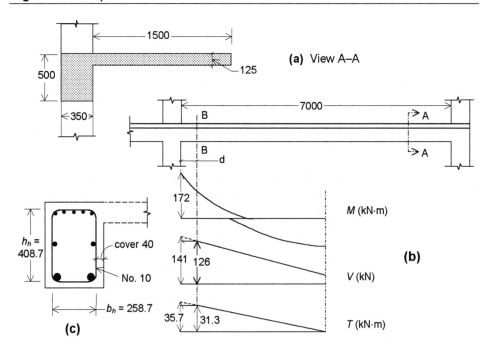

SOLUTION

1. Factored load effects (Fig. 7.10b)
   The loads on the overhanging portion of the slab per metre length of beam are:
   $DL = 1.25 \times 0.125 \times 1.50 \times 24 = 5.63$ kN/m
   $LL = 1.50 \times 1.5 \times 2.4 = 5.40$ kN/m
   Total = 11.0 kN/m
   Assuming a beam size of 350×500 mm, the torque due to the loads on slab is:

   $$T = 11.0 \left(\frac{1.5}{2} + 0.175 \right) = 10.2 \text{ kN} \cdot \text{m/m length}$$

   Dead load directly on the beam, including the beam weight, is:
   $1.25 (12 + 0.35 \times 0.5 \times 24) = 20.3$ kN/m
   Total load on beam = $11.0 + 20.3 = 31.3$ kN/m
   Maximum moment in the beam at the face of supporting column, using the CSA moment coefficients, is:

   $$M_f = \frac{w_f l_n^2}{10} = \frac{31.3 \times 7^2}{10} = 153 \text{ kN} \cdot \text{m}$$

   Maximum shear in the beam is:

   $$V_{max} = 1.15 \frac{w_f l_n}{2} = 1.15 \times \frac{31.3 \times 7}{2} = 126 \text{ kN}$$

   Maximum torque in beam, at face of support:

   $$T_{max} = 10.2 \times \frac{7}{2} = 35.7 \text{ kN} \cdot \text{m}$$

   The shear and torque vary linearly. The design shear and torque are taken at a section a distance d from face of the support. Assuming 40 mm cover, No. 10 stirrups and No. 25 bars, $d = 500 - 64 = 436$ mm. At the critical design section:
   Factored shear, $V_f = 126 - 35.0 \times 0.436 = 112$ kN
   Factored torque, $T_f = 35.7 - 10.2 \times 0.436 = 31.3$ kN·m

2. Check shear resistance and torsion resistance requirements and adequacy of overall section size
   Factored shear resistance provided by concrete, V_c, is (Eq. 6.22),

   $$V_c = 0.2 \lambda \phi_c \sqrt{f'_c} b_w d = 0.2 \times 1.0 \times 0.6 \times \sqrt{30} \times 350 \times 436 \times 10^{-3} = 100 \text{ kN}$$

   $V_f = 112$ kN $> 0.5 V_c = 50$ kN, shear reinforcement is required
   $V_s = V_f - V_c = 112 - 100 = 12$ kN
   $V_s \ll 0.8 \lambda \phi_c \sqrt{f'_c} b_w d$, and hence the size of section is adequate for shear.

232 *REINFORCED CONCRETE DESIGN*

$0.1\lambda\phi_c f_c' b_w d = 0.1 \times 1.0 \times 0.6 \times 30 \times 350 \times 436 \times 10^{-3} = 275$ kN

Since $V_f < 0.1\lambda\phi_c f_c' b_w d$, normal spacing requirements control. Pure torsional cracking resistance, T_{cr}, is obtained as (Eq. 7.12c),

$$T_{cr} = \frac{A_c^2}{p_c} 0.4\lambda\phi_c \sqrt{f_c'}$$

$A_c = 350 \times 500 = 175 \times 10^3$ mm^2, $p_c = 2(350+500) = 1700$ mm

$$T_{cr} = \frac{(175 \times 10^3)^2}{1700} \times 0.4 \times 0.6\sqrt{30} \times 10^{-6} = 23.7 \text{ kN} \cdot \text{m}$$

Since $T_f > 0.25\, T_{cr}$, torsion reinforcement is required.

The total nominal shear stress, considering V_f and T_f, must be such that Eq. 7.25 is satisfied, otherwise the section dimensions must be increased. Assuming 40 mm clear cover all around and No. 10 stirrups, the centreline dimensions of the stirrup cage (Fig. 7. 10c) are 258.7 × 408.7mm.

$A_{oh} = 408.7 \times 258.7 = 106 \times 10^3$ mm^2, $p_h = 2(408.7 + 258.7) = 1335$ mm

$$\frac{V_f}{b_w d} + \frac{T_f p_h}{A_{oh}^2} = \frac{112\,000}{350 \times 436} + \frac{31.3 \times 10^6 \times 1335}{(106 \times 10^3)^2} = 4.45 \text{ MPa}$$

$0.25\phi_c f_c' = 0.25 \times 0.6 \times 30 = 4.5$ MPa

Eq. 7.25 is satisfied and the cross-sectional dimensions are adequate.

3. Design of flexural reinforcement

$$K_r = \frac{M_f}{bd^2} = \frac{153 \times 10^6}{350 \times 436^2} = 2.30 \text{ MPa}$$

For this K_r, from Table 5.4:
$\rho = 0.0074 \ll \rho_{max} = 0.0243$
$A_{sf} = 0.0074 \times 350 \times 436 = 1129$ mm^2

4. Reinforcement for flexural shear
$V_s = V_f - V_c = 112 - 100 = 12$ kN
Using Eq. 6.23, area of vertical stirrup, A_v, is given by:

$$\frac{A_v}{s} = \frac{V_s}{\phi_s f_y d} = \frac{12000}{0.85 \times 400 \times 436} = 0.081 \text{ mm}^2/\text{mm}$$

Minimum transverse shear reinforcement is given by Eq. 6.26 as:

$$\left(\frac{A_v}{s}\right)_{min} = \frac{0.06\sqrt{f_c'}b_w}{f_y} = \frac{0.06\sqrt{30}}{400} \times 350 = 0228 \text{ mm}^2/\text{mm}$$

For satisfying the minimum A_v requirement, Code (Cl.11.2.8.5) permits inclusion of transverse torsion reinforcement. Hence, the checking of $A_{v,\min}$ will be done after designing torsion reinforcement. Choosing two-legged closed stirrups, area of stirrup required per leg, A_{v1}, is given by:

$$\frac{A_{v1}}{s} = \frac{A_v/s}{2} = \frac{0.081}{2} = 0.0405 \text{ mm}^2/\text{mm}$$

5. Torsional reinforcement
   Area of stirrup required for torsion, using Eq. 7.22 is:

   $$\frac{A_t}{s} = \frac{T_f}{2A_o \phi_s f_y}$$

   $$A_o \approx 0.85 A_{oh} = 0.85 \times 106 \times 10^3 = 90.1 \times 10^3 \text{ mm}^2/\text{mm}$$

   $$\frac{A_t}{s} = \frac{31.3 \times 10^6}{2 \times 90.1 \times 10^3 \times 0.85 \times 400} = 0.511 \text{ mm}^2/\text{mm}$$

   and total longitudinal torsional steel is (Eq. 7.23b):

   $$A_l = \frac{A_t}{s} p_h = 0.511 \times 1335 = 682 \text{ mm}^2/\text{mm}$$

6. Selection of stirrup and spacing
   Total area required for each leg, for resisting shear and torsion, is given by:

   $$\left(\frac{A_t}{s}\right)_{total} = \frac{A_{v1}}{s} + \frac{A_t}{s} = 0.0405 + 0.511 = 0.552 \text{ mm}^2/\text{mm}$$

   This is more than the minimum requirement of 0.114 mm²/mm and hence controls. Also note that the A_l/p_h required by design, = 0.511, is also greater than the $(A_v/s)_{\min}$ = 0.114 and hence adequate.
   Using No. 10 stirrups, A_v = 100 mm², and the required spacing is:

   $$s = \frac{A_v}{0.552} = \frac{100}{0.552} = 181 \text{ mm}$$

   Spacing must not exceed: $0.7d = 0.7 \times 436 = 305$ mm < 600 mm
   Select a spacing of 180 mm. The same spacing (180 mm) may be provided throughout the length of the beam even though the theoretical requirement is less.

7. Selection of longitudinal reinforcement
   Spacing of longitudinal reinforcement is limited to a maximum of 300 mm. Since h_h = 408.7 mm, longitudinal reinforcement will be required at mid-depth. Thus, A_l will be distributed in three layers. Providing two No. 10 bars

at mid-depth, one on each side (Fig. 7.10c), the balance of A_l to be provided between the top and bottom layers is 682 - 200 = 482 mm², at 482/2 = 241 mm² each.

Combined longitudinal reinforcement on the flexural tension side (top face) is:

$$A_s = A_{sf} + A_{l,top} = 1129 + 241 = 1370 \text{ mm}$$

Provide five No. 20 bars, giving A_s = 1500 mm². The longitudinal reinforcement on the flexural compression side is (bottom face here):

$$A'_s = A_{l,bottom} - \frac{M_f}{0.9 df_y} = 241 - \frac{153 \times 10^6}{0.9 \times 436 \times 400} = -734 \text{ mm}^2$$

The tension due to torsion is more than balanced by flexural compression. However, inside each corner of the stirrup, at least one longitudinal bar having a diameter not less than $s/16$ (180/16 = 11.25 mm) is a minimum requirement (Cl. 11.2.7). Hence, provide two No. 15 bars on the bottom side, as shown in Fig. 7.10c. In practice, the two longitudinal corner bars at the bottom may be provided by extending two of the positive moment bars from the midspan section into the support. The longitudinal bars required for torsion must be anchored into the support so as to develop the yield strength at the face of support.

The net longitudinal tension reinforcement, $A_s - A'_s$ = 1500 - 200 = 1300 mm², is well within the maximum allowable limit of ρ_{max} = 0.0243.

PROBLEMS

1. The critical section of a balcony beam has the cross-sectional dimensions shown in Fig. 7.11, and is subjected to the following actions:
   M_f = + 76 kN·m, T_f = 35 kN·m, V_f = 52 kN
   Check the adequacy of the section. Take f'_c = 30 MPa and f_y = 400 MPa.

2. A continuous bridge girder, curved in plan, has the following actions acting at the critical section, a distance d from the face of support: M_f = -39.5 kN·m, V_f = 53.5 kN, and T_f = 34.2 kN·m. Design the girder section and reinforcement. Take f'_c = 30 MPa, f_y = 350 MPa.

3. Figure 7.12 shows the overall dimensions of a sign post. The post is a reinforced concrete member of square cross section. The net wind load is estimated as 1.5 kN/m². Design the post, assuming:

Fig. 7.11 Problem 1

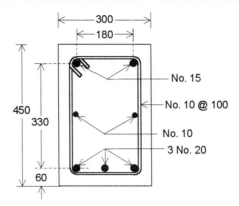

Fig. 7.12 Problem 3

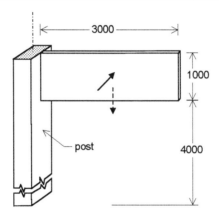

a) the effects of vertical loads are negligible compared to wind effects; and
b) the weight of the sign board is 5 kN and gravity load effects are to be included.

REFERENCES

7.1 Timoshenko, S., and Goodier, J.N., *Theory of Elasticity*, 3rd ed., McGraw-Hill, New York, 1970.

7.2 Popov, E.P., *Engineering Mechanics of Solids*, 2nd ed., Prentice-Hall, Englewood Cliffs, New Jersey, 1998, 864 pp.

7.3 Cowan, H.J., *Reinforced and Prestressed Concrete in Torsion*, Edward Arnold Ltd., London, 1965.

7.4   Hsu, T.T.C., *Torsion of Structural Concrete - Plain Concrete Rectangular Sections*, Torsion of Structural Concrete, ACI Publication SP-18, American Concrete Institute, Detroit, 1968, pp. 207-238.

7.5   Collins, M.P., *The Torque-Twist Characteristics of Reinforced Concrete Beams*, Inelasticity and Non-linearity in Structural Concrete, SM Study No. 8, University of Waterloo Press, Waterloo, 1972, pp. 211-232.

7.6   Hsu, T.T.C., *Torsion of Structural Concrete - Behaviour of Reinforced Concrete Rectangular Members*, ACI Publication SP-18, American Concrete Institute, Detroit, 1968, pp. 261-306.

7.7   ACI Committee 438, *Tentative Recommendations for the Design of Reinforced Concrete Members to Resist Torsion*, ACI Journal, Vol. 66, No. 1, Jan. 1969, pp. 1-8.

7.8   Lyalin, I.M., *The Ultimate Equilibrium Method*, Chapter 6 in Ref. 7.3, pp. 81-106.

7.9   Hsu, T.T.C., *Ultimate Torque of Reinforced Rectangular Beams*, ASCE Journal, Struct. Div., Vol. 94, No. ST2, Feb. 1968, pp. 485-510.

7.10 Lampert, P., and Collins, M.P., *Torsion, Bending, and Confusion - An Attempt to Establish the Facts*, ACI Journal, Vol. 69, Aug. 1972, pp. 500-504.

7.11 Mitchell, D., and Collins, M.P., *Diagonal Compression Field Theory - A Rational Method for Structural Concrete in Pure Torsion*, ACI Journal, Vol. 71, Aug. 1974, pp. 396-408.

7.12 Lampert, P., *Post Cracking Stiffness of Reinforced Concrete Beams in Torsion and Bending*, Analysis of Structural Systems for Torsion, ACI Publication SP-35, American Concrete Institute, Detroit, 1973, pp. 385-433.

7.13 Zia, Paul, *What Do We Know About Torsion in Concrete Members?* ASCE Journal, Struct. Div., Vol. 96, No. ST6, June 1970, pp. 1185-1199.

7.14 Collins, M.P. and Lampert, P., *Redistribution of Moments at Cracking - The Key to Simpler Torsion Design?* Analysis of Structural Systems for Torsion, ACI Publication SP-35, American Concrete Institute, Detroit, 1973, pp. 343-383.

7.15 Hsu, T.T.C., and Burton, K.T., *Design of Reinforced Concrete Spandrel Beams*, ASCE Journal, Struct. Div., Vol. 100, No. ST1, Jan. 1974, pp. 209-229.

7.16 Hsu, T.T.C., and Hwang, C.S., *Torsional Limit Design of Spandrel Beams*, ACI Journal, Vol. 74, American Concrete Institute, Detroit, Feb. 1977, pp. 71-79.

7.17 Mitchell, D. and Collins, M.P., *Detailing for Torsion*, ACI Journal, Vol. 73, American Concrete Institute, Detroit, Sept. 1976, pp. 506-511.

7.18 CEB-FIP, *Model Code for Concrete Structures* 1990, Comité Euro-International du Béton, Thomas Telford Publishing, London, 1993, 480 pp.

CHAPTER 8 Shear and Torsion Design - General Method

8.1 INTRODUCTION

As in the case of the Simplified Method dealt with in Chapter 6, the General Method for shear design also uses the truss analogy. However, unlike in the simplified method where the inclination of the diagonal cracks is taken as 45°, here, the angle of inclination, θ, of the diagonal compressive stresses to the longitudinal axis of the member is considered variable. Furthermore, the influence of diagonal tension cracking on the diagonal compressive strength of concrete, and the influence of shear on the design of longitudinal reinforcement are accounted for in a more direct manner. The General Method is based on the 'Compression Field Theory' (Ref. 8.1).

In order to identify the various parameters to be considered in analysing for shear strength, consider the compressive stress trajectories in a beam subjected to a bending moment, M, an axial force, N (considered positive if tensile), and a shear force, V, as shown in Fig. 8.1. Note that this is similar to the trajectories shown in Fig. 6.1d, except that in Fig. 8.1b the influence of axial force is also included. At any cross section such as section 1-1, the magnitude and direction of the principal compressive stresses and principal compressive strains will vary over the depth of the section. At the top face the inclination θ will be 0° and at the bottom face θ will be 90°. The shear stress distribution over the depth of section will also be non-uniform. Considering a small element such as at A at a depth y, the stresses, strains, and the corresponding Mohr's circles are as shown in Fig. 8.1h, j. In drawing these it is assumed that concrete has no tensile strength, and that the directions of principal stresses coincide with the directions of principal strains.

For a correct analysis, three parameters are required to be known/computed at each point over the depth of the section. These may be considered, for instance, as the principal strains ε_1 and ε_2 and the angle θ. The stress-strain relationship for concrete and reinforcing steel are also necessary. With these known, the principal stress, f_2, can be computed from the stress-strain relations for concrete and hence the normal stress, f_{cx}, and the tangential stress, v, at all points over the depth of the section. The longitudinal steel strain, ε_{sx}, and hence the stress in longitudinal steel, f_s, can also be computed assuming that reinforcing steel carries only axial forces. Thus, the distributions of axial and tangential stresses over the cross section can be obtained as shown in Fig. 8.1d and e. The resultant of the forces obtained as the product of these stresses and the cross-sectional area integrated over the entire depth of section gives the stress resultants M, N and V (Fig. 8.1f). The strain in the transverse direction, ε_t,

238 REINFORCED CONCRETE DESIGN

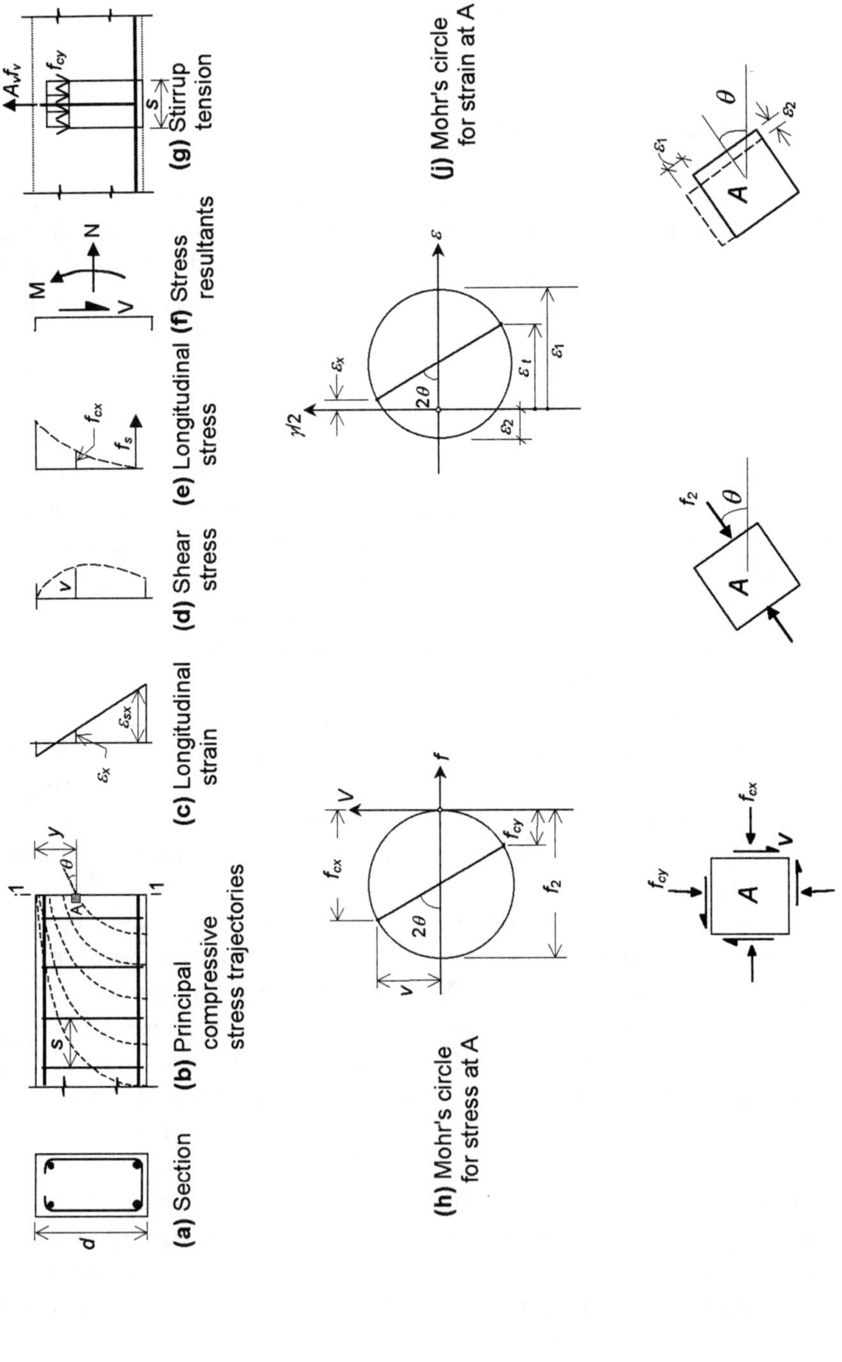

Fig. 8.1 Stress and strain under combined loads M, N, and V

determines the tensile stress, f_v, in the transverse shear reinforcement, and the tension in this reinforcement balances the transverse tensile stress in the concrete, f_{cy}, over the area tributary to it (Fig. 8.1g).

The strain distribution must be compatible with the geometry of deformation. Thus, assuming that beam deformations are such that plane sections of the beam remain plane, the distribution of ε_1, ε_2 and θ must be such that the associated longitudinal strain, ε_x varies linearly over the depth of the member, as shown in Fig. 8.1c.

An analysis problem may be the computation of the maximum shear that a given section can carry concurrently with given axial load and moment, or the computation of the axial load and moment capacity of the section in the presence of a given shear force. Because of the large number of unknowns involved, a direct solution to these problems is not possible, and a trial and error procedure has to be used. Two parameters that may be assumed initially are the shear stress distribution and the longitudinal strain distribution. This gives v and ε_x at all points. Taking trial values of a third parameter also, such as ε_1, and by successive iterations to satisfy equilibrium, compatibility, and stress-strain relations, the appropriate values of f_{cx} and v at element A can be found. Such a procedure to compute the shear strength of a given section acted on by given moment M and axial force N is presented in Ref. 8.2. However, such procedures are lengthy and tedious and seldom necessary in practice. A corresponding approximate procedure was specified in CSA A23.3-M84(11.4.1.5).

The procedure recommended in CSA A23.3-M84 neglected the contribution of tensile stresses in the cracked concrete, and hence was conservative. The revised procedure adopted in CSA A23.3-94(11.4) accounts for the contribution of tensile stresses in the concrete between cracks and is based on the 'modified compression field theory' (Ref. 8.1, 8.3). This is dealt with in Section 8.3.

8.2 STRESS-STRAIN RELATIONHIP FOR DIAGONALLY CRACKED CONCRETE

The necessity to know the stress-strain relationship for steel and concrete has already been mentioned. The state of stress in element A shown in Fig. 8.1h indicates that the maximum (principal) compressive stress in concrete, f_2, is inclined at an angle θ to the axis of the member. Associated with this stress is a compressive strain, ε_2, along f_2 and a tensile strain, ε_1, at right angles to the direction of f_2. Because of the low tensile strength of concrete (which is neglected here), tensile cracks will develop early along the direction of f_2, and the concrete in between these cracks acts as the parallel compression diagonals in the truss analogy. Therefore, the concrete carrying the diagonal compressive stress can be represented as shown in Fig. 8.2a.

Biaxially strained concrete, as in Fig. 8.2a, with compression in one direction

Fig. 8.2 *Stress-strain relationship for diagonally cracked concrete*

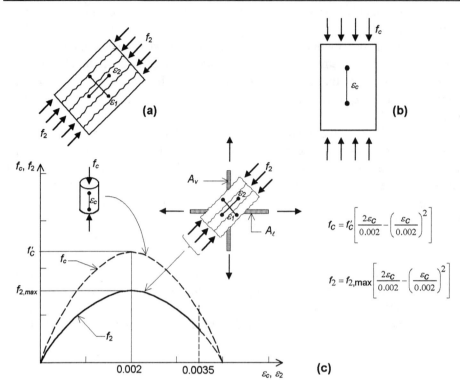

and a concurrent transverse tensile strain is weaker than concrete in uniaxial compression as in a cylinder test (Fig. 8.2b), where the lateral strain is only due to the Poisson effect. Based on results of tests on reinforced concrete panels subjected to in-plane shear and normal stresses (Ref. 8.3) the maximum compressive strength, $f_{2,max}$, of concrete in the presence of transverse tensile strain, ε_1, has been expressed by Eq. 8.1:

$$f_{2,max} = f'_c /(0.8 + 170\varepsilon_1)$$
$$\leq f'_c \tag{8.1}$$

where $f_{2,max}$ = compressive strength of concrete in presence of transverse tensile strain ε_1
f'_c = specified compressive strength of concrete
ε_1 = transverse tensile strain

The test panels forming the basis for the derivation of Eq. 8.1 were reinforced with uniform and closely spaced reinforcement along the longitudinal and transverse directions.

While Eq. 8.1 gives $f_{2,max}$, the *maximum strength* of concrete under transverse tensile strain ε_1, to compute the stress f_2 corresponding to a principal compressive strain ε_2 (concurrent with transverse tensile strain ε_1), a corresponding stress-strain relation for biaxially strained concrete (Fig. 8.2a) is also necessary. For this, it may be assumed that the general shape of this stress-strain relation remains the same as for uniaxial compression. One such parabolic relationship proposed in Ref. 8.4 is given in Eq. 8.2. Equation 8.2 is also shown in Fig. 8.2c where it is compared with the parabolic stress-strain diagram for uniaxial compression:

$$f_2 = f_{2,max}\left[\frac{2\varepsilon_2}{0.002} - \left(\frac{\varepsilon_2}{0.002}\right)^2\right] \qquad (8.2)$$

where f_2 is the compressive stress corresponding to compressive strain ε_2, in presence of transverse tensile strain ε_1, $f_{2,max}$ is given by Eq. 8.1, and the strain corresponding to the peak stress is taken as 0.002.

8.3 ANALYSIS BASED ON MODIFIED COMPRESSION FIELD THEORY

8.3.1 Assumptions and Equations

For simplicity, the case of a symmetrically reinforced beam under pure shear is considered initially. The effects of bending moment and axial force can be superimposed subsequently.

Before cracking, shear causes diagonal tensile and compressive stresses of equal magnitude, inclined at 45° to the beam axis. After diagonal cracks are formed, the tensile stress in concrete is reduced to zero at the cracks, while the concrete in-between cracks can sustain tensile stresses. Thus, the tensile stresses in concrete vary from zero at the cracks to a maximum value in between cracks, and, in deriving equilibrium equations, an *average* value, f_1, can be used. This average stress, f_1, is less than the maximum tensile stress reached prior to diagonal cracking. Furthermore, in deriving these equations, the following simplifying assumptions are made:

(a) The shear stress, v, is uniformly distributed over the web, having a width b_w and depth d_v, so that:

$$v = V/(b_w d_v) \qquad (8.3)$$

where d_v is the distance between the resultants of the tensile and compressive forces due to flexure.

242 REINFORCED CONCRETE DESIGN

(b) Under pure shear, and with symmetry, the longitudinal strain, ε_x, and the inclination, θ, of the principal compressive stress remain constant over the depth d_v.
(c) Compressive stress – strain relationship for concrete after diagonal cracking is as given be Eqs. 8.1 & 8.2.

With the above simplifying assumptions, the internal forces, stress and strain distributions and the stress resultants at a section subjected to shear only (such as at a point of contraflexure) are as shown in Fig. 8.3. The Mohr's circles for (average) stress and strain states at all points on the section are as shown in Fig. 8.3 (viii) and (ix). From the Mohr's circle of stress,

$$v = \frac{f_1 + f_2}{2} \sin 2\theta$$

or,
$$f_2 = \frac{v}{\sin\theta \cos\theta} - f_1 \tag{8.4}$$

$$= \frac{V}{b_w d_v \sin\theta \cos\theta} - f_1 \tag{8.5}$$

The force in the transverse reinforcement balances the vertical components of the concrete stresses f_1 and f_2. Considering equilibrium of stirrup forces and vertical components of f_1 and f_2 acting over the concrete area tributary to a stirrup, as shown in Fig. 8.3(x):

$$A_v f_v = b_w s (f_2 \sin^2\theta - f_1 \cos^2\theta) \tag{8.6}$$

Substituting for f_2 from Eq. 8.5,

$$\frac{A_v f_v}{b_w s} = \frac{V}{b_w d_v} \tan\theta - f_1 \tag{8.7}$$

and
$$V = \frac{A_v f_v d_v}{s} \cot\theta + f_1 b_w d_v \cot\theta \tag{8.8}$$

$$= V_s + V_c \tag{8.8a}$$

where,
$$V_s = \frac{A_v f_v d_v}{s} \cot\theta \tag{8.9}$$

and
$$V_c = f_1 b_w d_v \cot\theta \tag{8.10}$$

CHAPTER 8 SHEAR AND TORSION DESIGN

Fig. 8.3 Modified compression field theory–Analysis for shear force V

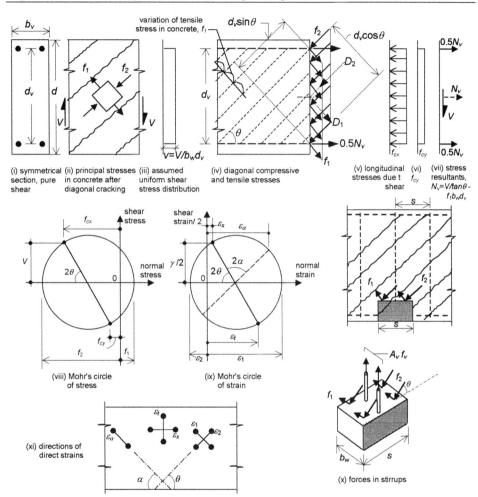

Equation 8.8 shows that the shear resistance consists of a part, V_s, contributed by the shear reinforcement, and a part, V_c, contributed by the concrete. The part V_c depends on the average tensile stress, f_1, in the diagonally cracked concrete. V_s is the same as derived earlier in Section 6.8 (Eq. 6.13a).

The stresses f_1 and f_2 over a cross section (Fig. 8.3(iv)) resulting from shear V has a net axial resultant, N_v, given by:

$$N_v = b_w\, d_v\, (f_2 \cos^2\theta - f_1 \sin^2\theta) \tag{8.11}$$

This has to be resisted by longitudinal reinforcement. Thus shear force introduces additional tensile stresses in the longitudinal reinforcements. If A_{sx} is the *total* area of such reinforcement and f_{sx} the tensile stress due to shear, $A_{sx} f_{sx} = N_v$, then substituting for f_2 from Eq. 8.5 into Eq. 8.11,

$$A_{sx} f_{sx} = N_v = V \cot\theta - f_1 b_w d_v \tag{8.12}$$

In Eqs. 8.8 and 8.12, f_1 is the *average* principal tensile stress carried by diagonally cracked concrete. Based on experimental results (Ref. 8.3), a relation between average tensile stress, f_1, and corresponding average tensile strain, ε_1, recommended in Ref. 8.1 is given by:

$$f_1 = E_c \varepsilon_1 \quad \text{for} \quad \varepsilon_1 \leq \varepsilon_{cr} \tag{8.13}$$

$$f_1 = \frac{\alpha_1 \alpha_2 f_{cr}}{1 + \sqrt{500 \varepsilon_1}} \quad \text{for} \quad \varepsilon_1 > \varepsilon_{cr} \tag{8.14}$$

where α_1 and α_2 are factors accounting for the bond characteristics of the reinforcement and the type of loading.

There are several other considerations in choosing the appropriate value for f_1. The equations given above are all based on *average stresses* and *average strains*. However, as already mentioned, because of the diagonal cracking there will be local variations, with the tensile stress in concrete reduced to zero at the crack and a corresponding local increase in the tensile stress in the transverse reinforcement, thus providing the required tensile stress component across the crack interface. The ability of the member to transmit tensile forces across the crack will control the shear strength of the member. Once the stress in the transverse reinforcement at the crack location reaches yield value, any increase in shear force can be resisted only by shear stresses, v_{ci}, transmitted along the crack interface (Fig. 8.4b,d). The magnitude of the shear stress, v_{ci} that can be transmitted between the two sides along the crack interface will depend primarily on the crack width, w, (Fig. 8.4b). The crack width, w, in turn depends on the average principal tensile strain, ε_1, and the average spacing, s_θ, of the diagonal cracks. Recommended limiting value of v_{ci} to avoid slipping along cracks is (Ref. 8.1):

$$v_{ci} = \frac{0.18\sqrt{f'_c}}{0.3 + \dfrac{24w}{a+16}} \tag{8.15}$$

where, $\quad w = \varepsilon_1 s_\theta \tag{8.16}$

and $\quad a$ = the maximum size of aggregate.

Fig. 8.4 *Transmission of forces across diagonal cracks*

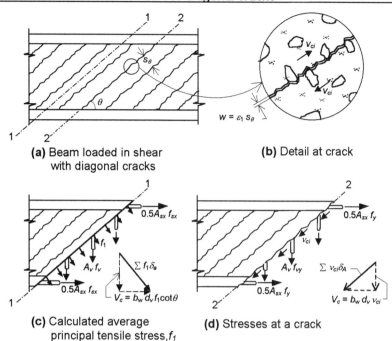

(a) Beam loaded in shear with diagonal cracks

(b) Detail at crack

(c) Calculated average principal tensile stress, f_1

(d) Stresses at a crack

The spacing of diagonal cracks, s_θ, depends on the type, amount and distribution of the longitudinal and transverse reinforcements. Expressions for estimating crack spacing are also given in Ref. 8.1. In Fig. 8.4c, the *average* tensile principal stress in concrete, f_1, is assumed to be developed midway between diagonal cracks. As one moves towards the crack, the concrete tensile stress reduces and the slack is taken up by increases in transverse reinforcement stress and/or the interface shear, v_{ci}. Yielding of transverse reinforcement at a diagonal crack and the limitation on the stress v_{ci} that can be mobilized can limit the magnitude of the effective *average* tension, f_1, that can be transmitted.

Equations 8.8 and 8.12 give the shear strength and the associated axial tensile reinforcement requirements in the case of pure shear in terms of f_1, f_2, θ and the steel stress. If three parameters, such as ε_1, f_2, and θ are known, f_1, f_2, and steel stress can be computed from the respective strains and the stress – strain relations, and the corresponding shear strength determined. Stresses f_1, f_2 and f_v must be within their respective upper limits. A trial and error procedure to predict the response of a beam loaded in shear is presented in Ref. 8.1. In this procedure, trial values are selected for ε_1, θ, and f_v and these are adjusted until equilibrium, compatibility, and limiting stress conditions are met.

In practice, shear occurs in combination with bending moments and, possibly, axial forces as well. The presence of bending moments and/or axial tension increases the axial tensile strain, ε_x, reducing the shear strength of the beam. When bending moment is present, the longitudinal strain, ε_x, and the inclination θ of the principal compressive stress vary over the depth of the section. Detailed procedure for analysis of a section subjected to combined shear and moment is complex and very laborious. Therefore recourse is made to simplifications. One such procedure is to consider the stresses and strains at just one level in the beam depth and to calculate corresponding θ, which is then considered applicable for the entire depth. Once again, a trial and error process is required for a solution. The strain profile over the depth (linear variation – plane section theory) and the value of θ are adjusted so that the stress limits are satisfied and the internal forces are in equilibrium with given bending moment, M, and axial force, N.

8.3.2 Simplified Design Procedure

Shear design involves checking the adequacy of the section selected from considerations of flexure and axial loads and computing the required transverse shear reinforcement and the additional longitudinal reinforcement to carry the applied shear. The nominal shear strength of the section can be expressed as (also Eq. 8.8 a):

$$V = V_c + V_s \tag{8.17}$$

where V_c is the part contributed be tensile stresses in the concrete and V_s is the part contributed by the transverse reinforcement. V_s is given by Eq. 8.9, in which, assuming that optimum amount of transverse reinforcement is provided so that they yield as ultimate strength is reached, f_v can be taken as the yield stress f_y, so that:

$$V_s = (A_v\, f_y\, d_v\, \cot\theta)\,/\,s \tag{8.18}$$

The part V_c is given by Eq. 8.10, which may be expressed as:
$$V_c = \beta\sqrt{f'_c}\, b_w d_v \tag{8.19}$$

where
$$\beta = f_1 \cot\theta\,/\,\sqrt{f'_c} \tag{8.20}$$

An evaluation of the shear strength (Eq. 8.17) now reduces to the determination of the appropriate values of θ and β to be used in Eqs. 8.18 and 8.19.

Substituting for f_1 from Eq. 8.14, and assuming the tensile stress at cracking, $f_{cr} = 0.33\sqrt{f'_c}$ and taking the factor $\alpha_1\alpha_2$ as equal to unity,

$$\beta = 0.33\cot\theta/(1+\sqrt{500}\,\varepsilon_1) \tag{8.21}$$

In a situation where the transverse reinforcement has yielded at failure, the shear contribution V_c of the tensile stresses f_1 has to be maintained at the diagonal cracks by the transverse component of the interface shear v_{ci} (Fig. 8.4a and b). This condition yields the equation:

$$V_c = f_1\, b_w\, d_v \cot\theta = v_{ci} b_w\, d_v$$

that is,
$$f_1 = v_{ci} \tan\theta \tag{8.22}$$

Substituting the limiting value for v_{ci} from Eq. 8.15, the corresponding limit on f_1 is:

$$f_1 = \frac{0.18\sqrt{f'_c}}{0.3 + \dfrac{24w}{a+16}}\tan\theta \tag{8.23}$$

Substituting this limiting value of f_1 in Eq. 8.20 yields:

$$\beta \le \frac{0.18}{0.3 + \dfrac{24w}{a+16}} \tag{8.24}$$

Thus the expressions for β are:

$$\beta = \frac{f_1 \cot\theta}{\sqrt{f'_c}} = \frac{0.33\cot\theta}{1+\sqrt{500\varepsilon_1}} \le \frac{0.18}{0.3 + \dfrac{24w}{a+16}} \tag{8.25}$$

where, w = $\varepsilon_1\, s_\theta$ is the crack width,
         a = maximum size of aggregate,
         s_θ = average spacing of diagonal cracks, and
         ε_1 = average principal tensile strain in concrete.

From the Mohr's circle of strain, the principal tensile strain, ε_1, may be expressed as:

$$\varepsilon_1 = \varepsilon_x + (\varepsilon_x + \varepsilon_2)\cot^2\theta \tag{8.26}$$

For diagonally cracked concrete, ε_2 is given by Eq. 8.2 as:

$$\varepsilon_2 = 0.002\left(1 - \sqrt{1 - f_2/f_{2,\max}}\right)$$

where,
$$f_{2,\max} = \sqrt{f_c'}/(0.8 + 170\varepsilon_1)$$

and f_2 may be taken conservatively (neglecting value of f_1 in Eq.8.4) as:

$$f_2 = v/(\sin\theta\cos\theta) = v(\tan\theta + \cot\theta)$$

Substituting these values in Eq. 8.26,

$$\varepsilon_1 = \varepsilon_x + \cot^2\theta\left[\varepsilon_x + 0.002\left(1 - \sqrt{1 - \frac{V}{\sqrt{f_c'}}(\tan\theta + \cot\theta)(0.8 + 170\varepsilon_1)}\right)\right] \quad (8.27)$$

If ε_x, $V/\sqrt{f_c'}$ and θ are known, ε_1 can be found and β can be computed from Eq. 8.25 (also assuming crack spacing s_θ and aggregate size a, if not known). An increase in the strain ε_x results in a decrease in the shear strength (decrease in both components, V_c and V_s, as β decreases and θ increases). The axial strain ε_x may be taken conservatively as the longitudinal strain in the flexural tension chord of the equivalent truss (Fig. 8.5). Accordingly, at a section subjected to a bending moment M and axial force N (assumed positive if tensile),

$$\varepsilon_x = \frac{0.5(N + V\cot\theta) + M/d_v}{E_s A_s} \quad (8.28)$$

The value of the parameter $V/\sqrt{f_c'}$ can be computed knowing the applied shear force V. As far as θ is concerned, a trial-and-error approach is needed. The appropriate value of θ is chosen such that:

(i) f_2 does not exceed $f_{2,\max}$,
(ii) strain in transverse reinforcement, ε_v, is at least equal to 0.002, and
(iii) the shear reinforcement is near minimum.

On the above basis, Tables/graphs have been prepared giving values of θ and β for different combinations of $V/\sqrt{f_c'}$ and ε_x. In computing the β values given in such tables, it is further assumed that, for members containing at least a minimum amount of transverse reinforcement, the average spacing of diagonal cracks is 300 mm and the aggregate size is about 19 mm.

Fig. 8.5 Longitudinal strain at flexural tension steel level

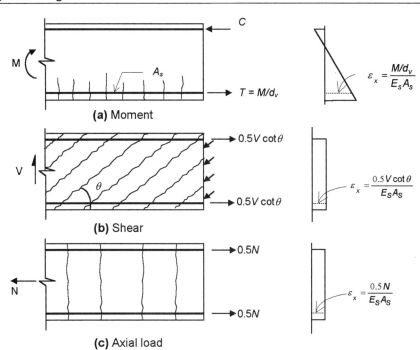

It was shown earlier that shear force also introduces tensile stresses in the longitudinal reinforcements (Fig. 8.3). The strain in the longitudinal reinforcement will have its peak value at the crack location. Considering the stress patterns given in Fig. 8.4, since the resultant horizontal force for the *average* stress conditions shown at (c) and for the conditions at a diagonal crack shown at (d) must be the same, if yielding of longitudinal bars at the crack is to be avoided,

$$A_{sx} f_y - v'_{ci} b_w d_v \cot\theta \geq A_{sx} f_{sx} + b_w d_v f_1 \tag{8.29}$$

When the transverse reinforcement yields at failure, v'_{ci} is given by Eq. 8.22. Further, for average stress conditions, $A_{sx} f_{sx}$ is given by Eq. 8.12. Substituting these values in the above equation and simplifying yields:

$$A_{sx} f_y \geq V \cot\theta + f_1 b_w d_v \cot^2\theta \tag{8.30}$$

Substituting $V_c = f_1 b_w d_v \cot\theta$ from Eq. 8.10

$$A_{sx} f_y \geq V \cot\theta + V_c \cot\theta$$

Since $V_c + V_s = V$,

$$A_{sx} f_y \geq (2V - V_s) \cot\theta$$

Considering the reinforcement on the flexural tension side only,

$$A_s f_y \geq (V - 0.5V_s) \cot\theta$$

Now considering stresses due to applied bending moment, M, and axial tension, N, to avoid yielding of the longitudinal reinforcement on the flexural tension side,

$$A_s f_y \geq \frac{M}{d_v} + 0.5N + (V - 0.5V_s) \tag{8.31}$$

For members without transverse reinforcements, the diagonal cracks will be more widely spaced and considerably greater than 300 mm assumed in the above case. For such cases also, tables have been prepared listing θ and β values for various combinations of longitudinal strain ε_x and a crack spacing parameter. In both cases, an over estimation of ε_x will give more conservative predictions of the shear strength.

This method attempts to arrive at more rational solutions by considering such aspects as influence of cracking on compressive strength, accounting for the tensile strength of cracked concrete, variable angle of inclination of principal stresses, influence of shear on stresses in longitudinal reinforcements, strain compatibility, etc. At the same time, to make the procedure tractable and suitable for a code format, a series of simplifying assumptions are made. These include neglect of the redistribution of shear stress, consideration of the stresses and strains at only one level in the cross section and applying the results to the entire section, taking the longitudinal strain at the flexural tension steel level, assumption of a constant crack spacing of 300 mm for all beams with shear reinforcement, etc. Despite the many such simplifying assumptions, a closed form solution is not possible and a trial-and-error approach involving complex equations, tables and charts are used. Even so, modification factors have to be applied in some cases, as with the constant 1.3 in Eq. 8.33 below. Unfortunately the objective for rigor is somewhat compromised by the need to introduce so many assumptions.

8.3.3 CSA Code Provisions for Shear Design by the General Method

The general method for shear design in CSA A23.3–94(11.4) follows the simplified procedure described in Section 8.3.2 above. The various load and resistance factors are also incorporated. The controlling design equation is:

$$V_{rg} = V_{cg} + V_{sg} \geq V_f \tag{8.32}$$

Where, V_{rg} is the factored shear resistance,
V_{cg} is the factored shear resistance attributed to concrete,
V_{sg} is the factored shear resistance provided by the shear reinforcement,
V_f is the factored shear force at the section, and

$$V_{cg} = 1.3\lambda\phi_c\beta\sqrt{f_c'}b_w d_v \tag{8.33}$$

for stirrups perpendicular to beam axis:

$$V_{sg} = \frac{A_v\phi_s f_y d_v \cot\theta}{s} \tag{8.34}$$

For transverse reinforcement inclined at an angle α to the longitudinal axis,

$$V_{sg} = \frac{A_v\phi_s f_y d_v (\cot\theta + \cot\alpha)\sin\alpha}{s} \tag{8.35}$$

The factor 1.3 in Eq. 8.33 compensates for the low value of ϕ_c and partially offsets the conservatism of the General Method. The expressions for V_{sg} are the same as those derived on the basis of the truss model in Section 6.8 (Eqs. 6.13 and 6.13a). To ensure that the transverse reinforcement will yield prior to the crushing of the concrete in the web in diagonal compression, V_{rg} is limited to:

$$V_{rg} \leq 0.25\phi_c f_c' b_w d_v \tag{8.36}$$

Tables and graphs are presented in the Code for determining values of β and θ for sections with and without the minimum amount of transverse reinforcements. For sections with transverse reinforcement, the table is in terms of parameters $v_f/(\lambda\phi_c\sqrt{f_c'})$, where v_f is the factored shear stress, and longitudinal strain at the tension steel level, ε_x. For evaluating these parameters,

$$v_f = V_v/(b_w d_v) \tag{8.37}$$

and

$$\varepsilon_x = [0.5(N_f + V_f \cot\theta) + M_f/d_v]/(E_s A_s) \tag{8.38}$$

$$\leq 0.002$$

For sections without the minimum transverse reinforcement, the parameters to be used are the crack spacing parameter, s_z, determination of which is as per Code Cl.11.4.7, and ε_x.

Longitudinal reinforcement is to be designed for the combined effects of flexure, axial load and shear. Accordingly, at all sections,

$$A_s f_y \geq M_f / d_v + 0.5 N_f + (V_f - 0.5 V_{sg}) \cot \theta \tag{8.39}$$

In the case of members not subjected to significant axial tension, the requirement of Eq. 8.39 may be satisfied by extending the flexural tension reinforcement a distance of $d_v \cot\theta$ beyond the location needed for flexure alone (compare with Fig. 6.9a). Similarly computation similar to development of Eq. 6.27 (Fig. 6.9b) with diagonal crack at angle θ will show that at exterior direct bearing supports, the bottom longitudinal reinforcement should be capable of resisting a tensile force T at the inside edge of the bearing area, given by:

$$T = (V_f - 0.5 V_{sg}) \cot \theta + 0.5 N_f \tag{8.40}$$

Shear failure by yielding of transverse reinforcements involves the reinforcements over a length of about $d_v \cot\theta$, as can be seen in Fig. 8.6. Accordingly, the Code (Cl. 11.4.8) permits the provision of transverse reinforcement over such lengths based on the average requirement for this length.

Fig. 8.6 Design for average shear over length $d_v \cot\theta$

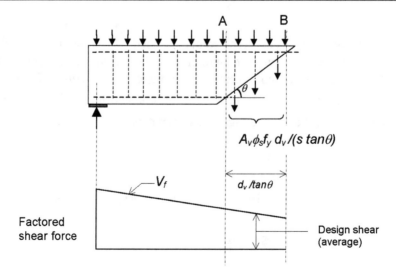

8.4 COMBINED SHEAR AND TORSION

The shear stress due to torsion and the shear stress due to transverse shear are additive on one side of the cross section and they counteract on the opposite side. The transverse reinforcement is designed considering the side where the stresses are additive. The Code requires that the transverse reinforcement provided shall be at least equal to the sum of that required for the shear and the coexisting torsion. While the equations presented in Section 8.3 above are used to compute the area of transverse reinforcement required for shear, equations derived on the basis of the *space truss analogy,* with the spiral cracks making an angle θ with the longitudinal axis, and assuming that cracked concrete carries no tension, are used for the computation of the transverse reinforcement required for torsion (Section 7.3.2 b and Fig. 7.7 for relevant derivations). On this basis, the factored torsional resistance of the section, T_{rg}, is given by (see Eqs. 7.15 & 7.16):

$$T = 2A_o \frac{\phi_s A_t f_y}{s} \cot\theta \qquad (8.41)$$

Here A_t is the area of one leg of closed transverse torsion reinforcement. In this equation A_o, the area enclosed by shear flow path, is to be taken as $0.85 A_{oh}$, where A_{oh} is the area enclosed by centerline of exterior closed transverse torsion reinforcement; and θ is to be determined from tables and graphs given in the Code, as explained in Section 8.3.3. To determine θ, the factored shear stress, v_f, and the longitudinal strain, ε_x, are required.

For thin walled tubular sections, the torsional shear is uniform over the thickness as explained in Chapter 7. Hence for box type sections, the factored shear stress due to combined shear, V_f, and torsion, T_f, is given by:

$$v_f = \frac{V_f}{b_w d_v} + \frac{T_f p_h}{A_{oh}^2} \qquad (8.42)$$

For other cross sectional shapes, such as a rectangle, torsional shear stress at first (diagonal) cracking varies over the section from zero to a maximum, with the possibility for considerable redistribution. To allow for this, the factored shear stress is taken as the square root of the sum of the squares of the shear stresses calculated individually for shear and torsion. Accordingly,

$$v = \sqrt{\left(\frac{V_f}{b_w d_v}\right)^2 + \left(\frac{T_f p_h}{A_{oh}^2}\right)^2} \qquad (8.43)$$

where p_h is the perimeter of the centerline of the closed transverse torsion reinforcement. The longitudinal strain, ε_x, may be taken as 0.002, or alternatively computed from Eq. 8.44 below, which is also based on a root-mean-square approach.

$$\varepsilon_x = \frac{0.5N_f + 0.5\cot\theta\sqrt{V_f^2 + \left(\frac{0.9p_h}{2A_0}\right)^2} + \frac{M_f}{d_v}}{E_s A_s} \geq 0 \qquad (8.44)$$

Similarly, allowing for torsion also, the longitudinal reinforcement is to be proportioned such that:

$$A_s \phi_s f_y \geq \frac{M_f}{d_f} + 0.5N_f + \cot\theta\sqrt{(V_f - 0.5V_{sg})^2 + \left(\frac{0.45p_h T_f}{2A_0}\right)^2} \qquad (8.45)$$

8.5 DESIGN USING STRUT-AND-TIE MODEL

Reinforced concrete members or portions of them can be analyzed, designed and detailed by idealizing them as an appropriate truss consisting of interconnected reinforcing steel tensile ties and concrete compressive struts, as shown in Fig. 8.7. It is a basic concept in structural design that, for transferring a system of loads to the supports, any stable skeletal framework, such as a truss, grid, or arch, compatible with the actual deformation pattern, may be delineated and the members and their joints designed for the resulting forces thereon. Designers have always used this concept for design in special situations, such as openings in webs of beams, corbels, end blocks in prestressed concrete beams, etc. The skeleton (or truss) may be implicit and embedded within a member, as in the case of the truss analogy for shear design of concrete beams and the truss analogy for plate girder design, or it may be explicit and externally visible as in a real truss. For a given structure and loading, a number of different strut-and-tie models may be feasible. Thus for a deep beam, for instance, it is possible to conceive a truss, a tied arch, a strutted catenary, and several other combinations (Fig 8.8) as the skeletal load transfer forms within the member and design accordingly. The optimum design is the one which is economic, has stability, adequate reserve strength and ductility, and meets the serviceability conditions satisfactorily. Usually, the model involving the most direct load path to supports and relatively large angles between the strut and tie at nodes will be more efficient.

Truss models for load transfer schemes for the cases shown in Fig. 8.7a are shown in Fig. 8.7b. The forces in the truss members can be computed from statics. The

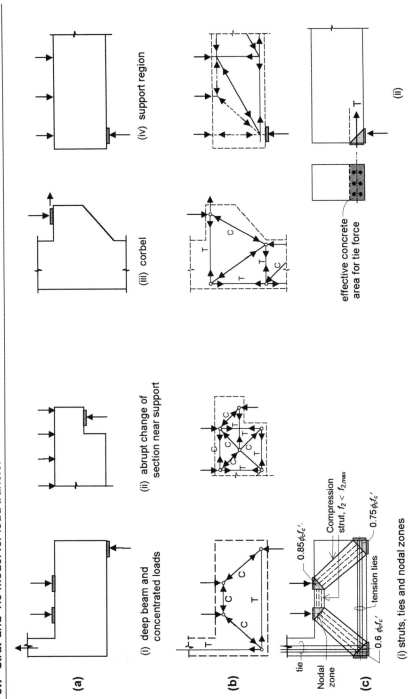

Fig. 8.7 Strut-and-Tie model for load transfer

Fig. 8.8 Alternative load transfer schemes

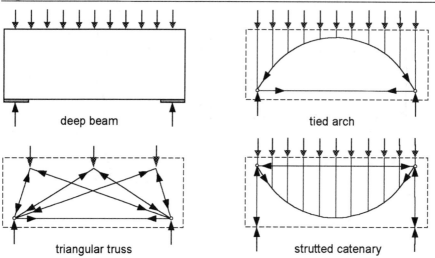

truss members and their joints (node regions) must have adequate strength to carry these forces. In the case of a real truss, the identification of the member areas and their design is fairly straightforward. However, in the case of an implicit truss embedded in concrete such as the one shown in Fig. 8.7c, the evaluation of appropriate member cross sectional areas is not so simple, especially for the compressive struts. The following general guidelines may be helpful here:

1. The stress distribution in the cross section of a truss member may be assumed to be uniform, so that the member force will act along the member centreline.
2. All the member forces, loads, and reactions meeting at a node must form a system of concurrent forces.
3. The node region is bounded by sections of the members meeting at the node and the load or reaction bearing area, if applicable.
4. The member section areas and the node dimensions should be adequate to carry the loads without exceeding the stress limits applicable.

CSA A23.3-94(11.1.1) permits the design of flexural members (and regions) for shear and torsion using the strut-and-tie model. Also, regions of members where the assumption "plane sections remain plane" is not applicable have to be proportioned for shear and torsion using the strut-and-tie model (Cl. 11.5). Such regions include areas of static or geometric discontinuities, deep beams and corbels, (Fig. 8.7).

As the tension reinforcement forms the effective tension tie, its area, A_{st}, is obtained as the tie force divided by $\phi_s f_y$. This reinforcement has to be so distributed as to give adequate dimensions for the nodes and joining compressive struts to carry their respective forces satisfactorily (Fig 8.9). This tension tie reinforcement must be

Fig. 8.9 Compressive strut dimensions

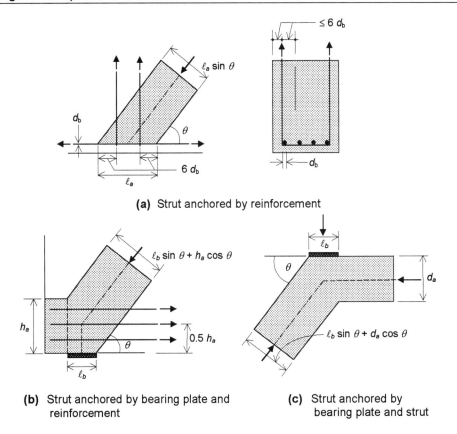

(a) Strut anchored by reinforcement

(b) Strut anchored by bearing plate and reinforcement

(c) Strut anchored by bearing plate and strut

anchored by appropriate embedment lengths, hooks, or mechanical devices so that it is capable of developing the required stress at the inner edge of the node region (Chapter 9). For straight bars, if the extension beyond the inner edge of the node region, x, is less than the development length, l_d, of the bar, the bar stress has to be limited to $f_y(x/l_d)$.

The cross sectional area of the compressive strut has to be computed based on the guidelines given above and also considering the concrete area available, as well as the anchorage conditions at the end of the strut. A few typical cases are depicted in Fig. 8.9. The compressive strut (Fig. 8.10) is likely to have tension cracks developed parallel to its axis. The allowable compressive stress in the strut has to take this into account (Section 8.2). If a tension tie is crossing a compression strut and has a strain $\varepsilon_s = f_s/E_s$, along the tie, the principal strains in the concrete at this location, ε_1 and ε_2, must be compatible with ε_s. On this basis, and further assuming that the maximum principal compressive strain, ε_2, in the direction of the strut equals 0.002, the Code

Fig. 8.10 Strain conditions in concrete strut

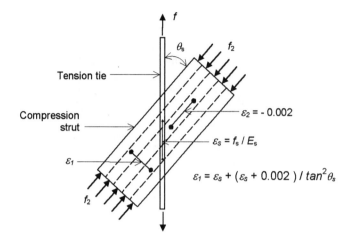

recommends the following equations for the limiting compressive stress in concrete strut, f_{cu}.

$$f_{cu} = \frac{f_c'}{0.8 + 170\varepsilon_1} \leq 0.85 f_c' \qquad (8.46)$$

where,

$$\varepsilon_1 = \varepsilon_s + (\varepsilon_s + 0.002)\cot^2\theta_s \qquad (8.47)$$

and θ_s is the smallest angle between the compressive strut and the adjoining tensile tie and ε_s is the tensile strain in this tie.

If the compressive strut is reinforced for compression with bars having an area A_{ss}, placed parallel to the strut axis and detailed so as to develop its yield strength, f_y, the limiting strength of the strut is given by :

$$A_c \phi_c f_{cu} + A_{ss} \phi_s f_y \qquad (8.48)$$

The nodal zones are the regions of concrete subjected to multidirectional compression, where the struts and ties of the truss meet. The allowable compressive stress in these regions are dependant on the degree of confinement and the adverse effects of tensile straining caused by anchoring of tension ties in this region, if any. The effective area of the node zone can be increased and the stresses in the region reduced by increasing the size of the bearing plates, by increasing the section dimensions of the compressive struts, and by increasing the effective anchorage area of tension ties. Unless special confining reinforcement is provided, the compressive stresses in the concrete in the node regions are limited to the maximum values given below (CSA A23.3-94(11.5.4)):
(a) $0.85 \phi_c f_c'$ in node regions bounded by compressive struts and bearing areas;

(b) $0.75\phi_c f_c'$ in node regions anchoring a tension tie in only one direction; and
(c) $0.65\phi_c f_c'$ in node regions anchoring tension ties in more than one direction.

Examples of zones to which each of these limits apply are also identified in Fig. 8.7c. In most situations, since compressive stress in the compression strut is limited to a maximum of $f_{2,max}$ (Eq. 8.46), the compressive stress on the face of the nodal region bearing against a compression strut will be within safe limits. The stress limits in the node region may be considered satisfied if:
1. The bearing stress due to concentrated loads or reactions does not exceed the limits given above, and
2. The tie reinforcement is uniformly distributed over an effective area of concrete at least equal to the tie force divided by the stress limits given above.

The strut-and-tie model discussed so far leaves large areas of concrete in the member, outside the truss, unreinforced. In order to control crack widths and to impart some ductility to the member, the Code also requires provision of an orthogonal grid of reinforcing bars near each face. Such reinforcement should be not less than 0.002 times the gross concrete area in each direction, with a maximum spacing of 300 mm.

EXAMPLE 8.1

Design the shear reinforcement for the T – beam in Example 6.2 using the General Method.

SOLUTION

1. Shear force diagram
   The factored shear force envelope (Example 6.2) is shown in Fig. 8.11b.

2. Check adequacy of the section
   Near the support section, where the shear is maximum, $d = 434$ mm, $b_w = 300$ mm,
   $d_v = 0.9\ d = 391$ mm

   Admissible maximum V_{rg} is given by Eq. 8.36
   $V_{rg,max} = 0.25\ \phi_c f_c' b_w d_v = 0.25 \times 0.6 \times 20 \times 300 \times 391 \times 10^{-3} = 351.9$ kN
   The maximum shear at the critical section near the support, distant d_v from the face of the support is

260 REINFORCED CONCRETE DESIGN

Fig. 8.11 Example 8.1

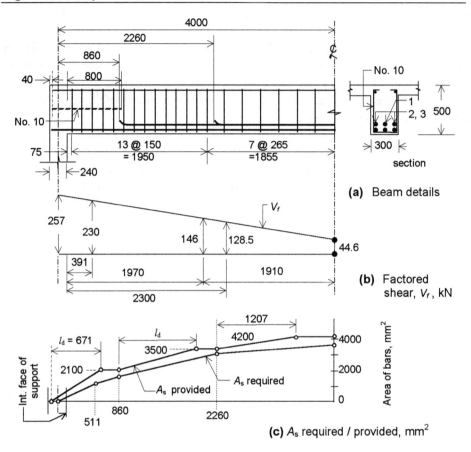

(a) Beam details

(b) Factored shear, V_f, kN

(c) A_s required / provided, mm²

$$V_f = 44.6 + (257 - 44.6)(4 - 0.12 - 0.391)/4 = 230 \text{ kN}$$
$$V_f < V_{rg,max} \qquad \text{OK}$$

3. Different zones for shear reinforcement and spacing
   (a) Where $V_f \leq 0.5 V_c$, no stirrups are required.
   For this, V_c may be taken as (Cl. 11.2.8.2)

$$V_c = 0.2 \lambda \phi_c \sqrt{f'_c} b_w d_v = 0.2 \times 1.0 \times 0.6 \times \sqrt{20} \times 300 \times 434 \times 10^{-3} = 69.87 \text{ kN}$$

Since the minimum $V_f = 44.6$ kN is greater than $0.5 V_c$ calculated above, shear reinforcement is required throughout the length of the beam.

(b) Spacing limitations
The normal spacing requirements (Cl. 11.2.11(a) - 600mm and 0.7 d) applies where
$$V_f < 0.1\, \lambda\, \phi_c f_c'\, b_w\, d_v = 0.1 \times 1.0 \times 0.6 \times 20 \times 300 \times 406 \times 10^{-3} = 146 \text{ kN}$$

The location where $V_f = 146$ kN is given by
$(146 - 44.6) \times 4000 /(257 - 44.6) = 1910$ mm from midspan, or a distance of 1970 mm from face of support.
Where $V_f > 146$ kN, spacing limits are 300 mm and $0.35d$

4. Design of stirrups
(a) Critical section at $d_v = 391$ mm from face of support, where $V_f = 230$ kN
The parameters θ and β are to be determined from Table 11.1 of Code. For this the factored shear stress is
$$v_f = V_f/(b_w\, d_v) = 230 \times 10 /(300 \times 391) = 1.96 \text{ MPa}$$
Factored shear stress ratio, $v_f /(\lambda\, \phi_c f_c') = 1.96 / (1.0 \times 0.6 \times 20) = 0.163$

Longitudinal strain is (Eq. 8.38), $\varepsilon_x = (0.5\, V_f \cot\theta + M_f/d_v) / (E_s A_s)$, where M_f is the bending moment at the critical section at 391 mm from the support, corresponding to the load causing maximum shear at this section. However, for convenience and to be conservative, this moment is taken here as the moment with full load on the entire span. Accordingly,
$M_f = 64.2 \times (4 \times 0.511 - 0.511^2/2) = 122.8$ kN·m
$\varepsilon_x = (0.5 \times 230 \times 10^3 \cot\theta + 122.8 \times 10^6 / 391) / (200\,000 \times 2100)$
    $= (0.274 \cot\theta + 0.748) \times 10^{-3}$

Since θ is not yet known, a trial and error procedure is needed. It is conservative to overestimate ε_x. For $v_f/(\lambda\, \phi_c f_c') \leq 0.200$ and $\varepsilon_x \leq 0.001$, Table 11.1 of the Code gives $\theta = 34.5°$. With this value of θ, ε_x is calculated as:

$$\varepsilon_x = (0.274 \cot 34.5° + 0.748) \times 10^{-3} = 1.147 \times 10^{-3}$$

which is greater than the value 0.001 assumed. Choosing from the table the value for $\varepsilon_x \leq 0.0015$, $\theta = 35°$ and corresponding $\varepsilon_x = 0.001139 < 0.0015$ assumed and hence OK. For this, from the Code, Table 11.1, $\beta = 0.100$.

The factored shear resistance contributed be concrete is (Eq. 8.33)
$$V_{cg} = 1.3 \times 1.0 \times 0.6 \times 0.100 \times \sqrt{20} \times 300 \times 391 \times 10^{-3} = 40.9 \text{ kN}$$

The factored shear resistance to be provided by stirrups is
$V_{sg} = V_f - V_{cg}$ = 230 – 40.9 = 189.1 kN

Assuming No. 10 U stirrups placed perpendicular to beam axis, the required spacing is given by (Eq.8.34)
$s = (\phi_s A_v f_y d_v \cot\theta)/V_{sg}$
$= (0.85 \times 200 \times 400 \times 391 \cot 35°) / 189\,100 = 201$ mm

As V_f here is > 146 kN, the limiting spacing is given by
300 mm or 0.35 d = 0.35 × 434 = 152 mm

Hence, the limiting spacing controls, and a spacing of 150 mm is selected. The shear force is 146 kN at a distance of 1970 mm from the face of support, and the limiting spacing is applicable upto this location. Therefore, provide the first stirrup at a distance of 75 mm from the face of the support, followed by 13 more stirrups at 150 mm, covering a total length of 2025 mm from the face of the support.

(c) Section at 2.3 m from face of support
The Code permits the design of stirrups for the average shear over a length of $d_v \cot\theta$. Here, $d_v \cot\theta$ is in the range of 391cot35° = 558 mm (although both d_v and θ will vary slightly along the span). In practice, designing for every discrete lengths of $d_v \cot\theta$ is not warranted. In this example, the next section for design is taken at a distance of 2 m from face of support (which is approximately 558/2 mm from the location of the last stirrup designed.

The shear at 2.3 m from face of support is
V_f = 44.6 + (257 – 44.6) (4 –2.42) / 4 = 128.5 kN.

The bending moment corresponding to this shear is,
M_f = 19.6 × 4 × 2.42 – 19.6 × 2.42² / 2 + 44.6 × 5.58² × 2.42 /(8 × 2)
= 342.4 kN·m.

At 2.3 m from face of support, there are 5 No. 30 bars and effective depth is different from at support. Here, conservatively, the effective depth at midspan, equal to 406 mm will be used for this section also. Corresponding d_v = 0.9 × 406 = 365 mm.

Factored shear stress ratio,
$v_f/(\lambda \phi_c f_c')$ = 128.5 × 10³ /(300 × 365 × 1.0 × 0.6 × 20) = 0.098

Longitudinal strain is
$$\varepsilon_x = (0.5 \times 128.5 \times 10^3 \cot\theta + 342.4 \times 10^6 /365)/ (200\,000 \times 3500)$$
$$= (0.092 \cot\theta + 1.340) \times 10^3$$
For $v_f/(\lambda\,\phi_c f_c') < 0.100$ and $\varepsilon_x < 0.0015$, from Table 11.1 of Code, $\theta = 38°$. Corresponding to this, $\varepsilon_x = 1.458 \times 10^{-3}$ which is less than the 0.0015 assumed and hence OK. Corresponding $\beta = 0.143$.

Shear strength due to concrete,
$$V_{cg} = 1.3 \times 1.0 \times 0.6 \times 0.143 \times \sqrt{20} \times 300 \times 365 \times 10^{-3} = 54.62 \text{ kN}$$
Shear due to stirrups is $128.5 - 54.62 = 73.9$ kN.

Spacing of stirrups is,
$$s = (0.85 \times 200 \times 400 \times 365 \cot 38°)/ 73900 = 430 \text{ mm}$$
The maximum spacing in this region is given by
600 mm or $0.7\,d = 0.7 \times 406 = 284$ mm.

Hence, the limiting spacing controls for the remaining portions of the beam. Here, from the last stirrup already provided earlier, 7 more stirrups may be provided at a uniform spacing of 265 mm, which results in the last stirrup being placed at mid-span. The arrangement of stirrups is shown in Fig. 8.11a.

5. Check adequacy of longitudinal reinforcements
   Out of the 6 – No. 30 bars at mid-span, one is terminated at a distance of 2260 mm and two more at a distance of 860 mm from the center of support, respectively. Allowing for the effects of shear, the required factored resistance of tension reinforcement is given by

   $$N_s = M_f / d_v + (V_f - 0.5\,V_{sg})\cot\theta$$

   or, with the stress in steel fully developed to f_y, the required area of steel is given by $A_s = N_s / f_y$. The stress f_y will be developed in a bar at a distance equal to the development length, l_d, from the free end.

The loading conditions for the maximum moment and the maximum shear force at a section are different, and the combination to be checked is the maximum M_f and concurrently occurring V_f and vice versa. However, in this example, the maximum moment M_f and the shear V_f determined from the shear envelope in Fig. 8.11b will be taken together for the checking. The section at a distance d_v from the face of the support, the sections were bars are terminated and the mid-span section will be investigated. The calculations are shown in Table 8.1 below.

Distance of section from upport $^C/_L$ (m)	M_f (kN·m)	V_f (kN)	d_v (mm)	V_{sg} (kN)	θ (degrees)	Required A_s (mm²)
0.511	122.8	230	391	189.1	35	1269
0.860	197.1	211	391	168.9	37	1680
2.260	416.4	137	371	90.4	39	3089
4.000	513.6	44.6	365.4	Nil	43	3633

The terminated bars will be fully effective only at a distance l_d from the free end. For No. 30 bottom bars in regions containing minimum stirrups, from Table 12.1 of Code,

$l_d = 0.45 k_1 k_2 k_3 k_4 f_y d_b / \sqrt{f_c'}$

$= 0.45 \times 1.0 \times 1.0 \times 1.0 \times 1.0 \times 400 \times 30 / \sqrt{20} = 1207$ mm

The required A_s and the actual area provided are presented graphically in Fig. 8.11c. The longitudinal reinforcement provided are OK.

6. Check adequacy of reinforcement area at exterior support (Code Cl. 11.4.9.4)
Tensile force to be resisted at the inside edge of bearing area is (Eq. 8.40)
$T = (V_f - 0.5 V_{sg}) \cot\theta = (230 - 0.5 \times 189.1) \cot 35° = 193.4$ kN

If the bar is provided straight and the cover at the edge is 40 mm, the available development length up to the inside edge of the bearing area is = 240 − 40 = 200 mm. The stress that can be developed at the inside edge is
$f_s = (200 / 1207) \times 400 = 66.3$ MPa.

Hence, the force that can be resisted is $A_s f_s = 2100 \times 66.3 \times 10^{-3} = 139$ kN. This is inadequate. Hence the bars may be provided with hooks so as to develop a stress of at least $193.4 \times 10^3 / 2100 = 92$ MPa.
Providing the bars continued over the support region with standard 90° hooks, the development length, l_{dh}, is given by (Cl. 12.5)

$l_{dh} = 100 d_b / \sqrt{f_c'} = 100 \times 30 / \sqrt{20} = 671$ mm

The stress developed in the bars at the inside edge of bearing area is
$400 \times (200 / 671) = 119$ MPa > 92 MPa, OK.
Note that in Fig. 8.11c, the stresses in the bars are taken as fully developed over a length of 671 mm for the bars at the support and over a length of 1207 mm for the terminated bars.

PROBLEMS

See Chapters 6 and 7 for problems on shear and torsion design, to be solved using the General Method.

REFERENCES

8.1 Collins, M.P., Mitchell, D., *Prestressed Concrete Structures,* Prentice Hall, Englewood cliffs, NJ, 1991, pp 338-411.
8.2 *Concrete Design Handbook*, Part II, Chapter 4, Canadian Portland Cement Association, Ottawa, 1985.
8.3 Vecchio, F.J. and Collins, M.P., *The Modified Compression-Field Theory for Reinforced Concrete Elements Subjected to Shear*, J. ACI, Vol. 83, March-April 1986, pp. 219-231.
8.4 Park, R. and Paulay, T., *Reinforced Concrete Structures*, John Wiley & Sons, Inc., New York, 1975, 769 pp.

CHAPTER 9 Bond and Development

9.1 BOND STRESS

The transfer of axial force from a reinforcing bar to the surrounding concrete results in the development of tangential stress components along the contact surface (Fig. 9.1d). This stress acting parallel to the bar along the interface is called *bond stress* and will be denoted by u per unit area of bar surface. Bond stress results from a change in the bar force, T, along its length.

Bond resistance is made up of chemical adhesion, friction, and mechanical interlock between the bar and surrounding concrete. In plain bars, only the first two of these components contribute to the bond strength. In contrast, with deformed bars, the surface protrusions or ribs (oriented transversely to the bar axis) interlocking with and bearing against the concrete key formed between the ribs contributes more positively to bond strength, and is the major reason for their superior bond effectiveness. (CSA A23.3-94(3.1.2) permits only deformed bars as reinforcement except for spirals, and for stirrups and ties smaller than 10 size.)

9.2 FLEXURAL BOND

Fig. 9.1b shows the flexural stresses at two adjacent sections of a beam, dx apart, between which the moment changes by dM. With the usual assumptions made in flexural design, the change in the bar force over the length dx is:

$$dT = \frac{dM}{z}$$

This unbalanced bar force is transferred to the surrounding concrete through the bond stress, u, developed along the interface (Fig. 9.1d). The bond stress, u per unit of bar surface, assuming it to be uniformly distributed over the surface, is obtained from equilibrium of forces as:

$$u \sum o\, dx = dT$$

$$u = \frac{1}{\sum o} \frac{dT}{dx} \qquad (9.1a)$$

$$= \frac{1}{\sum o} \frac{dM/dx}{z}$$

$$u = \frac{V}{\sum o\, z} \qquad (9.1b)$$

Fig. 9.1 Bond stress in a beam

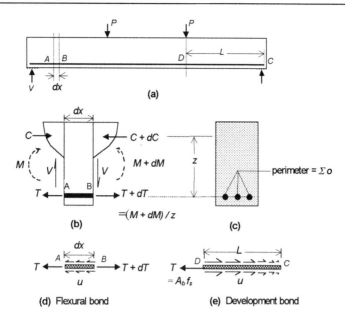

where Σo is the total perimeter of the bars at the section, and $V = dM/dx$ is the transverse shear force. From Eq. 9.1a, the bond stress is proportional to the change of the bar force.

The bond stress expressed by Eq. 9.1b is associated with a shear or variation of bending moment and is called *flexural bond*. Although Eq. 9.1b appears to give the exact value of the local bond stress, because of variations along the bar due to the influence of flexural cracking, local slip, splitting, and other secondary effects, in reality it represents only a measure of an average local bond stress.

9.3 ANCHORAGE OR DEVELOPMENT BOND

The anchorage zone CD of the tension reinforcement of a cantilever beam is shown in Fig. 9.2. The design bending moment and, hence, the reinforcement stress, f_s, are maximum at section D. At the end C of the bar, the stress is zero. The change in force between C and D, equal to $T = A_b f_s$, is once again transferred to the surrounding concrete through bond. If the bond stress is assumed to have an average uniform value of u, equilibrium of forces gives:

$$(\pi d_b L)u = A_b f_s$$

$$(\pi d_b L)u = \frac{\pi d_b^2}{4} f_s$$

$$u = \frac{A_b f_s}{\pi d_b L} = \frac{d_b f_s}{4L} \tag{9.2}$$

where
A_b = area of bar
d_b = bar diameter
L = length of embedment CD
f_s = maximum stress in bar (at D)

The *average* bond stress so calculated using the nominal bar diameter d_b gives the nominal bond stress. A similar situation exists in the segment CD of the reinforcement in Fig. 9.1, between the point of maximum tension and the end of the bar (Fig. 9.1e). The bond stress in these situations can be viewed as providing anchorage for a critically stressed bar, or alternatively as helping develop over the length L the maximum stress f_s needed at the critical section, and is referred to as *anchorage bond* or *development bond*.

9.4 VARIATION OF BOND STRESS

Actual bond stress in a bar is not uniform along its length. In anchorage bond, the stress may be high near the loaded end and much lower, even zero, near the outer end. In flexural members, in a constant moment region the shear force, V, is zero and Eq. 9.1b

Fig. 9.2 Anchorage bond stress

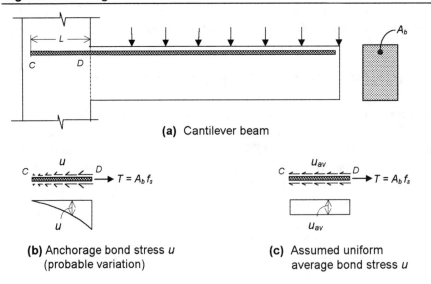

(a) Cantilever beam

(b) Anchorage bond stress u
(probable variation)

(c) Assumed uniform
average bond stress u

would also indicate zero bond stress. However, the steel force, T, varies between flexural cracks, and bond stresses are developed (Fig. 9.3). At the flexural cracks, the tension is carried by the reinforcement alone, while in between the cracks concrete carries some tension, thereby partially relieving the steel bar force. Since the bond stress is proportional to the rate of change of bar force (Eq. 9.1a), local bond stresses develop, following a distribution similar to that shown in Fig. 9.3d, with the direction of bond stress reversing between the cracks (Ref. 9.1). The net bond force between the cracks will be, of course, zero. With large size bars, these local peak bond stresses may cause local slip and/or splitting. In regions of flexural cracking where the moment varies, the distribution of bond stress will differ from that in Fig. 9.3d, as now the net bond force between the cracks must equal the unbalanced bar force, dT. Beam tests show that longitudinal splitting cracks tend to initiate at flexural cracks.

Unlike in tension bars, transverse cracking is generally not present when the bar is in compression. Furthermore, the bar end bearing against the concrete will transfer part of the force to the concrete. Hence, bond resistance of a compression bar is generally greater than that of a tension bar.

Fig. 9.3 *Effect of flexural cracks on bond stress in constant moment region*

(a) Constant moment region between flexural cracks

(b) Probable variation of bar tension T

(c) Bond stress

(d) Probable variation of bond stress u

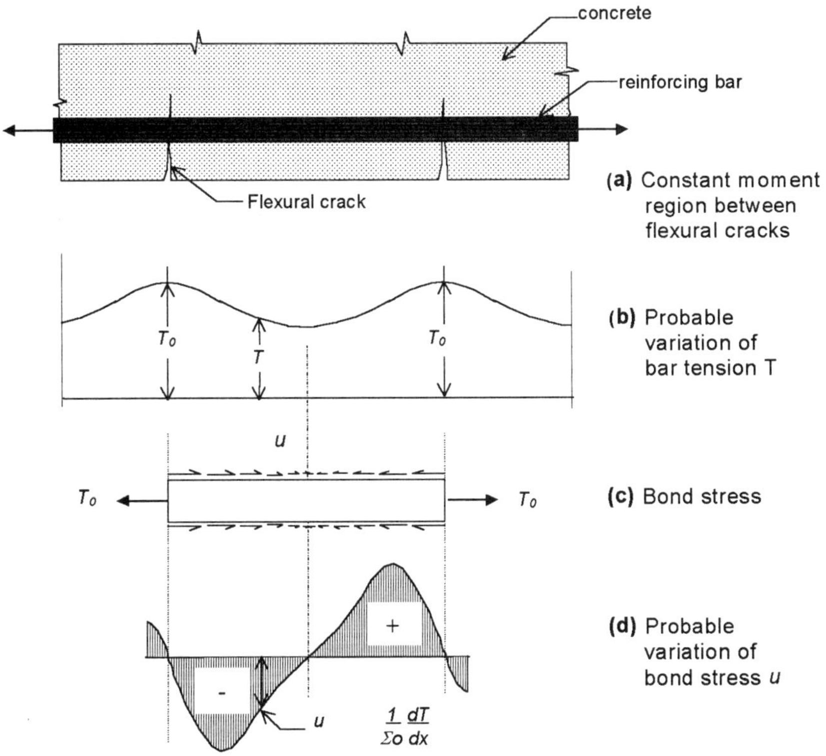

9.5 BOND FAILURE

The mechanisms that initiate bond failure may involve a breakup of adhesion between the bar and the concrete, longitudinal splitting of the concrete around the bar, crushing of the concrete in front of the bar ribs, shearing of the concrete keyed between the ribs along a cylindrical surface surrounding the ribs, or any combinations of these. Bond failures in beams occur generally when the reinforcement bar pulls loose, following longitudinal splitting of the concrete along the bar embedment (Fig. 9.4). Occasionally failure can occur when the bar pulls out of the concrete, leaving a circular hole and without extensive splitting of the concrete. The latter type of failure may occur with plain smooth bars placed with large cover, and with very small diameter deformed bars (wires) having large concrete covers. This is particularly true with top bars cast in low density concrete (Ref. 9.1). However, with deformed bars and with the normal amount of concrete cover used in ordinary beams, bond failure is usually a result of longitudinal splitting. Because of the inclination of the rib face, the bearing pressure between the rib and concrete is at an angle to the bar axis, which introduces radial forces in the concrete (commonly termed *wedging action*), as shown in Fig. 9.5. The radial forces, p, cause circumferential tensile stresses in the concrete surrounding the bar (similar to the stresses in a pipe subjected to internal pressure) and tend to split the concrete along the weakest plane. Although the splitting tendency is present even in plain bars, it is apparent that splitting is generally more critical with deformed bars. If the analogy of the bursting of a pipe is applied to a failure cylinder, as shown in Fig. 9.5c, it can be shown (Refs. 9.2, 9.3, 9.4) that for a bar subjected to an axial force only and not

Fig. 9.4 Typical bond splitting crack patterns

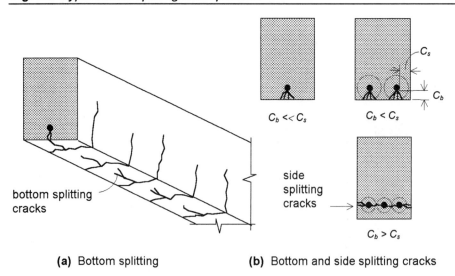

(a) Bottom splitting   (b) Bottom and side splitting cracks

Fig. 9.5 *Splitting forces with deformed bars*

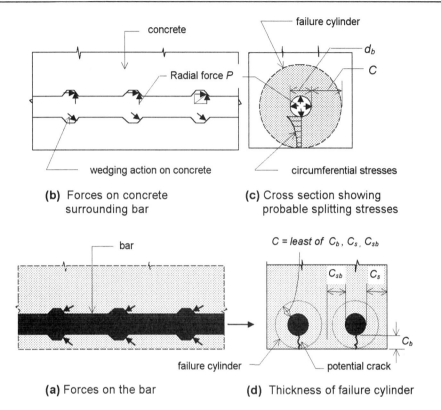

(a) Forces on the bar
(b) Forces on concrete surrounding bar
(c) Cross section showing probable splitting stresses
(d) Thickness of failure cylinder

enclosed by stirrups, the average uniform bond stress at which splitting occurs is a function of the tensile strength of concrete (or $\sqrt{f'_c}$), the ratio C/d_b, and ratio d_b/L. Here, C is the thickness of the idealised failure cylinder, equal to the least of the bottom clear cover, side cover, and one-half the clear spacing between bars (Fig. 9.5c, d); d_b is the bar diameter; and L the embedment length. Splitting occurs along the thinnest surrounding concrete section and the direction of the splitting crack (bottom splitting versus side splitting) will depend on the relative values of the bar spacing and concrete cover, as shown in Fig. 9.4b.

In flexural members, splitting cracks usually appear on the surface as starting from flexural and diagonal tension cracks, beginning in regions of high steel stress (regions where bond stress has local high values, Fig. 9.3), and with increased loads, progress along the length of embedment. Since the distribution of bond stress along the length of the bar is non-uniform, splitting is usually a gradual and progressive phenomenon, with local splitting at regions of high bond stress resulting in redistribution of bond stresses along the length of embedment. Local splitting does not affect the load carrying capacity of a beam, although it indicates bond distress. Splitting

Fig. 9.6 *Tied-arch action with bar anchorage only*

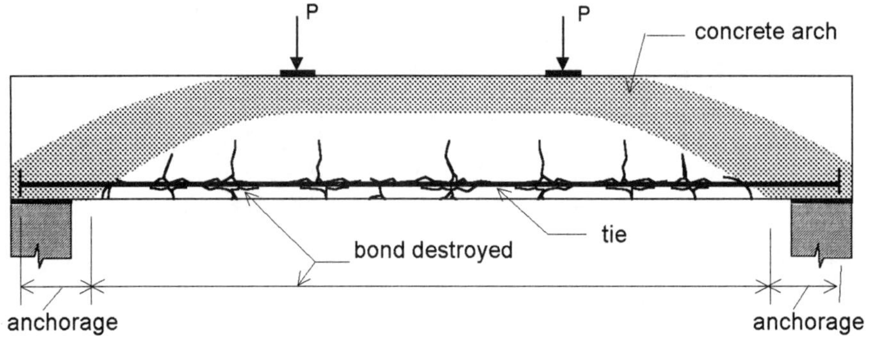

may develop over 60-75 percent of the bar length without loss of average bond strength. In beams without stirrups, however, the final failure is sudden, as the longitudinal split runs through to the end of the bar (Ref. 9.1).

A simple beam can act as a two-hinged arch and carry substantial loads even if the bond is destroyed over the length of the bar, as long as the tie bar is anchored at its ends (Fig. 9.6). Such end anchorage may be realised through anchorage bond over an adequate development length or by special devices (such as hooks and bends, nuts and bolts, welded plates, etc.), or by a combination of both. However, the deflection and crack widths of such a beam may be excessive.

9.6 FACTORS INFLUENCING BOND STRENGTH

The bond strength and mode of bond failure are affected by several variables. The importance of the parameters C, d_b, L, and the tensile strength of the concrete was indicated in the previous section. Any increase in confinement of the bar by the surrounding concrete, transverse reinforcement (stirrups, ties, and spirals) or bearing reaction increases bond strength and minimises splitting. Confinement by the concrete is dependent on the cover and bar spacing, and recent studies indicate that it can be appropriately accounted for by the parameter C/d_b (Refs. 9.2, 9.3). Stirrups and ties cross potential longitudinal splitting cracks at right angles (Fig. 9.7), and hence can resist, partially, the tensile force that causes splitting, thereby delaying initiation of splitting. They can also restrain the width and propagation of splitting cracks, once formed. Statistical analysis of test results indicates good correlation between the increase in bond stress due to stirrups and the parameter $A_t f_y/s d_b$, there being an upper limit for the increase that can be so obtained (Ref. 9.2, 9.3). Here, A_t, f_y, and s are the area, yield stress, and spacing of the transverse reinforcement, respectively, and d_b is the diameter of the longitudinal reinforcement. The beneficial effects of confinement

Fig. 9.7 Stirrup forces due to bond splitting

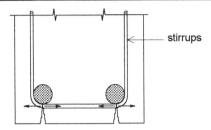

provided by transverse reinforcement in preventing/delaying bond splitting failure is recognised in the Code provisions for calculation of development length requirements.

Formation of diagonal tension cracks leads to an increase in the stress in the tension reinforcement where it intercepts the cracks (Section 5.10.3 and Fig. 5.12) and, hence, an increase in bond stresses on both sides of the crack. In addition, the dowel force that may develop in the tension reinforcement at a diagonal crack due to shear adds to the splitting force due to bond. One conservative estimate, based on test results, puts the loss of bond capacity due to relatively high values of dowel forces at 20 percent (Ref. 9.3). However, closely spaced stirrups will reduce the adverse effect of dowel forces and improve the bond strength.

When flexural reinforcement is terminated in a tension zone, the discontinuity results in a greater rate of change of bar forces and, hence, larger bond stresses. In addition, premature flexural cracks develop at such points, increasing the local bond stress concentrations.

Another factor that influences the bond strength of bars is the depth of fresh concrete below the bar during casting. Water and air, which rises towards the top of the concrete mass, tend to get entrapped beneath horizontally placed bars and weaken the bond as well as the concrete on the underside of the bar. Hence, top reinforcement, which is classified in CSA A23.3 as horizontal bars placed with more than 300 mm of concrete cast in the member below the bar, has a lower bond resistance.

Sometimes epoxy-coated reinforcements are used as a protection against corrosion. Studies have shown that such coating reduces significantly both adhesion and friction between the bars and concrete and the bond strength.

As already indicated, bond strength is proportional to the tensile strength of concrete and the latter, for normal density concretes, can be assumed to vary as the square root of the cylinder strength f_c'. (These correlations have been arrived at based on results of tests on concrete having compressive strengths not exceeding about 65 MPa. The validity of the correlation for concrete with f_c' exceeding the above value has not been well established. Therefore, the Code, while permitting f_c' up to 80 MPa (and even higher), limits the value of $\sqrt{f_c'}$ for use in bond and anchorage calculations to a maximum of 8 MPa). However, for low density concrete the tensile strength is somewhat less than that of normal density concrete having the same compressive

274 REINFORCED CONCRETE DESIGN

strength. Also, bond failure by pull out of the bar, following crushing or shear failure of the concrete between bar ribs, has been observed more often in low density concrete than normal density concrete.

The dimensions and shape of the ribs (deformations) of deformed bars may be expected to have some influence on bond resistance. Experiments show that several deformation patterns, all meeting the specifications for bar deformations, yield substantially the same bond resistance (Refs. 9.3, 9.5).

Because of the extreme non-uniformity in the bond stress distribution, and the large number of variables and uncertainties involved, there is, as yet, no theoretical method for evaluating the bond strength of a reinforcing bar embedded in concrete. Hence, specifications and design practice have depended on bond test results to arrive at maximum allowable *nominal* bond stresses, computed as the bar force divided by the nominal surface area of the embedment length of the bar ($u = P/(\Sigma oL)$), or alternatively to arrive at the minimum development length required to develop a required stress in the reinforcement.

9.7 BOND TESTS

Bond strength is determined with either pull-out tests or some sort of beam tests. In a typical pull-out test, shown schematically in Fig. 9.8a, a bar embedded in a concrete cylinder or prism is pulled until failure by splitting, excessive slip, or pull out. The nominal bond stress is computed as $u = P/(\Sigma oL)$, where P is the pull at failure. Bond conditions in a pull-out test do not ideally represent those in a flexural member. Factors such as flexural cracks, shear, diagonal tension cracks, and dowel forces, which lower the bond resistance of a flexural member are not present in a concentric pull-out test. The concrete in the pull-out specimen is in compression, and the friction at the bearing on the concrete offers some restraint against splitting.

Several forms of beam tests for bond have been developed, all designed to duplicate, as best as practicable, the actual bond conditions in a flexural member (Ref. 9.6). The set-up for one such test is shown in Fig. 9.8b (Ref. 9.3). In addition to simulating the bond stress conditions in a beam, this set-up is capable of allowing for dowel force effect at diagonal shear cracking.

Based on early bond test results, the maximum allowable ultimate bond stress was expressed as $u_u = K \sqrt{f'_c}/d_b$ (and $K' \sqrt{f'_c}$ for special large size deformed bars) in the 1970 edition of CSA A23.3 (also in Ref. 9.7), where K and K' are appropriate constants specified in the Code. Although from the 1973 edition of the CSA A23.3 onwards the emphasis on design has shifted from an allowable bond stress to a required development length, l_d, the basis for l_d is still essentially an allowable bond stress value.

Fig. 9.8 Bond test specimens

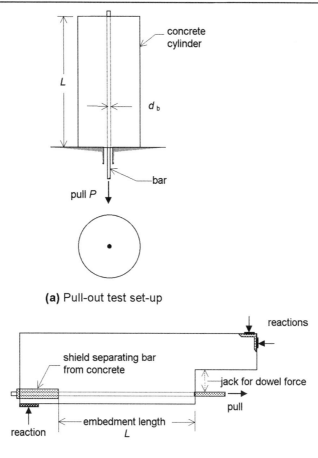

(a) Pull-out test set-up

(b) Modified cantilever bond test specimen

9.8 DEVELOPMENT LENGTH

The *development length* is the length necessary to introduce or develop a given stress, f_s, into the bar through bond or, viewed in reverse, the length necessary to take the maximum stress, f_s, out of the bar (length *CD* in Fig. 9.1 and 9.2). Sometimes the term *anchorage length* is used for the development length, in situations where the embedment portion of the bar is not subjected to any flexural bond (Fig. 9.2). However, the distinction should be recognised that *anchorage* can be obtained by methods other than bond development, such as by hooks and bends, mechanical fasteners, etc. The concept of development length is that a certain minimum length of bar is needed on either side of a point of maximum steel stress to prevent the bar from pulling out. However, when the bar cannot be provided with the required extension due to placement difficulties, or when it is more convenient to do so, the required development length may be provided by a combination of the available embedment length plus additional equivalent embedment length of a hook or mechanical anchorage.

Prior to the 1973 edition of CSA A23.3, design for bond required the consideration of both flexural bond stress and anchorage bond stress. Because of the unpredictable and extreme non-uniformity in the distribution of the actual bond stress, the flexural bond stress computed by Eq. 9.1b gives only an unreliable index. With such variations in bond stress, localised bond failures may occur at loads well below ultimate; however, these local failures do not impair the strength of the beam, provided the bars are still anchored at their ends (Fig. 9.6). Further, the allowable bond stress determined from bond tests represents an average bond resistance over the full embedment length. Thus, the development of bar stresses through adequate anchorage is the primary requirement rather than the control of local bond stresses, whose predicted values are unrealistic. Thus, the 1973 and subsequent editions of CSA A23.3 consider essentially the development bond only in the form of a specified minimum required development length. This length is derived on the basis of a maximum permissible average bond stress over the length of embedment. Thus, the *development length* specified in CSA A23.3-94 is the length required to develop a bar stress equal to f_y with a permissible average anchorage bond stress u_u. This value of l_d may be obtained from Eq. 9.2 as:

$$l_d = \frac{f_y d_b}{4 u_u} \tag{9.3}$$

Several studies have been made for predicting the average bond strength at splitting, accounting for the influence of bar cover and spacing, shear force, dowel force and transverse reinforcement (stirrups) (Refs. 9.2, 9.3, 9.4 and 9.8). Based on these, Refs. 9.2 and 9.3 have recommended Eqs. 9.4 and 9.5, respectively, for the average ultimate bond strength, u:

$$u = \left[0.1 + 0.25 \frac{C}{d_b} + 4.2 \frac{d_b}{l_s} + 0.024 \frac{A_{tr} f_{yt}}{s d_b} \right] \sqrt{f_c'} \tag{9.4}$$

where the last term within the bracket is limited to 0.25

$$u = \left[0.55 + 0.24 \frac{C}{d_b} \right] \sqrt{f_c'} + 0.191 \frac{A_t f_{yt}}{s d_b} \tag{9.5}$$

In Eqs. 9.4 and 9.5, C is the lesser of clear cover and one-half bar spacing in mm, d_b

is the diameter of the main bar in mm, A_{tr} is the effective area of transverse reinforcement (stirrups) per main bar in mm², A_t is the area of stirrup in mm², f_{yt} is the yield strength of stirrups in MPa, s is the spacing of stirrups in mm, and l_s is the splice length or development length in mm. Reference 9.2 also concludes that the basic requirements for development length and lap splice length are identical. Substituting the expression in Eq. 9.4 for u_u in Eq. 9.3 yields:

$$l_d = \frac{4}{\pi} \frac{A_b f_y}{(0.4d_b + C + 16.8 d_b^2/l_s + 0.096 A_{tr} f_{yt}/s)\sqrt{f_c'}} \tag{9.6}$$

9.9 CSA CODE DEVELOPMENT LENGTH

The expression for development length of deformed bars in tension given in CSA A23.3-94(12.2) is a modified form of Eq. 9.6. In development length computations, *no allowance need be made for the material resistance factors*. The area of steel being developed is already computed during its design with an allowance for the ϕ factors. Hence, the only requirement is to provide adequate development for the steel area so designed. The required development length must be provided on both sides of peak stress points. The Code (Clause 12) requirements are summarised below.

These requirements are used to compute the development length, l_d, necessary to develop a stress of f_y in the bar at one end of this length. Conversely, these relationships may also be used to compute the reduced stress $f_s < f_y$ obtainable at one end of a given development length, which is less than the minimum required for f_y.

9.9.1 Tension Bars

For deformed bars and deformed wires in tension, the Code recommends Eq. 9.7 for the computation of the development length, l_d, in mm. However, l_d shall not be less than 300 mm. Equation 9.7 is a modified form of the general expression recommended by ACI Committee 408 (Ref. 9.9) and is based on the studies reported in Ref. 9.2 (see also Eq. 9.6). The factors k_1 to k_4 are introduced to allow for the influence of various parameters discussed in Section 9.6. The factor K_{tr} represents the contribution due to confinement afforded by the transverse reinforcement across potential planes of splitting. As confinement increases, bond failure changes from bond-splitting type to a pullout type failure. Hence, there is a limit to the confinement that can be considered effective, and the Code limits $(d_{cs} + K_{tr})$ to a maximum of $2.5d_b$.

$$l_d = 1.15 \frac{k_1 k_2 k_3 k_4}{(d_{cs} + K_{tr})} \tag{9.7}$$

where
$$K_{tr} = \frac{A_{tr} f_{yt}}{10.5sn}$$

The term $(d_{cs} + K_{tr})$ in Eq. 9.7 is limited to a maximum of $2.5d_b$. Also, the value of $\sqrt{f'_c}$ to be used in bond and anchorage calculations is limited to a maximum of 8 MPa. In Eq. 9.7,

A_b = area of individual bar in mm²,
A_{tr} = area of reinforcement within l_d which crosses the potential bond splitting crack,
d_b = nominal diameter of bar or wire,
d_{cs} = the smaller of (a) the distance from the closest concrete surface to the centre of the bar being developed, or (b) two-thirds the centre-to-centre spacing of the bars being developed,
f_{yt} = specified yield strength of transverse reinforcement,
n = number of bars or wires being developed along potential plane of bond splitting,
s = maximum centre-to-centre spacing of transverse reinforcement within l_d, and the modification factors are:

Bar location factor, k_1
 = 1.3 for horizontal reinforcement so placed that more than 300 mm of fresh concrete is cast in the member below the development length or splice.
 = 1.0 for other cases.

Coating factor, k_2
 = 1.5 for epoxy-coated reinforcement with clear cover < $3d_b$, or with clear spacing between bars being developed < $6d_b$.
 = 1.2 for all other epoxy-coated reinforcement.
 = 1.0 for uncoated reinforcement.

Concrete density factor, k_3
 = 1.3 for structural low-density concrete.
 = 1.2 for structural semi-low-density concrete.
 = 1.0 for normal-density concrete.

Bar size factor, k_4
 = 0.8 for No. 20 and smaller bars and deformed wires.
 = 1.0 for No. 25 and larger bars.

Note that the product k_1k_2 need not be taken greater than 1.7.

Where the reinforcement provided in a flexural member is in excess of that required by analysis, the stress in it will be correspondingly low. Therefore, the development length, l_d, may be multiplied by the factor $(A_{s,required})/(A_{s,provided})$.

As an alternative and simpler format, for deformed bars and deformed wires in tension and satisfying the clear cover and clear spacing requirements of CSA A23.1-94, the Code permits the calculation of l_d using the expressions given in Table 9.1.

Table 9.1 *Development Length, l_d, in mm, of Deformed Bars and Deformed Wires in Tension*

Cases	Minimum development length, l_d
(a) Member containing minimum stirrups or ties (Code Cl. 11.2.8.4 or Cl. 7.6.5) within l_d, or Slabs, walls, shells, or folded plates having clear spacing between bars being developed not less than $2d_b$	$0.45\, k_1 k_2 k_3 k_4\, \dfrac{f_y}{\sqrt{f'_c}}\, d_b$
(b) Other cases	$0.60\, k_1 k_2 k_3 k_4\, \dfrac{f_y}{\sqrt{f'_c}}\, d_b$

9.9.2 Compression Bars

For deformed bars in compression the development length, l_d, is computed as the product of the basic compression development length, l_{db}, and the applicable modification factors. The basic compression development length, $l_{db} = 0.24\, d_b f_y / \sqrt{f'_c}$, but not less than $0.044\, d_b f_y$. The modification factors, with a cumulative value of not less than 0.6, are:

(a) When reinforcement in excess of that required by analysis is provided: $(A_{s,required})/(A_{s,provided})$

(b) Reinforcement enclosed within spiral reinforcement having a diameter of at least 6 mm and having a pitch no greater than 100 mm or within No. 10 ties satisfying Code Cl. 7.6.5 and spaced at not more than 100 mm on centre: 0.75

9.9.3 Bundled Bars

In bundled bars, the perimeter of individual bars exposed to concrete is reduced, and the bond resistance may not be fully mobilised in the core of the bundle. Hence, the development length computed for the individual bars in a bundle is increased by 10 percent for a 2-bar bundle, 20 percent for a three-bar bundle, and 33 percent for a four-bar bundle. Also, the cut-off points within the span of flexural members of individual bars in a bundle are staggered by at least forty bar diameters.

9.9.4 Welded Wire Fabric in Tension

For smooth WWF, the yield strength is considered to be developed with the embedment of at least two cross wires, with the closer one at least 50 mm from the critical section (Fig. 9.9). However, the length, l_d, measured from the critical section to the outermost cross wire must be not less than:

Fig. 9.9 Combination development length with hooks and embedment

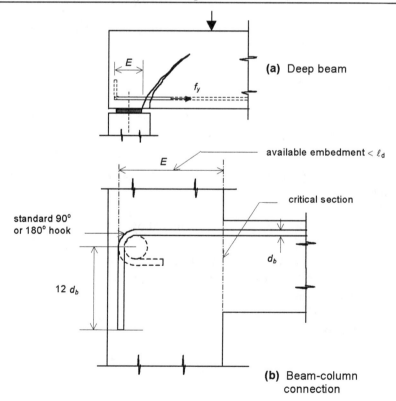

(a) Deep beam

(b) Beam-column connection

$$l_d = 3.3 \frac{A_w}{s_w} \frac{f_y}{\sqrt{f'_c}}$$

where A_w is the area and s_w is the spacing of the individual wires to be developed. The length given by the above equation may be modified by the excess reinforcement factor given in Sections 9.9.1 in case of any excess area provided. However l_d must be not less than 150 mm.

For deformed WWF, the development length, l_d, measured from the critical section to the end of the wire is taken as the product of the development length, l_d, as given in Section 9.9.1 above and the applicable wire fabric factor(s) given below. However, l_d must not be taken less than 200 mm except in the development of web reinforcement.

The wire fabric factors are:
(a) when there is at least one cross wire within the development length and not less than 50 mm from the point of the critical section, the wire fabric factor shall be taken as $= (f_y - 240)/\sqrt{f'_c}$, or $5d_b/s_w$, which ever is larger, but need not be taken greater than 1.0.
(b) when there are no cross wires within the development length, or with a single cross wire less than 50 mm from the point of the critical section, the wire fabric factor shall be taken as 1.0.

9.10 HOOKS AND MECHANICAL ANCHORAGES

There are situations where the straight distance available beyond a critical section may be inadequate to provide the required development length, l_d (Fig. 9.9). Figure 9.9a shows a deep beam where inclined cracks cause the beam to behave like a tied-arch. The full yield strength may have to be developed at the face of the support and the available straight embedment length, E, may be less than the length, l_d, required. A similar situation exists at the top bars in a beam or bracket where it frames into a column (Fig. 9.9b). The column face is the critical section for the negative moment reinforcement, and again the straight embedment length available within the column may be less than l_d. In such situations CSA A23.3-94 allows the development of the tension in the reinforcement through a combination of the straight lead-in length, l_{li}, available, plus a standard hook or mechanical anchorage. The effectiveness of mechanical anchorage devices must be ascertained by tests. Hooks can be used in developing tensile stresses only and are ineffective in compression.

The proportions of a standard hook as defined in CSA 23.1-94 are shown in Fig. 9.10 (proportions for stirrup and tie hooks were given in Chapter 6, Fig. 6.12). The development length, l_{dh}, of a deformed bar terminating in a standard hook is the length measured from the outside of the hook to the critical section where the

Fig. 9.10 Development length of standard hooks

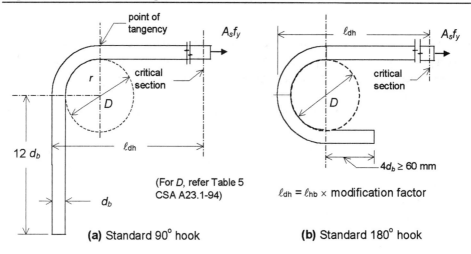

(a) Standard 90° hook
(b) Standard 180° hook

reinforcement yield stress, f_y, is developed, as shown in Fig. 9.10. The length, l_{dh}, is computed as the product of a basic development length of a standard hook in tension, l_{hb}, and applicable modification factors given below. However l_{dh} is taken as not less than $8d_b$ or 150 mm, whichever is greater.

The basic development length for a hooked bar with $f_y = 400$ MPa is

$$l_{hb} = 100 d_b / \sqrt{f'_c}$$

The values of l_{hb} for various size bars and concrete strengths are given in Table 9.2. The modification factors for hook development length are:

(a) for f_y other than 400 MPa: $\qquad f_y/400$
(b) for No. 35 and smaller bars, with side cover (normal to plane of hook) not less than 60 mm, and for 90E hooks with cover on the bar extension beyond the hook of not less than 50 mm (Fig.9.10c): $\qquad 0.7$
(c) for No. 35 and smaller bars, with hook enclosed vertically or horizontally within at least three ties spaced along a length at least equal to the inside diameter of the hook, at a spacing not greater than $3d_b$, d_b being the diameter of the hooked bar (Fig. 9.10c): $\qquad 0.8$
(d) for reinforcement provided in excess of that required by analysis, for which development of f_y is not required: $\qquad (A_{s,required})/(A_{s,provided})$
(e) for structural low density concrete: $\qquad 1.3$
(f) for epoxy-coated reinforcement: $\qquad 1.2$

For bars being developed by a standard hook at the ends of members where the side cover and the top (or bottom) cover over the hook is less than 60 mm, the hook

CHAPTER 9 BOND AND DEVELOPMENT

must be enclosed within a minimum of three ties spaced along a length at least equal to the inside diameter of the hook and at a spacing not greater than $3d_b$. In this case the factor of 0.8 at (c) above will not apply (Fig. 9.11(c)(iii)).

Table 9.2 *Basic Development Length, l_{hb}, for Standard Hooks in Tension*

Bar no.	Diameter of bar, d_b mm	l_{hb}				
		$f_c' = 20$ MPa	$f_c' = 25$ MPa	$f_c' = 30$ MPa	$f_c' = 35$ MPa	$f_c' = 40$ MPa
10	11.3	253	226	206	191	179
15	16.0	358	320	292	270	253
20	19.5	436	390	356	330	308
25	25.2	563	504	460	426	398
30	29.9	669	598	546	505	473
35	35.7	798	714	652	603	564
45	43.7	977	874	798	739	691
55	56.4	1261	1128	1030	953	892

$l_{hb} = 100 d_b / \sqrt{f_c'}$ for $f_y = 400$ MPa

Development length, $l_{dh} = l_{hb} \times$ modification factors, but not less than $8d_b$ and 150 mm.

Tail extensions beyond those of a standard hook are not effective in providing development length (Ref. 9.10). Hooks are used for anchorage or development and should not be used to replace the bar extensions discussed in Section 5.10.4.

Fig. 9.11 *Bar size limitation at points of inflection*

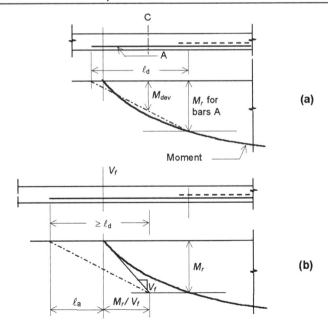

9.11 CSA CODE DEVELOPMENT REQUIREMENTS

The computed stress at *every section* of a reinforcing bar must be developed on both sides of the section. Flexural tension reinforcement can be developed by bending it over to the compression face of the member and anchoring it, or making it continuous with the compression reinforcement at this face. In flexural members, the critical sections for development are: (1) the points of maximum stress in the bar, such as section $C1$ for positive moment reinforcement and section $C2$ for negative moment reinforcement (see Fig. 5.14), and (2) when part of the reinforcement is terminated (or bent) at points within the span, for the continuing reinforcement, the point where parts of the bars are terminated (or bent) (such as sections, X_1, Y_1 in Fig. 5.14). The development length and bar extension requirements at these critical sections in flexural members, specified in CSA A23.3-94(12.10, 12.11 & 12.12), were detailed in Sections 5.10.4 and 5.10.5 and Figs. 5.14 and 5.15. When the flexural member is part of the primary lateral load resisting system, the minimum positive reinforcement required to be continued into the support (CSA A23.3-94 (12.11)) must be anchored to develop its yield strength in tension at the face of the support (at section $C2$ in Fig. 5.14). Likewise, if the bottom reinforcement at the face of the support is counted on as compression reinforcement at this section, it must be anchored to develop the stress, f_y.

To ensure that the computed bar stress is developed at every section, the positive moment region of beams with distributed loading requires special consideration. In such regions, the moment diagram is nonlinear (and convex), but the bar stress development over the length l_d is assumed to be linear. Hence, providing the minimum l_d beyond a critical section may not always ensure the development of required stresses at other sections. Such a situation is illustrated in Fig. 9.11a. In this figure, assuming a linear variation of bar stress over the length, l_d, the moment of resistance corresponding to the developed bar stress varies as shown by the chain line. At a section such as C, the moment that is developed (M_{dev}) is less than the applied moment. To avoid such a situation, the Code stipulates that (Fig. 9.11b):

$$l_a + \left(\frac{M_r}{V_f}\right) \geq l_d \qquad (9.7)$$

where M_r is the factored moment corresponding to the reinforcement extended beyond the section of zero moment, V_f is the factored shear force at the section of zero moment, and l_a is limited to the larger of d or $12d_b$. V_f represents the slope of the moment diagram at the point of zero moment and the length $l_a + M_r/V_f$ is indicated in Fig. 9.12b. The moment developed with this requirement is shown by the dotted line in Fig. 9.12b. The requirement in Eq. 9.7 is to be satisfied at supports of simply supported members and at points of inflection of continuous beams. Although a similar situation could arise at bar cut-off points, if the bar size satisfies the requirement at the point of zero moment, usually the condition will be satisfied at cut-off points as well. When the

Fig. 9.12 Example of special end anchorage requirement

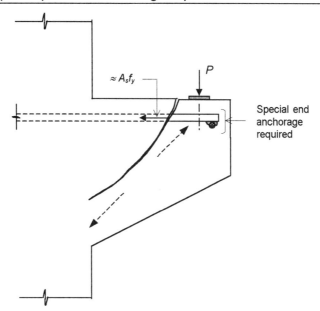

General Method of shear design in CSA A23.3-94 (11.4) is followed, the requirement given in Eq. 9.7 can be waived.

Special members, such as sloped, stepped or tapered footings, deep beams, brackets, and members of variable depth, in which the steel stress is not directly proportional to the moment, require special consideration for the development of tension reinforcement. For the cases shown in Figs. 9.10a and 9.13, the steel stress is relatively high close to the end of the member, and the development of the bar depends on providing adequate end anchorage.

For web reinforcement, the critical section for development is considered to be at mid-depth, $d/2$, of the beam. The anchorage requirements for web reinforcement were detailed in Section 6.11 and Fig. 6.12.

9.12 FACTORED MOMENT RESISTANCE DIAGRAMS

The theoretical cut-off or bend points may often be determined by scaling from the factored moment envelope. Once the actual cut-off details are worked out, a convenient method to check their adequacy is to construct the corresponding factored moment resistance diagram (Fig. 9.13c). In determining the latter, a bar is assumed to develop its full yield strength, and hence its share of the moment resistance, linearly over a length l_d from the *effective* point of cut-off (Fig. 9.13b). The Code requirement in Clause 12.10.3 (and also 11.3.8) implies that at cut-off points the factored moment

286 REINFORCED CONCRETE DESIGN

Fig. 9.13 Factored moment resistance diagram

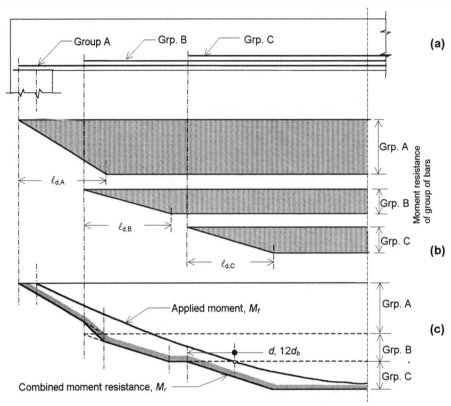

resistance diagram must have a horizontal offset from the applied moment diagram by the larger of d or $12d_b$, except at simple or exterior supports and at free ends of cantilevers.

9.13 DESIGN EXAMPLES

Calculations for the development requirements of flexural reinforcement in simply supported beams were illustrated in design Examples 5.9 and 5.10 in Chapter 5. Examples of a continuous beam and a beam-column connection are given below.

EXAMPLE 9.1

The moment envelope for an interior span of a continuous beam is shown in Fig. 9.14a. With a beam size of 300 × 450 mm and an effective depth of 390 mm, the flexural design gives three No. 25 bars as positive moment reinforcement at midspan and four

CHAPTER 9 BOND AND DEVELOPMENT 287

No. 25 bars as negative moment reinforcement at the support. The beam is designed for shear using the Simplified Method in CSA A23.3-94(11.3). That is, the effect of shear on longitudinal reinforcement has not been accounted for in flexural design.

One-third of the positive moment reinforcement and one-half of the negative moment reinforcement are to be terminated, as they become no longer necessary for flexure. Design the bar cut-off details and check the development requirements by the Code. Take $f_y = 400$ MPa, and $f_c' = 30$ MPa. The factored load on the span is 48 kN/m.

SOLUTION

(See Fig. 9.14b)
1.   Positive Moment Reinforcement

    (a) Development length required: The required development length for the bottom reinforcement, using the simpler equation given in Table 9.1 (case a) is:

Fig. 9.14 Example 9.1

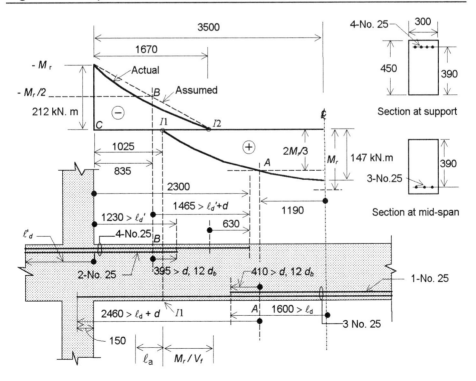

$$l_d = 0.45 k_1 k_2 k_3 k_4 (f_y / \sqrt{f'_c}) d_b \, l_d$$
$$= 0.45 \times 1 \times 1 \times 1 \times 1 \times (400/\sqrt{30}) \times 25.2 = 828 \text{ mm}.$$

(b) Location of bar cut-off: The reinforcement ratio provided at midspan is $\rho = 3 \times 500/(300 \times 390) = 0.0128$. For this, from Table 5.4, $K_r = 3.7$ and the factored moment of resistance is $M_r = 3.7 \times 300 \times 390^2 \times 10^{-6} = 169$ kN·m. Then, one of the three bars is no longer needed for bending moment, at a section where

$$+M = \frac{2}{3}(+M_r) = \frac{2}{3} \times 169 = 113 \text{ kN·m}$$

From the moment envelope in Fig. 9.15a, this section is located at A, a distance of 1190 mm from midspan. The bar must be continued beyond this point, a minimum distance equal to the larger of $d = 390$ mm or $12 d_b = 302$ mm (CSA A23.3-94(11.3.8)). Hence, the one bar is continued a distance $= 1190 + 390 \approx 1600$ mm from midspan.

(c) Check development

(i) The cut-off bar has its critical section at midspan. The actual length of embedment beyond the critical section is:

    1600 mm $> l_d = 828$ mm   OK

(ii) The continuing bar has the critical section for development at section A, where the terminated bar is no longer required to resist flexure. With the two bars continued into the support a distance of 150 mm (CSA A23.3-94(12.11.1)), the length of embedment provided is:

    $3500 + 150 - 1190 = 2460$ mm
    $> l_d + (d, \text{ or } 12d_b) = 828 + 390 = 1218$ mm   OK

(iii) To check bond stresses at the point of inflection (bar size limitation - CSA A23.3-94(12.11.3)), the maximum additional embedment length beyond the point of inflection, $l1$, that can be considered effective, is

l_d = the larger of d or $12d_b$
    = 390 mm
M_r = flexural strength for the 2 bars
    $\approx \frac{2}{3} \times 169 = 113$ kN·m

Shear at face of support $= 48 \times 3.5 = 168$ kN
V_f = shear at point of inflection
    $= 168 - 48 \times 1.025 = 119$ kN

$$l_d + \frac{M_r}{V_f} = 390 + \frac{113 \times 10^6}{119 \times 10^3} = 1340 \text{ mm} > l_d \qquad \text{OK}$$

(iv) In addition, since a flexural reinforcement bar has been terminated in a tension zone, the shear strength requirements at the cut-off location (CSA A23.3-94(12.10.5)) must be checked. This was illustrated in Chapter 6 in Examples 6.1 and 6.2.

Note: the minimum positive moment reinforcement that must be continued into the support is only $0.25A_s$, which in this example is one bar. If the beam is part of a primary lateral load resisting system (by frame action, either alone or in combination with shear-walls, etc.), this minimum required reinforcement must be anchored into the support to develop stress f_y at the column face, in the event of possible stress reversals and to ensure ductility by yielding of bars at the connection. It is *not* sufficient to develop the larger area continued into the support for a proportionately lower stress in lieu of providing full anchorage for the minimum required reinforcement.

2. Negative Moment Reinforcement
   (a) Development length l_d' for top reinforcement
   The negative moment reinforcement has more than 300 mm of concrete below it and hence will classify as "top reinforcement" and the modification factor of $k_1 = 1.3$ is applicable.
   $l_d' = 1.3 \times 828 = 1076$ mm

   (b) Location of cut-off
   With four No. 25 bars, $A_s = 2000$ mm², $\rho = 1.71\%$, and from Table 5.4, $K_r = 4.64$, $M_r = 4.64 \times 300 \times 390^2 \times 10^{-6} = 212$ kN·m. Two of the four bars are not needed for flexure at the section where
   $$-M = \frac{1}{2}(-M_r) = 106 \text{ kN·m}$$
   Assuming for simplicity (and slightly conservatively) that the negative moment envelope is a straight line, $-M = 106$ kN·m at section B, a distance 835 mm from the face of the support. Continuing the bars for $d \geq 12d_b$ beyond B, two of the four bars can be terminated at $835 + 390 \approx 1230$ mm from the face of the support.
   The remaining two bars (which is greater than the minimum of 1/3 required by the Code) are continued beyond the point of inflection $I2$ for a distance equal to the greater of d, $12d_b$, or 1/16 of clear span, = 440 mm. Hence these bars must be continued to $1670 + 440 = 2110$ mm, at least,

290 REINFORCED CONCRETE DESIGN

from the face of the support.

(c) Check development
   (i) For the two bars cut off, the actual embedment length beyond the critical section C (face of support) is:

$$1230 \text{ mm} > l_d' = 1076 \text{ mm} \quad \quad \text{OK}$$

   (ii) For the two bars continued, the critical section for development is at B, where the terminated bars are no longer required to resist flexure. Length provided beyond B, determined above, is: 2110 - 835 = 1275 mm
   Length required in accordance with CSA A23.3-94(12.10.4) is,
$$l_d' + (d \text{ or } 12d_b) = 1076 + 390 = 1467 \text{ mm}$$
   This is more than the length determined above = 1275 mm. Hence, the two bars continued must be taken to a length of $835 + 1467 \approx 2300$ mm from the face of the support.

   (iii) All the negative moment reinforcement must also be fully developed to the left of section C by anchorage in or through the column with embedment length, hooks, or mechanical anchorage. At an interior column this is usually done by continuing the reinforcement into the adjacent span as part of the reinforcement in that span.

   (iv) At the bar cut-off in the tension zone, the shear strength requirements must be checked as before.

EXAMPLE 9.2

Design the anchorage details for the negative moment reinforcement in the beam at the exterior beam-column connection shown in Fig. 9.15, using $f_y = 400$ MPa, and $f_c' = 20$ MPa.

SOLUTION

Required development length in tension, allowing the 1.3 factor for top reinforcement, is

$$l_d = 0.45 k_1 k_2 k_3 k_4 d_b f_y / \sqrt{f_c'}$$
$$= 0.45 \times 1.3 \times 1 \times 1 \times 1 \times 25.2 \times 400 / \sqrt{20} = 1319 \text{ mm}$$

Fig. 9.15 Example 9.2

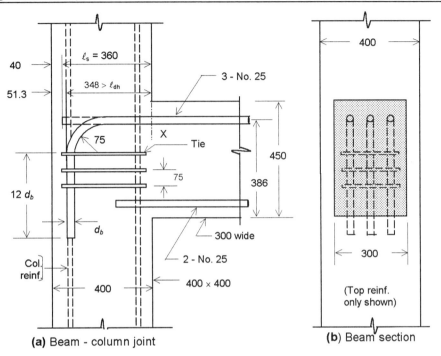

(a) Beam - column joint (b) Beam section

Allowing the minimum clear cover of 40 mm, the maximum available straight length, l, is 400 - 40 = 360 mm < l_d. Since a straight embedment is inadequate, anchorage through the provision of a standard 90° hook with a lead-in embedment length is considered. Basic development length for a standard hooked bar is

$$l_{hb} = 100 d_b / \sqrt{f'_c} = 100 \times 25.2 / \sqrt{20} = 563 \text{ mm}$$

The critical section is x at the face of support. With beam width = 300 mm and column width = 400 mm, the side cover is in excess of 60 mm. The hook will be placed with a cover of 51.3 mm on the vertical bar extension beyond the hook (that is, within column ties). Furthermore, the hook will be enclosed horizontally within at least three ties conforming to Code Clause 12.5.3(c). With these, the modification factors applicable are 0.7 (as per Cl. 12.5.3(b)) and 0.8, and the development length of the hook will be $l_{dh} = 0.7 \times 0.8 \times 563 = 315$ mm. The length available with the 51.3 mm cover is 400 - 51.3 = 348 mm > 315 mm   OK

Using No. 10 closed horizontal ties to confine the concrete in the anchorage zone, the maximum spacing is $3d_b = 3 \times 25.2 = 75.6$ mm. Use a spacing of 75 mm.

The minimum length over which the stirrups are to be placed is equal to the inside diameter of hook which for No. 25 bars with $f_y = 400$ MPa is 150 mm (CSA A23.1-94, Table 5). Three No. 10 horizontal ties at a spacing of 75 mm may be provided as shown.

EXAMPLE 9.3

In the beam-column connection shown in Fig. 9.15, if the positive moment reinforcement continued into the support is considered as compression reinforcement in the design for negative bending moment at the column face, check the anchorage requirements at the support. Use $f_y = 400$ MPa and $f_c' = 20$ MPa.

SOLUTION

The basic development length for compression is:
$$l_{db} = 0.24 f_y d_b / \sqrt{f_c'} \geq 0.044 f_y d_b \text{ or } 200 \text{ mm}$$
$$= 0.24 \times 400 \times 25.2 / \sqrt{20} = 541 \text{ mm}$$

As no modification factors are applicable in this case, $l_d = l_{db} = 541$ mm

The maximum available straight length is 360 mm (Fig. 9.15), and hooks are ineffective for development of compressive stress. Hence, the compressive stress in the bar at the column face has to be limited to the value that can be developed or, alternatively, smaller diameter bars must be used.

The stress that can be developed in the No. 25 bar with the available development length of 360 mm may be computed from:
$$l_{dc} = 0.24 f_s d_b / \sqrt{f_c'}$$
$$360 = 0.24 \times f_s \times 25.2 / \sqrt{20}, \; f_s = 266 \text{ MPa}$$

If the No. 25 bars are replaced by smaller size bars, the maximum size of bar which can develop a compressive stress of $f_y = 400$ MPa at the column face with the available development length may be computed from the above equation as:
$$360 = 0.24 \times 400 \times d_b / \sqrt{20}, \; d_b = 16.8 \text{ mm}$$

Thus No. 15 size bottom bars can develop a stress of f_y in compression at the column face.

EXAMPLE 9.4

The reinforcement details and the moment envelope for a continuous T-beam are given in Fig. 9.16. Draw the moment capacity diagram and check the development requirements at the inflection points. This beam was designed with $\sqrt{f'_c} = 20$ MPa, and $f_y = 300$ MPa. Assume that for bar extension requirements bent bars are effective from mid-depth of the beam. The beam is 300 mm wide and 600 mm deep overall, the effective width of flange is 1850 mm, and $h_f = 110$ mm. The maximum shear at the exterior face of the first interior support is 199 kN. Assume that design for shear is by the Simplified Method in CSA A23.3-94(11.3).

SOLUTION

The moment capacities are first computed. The reinforcement ratios at all sections are below the maximum permissible (for the given material properties) of $\rho_{max} = 0.02485$.

1. Section at exterior support:
    (i) With full reinforcement, $A_s = 1500$ mm^2
    $d = 536$ mm, $\alpha_1 = 0.82$
    $$a = \frac{1500 \times 0.85 \times 300}{0.82 \times 0.6 \times 20 \times 300} = 130 \text{ mm}$$
    $M_r = 0.85 \times 1500 \times 300 \times (536 - 130/2) \times 10^{-6} = 180$ kN·m
    With two No. 25 bars, $M_r \approx \frac{2}{3} \times 180 = 120$ kN·m
    With one No. 25 bar, $M_r \approx \frac{1}{3} \times 180 = 60$ kN·m
    l_d for No. 25 top bars is $= 0.45 \times 1.3 \times 300 \times 25.2/\sqrt{20} \approx 989$ mm. The moment resistance diagram for the negative moment in this region is drawn in Fig. 9.16b.

2. Section at interior support:
    (i) For full reinforcement
    $A_s = 2200$ mm^2, $d = 538$ mm, $a = 190$ mm, $M_r = 249$ kN·m
    (ii) With four No. 20 and one No. 25,
    $$A_s = 1700 \text{ mm}^2, M_r = \frac{1700}{2200} \times 249 = 192 \text{ kN·m}$$
    (iii) With four No. 20 bars only
    $A_s = 1200$ mm^2, $M_r = 136$ kN·m

294 *REINFORCED CONCRETE DESIGN*

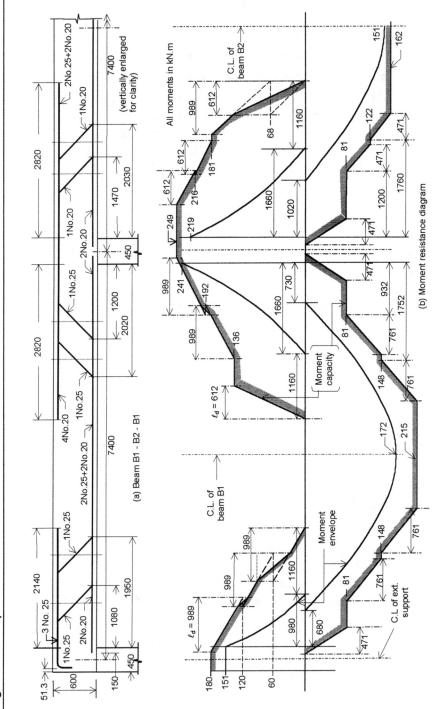

Fig. 9.16 Example 9.4

l_d for No. 20 top bars = $0.45 \times 1.3 \times 1 \times 1 \times 0.8 \times 300 \times 19.5 / \sqrt{20}$
= 612 mm
l_d for No. 20 bottom bars = 612/1.3 = 471 mm
l_d for No. 25 top bars = 989 mm as before.

(iv) For interior span at this section
For two No. 25 and three No. 20 bars, A_s = 1900 mm²,
M_r = 215 kN·m
For two No. 25 bars and two No. 20 bars, A_s = 1600 mm²,
M_r = 181 kN·m
For No. 20 bars only, A_s = 600 mm², M_r = 68 kN·m

3. Positive moment region, exterior span:
   (i) Width of T-beam flange = 1850 mm, and d = 537 mm
   (ii) With full reinforcement, A_s = 1600 mm²
   $$a = \frac{1600 \times 0.85 \times 300}{0.82 \times 0.6 \times 20 \times 1850} = 22.4 \text{ mm} < h_f$$
   $M_r = 0.85 \times 1600 \times 300 \times (537 - 22.4/2) \times 10^{-6}$ = 215 kN·m
   (iii) With one No. 25 bent, A_s = 1100 mm², M_r = 148 kN·m
   (iv) With two No. 25 bent, A_s = 600 mm², M_r = 81 kN·m
   (v) l_d for No. 20 bottom bars = 471 mm
   (vi) l_d for No. 25 bottom bars = 989/1.3 = 761 mm

4. Positive moment region, interior span:
   (i) With full reinforcement
   A_s = 1200 mm², d = 539 mm, a = 16.8 mm, M_r = 162 kN·m
   (ii) With one No. 20 bent, M_r = (3/4) × 162 = 122 kN·m
   (iii) With two No. 20 bent, M_r = (1/2) × 162 = 81 kN·m
   The complete moment capacity diagram is shown in Fig. 9.16b. The factored moment resistance developed at all sections exceeds the factored moment, M_f. The requirement of bar extension by the larger of d or $12d_b$ (= 539 mm here) specified in Code Clauses 12.10.3 and 12.10.4 implies that the moment resistance diagram must have a horizontal clearance of at least d or $12d_b$ from the factored moment diagram at points of bar cut-off. The moment resistance diagram in Fig. 9.16b meets this requirement. Also at least one-third of the negative moment reinforcement must extend beyond the point of inflection for a length equal to the greatest of d, $12d_b$ or $l_n/6$, equal to 1159 mm in this case. The extension provided is 1160 mm.
   The No. 25 bars used as negative reinforcement at the exterior

support must have adequate anchorage within the beam. Providing a standard 90° hook, the basic development length of hook is:
$$l_{hb} = 100 d_b / \sqrt{f'_c} = 563 \text{ mm}$$

Here the yield strength of reinforcement is f_y = 300 MPa. Furthermore, the side cover to the hook is greater than 60 mm and the cover on the bar extension beyond the hook is 51.3 mm > 50 mm. The applicable modification factors are $f_y/400$ and 0.7 (CSA A23.3-94(12.5.3)). Hence, the development length of the hook is:
l_{dh} = 563 × 0.7 × 300/400 = 296 mm

Length available within the girder is:
450 - 51.3 = 399 mm > 296 mm   OK

To check the bar diameter restriction of positive moment reinforcement at inflection points (Code Cl. 12.11.3), the critical section will be at the inflection point closest to the interior support of the outer span as the shear is greatest here. The shear at the inflection point may be taken conservatively as equal to the shear at the support = 199 kN. $l_a = d = 537$ mm.

$$\frac{M_r}{V_f} + l_a = \frac{81 \times 10^6}{944 \times 10^3} + 537 = 944 \text{ mm} > l_d = 471 \text{ mm} \qquad \text{OK}$$

9.14 SPLICING OF REINFORCEMENT

9.14.1 General

Splices are required when reinforcement has to be extended beyond the available length of individual bars or when bars are placed short of the required length, in several stages (such as in columns) for convenience in construction. Splicing of reinforcement is done by butt-welding, mechanical connection, or overlapping the bars to develop their strength through bond.

CSA A23.3 requires that for a full welded splice, the bars be butted and welded to develop in tension at least 120 percent the specified yield strength of the bar, but not less than 110 percent of the actual yield strength of the bar used in the test of the welded splice. The 20 percent increase above f_y ensures yielding and a ductile failure and is adequate for a compression splice as well. All welding of reinforcing steel must conform to CSA Standard W186 - *Welding of Reinforcing Bars in Reinforced Concrete Construction*.

A mechanical connection must also be capable of developing in tension or compression, at least 120 percent of the specified yield strength of the bar, and 110 percent of the actual yield strength of the bar used in the connection test. However, in regions where the area of reinforcement provided is at least twice that required by analysis, welded or mechanical splices of strength less than 1.20 f_y are permitted as specified in CSA A23.3-94 (12.15.4).

9.14.2 Lap Splice

In general, No. 35 and smaller size reinforcing bars are spliced by lapping the bars (Fig. 9.17a). The cracking and splitting behaviour observed in lap splice tests are similar to those in development length tests (compare Fig. 9.5 and Fig. 9.17b). The discontinuity and stress concentrations at each end of a tension lap splice tend to induce early cracking and splitting from both ends. Evaluation of results of both development length and splice tests shows that the same parameters influence the length requirements in both cases and, the same expression (Eq. 9.4) can be used to determine both development and splice lengths (Ref. 9.2). The maximum reinforcement stress that is allowed in design is the yield strength. To avoid sudden failure at the splice, a full strength lap splice should develop more than the yield strength of the reinforcement. This will ensure yielding of bars and, thereby, a ductile behaviour of the member.

The length of splice required depends on the computed stress in the bar at the

Fig. 9.17 *Spliced bars*

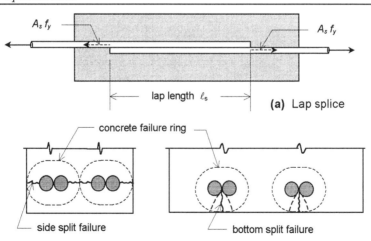

splice location and the proportion of the total number of bars that are spliced in the same region. As far as possible, splices should be located away from points of maximum stress in the reinforcement and should be staggered. The Code requirements for splice length are based on the development length, l_d, the length being increased depending on the severity of the stress condition. However, the comparisons in Ref. 9.2 indicate that the Code requirements are very conservative for lap splices in regions of high stress.

9.14.3 Code Requirements for Lap Splices

CSA A23.3-94 (12.14) permits lap splices for No. 35 and smaller size bars only except as provided in Clauses 12.16.2 and 15.8.2.5. For larger bars, splice length requirements have not been established. Lap splices are generally made with the bar in contact, and in the case of noncontact splices the transverse bar spacing is limited to a maximum of one-fifth the splice length or 150 mm, whichever is less.

(a) Tension Splices of Deformed Bars and Deformed Wires

Tension lap splices are classified as A or B, with corresponding minimum specified lap lengths. The classification as A and B depends on the design stress in the bar in the splice region and the fraction of bars spliced within a lap length, as detailed in Table 9.3 wherein the corresponding minimum required lap lengths are also indicated. In this table, l_d is the development length for a tensile stress of f_y computed as given in Section 9.9.1, but *without* the modification factor for excess reinforcement. The minimum limit of 300 mm on l_d need not be applied; however, the lap length computed from Table 9.3 is subject to a minimum of 300 mm.

Table 9.3 Tension Lap Splice Classification and Specified Minimum Lap Lengths, l_s

$\left(\dfrac{A_{s\ provided}}{A_{s\ required}}\right)$	Maximum Percent of A_s Spliced Within Required Lap Length	
	50	100
≥ 2.0	Class A $l_s = 1.0\ l_d$	Class B $l_s = 1.3\ l_d$
< 2.0	Class A $l_s = 1.3\ l_d$	Class B $l_s = 1.3\ l_d$

l_s shall not be taken less than 300 mm.

(b) Compression Splices of Deformed Bars

The minimum lap length is specified as not less than: $0.073\ f_y\ d_b$ for $f_y \leq 400$ MPa and $(0.133\ f_y - 24)\ d_b$ for $f_y > 400$ MPa, or 300 mm.

When the splice is enclosed throughout its length by minimum ties or spirals, reduced lap lengths are permitted (CSA A23.3-94(12.17)), subject to a minimum requirement of 300 mm.

In vertical bars required for compression only, end bearing splices are also permitted (Code Cl. 12.16.4).

(c) Other Cases

The Code gives specifications for the splicing details for bundled bars, welded wire fabric (both plane wire and deformed wire), and columns (CSA A23.3-94 Clause 12). For pairs of U-stirrups turned over to form a closed unit, the lap length required is $1.3l_d$. However, in members at least 450 mm deep, such splices with $A_b f_y$ not more than 40 kN per leg may be considered effective if the legs extend the full available depth of the member.

REFERENCES

9.1 ACI Committee 408, *Bond Stress - The State of the Art*, J. ACI, Vol. 63, Oct. 1966, pp. 1161-1190.

9.2 Orangun, C.O., Jirsa, J.O., and Breen J.E., *A Reevaluation of Test Data on Development Length and Splices*, J. ACI, Vol. 74, Mar. 1977, pp. 114-122.

9.3 Kemp, E.L., and Wilhelm, W.J., *Investigation of the Parameters Influencing Bond Cracking*, J. ACI, Vol. 76, Jan. 1979, pp. 47-71.

9.4 Jimenez, R., White, R.N., and Gergely, P., *Bond and Dowel Capacities of Reinforced Concrete*, J. ACI, Vol. 76, Jan. 1979, pp. 73-92.

9.5 Soretz, S., and Holzenbein, H., *Influence of Rib Dimensions of Reinforcing Bars on Bond and Bendability*, J. ACI, Vol. 76, Jan. 1979, pp. 111-125.

9.6 Ferguson, P.M., *Reinforced Concrete Fundamentals*, 4th ed., John Wiley and Sons, New York, 1979.

9.7 ACI Standard 318-95, *Building Code Requirements for Structural Concrete*, and *Commentary* (ACI 318R-95), American Concrete Institute, Detroit, Michigan, 1995, 369 pp.

9.8 Untrauer, R.E., and Warren, G.E., *Stress Development of Tension Steel in Beams*, J. ACI, Vol. 74, Aug. 1977, pp. 368-372.

9.9 ACI Committee 408, *Suggested Development, Splice, and Standard Hook Provisions for Deformed Bars in Tension*, Concrete International, ACI, V.1, N.7, July 1979, pp. 44-46.

9.10 Marques, J.L.G., and Jirsa, J.O., *A Study of Hooked Bar Anchorages in Beam-Column Joints*, J. ACI, Vol. 72, May 1975, pp. 198-209.

CHAPTER 10 Continuity in Reinforced Concrete Construction

10.1 GENERAL

Design of flexural members for a given bending moment and shear force was explained and illustrated in Chapters 5 to 8. The detailing of reinforcement in members with known variations of bending moment was presented in Chapter 9. Most of the examples in these chapters dealt with simple beams, in which the bending moments and shear forces were determined from statics. Where a continuous beam example was used, it was assumed that the moment variation was known.

Reinforced concrete structures are usually cast-in-place, with all the structural elements such as slabs, beams, girders, columns, and foundations cast monolithically. Such a continuous structure is statically indeterminate, and before the individual member dimensions and reinforcement can be determined by the methods described in Chapters 5 to 9, the distribution of design moments and shear forces along the individual members need to be determined. This is a problem of structural analysis. However, the analysis of a complete three-dimensional space frame subjected to various combinations of loading and different patterns of loading is very complex. Simplifying approximations, which are permitted design codes are usually made. The basic approximation involves the subdivision of the structure into smaller, two-dimensional subframes (floors or roof, walls and plane frames; Fig. 2.3), which in most instances can be further simplified into continuous beams or partial frames for the analysis of *gravity load* effects (Section 10.2).

Even in a detailed analysis of the complete frame, as for an unusual structure or structures of major consequence, the approximate analysis methods provide a handy tool to (1) proportion members in the preliminary design stage to obtain relative member stiffnesses needed for the detailed analysis, and (2) provide a rough check on the results of the detailed analysis.

The maximum effects (moment, shear, etc.) on members of continuous structures due to the design loads (factored loads) are usually determined by elastic analysis (CSA A23.3-94(9.3)). There is an apparent inconsistency between the analysis and design procedures, as the former uses elastic theory, while the latter is based on inelastic section behaviour. However, the procedure leads to safe and satisfactory designs, and the procedure is justified if a linear moment-curvature (M - ψ) relationship is assumed up to a moment of $M = M_r$. Since the design moments at various critical sections are determined for different loading patterns (Section 10.2), for any one loading pattern most sections will have $M < M_r$, and a few sections will just reach $M = M_r$, so that the entire structure is within the linear range of the M - ψ relationship. Under-reinforced sections, as permitted by the Code, have a nearly linear M - ψ relationship up to $M = M_n$ (Fig. 4.7a), and the inelastic analysis procedures such as "limit design" and "yield-line theory" for slabs usually assume such a linear M - ψ variation up to $M = M_n$. The Code recognises some inelastic response by permitting a limited redistribution of the moments in continuous flexural members calculated by the elastic theory (CSA A23.3-94(9.2.4), Section 10.7).

For a first order elastic analysis, where the effects of deflection and axial deformation are not included, the gravity load effects and lateral load (wind or earthquake) effects can be determined separately and superimposed. As is shown in Section 10.2, the gravity load effects can often be determined by considering one floor at a time, together with the columns framing into that floor, and analysing this partial frame. For lateral loads, however, each bent, consisting of the columns in *all* storeys in the bent together with the horizontal floor members (beam and/or slabs) connecting these columns, is analysed as a plane frame, with the horizontal loads generally applied at the joints (Fig. 2.3). Again, for buildings of moderate height, approximate methods of analysis may be used to determine wind effects on frame members.

To simplify the analysis, a three-dimensional framed structure is treated as a series of independent parallel plane-frames (bents) along the column lines in the longitudinal and transverse directions (Fig. 10.1). For gravity load effects, these plane frames can be further simplified into continuous beams or partial frames, and analysed with the loads so placed as to produce the worst effects at the design section considered. The critical loading patterns and the simplifications applicable for continuous beams and plane frames under vertical loads are considered next.

10.2 LOADING PATTERNS FOR CONTINUOUS BEAMS AND PLANE FRAMES

While the dead loads, by their nature, act throughout the structure at all times, the live load may act on any or all spans. In order to determine the maximum (and minimum) effects of the design load at any section in a member of a continuous beam or frame, it is first necessary to identify the spans to be loaded with live loads so as to create the worst effects. This is conveniently done by sketching, qualitatively, the shape of the influence line for the effect in question, using the Müller-Breslau principle (Ref. 10.1). The principle is that the influence line for any action (moment, shear force, reaction, etc.) at a section is given by the deflected shape of the structure resulting from a unit displacement *corresponding*

302 REINFORCED CONCRETE DESIGN

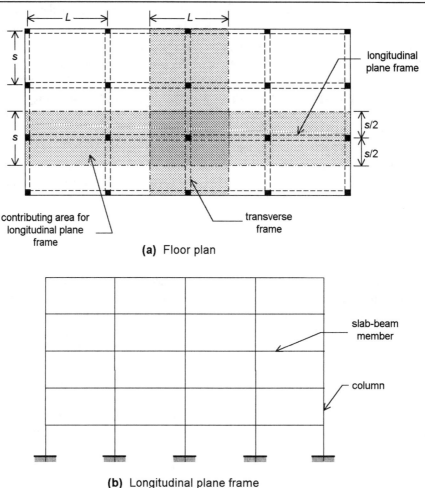

Fig. 10.1 Building frame approximated as series of plane frames

to that action. Influence lines for the moments and shear forces at two sections in a continuous beam, and for the moments at two sections in a plane frame are shown in Figs. 10.2 and 10.3, respectively.

In continuous beams, the dead load which acts on all spans normally produces positive moment (that is, moment causing tension at the bottom fibres of the beam) in midspan regions and negative moments at support sections. Regarding the positioning of live loads, from the shape of the influence lines in Fig. 10.2 it can be concluded that:

1. the maximum positive moment in a span occurs when live load is placed on that span and every alternate span (Fig. 10.2 b(ii));

CHAPTER 10 CONTINUITY IN CONSTRUCTION 303

Fig. 10.2 Influence lines and load (gravity) patterns for continuous beams

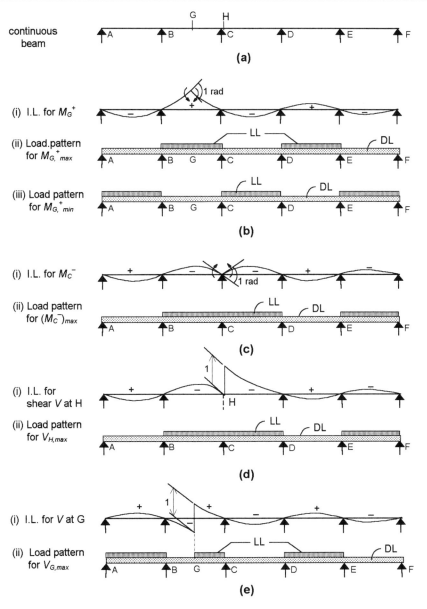

2. the minimum positive moment in a span occurs when live load is placed on alternate spans, with the span under consideration unloaded (Fig. 10.2b(iii));

304 REINFORCED CONCRETE DESIGN

3. the maximum negative moment at the support section occurs when live load is placed on the two neighbouring spans and then every alternate span (Fig. 10.2c(ii)); and
4. the influence on the desired effect at a section, of loads on spans far removed from the section under consideration, is relatively small.

Indeed, the alternating loading pattern shown in Fig. 10.2b gives the maximum positive moments, in *all* spans with live loads and the minimum positive moments in *all* spans with dead load only. With certain span proportions and live load/dead load ratios, the "minimum positive" span moment may be negative. For instance, a lightly loaded short span, flanked on either side by heavily loaded long spans, may have a net negative moment at midspan under this loading condition. Such a beam is subjected to positive or negative moment at midspan, depending on the loading pattern, and has to be reinforced with tension reinforcement placed at the top as well as at the bottom in the midspan regions. Minimum negative moments at support sections are not usually determined, as there is seldom any chance of reversal of moment at this section under gravity loading. (Reversal of moment at support sections could, however, occur under lateral loading or settlement of supports.)

The fourth conclusion above indicates that, when the spans and loadings do not vary a great deal, there is no great loss of accuracy if the influence of spans more than two supports away from the section under consideration are ignored. Thus, in Fig. 10.2, the positive moment in span *BC* can be determined with reasonable accuracy by considering the portion *ABCD* only, and assuming that the beam is fixed at support *D* (beyond which the beam is, in fact, continuous). Similarly, the negative moment at support *C* may be obtained by analysing the part *ABCDE* of Fig. 10.2c(ii), with support *E* assumed as fixed.

Figures 10.3a and b show the influence lines and loading patterns for the maximum positive moment at midspan, and the maximum negative moment near the right end, for the interior beam *BC* of a plane frame, respectively. The maximum positive moment at midspan of *BC* is obtained for the checkerboard pattern of live load placement, with the total load (dead + live) applied on the span in question and on alternate spans, and the dead load only on the remaining spans (Fig. 10.3a). This loading pattern produces the maximum positive span moments in all spans with total load and minimum positive span moments in all spans with dead load only.

In Fig. 10.3b, the influence line is partially positive and partially negative along the beams above and below beam *BC*. Correctly, these two beams should receive live load only partially. However, for simplicity it is common practice in influence lines decrease as one moves away from the spans under consideration, and particularly so as one moves on to floors at other levels. Therefore, in analysing for gravity load effects on flexural members on a floor, the effects of loads on other floors may be neglected. Also, in usual building frames of regular

Fig. 10.3 *Gravity load patterns for plane frame*

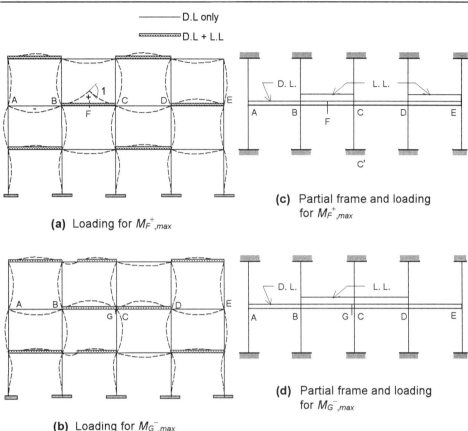

(a) Loading for $M_{F,max}^{+}$

(b) Loading for $M_{G,max}^{-}$

(c) Partial frame and loading for $M_{F,max}^{+}$

(d) Partial frame and loading for $M_{G,max}^{-}$

shapes with no pronounced asymmetry in the geometry and loading, the sidesway caused by gravity loads is small and can be neglected. With these approximations, the analysis for gravity load effects on floor members may be made with reasonable accuracy by considering one floor at a time together with the columns framing into that floor, with the far ends of these columns assumed as fixed. This limited frame can be analysed with the loads placed so as to produce the worst effect at critical sections. Thus, the analysis for the cases in Fig. 10.3a and b may be simplified to the analysis of the corresponding partial frames shown in Fig. 10.3c and d, respectively.

For a column, the loading pattern producing the maximum moment and that producing the maximum axial load are different. Therefore, the maximum axial load and maximum moment do not occur simultaneously. The loading pattern for maximum moment in columns, shown in Fig. 10.4a for column CC',

Fig. 10.4 Gravity load pattern for maximum column moment

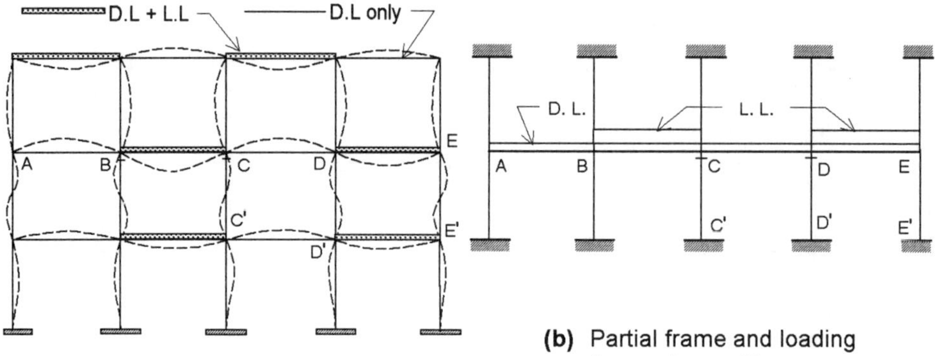

(a) Loading pattern for column CC'

(b) Partial frame and loading for maximum $M_{cc'}$

can once again be determined by sketching the influence line. However, although the loading in Fig. 10.4a gives the maximum moments at the ends of column CC', the column bends in reverse curvature. In a relatively long column, a checkerboard pattern of loading causing single curvature bending of the column may be more critical than the pattern shown in Fig. 10.4a. Again the moment at end C of column CC' in Fig. 10.4a is approximately given by the moment obtained from the analysis of the corresponding partial frame, shown in Fig. 10.4b. For the partial frame, the total column moment at a support such as C is distributed between the two columns above and below C in proportion to their stiffnesses. Somewhat greater column moments will be obtained, particularly in the exterior columns (column EE' in Fig. 10.4), if the influence of loading on the adjacent floors is also taken into account; however, even for these columns the partial frame analysis is generally satisfactory, except for highly irregular spans or loadings. In general, column moments are much more sensitive to changes in the assumption and cannot be determined with the same degree of accuracy as beam moments.

10.3 APPROXIMATIONS PERMITTED BY CODE FOR FRAME ANALYSIS

CSA A23.3-94(9.1) permits the use of approximate methods of frame analysis for buildings of usual types of construction, spans and storey heights. Specific approximations suggested are summarised below.

10.3.1 Floor Members

In accordance with the general frame behaviour discussed in Section 10.2, CSA A23.3-94(9.3) allows the approximate frame analysis for vertical loading of floor (or roof) members of continuous construction by considering live load applied only to the floor (roof) under consideration and the far ends of the columns assumed as fixed. The loading patterns to be considered are (1) factored dead load on all spans plus factored live load on two adjacent spans (Fig. 10.3d), giving the design negative moment at supports, and (2) factored dead load on all spans with factored live load on alternate spans (Fig. 10.3c), giving the maximum and minimum positive span moments on spans with and without live loads, respectively. In determining the bending moment at a support, the beam member may be assumed as fixed at any support two panels distant, provided the beam continues beyond that point (CSA A23.3-94(13.9.1.3)). In analysing for maximum and minimum positive moments in spans, the far ends of adjacent spans may similarly be considered as fixed.

10.3.2 Columns

For columns (CSA A23.3-94(10.10.1)), the loading pattern with factored live load on all floors above, which results in the maximum axial force (subject to the reduction allowed when the total tributary floor area supported by the column is large), and factored live load on a single (usually the larger and heavier loaded) adjacent span of the floor under consideration, which produces the maximum moment that can accompany the maximum axial force, is the most general combination for gravity loading. In addition, because the interaction between axial strength and flexural strength of columns is non-linear (Chapter 15), it is also necessary in column design to consider the loading condition which produces the maximum ratio of bending moment to axial force. This condition generally occurs under the checkerboard pattern of loading. In computing gravity load moments in columns by analysing partial frames, the far ends of columns monolithic with the structure may be considered fixed as shown in Fig. 10.4b.

10.3.3 Moment Coefficients

For the design of continuous beams and one-way slabs with two or more approximately equal spans (the longer of two adjacent spans not exceeding the shorter by more than 20 percent) and with loads uniformly distributed, where the factored live load does not exceed two times the factored dead load, CSA A23.3-94(9.3.3) specifies approximate moment and shear values which may be used in lieu of more accurate analysis. These values are given in Table 10.1 and are generally conservative. The load patterns that produce the critical values for column moments (Section 10.3.2) are different from those for maximum negative moments in beams. Therefore, the moment values in Table 10.1 should not be used to evaluate column moments.

Table 10.1 Moments and Shears in Continuous Beams Using CSA Coefficients (CSA A23.3-94(9.3.3))

(a) **POSITIVE MOMENTS**
  (i) End spans
    (1) if discontinuous end is unrestrained $\quad \dfrac{1}{11} w_f l_n^2$
    (2) if discontinuous end is integral with support $\quad \dfrac{1}{14} w_f l_n^2$
  (ii) Interior spans: $\quad \dfrac{1}{16} w_f l_n^2$

(b) **NEGATIVE MOMENTS**
  (i) At exterior face of first interior support:
    (1) two spans $\quad \dfrac{1}{9} w_f l_n^2$
    (2) more than two spans $\quad \dfrac{1}{10} w_f l_n^2$
  (ii) At other faces of interior supports $\quad \dfrac{1}{11} w_f l_n^2$
  (iii) At interior faces of exterior supports for members built integrally with their supports
    (1) where the support is a spandrel beam or girder $\quad \dfrac{1}{24} w_f l_n^2$
    (2) where the support is a column $\quad \dfrac{1}{16} w_f l_n^2$

(c) **SHEAR**
  (i) in end members at face of first interior support $\quad 1.15 \dfrac{w_f l_n^2}{2}$
  (ii) at face of all other supports $\quad \dfrac{w_f l_n^2}{2}$

Notes: w_f = factored load per unit length of beam

l_n = clear span for positive moment, negative moment at an exterior support, or shear, and the average of adjacent clear spans for negative moment at interior supports.

10.3.4 Moment Envelopes for Use with CSA Moment Coefficients

To facilitate the detailing of bending or cut off of flexural reinforcement for members designed using the CSA moment coefficients, approximate moment envelopes and inflection points corresponding to these moment coefficients may be obtained based on the following considerations:

1. In the exterior span, the maximum positive moment in the span and the maximum negative moment at the exterior support occur simultaneously under the same loading pattern (Fig. 10.5a).
2. In interior spans, when the span moment is maximum, the negative moments at the two supports are assumed equal (Fig. 10.5b).
3. The specified positive moment in spans is assumed to be at the midspan section. Although the actual maximum positive moment may be slightly greater than the midspan moment, the latter is considered as the design moment. (An alternative procedure is to take the maximum span moment as the specified value.)
4. The loading patterns for the maximum negative moments at each end of a given span are different. However, to obtain the negative moment envelopes, it is conservatively assumed that these maximum negative moments occur simultaneously. Two typical cases are shown in Fig. 10.5c.

Based on these approximations, the moment diagrams and the points of inflection may be computed using statics. The moment diagrams so obtained for the various moment coefficients are shown non-dimensionally in Fig. 10.6, which may be used to determine locations where tension reinforcement is no longer required to resist flexure.

10.4 ANALYSIS PROCEDURES

Several methods of analyses are specified in the Code. These are elastic frame analysis approximate frame analysis, analysis by strut-and-tie models, elastic stress analysis, elastic plate analysis, and plastic analysis (Refs. 10.1, 10.2, 10.3). For the analysis of partial frames, as detailed in Section 10.2, the method of moment distribution is convenient and still widely used. Ref. 10.4 shows that for building frames analysed for gravity loads by considering each floor with the far ends of the columns as fixed, two-cycles of moment distribution are adequate to yield results of sufficient accuracy.

310 REINFORCED CONCRETE DESIGN

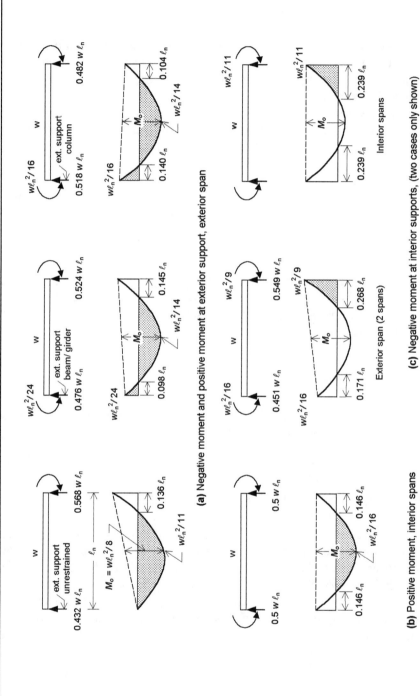

Fig. 10.5 Moment diagrams for CSA coefficients (typical cases)

Convenient tabular arrangements for computing maximum and minimum positive span moments, negative moments at beam ends, and column moments are also given in Ref. 10.4 and 10.5.

10.5 STIFFNESS OF MEMBERS

Regardless of the procedure used for the analysis of indeterminate frames, the stiffnesses of the members must be established first. For a prismatic member EI and GJ give the flexural and torsional stiffnesses, where I and J are, respectively, the moment of inertia and the torsional inertia (cross-sectional property analogous to the polar moment of inertia of a circular section) of the member

Fig. 10.6 Non-dimensionalized moment diagrams for CSA moment coefficients

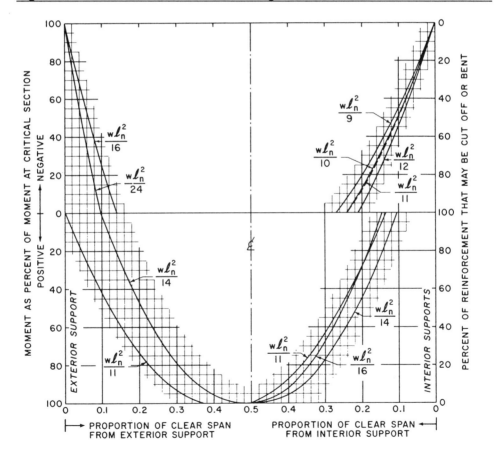

cross section. The modulus of elasticity $E = E_c$ is given by Eq. 1.1, and the shear modulus G may be taken as $E_c/2 (1 + \mu) \approx E_c/2$, (assuming Poisson's ratio, μ, for concrete as zero). To compute I and J, the cross-sectional dimensions, which are not initially known, must be assumed or estimated based on preliminary design. The relative stiffnesses and the analysis may have to be back-checked and perhaps revised, if the final dimensions selected are markedly different from the preliminary values.

Moment of inertia, I, of a reinforced concrete section is usually computed on the basis of either the gross concrete section (neglecting reinforcement) or the cracked section transformed to concrete. Ideally, the effective member stiffness values used in the frame analysis should reflect the degree of cracking, inelastic action and the amount of reinforcement along the member; however, it is difficult to account accurately for all such influences. CSA A23.3-94(9.2.1.1) allows the use of any reasonable and "consistent" assumptions for computing the flexural and torsional stiffnesses.

For braced frames, only relative values of stiffnesses are required. In such situations, the assumption used to compute I does not significantly affect the analytical results, as long as the same assumption is used for all members. To account for the higher degree of flexural cracking in beams relative to columns, the I values used for braced frames are one-half of the gross section I for beams and the gross section I for columns (Ref. 10.6). For unbraced or lightly braced frames, CSA A23.3-94(10.14.1) prescribes that the I values used be 0.35 times the gross section I for beams and 0.7 times the gross section I for columns. For T-beams, the gross section I may be approximated as two times the gross section I for the web.

The precracking torsional stiffness of a rectangular beam is only about 10 to 20 percent of its flexural stiffness. Torsional cracking reduces the torsional stiffness to about 10 to 20 percent of its precracking value, so that the torsional stiffness becomes almost negligible compared to the corresponding flexural stiffness. Therefore, in indeterminate structures where torsion arises because the member must twist to maintain compatibility, it is reasonable to neglect torsional stiffness in the analysis of frames. Thus, the torsional stiffness of beams transverse to the frame being analysed is usually neglected except in two-way slab floor systems (Ref. 10.6). If the torsional stiffness is required for an analysis, the J values may be taken as 0.2 times the gross section J for service loads and 0.15 times the gross section J for ultimate loads.

Another question that arises in computing the stiffness of beams is the effect of flanges, and the flange width to be considered as contributing to the stiffness. In a one-way floor system, such as shown in Fig. 10.7a, for the analysis of a plane frame along the columns in the east-west direction, the floor loading to be considered is that placed over the width b. It would appear that the slab-beam member whose flexural stiffness is to be considered in the analysis of the frame is the T-beam with a flange width of b. However, with large values of b, the assumption that the entire flange width is fully effective over the entire span is questionable, particularly with the flanges in the tension zone in regions of negative moment. When flanges are in tension, the Code (Cl. 10.5.3) does provide for the distribution of main tension reinforcement over the "effective flange width" used in the design of sections. Therefore, it would be appropriate to use the same effective flange width to compute the moment of inertia as is used in design computations according to CSA A23.3-94(10.3), and the authors prefer this procedure.

An alternative procedure that has been suggested in Ref. 10.4 is to use twice the moment of inertia of the gross web section ($I - 2 \times b_w h^3/12$). This closely corresponds to an effective flange width of six times the web width for T-sections with a flange thickness to overall depth ratio (h_f/h) of 0.2 to 0.4. One advantage of this procedure is that it eliminates the need to compute the moment of inertia of the T-shaped section.

Yet another common situation, where the assumptions regarding the moment of inertia adopted for the floor members vary amongst designers, is the one-way slab-beam-girder floor system shown in Fig. 10.7b. One procedure is to treat the interior beam marked $B1$ as a continuous T-beam supported on girders and carrying the gravity load on its tributary area of width b'. In the analysis of this continuous beam, the I of the effective T-section of the beam $B1$ is used for computing flexural stiffness. The torsional restraint offered by the girders may be neglected and the girders assumed to offer only vertical supports, except that some allowance is made for the restraint (and resulting negative moment) at the exterior support. A separate analysis is made for the beam $B2$ along the column line, considering it as a continuous T-beam supported and restrained by the columns. This procedure will usually result in slightly different designs for beams $B1$ and $B2$. An alternative procedure is to make all the beams within a panel width b tributary to a bent identical, and take their combined stiffness as the beam-member stiffness in the analysis of the frame along the column line. This procedure will be more appropriate when the girders are torsionally very stiff. In the case of two-way slab systems with or without beams along column lines (Figs. 2.5, 2.6, and 2.9b), CSA A23.3-94(13.9) specifies the use of the gross moment of inertia of the full panel width to compute the stiffness of the slab-beam member for frame analysis.

Sometimes beams are provided with haunches in order to increase their depth at supports in continuous spans, where the negative moment and shear force are greater than in the span. Similarly, drop panels are frequently used around columns in flat slabs (Chapter 14). In such cases the effects of the varying depth must be included in the analysis. CSA A23.3-94(13.9) gives approximate methods of calculating member stiffness of non-prismatic members.

Fig. 10.7 Definition of slab-beam member for analysis of bent in east-west direction

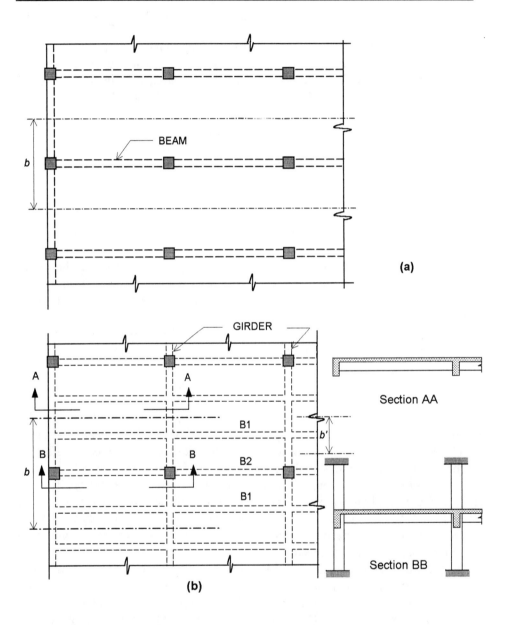

Finally, to calculate gravity load deflections, CSA A23.3-94(9.8.2.3 and 9.8.2.4) prescribes the calculation of I to be used. This is discussed in detail in Chapter 11.

10.6 USE OF CENTRE-TO-CENTRE SPAN AND MOMENT AT SUPPORT FACE

In conventional analysis of frames, the members are represented by their centrelines and the resulting member lengths used (Fig. 10.8a,c). CSA A23.3-94(9.2.2) also specifies the use of centre-to-centre distances for span lengths to determine the moment diagram. The use of centreline dimensions implies the following approximations:

1. The support reaction is concentrated at the support centrelines instead of being distributed in some fashion over the width of support (width of column).
2. The beam is prismatic up to the support centreline, and the beam stiffness is unaffected by the increased effective beam section between the column face and column centreline where the beam merges into the column, (portion BC in Fig. 10.8a,b).
3. The beam centreline can deflect a column face (point B', in Fig. 10.8c) rather than such movement being restricted by the column.

The effect of the first assumption on the moment diagram is usually insignificant and can be ignored (Ref. 10.4). The neglect of the increased stiffness and restraint of the beam within the column width results in a slight over-estimation of the positive moment at the midspan and a corresponding under-estimation of the negative moment at the support. These effects can be reasonably accounted for by assuming that the moment of inertia of the beam section is infinite over the width of the column (Fig. 10.8b). Such an adjustment will result in an upward shift of the theoretical moment diagram in Fig. 10.8d by $Vb/6$, where V is the computed beam shear at the column centreline and b is the width of the column (Ref. 10.4). Since the effective beam depth is greatly increased once it merges into the column, the critical section of the beam for negative moment is generally at the column face. The Code (9.2.2) also permits beams and girders to be designed for the moments at the faces of supports. In Fig. 10.8d, $M_c - Vb/3$ gives the adjusted design moment at the face of the support, where M_c is the theoretically computed moment at the support centreline. As far as the midspan section is considered, it is satisfactory and conservative to design for the theoretical moment, ignoring the correction $Vb/6$ which is usually small. For the analysis of frames with two-way slab systems, CSA A23.3-94(13.9) specifies that account be taken of the enhanced moment of inertia of both slab-beam members and columns at their intersections (Chapter 14). When such allowance is made in the analysis, the shift of the moment diagrams indicated in Fig. 10.8d need not be made, and the design moment at the face of the support is obtained directly as $\approx M_c - Vb/2$.

Fig. 10.8 Moment correction to account for support width in continuous beams

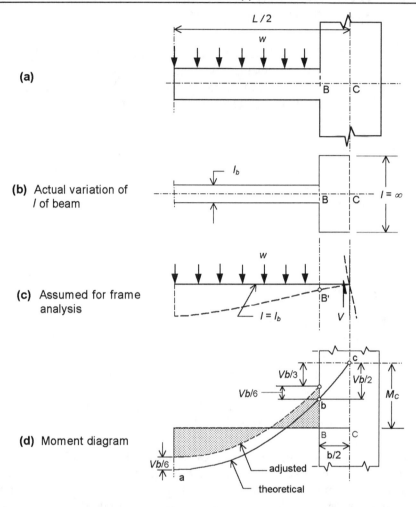

(a)

(b) Actual variation of *I* of beam

(c) Assumed for frame analysis

(d) Moment diagram

The conditions described above and in Fig. 10.8, with reference to a beam built into the supporting member, also exist in a column (Fig. 10.9). Generally for the analysis, the length of the column is taken as equal to the centre-to-centre distance or storey height, and the increase in moment of inertia of the column section within the beam depth is neglected (except for two-way floor systems presented in Chapter 14). With the lower moment gradient that usually exists in columns, the adjustments to be made to the centreline moment to obtain the column moment at the beam face is small (Fig. 10.9b). Therefore, the column moment may be taken as that at the centreline of the beam, as obtained from the analysis.

Fig. 10.9 Column moments from frame analysis

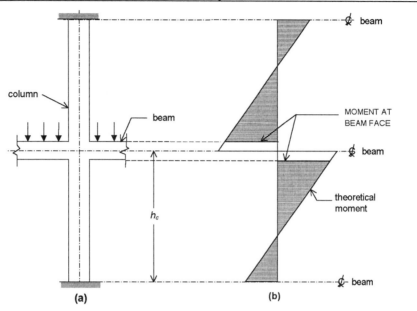

(a) (b)

Fig. 10.10 Example 10.1

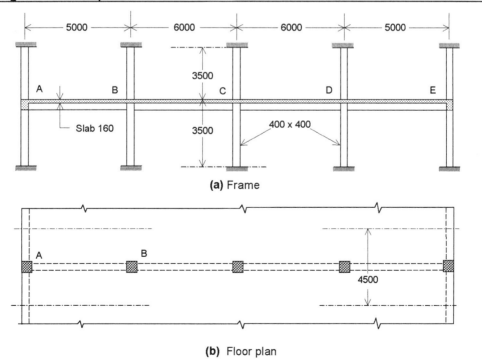

(a) Frame

(b) Floor plan

EXAMPLE 10.1

Centreline dimensions of a partial frame, consisting of one floor (slab-beam member) and the columns immediately above and below the floors, are given in Fig. 10.10a. The slab is one-way, spanning the frames spaced at 4.5 m (Fig. 10.10b), and has been designed as 160 mm thick. The specified live load on the floor is 5.44 kN/m^2, and there is a dead load of 0.5 kN/m^2 due to floor finish, ceiling, etc. The column size is estimated as 400 mm × 400 mm. Beams are provided only along the column lines in the longitudinal directions, as shown in Fig. 10.10b. Use $f_c' = 30$ MPa and $f_y = 400$ MPa.

Analyse the frame for gravity loads and determine the moment envelope.

SOLUTION

The load per metre length of the beam is that coming from a width of 4.5 m (equal to the sum of one-half panel width on either side of the frame).

Dead load from slab = $0.16 \times 4.5 \times \dfrac{2400 \times 9.81}{1000} = 17.0$ kN/m

Dead load of beam stem, assuming overall beam size of 300 mm × 450 mm,
= $0.3 \times (0.45 - 0.16) \times 2.4 \times 9.81 = 2.05$ kN/m
Dead load due to floor finish, etc. = $0.5 \times 4.5 = 2.25$ kN/m
Live load from slab = $5.44 \times 4.5 = 24.5$ kN/m
Factored dead load = $1.25(17.0 + 2.05 + 2.25) = 26.7$ kN/m
Factored live load = $1.5 \times 24.5 = 36.7$ kN/m

To estimate beam size, the maximum negative moment in the beam may be taken approximately as:

$$\dfrac{w_f l_n^2}{10} = \dfrac{(26.7 + 36.7)(6.0 - 0.4)^2}{2} = 199 \text{ kN} \cdot \text{m}$$

and maximum shear as:

$$V_f = 1.15 \times \dfrac{w_f l_n^2}{2} = 1.15 \times \dfrac{(26.7 + 36.7)(6.0 - 0.4)}{2} = 204 \text{ kN}$$

For negative moment, the T-beam acts as a rectangular beam. Assuming a reinforcement ratio of about $0.65 \rho_{max}$, from Table 4.1, $K_r = 4.37$.

Required $bd^2 = \dfrac{M_r}{K_r} = \dfrac{199 \times 10^6}{4.37} = 45.5 \times 10^6 \text{ mm}^3$

Taking b = 300 mm, required $d = \sqrt{45.5 \times 10^6 / 300} = 390$ mm.
An overall depth of h = 450 mm initially assumed should be adequate for moment, although the reinforcement ratio may be slightly higher than $0.65\rho_{max}$. The shear resistance, V_c, with this depth ($d \approx 450 - 65 = 385$) is:

$V_c = 0.2\lambda\varphi_c\sqrt{f'_c}b_w d = 0.2 \times 1.0 \times 0.6 \times \sqrt{30} \times 300 \times 385 \times 10^{-3} = 75.9 \text{ kN}$

$V_s = V_f - V_c = 178 - 75.9 = 102 \text{ kN} < V_{s,\lim} = 0.8\lambda\varphi_c\sqrt{f'_c}b_w d = 304 \text{ kN}$

Depth is adequate for shear

Minimum depth for deflection control (Table 5.2) is, for end spans, (5000-400)/18.5 = 249 mm, and for interior spans, (6000-400)/21 = 267 mm. The preliminary design of 300 × 450 mm for the beam section is OK.

The frame will be analysed using the moment distribution method (Ref. 10.1)

Member Stiffness
Stiffnesses will be computed based on gross concrete area. For the slab/beam member, the moment of inertia of the effective T-section used in design (Section 5.9) will be used.
For columns,

$$I_c = 400 \times 400^3 / 12 = 2.13 \times 10^9 \text{ mm}^4$$

$$K_c = \dfrac{4E_c I_c}{h_c} = \dfrac{4E_c \times 2.13 \times 10^9}{3500} = 2.43 E_c \times 10^6$$

For the T-beam, the effective width of flange is the least of:

$\left. \begin{array}{l} 0.1 \text{ span } \times 2 + b_w = 1300 \text{ mm for exterior span} \\ \qquad\qquad\qquad\quad = 1500 \text{ mm for interior spans} \end{array} \right\}$ controls

$24 h_f + b_w \qquad = 24 \times 160 + 300 = 4140 \text{ mm}$

spacing of beams = 4500 mm

For exterior span, the section is shown in Fig. 10.11a

320 *REINFORCED CONCRETE DESIGN*

$$\bar{y} = \frac{1000 \times 160 \times 80 + 300 \times 450 \times 225}{1000 \times 160 + 300 \times 450} = 146 \text{ mm}$$

$$I_{b1} = \frac{1300 \times 146^3}{3} + \frac{1000 \times 14^3}{3} + \frac{300 \times 304^3}{3} = 4.16 \times 10^9 \text{ mm}^4$$

$$K_{b1} = \frac{4EI_{b1}}{L_1} = \frac{4E_c \times 4.16 \times 10^9}{5000} = 3.33 E_c \times 10^6$$

(Note that if I_b is taken as twice I of the stem of the Tee, $I_b = 2 \times 300 \times 450^3/12 = 4.56 \times 10^9$ mm^4. This is close to the I_{b1}, computed by detailed calculation (4.16×10^9), and is sufficiently accurate for relative stiffness computations).

Similarly, for interior beams *BC* and *CD* (Fig. 10.11b);

$$I_{b2} = 4.35 \times 10^9 \text{ mm}^4$$

$$K_{b2} = 2.90 \, E_c \times 10^6$$

The torsional stiffness of the edge beams at *A* and *E* will be ignored.

The frame analysis will be done by the method of moment distribution. The distribution factors at each joint for the beam members are:

$$\text{Joint } A : D_{AB} = \frac{K_{b1}}{\Sigma K} = \frac{3.33}{2 \times 2.43 + 3.33} = 0.407$$

$$\text{Joint } B : D_{AB} = \frac{K_{b1}}{\Sigma K} = \frac{3.33}{2 \times 2.43 + 3.33 + 2.90} = 0.300$$

$$D_{BC} = \frac{2.90}{2 \times 2.43 + 3.33 + 2.90} = 0.261$$

$$\text{Joint } C : D_{CD} = D_{CD} = \frac{290}{2 \times 2.43 + 2 \times 2.90} = 0.272$$

Fig. 10.11 *Effective section for slab-beam member*

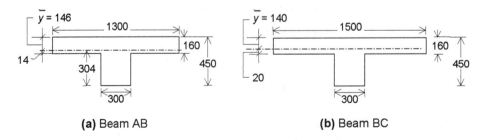

(a) Beam AB **(b)** Beam BC

The fixed-end moments ($w_f L^2/12$) and simple span moments ($w_f L^2/8$) needed for moment calculations are tabulated in Table 10.2.

Table 10.2 Loading patterns for critical moments in beam

	Exterior span		Interior span	
	FEM (kN·m)	S.S. moment (kN·m)	FEM (kN·m)	S.S. moment (kN·m)
Dead load only (26.7 kN/m)	55.6	83.4	80.1	120
Live load only (36.7 kN/m)	76.5	115	110	165
Total load (63.4 kN/m)	132	198	190	285

The load patterns for design moments at various locations are different, as shown in Fig. 10.12. Instead of analysing the frame for every loading pattern individually, all required cases can be obtained by appropriately combining the moment diagrams for (1) dead load only on all spans, and (2) live load only on each individual span, in turn. To get the latter, only two analyses need be done in this example, namely live load in spans *AB* and *BC* in turn. Because of symmetry, moment diagrams for live load in spans *CD* and *DE* will be the mirror image (end to end) of those due to loads in *BC* and *AB*, respectively. (Also, separation of dead load effects and live load effects may be required if gravity load effects have to be combined with lateral load effects.) Thus in all, only three moment distributions are required, namely:

1. dead load on all spans
2. live load on span *AB* only
3. live load on span *BC* only

These three distributions are set up in Table 10.3 at *a*, *b*, and *c*, respectively. (In these tables the distribution is set up in the conventional pattern for four cycles.)

In practice, the part of the frame beyond the support second removed from the section under consideration can be ignored, as previously explained, and the distribution may be limited to two cycles. Detailed explanation of such a set up is presented in Ref. 10.4. In Table 10.3, moments on only the beam members are presented with clockwise moment taken as positive. For any loading case, the total column moment at a joint can be obtained as the net unbalanced beam moment at the support, and the individual column moments can be obtained by distributing the total column moment between the columns immediately above and below the floor in proportion to their relative stiffnesses.

Table 10.3 Moment Distributions

Distribution Factor	0.407	½	0.300	0.261	½	0.272	0.272	½	0.261	0.300	½	0.407
C.O.F.	**A**		**B**			**C**			**D**			**E**
(a) DL on All Spans												
FEM	-55.6		55.6	-80.1		80.1	-80.1		80.1	-55.6		55.6
Bal.	22.6		7.4	6.4		—	—		6.4	-7.4		-22.6
C.O.	3.7		11.3	-2.9		3.2	-3.2		2.9	-11.3		-3.7
Bal.	-1.5		-3.4	0.2		-1.5	1.5		-0.2	3.4		1.5
C.O.	-1.7		-0.8	-0.1		0.1	-0.1		0.1	0.8		1.7
Bal.	0.7		0.2	—		—	—		—	-0.2		-0.7
C.O.	0.1		0.3	—		—	—		—	-0.3		-0.1
Bal.	—		-0.1	-0.1		—	—		0.1	0.1		—
Total	-31.7		70.5	-76.5		81.9	-81.9		76.5	-70.5		31.7
(b) LL on Span AB												
FEM	-76.5		76.5	-20.0								
Bal.	31.1		-23.0									
C.O.	-11.5		15.6	-4.1		2.7						
Bal.	4.7		-4.7	1.4								
C.O.	-2.4		2.4	-1.0		-2.0	2.7		1.4			
Bal.	1.0		-1.1	0.3		0.5	-0.2		-0.4			
C.O.	-0.5		0.5	-0.2		-0.5	-0.2		0.3	-0.4		-0.2
Bal.	0.2		-0.2	0.2		0.2	0.2		-0.1	-0.1		0.1
Total	-53.9		66.0	-23.6		-9.1	3.2		1.2	-0.5		-0.1
(c) LL on Span BC												
FEM				-110		110	-29.9					
Bal.			33.0	28.7		-29.9			-15.0			
C.O.	16.5		-15.0	-15.0		14.4	-3.9		3.9	4.5		
Bal.	-6.7		3.9	3.9		-3.9	-1.1					
C.O.	2.2		-3.3	-2.0		2.0	2.0		-2.0			2.2
Bal.	-0.9		1.6	1.4		-1.1	-1.1		0.5	0.6		-0.9
C.O.	0.8		-0.4	-0.5		0.7	0.3		-0.5	-0.4		0.3
Bal.	-0.3		0.3	0.2		-0.3	-0.3		0.2	0.3		-0.1
Total	11.6		35.7	-93.3		91.9	-32.9		-12.9	5.0		1.5

Table 10.4

(a) Loading 1 of Fig. 10.12

	A		B		C		D		E
w, kN/m		63.4		26.7		63.4		26.7	
$wL^2/8$		198		120		285		83.4	
DL (Table 10.3a)	-31.7		70.5	-76.5	81.9	-81.9	76.5	-70.5	31.7
LL on AB (T. 10.3b)	-53.9		66.0	-23.6	-9.1	3.2	1.2	-0.5	-0.1
LL on CD (T. 10.3c)	-1.5		-5.0	12.9	32.9	-91.9	93.3	-35.7	-11.6
Total	-87.1	88.5	132	-87.2	23.4	106	-171	114	171	-107	19.9	20.0

(b) Loading 3 of Fig. 10.12

	A		B		C		D		E
$w=$		63.4		63.4		26.7		63.4	
$wL^2/8$		198		285		120		198	
DL (Table 10.3a)	-31.7		70.5	-76.5	81.9	-81.9	76.5	-70.5	31.9
LL in AB (T. 10.3b)	-53.9		66.0	-23.6	-9.1	3.2	1.2	-0.5	-0.1
LL in BC (T. 10.3c)	11.6		35.7	-93.3	91.9	-32.9	-12.9	5.0	1.5
LL in DE (T. 10.3b)	0.1		0.5	-1.2	-3.2	9.1	23.6	-66.0	54.2
Total	-73.9	76	173	-195	107	162	-103	24.3	88.4	-132	88.3	87.5

(c) Loading 4 of Fig. 10.12

	A		B		C		D		E
$w=$		26.7		63.4		63.4		26.7	
$wL^2/8$		83.4		285		285		83.4	
DL (T. 10.3a)	-31.7		70.5	-76.5	81.9	-81.9	76.5	-70.5	31.7
LL in BC (T. 10.3c)	11.6		35.7	-93.3	91.9	-32.9	-12.9	5.0	1.5
LL in CD (T. 10.3c)	-1.5		-5.0	12.9	32.9	-91.9	93.3	-35.7	-11.6
Total	-21.6	22.1	101	-157	103	207	-207	103	157	-101	22.1	21.6

Fig. 10.12 Loading patterns for critical moments in beam

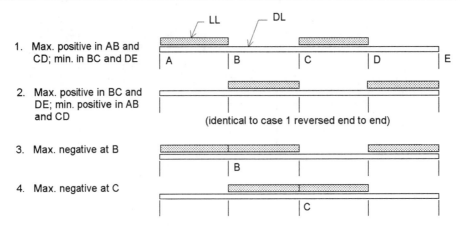

Combining the Distributions

The loading patterns for critical moments at the design sections in the beams are shown in Fig. 10.12. The moments for loading pattern No. 1 in Fig. 10.12 can be obtained by superposition of the moments for dead load on all spans (Table 10.3a), live load on span *AB* (Table 10.3b) and live load on span *CD* (Table 10.3c reversed end-to-end). This is done in Table 10.4 at *a*. This loading case gives the maximum span moment in *AB* and the minimum span moment in *BC*, as well as the maximum moment in span *CD* and the minimum moment in span *DE*. The latter two values, because of symmetry, are also equal to the maximum span moment in *BC* and the minimum span moment in *AB*. Thus from the single combination in Table 10.4a, the maximum and the minimum span moments can be obtained for both spans *AB* and *CD*. The complete moment diagram for all the spans obtained from Table 10.4a is plotted in Fig. 10.13.

Knowing the end moments in each span, the end shears and points of inflection in all spans may also be computed from statics. These computations are evident from the freebody diagram of each span, also given in Fig. 10.13. The moment diagram in Fig. 10.13 is plotted following the usual convention that sagging moment is positive and hogging moment is negative, and moments are drawn on the side on which each produces tension. When the end moments in a span are unequal, as they usually are, the peak positive span moment will not be at midspan. The location of peak moment (zero shear) and the magnitude of the peak span moment can be computed from the freebody diagram for each span. However, for regular spans and loadings (and particularly in interior spans) the location of peak span moment is not far removed from midspan, and the peak moment is only slightly greater than the midspan moment. In such cases, the midspan moment may be taken as the design moment for the span. For instance, with the moment diagram in Fig. 10.13, for span *AB*, the moment is maximum

Fig. 10.13 Moment diagram for load pattern 1 of Fig. 10.12

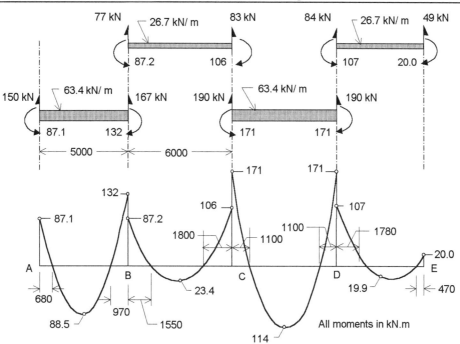

Fig. 10.14 Maximum and minimum span moments

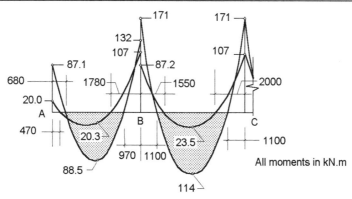

(that is, shear is zero) at a distance from A of 150/63.4 = 2.37 m compared to 2.5 m to midspan. The maximum moment at 2.37 m is 90.3 kN·m, which is only 2 percent in excess of the moment at midspan. The points of inflection are needed to detail bar extensions. Once again, for regular spans and loadings, exact calculations for location of inflection points are seldom needed, and bar extensions can be detailed in terms of span proportions (see Section 5.7)

326 REINFORCED CONCRETE DESIGN

Combining the two halves of Fig. 10.13, the combined maximum and minimum span moment diagrams for spans AB and BC are drawn in Fig. 10.14. The shaded part of the moment diagram in Fig. 10.14 indicates the envelope for positive moment.

Following the procedure described above, the bending moments for loading cases 3 and 4 of Fig. 10.12 (giving the maximum negative moments at supports B and C, respectively) can be obtained by combining appropriately the results of the three moment distributions in Table 10.3. These combinations are also set up in Table 10.4 at b and c. The resulting moment diagrams are shown in Fig. 10.15, where only the spans on either side of the support considered are shown, as these loading cases have no relevance beyond the respective negative moment regions. For the exterior support A, the loading pattern for the peak negative moment is the same as at 1 in Fig. 10.12, for which the moment diagram is shown in Fig. 10.13. Combining these peak negative moment diagrams with the peak positive moment diagram in Fig. 10.14, the complete moment envelope shown in Fig. 10.16 can be obtained.

EXAMPLE 10.2

For the problem in Example 10.1, determine the design moments, both positive and negative, for the spans AB and BC and compare with the approximate values recommended by the Code.

SOLUTION

The theoretical moment envelope, using centreline dimensions, was determined in Example 10.1 and is shown in Fig. 10.16. Adjustment to this diagram must be made to account for the increased beam stiffness within the column width, and the design negative moment must be determined at the column face. As explained in Section 10.6, these adjustments will result, approximately, in reductions in the span moment by $Vb/6$ and in the support moment at centreline by $Vb/3$; where V is the shear force at the support for the relevant loading case, and b is the width of column. The reduction in span moment is usually ignored so that the final design positive moments are:

for span AB, = 88.5 kN·m
for span BC, = 114 kN·m

For negative moments at support B, the centreline moments and shears were presented in Fig. 10.15a. Making the adjustments of $Vb/3$, as shown in Fig. 10.17,

Fig. 10.15 Moment diagram for maximum negative moments

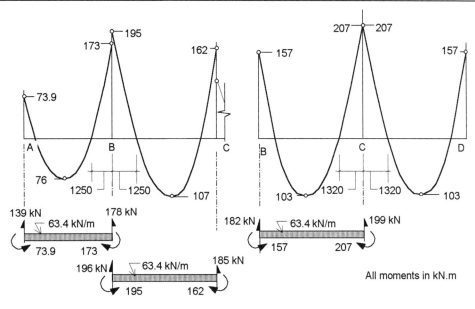

(a) Maximum negative moment at B   (b) Maximum negative moment at C

Fig. 10.16 Moment envelops for spans AB and BC

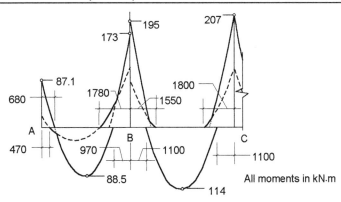

$$M_{BA} = 173 - \frac{178 \times 0.40}{3} = 149 \text{ kN·m}$$

$$M_{BC} = 194 - \frac{196 \times 0.40}{3} = 168 \text{ kN·m}$$

Similar computations for the peak negative moments at A and C give:

328 REINFORCED CONCRETE DESIGN

Fig. 10.17 Moments at column face

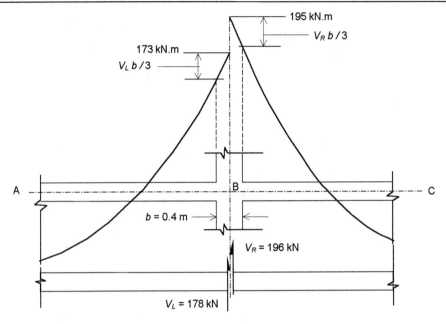

$$M_{AB} = 87.1 - \frac{150 \times 0.40}{3} = 67.1 \text{ kN·m}$$

$$M_{CB} = 207 - \frac{199 \times 0.40}{3} = 180 \text{ kN·m}$$

The final design moments are presented in Fig. 10.18, for midspan sections and face-of-support sections.

The continuous beam in Example 10.1 satisfies the limitations in CSA A23.3-94(9.3.3). Therefore, the moment values specified therein and summarised in Table 10.1 may be used for design. These values are computed below:

For span *AB*, l_n = 5 - 0.4 = 4.6 m, and for span *BC*, l_n = 5.6 m, w_f = 63.4 kN/m²

Positive span moments are:

(i) for span *AB* (discontinuous end *A* integral with support),

$$\frac{1}{14} w_f l_n^2 = \frac{1}{14} \times 63.4 \times 4.6^2 = 95.8 \text{ kN·m}$$

(ii) for span BC,

$$\frac{1}{16}w_f l_n^2 = \frac{1}{16} \times 63.4 \times 5.6^2 = 124 \text{ kN·m}$$

Negative moments are:

(i) at interior face of exterior support (member built integrally with supporting column),

$$M_{AB} = \frac{1}{16}w_f l_n^2 = \frac{1}{16} \times 63.4 \times 4.6^2 = 83.8 \text{ kN·m}$$

(ii) at exterior face of first interior support (more than two spans),

$$M_{BA} = \frac{1}{10}w_f l_n^2 = \frac{1}{10} \times 63.4 \times \frac{(4.6 \times 5.6)^2}{2} = 165 \text{ kN·m}$$

(iii) at other faces of interior supports:

$$M_{BC} = \frac{1}{11}w_f l_n^2 = \frac{1}{11} \times 63.4 \times \frac{(4.6+5.6)^2}{2} = 150 \text{ kN·m}$$

$$M_{CB} = \frac{1}{11}w_f l_n^2 = \frac{1}{11} \times 63.4 \times \frac{(5.6+5.6)^2}{2} = 181 \text{ kN·m}$$

Fig. 10.18 *Comparison of design moments by analysis and CSA coefficients*

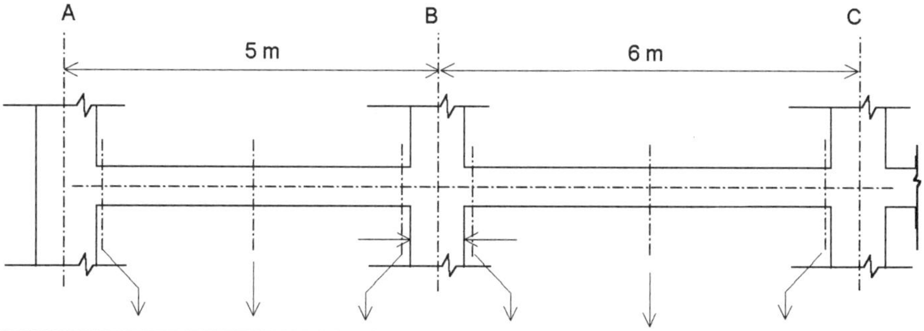

(a) Analysis	67.1	88.5	149	168	114	180
(b) CSA cl. 9.3.3	83.8	95.8	165	150	124	181
Ratio (a) / (b)	0.83	0.92	0.90	1.12	0.92	0.99

All moments in kN·m

These values obtained by the Code equations are also presented in Fig. 10.18. The Code coefficients are, in general, conservative. This is so because they make allowances for several influences such as live load to dead load ratios of up to two and differences between all adjacent spans of up to 20 percent.

EXAMPLE 10.3

For the frame in Example 10.1, determine the design moments due to gravity loads for the columns framing into A and B at the floor level considered.

SOLUTION

The load pattern for the maximum bending moment in the column consists of live load placed on a single adjacent span and every alternate span. For the column at A, this is shown as loading pattern 1 in Fig. 10.12 for which the moments were determined in Table 10.4a. The total column moment at A is equal to the unbalanced beam moment = 87.1 kN·m. Distributing this between the columns immediately above and below the floor, in proportion to their relative stiffnesses (the column stiffnesses are equal in this Example), at A,

$$M_{COL,ABOVE} = M_{COL,BELOW} = \frac{1}{2}(87.1) = 43.6 \text{ kN·m}$$

At support B, the live load must again be placed on a single adjacent span. Loadings 1 and 2 in Fig. 10.12 correspond to live load being placed on the span to the left or right of B, respectively. From the moment values given in Table 10.4a, the larger unbalanced moment at B occurs when the load is placed to the right of B (loading 2 of Fig. 10.12), moments for which can be obtained as the end-for-end reversed values in Table 10.4a.

Total column moment at B = 171 - 107 = 64 kN·m.
Moments on columns above and below B are each ½ (64) = 32.0 kN·m.

The column moments are indicated in Fig. 10.19. Note that no adjustment has been made for the beam depth and the increased column stiffness within the beam depth.

Fig. 10.19 Column moments, Example 10.3

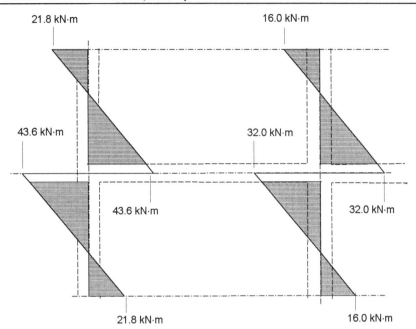

10.7 INELASTIC ANALYSIS AND MOMENT REDISTRIBUTION

10.7.1 Limit Analysis

Reinforced concrete frames may be analysed by inelastic theory (Refs. 10.8, 10.9). The procedure generally applied to reinforced concrete is termed *Limit Analysis* and is very similar to the *Plastic Analysis* procedure applied to steel structures. Limit analysis is based on the moment curvature (M-ψ) relationship of under-reinforced sections.

The M-ψ relationship for under-reinforced sections with steel bars having a well defined yield plateau (Fig. 4.7a) can be idealised as a bilinear elastic-plastic relation, as shown in Fig. 10.20. This idealisation implies that in a section subjected to flexure, the moment increases linearly with curvature until the applied moment reaches the value M_n. The latter can be taken equal to the nominal ultimate strength of the section developed in Eq. 4.17, or using Eq. 4.36 with ϕ_s and ϕ_c taken equal to unity, as:

Fig. 10.20 Idealised moment-curvature relation for under-reinforced beams

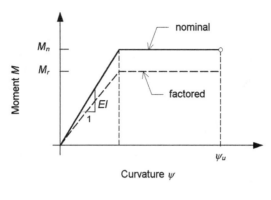

$$M_n = \rho f_y \left(1 - \frac{\rho f_y}{2\alpha_1 f_c'}\right) bd^2$$

On further straining, the moment at the section cannot increase; however, the section "yields", and the curvature will continue to increase at constant moment. The section is then said to have developed a "plastic hinge." The section fails when the curvature reaches its ultimate value ψ_u. If the material resistance factors, ϕ_s and ϕ_c, are also taken into account, the factored moment resistance is M_r, (given by Eq. 4.36) instead of M_n, and the M-ψ relationship can be idealised as shown by the broken lines in Fig. 10.20. The plastic hinge will form when the moment reaches the value M_r.

A fixed-ended beam of uniform strength, subject to an increasing uniform load of w per unit length, is shown in Fig. 10.21a. As long as the maximum moment in the beam is less than the factored resistance of the section, M_r, the beam behaves elastically. When the maximum moment, which is at the support sections A and B and equal to $wL^2/12$, reaches the value M_r at a load of, say, w_1, the limit of elastic behaviour is reached, and these sections are about to develop plastic hinges. The moment diagram at this load is as shown in Fig. 10.21b, and gives:

$$w_1 \frac{L^2}{12} = M_r$$

$$w_1 = \frac{12 M_r}{L^2}$$

Any further loading will not increase the support moment, but will increase the curvature at these sections so that the beam now behaves like a pin-ended beam with constant end moments equal to M_r (Fig. 10.21c). Additional

Fig. 10.21 Inelastic behaviour of fixed-ended beam

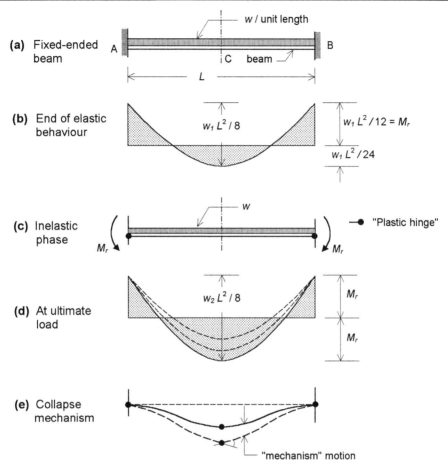

loading causes an increase in span moment only, as indicated by the broken lines in Fig. 10.21d. This increase can continue until the midspan moment also reaches M_r, provided of course that the curvature at the support does not exceed the ultimate value ψ_u. At this load, say, w_2, a plastic hinge will form at midspan as well. The bending moment diagram will now have the distribution shown by the solid line in Fig. 10.21d. The beam, now connected by three plastic hinges, behaves like a hinged "mechanism" which can deform with no further increase in load (Fig. 10.21e). It is also evident from the moment diagram in Fig. 10.21d that the load cannot increase above w_2 without an increase in moment at the supports or midspan, which is not possible. Thus, for the fixed ended beam in Fig. 10.21a, a "collapse mechanism" is formed as the third plastic hinge develops, and the corresponding collapse load (or ultimate load) is obtained from Fig. 10.21d as:

$$2M_r = \frac{w_2 L^2}{8}$$

$$w_2 = \frac{16 M_r}{L^2}$$

In this example the collapse load $w_2 = 16 M_r/L^2$ is one-third greater than the load at the end of elastic behaviour, $w_1 = 12 M_r/L^2$. This simple example illustrates that in an indeterminate structure the attainment of the ultimate strength at one section does not necessarily mean that the maximum strength of the structure is reached. On the contrary, with additional loading, such highly stressed sections will yield locally (develop plastic hinges) which could lead to a *redistribution* of moments to lower stressed sections resulting in substantially increased capacity.

Referring to factored moment, a fixed ended beam with a uniform load w_f analysed by elastic theory (Fig. 10.22a) will have to be designed for a moment $M_f = w_f L^2/12$. The maximum moment at midspan will be only $w_f L^2/24$. If the beam has uniform strength throughout, this will mean that the strength at the midspan section is not fully utilised. The same beam analysed by limit analysis procedure with full moment redistribution developed, will have a design moment of $M_f = w_f L^2/16$ (Fig. 10.22b), which is only 75 percent of the design moment based on elastic analysis. Thus economy could result from recognising inelastic behaviour. In steel structures, prismatic members of uniform strength are, in fact, mostly used, and substantial savings result by adopting plastic analysis and design. However, in reinforced concrete, even when a prismatic member is used, it is possible to design the amount of flexural reinforcement to suit the actual moment at any section. The savings in this case may not be so significant. Thus, the positive moment reinforcement for the beam in Fig. 10.22a could be designed for the moment $w_f L^2/24$, in which case both elastic analysis and limit analysis will give identical results. In continuous structures, the situation is somewhat different in that the loading conditions for critical design moments at support sections and at midspan are usually different. Therefore, for a specific loading pattern causing maximum moment at a chosen section (such as a support), there will be spare capacity elsewhere in the structure (such as at midspan), and advantage could be taken of the economy resulting from possible inelastic action and consequent moment redistribution. Despite this, the major advantage to the designer of recognising moment redistribution is that it enables him or her to slightly "adjust" the moment diagram, reducing (or increasing) the moment at some sections, and correspondingly in increasing (or decreasing) the moment elsewhere, leading to the design of an economical structure of more balanced proportions.

Fig. 10.22 *Design based on elastic and limit analysis*

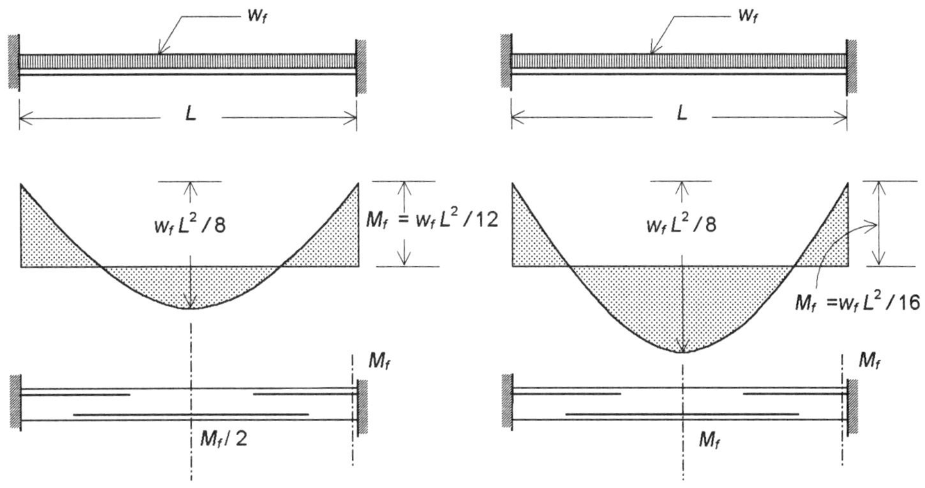

(a) Elastic analysis

(b) Limit analysis

One other consideration that has somewhat limited the use of limit analysis is the restricted rotation capacity of hinging regions in reinforced concrete members. Because of the large inelastic strain which steel can sustain, plastic hinges in steel structures can usually develop the rotations necessary to effect full redistribution of moment leading to the formation of a collapse mechanism. However, the maximum compressive strain which concrete can sustain is limited, as is the maximum inelastic rotation that a plastic hinge in a reinforced concrete member can develop before failure. Therefore, full redistribution cannot be taken for granted, and in using limit analysis it is necessary to evaluate the rotational capacities of hinging regions to ensure that these are adequate to meet the respective rotational requirements implied in the analysis.

10.7.2 Redistribution of Moments Permitted by Code

Recognising the inelastic behaviour of under-reinforced flexural members CSA A23.3-94(9.2.4) permits a limited amount (up to 20 percent) of redistribution of negative moments in continuous beams, as calculated by elastic analysis. Accordingly, the negative moments at supports computed by elastic theory for any loading pattern may be increased or decreased by up to

$$(30 - 50\ c/d\ ) \leq 20 \text{ percent}$$

and the negative moments so modified used for calculation of the moments at sections within the span. Here c/d is the neutral axis depth ratio. Sections having

336 REINFORCED CONCRETE DESIGN

low percentages of tension reinforcement and/or containing compression reinforcement have low c/d ratios and can effect larger redistributions. Under-reinforced continuous beams possess adequate rotation capacity to effect the limited moment redistribution permitted above.

EXAMPLE 10.4

Assuming that the reinforcement ratios are such that a 15 percent adjustment of support moments is permitted for all spans, determine a more economical and balanced distribution of design moments for the continuous beam analysed in Example 10.1.

SOLUTION

The moment envelope obtained by elastic analysis is given in Fig. 10.16. For flexure, the overall beam size is controlled by the maximum negative moment at supports, where the continuous T-beam acts as a rectangular beam. Therefore, it may be desirable to reduce the maximum negative moment, which is 207 kN·m at support C for the loading case shown in Fig. 10.15b. The corresponding moment diagram obtained by elastic analysis is shown in Fig. 10.23a for span BC. Reducing the support moment by the full 15 percent brings it down to 207 × 0.85 = 176 kN·m. In order to cause the least increase in the span moment, the negative moment at B in span BC for this loading is increased to 176 kN·m (an increase of only 12 percent) so that both support sections will now have the same modified design moment. These two adjustments cause the span moment to increase to 285 − (176 + 176) = 109 kN·m which is still less than the maximum span moment of 114 kN·m in the analytical moment envelope (Fig. 10.16). The broken lines in Fig. 10.23a show the modified moment diagram.

    The moment diagram giving the maximum negative moment at B obtained in Example 10.1 is shown in Fig. 10.23b. Having seen that the lowest adjusted negative moment at C is 176 kN·m, and decided that both supports B and C will be designed for the same adjusted moment, once again the negative moments at B and C in Fig. 10.23b are adjusted to 176 kN·m each, corresponding to a decrease of 9.7 percent at B and an increase of 8.7 percent at C. The modified diagram is also shown in Fig. 10.23b and is identical to the modified diagram in Fig. 10.23a.

    The elastic analysis distribution giving the maximum span moment for BC is presented in Fig. 10.23c. In order to reduce the span moment, the support

Fig. 10.23 Redistribution of moments, Example 10.4

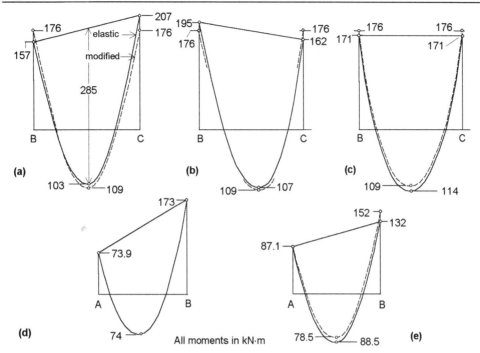

All moments in kN·m

moments are adjusted upward to the previously established value of 176 kN·m, an increase of 3 percent. The adjusted span moment is again 109 kN·m, as shown in Fig. 10.23c.

Considering span AB and the moment diagram for maximum negative moment at B (Fig. 10.23c), there is little to be gained by adjusting any of the moments, as the support section B is conveniently reinforced for the moment of 176 kN·m by continuing the bars designed for the negative moment at the left end of span BC. The moment diagram giving the maximum negative moment at support A and maximum span moment in AB is shown in Fig. 10.23e. Since there is reserve capacity at support B, this moment may be adjusted upwards by 15 percent to a value of $132 \times 1.15 = 152$ kN·m. This results in a reduction of span moment to 78.5 kN·m.

The moment at all critical sections obtained from elastic analysis is presented along with the corresponding moments obtained by allowing moment redistribution as permitted by the Code in Table 10.5 for comparison. It should be noted that the redistribution of moments may be effected in other ways as well, depending on specific requirements.

Table 10.5 Moment Redistribution as Permitted by Code

	A			B		C
	Span AB			Span BC		
	-ve	+ve	-ve	-ve	+ve	-ve
Moment from elastic analysis	87.1	88.5	173	195	114	207
Adjusted moments	87.1	78.5	173	176	109	176

All moments in kN·m

10.8 PRELIMINARY DESIGN

Before a structure can be analysed for load effects and its members designed, it is necessary to estimate cross-sectional dimensions, to assess the dead loads due to self weight and to compute the moment of inertia of sections and member stiffness properties needed for the analysis. This may be done by a preliminary design, based on member loads estimated using tributary areas supported by each member, and approximate moment factors, although with experience the member sizes can be estimated without such calculations. For preliminary designs, the floor areas tributary to beams, girders, and columns may be taken approximately as shown in Fig. 10.24. These areas are based on the assumption that if a two-way slab is divided into four areas by lines midway between lines of support as

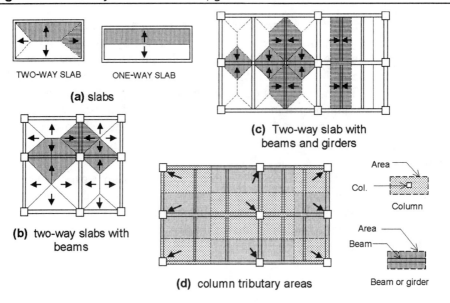

Fig. 10.24 Tributary areas for beams, girders and columns

(a) slabs — TWO-WAY SLAB, ONE-WAY SLAB

(b) two-way slabs with beams

(c) Two-way slab with beams and girders

(d) column tributary areas

shown in Fig. 10.24a, the load on each area is transferred to the adjacent support. One-way slabs are assumed to transfer the loads only to the longer (supported) sides. Girders have a tributary area equal to the floor area directly supported by them, if any, plus one-half of the sum of the tributary areas of beams supported by them. For columns, the tributary area per floor is one-half the sum of the tributary area of the floor members framing into it. The moments and shears may be estimated using the Code coefficients (Section 10.3.3); or alternatively, the moment may be taken equal to the fixed-end moment for the span and loading, and the shears as those under simple support conditions. For continuous beams, the moment and shears at the support sections govern cross-sectional dimensions. For columns, the tributary area is the sum of the tributary areas for the column from all levels above the storey under consideration. The axial loads on them primarily determine sizes of interior columns, since the unbalanced moments are not large for regular column spacings. The exterior columns with the floor extending on one side only are generally subjected to greater unbalanced moments, and their size, determined on the basis of axial load, must be increased to allow for these moments, particularly in the upper storeys where the axial loads are relatively low compared to the moments.

PROBLEMS

1. Using elastic analysis, determine the moment envelope for the slab-beam member for the plane frame shown in Fig. 10.25. The load on the slab-

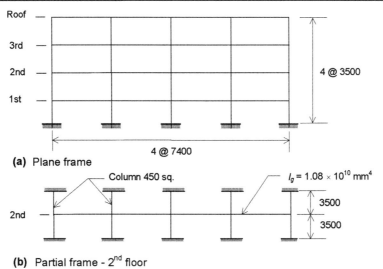

Fig. 10.25 Problems 1, 2 and 3

beam member consists of a dead load (including self weight of slabs and beams) of 19.8 kN/m and a live load of 16.8 kN/m.

2. For the floor member in the partial frame in Problem 1, determine the design moments at all critical sections and compare with the moments obtained by the Code coefficients.

3. Assuming that the reinforcement ratio is such to allow a moment redistribution of 10 percent, determine optimum redistributed design moments at all critical sections for the floor member in Problem 1.

REFERENCES

10.1 Wang, C.K. *Intermediate Structural Analysis*, McGraw Hill, New York, 1983, 656 pp.

10.2 Sennett, R.E., *Matrix Analysis of Structures*, Prentice Hall, Inc., New Jersey, 1993, 228 pp.

10.3 Muttoni, A., Schwartz, J., and Thurlimann, B., *Design of Concrete Structures with Stress Fields*, Springer-Verlaag, New York, 1997, 152 pp.

10.4 *Continuity in Concrete Building Frames*, 4th Ed., Portland Cement Association, Chicago, Illinois, 1959, 56 pp.

10.5 *Frames and Continuous Structures - Analysis by Moment Distribution*, Concrete Information IS 210.01D, Portland Cement Association, Skokie, Illinois, 1980.

10.6 ACI Standard 318-95, *Building Code Requirements for Structural Concrete* (ACI 318R-95) and *Commentary* (ACI 318R-95), American Concrete Institute, Detroit, Michigan, November 1995, 369 pp.

10.7 *Handbook of Frame Constants*, Engineering Bulletin, Portland Cement Association, Skokie, Illinois, available through Canadian Portland Cement Association, Ottawa, Ont., 1958, 33 pp.

10.8 ACI-ASCE Committee 428, *Limit Design of Reinforced Concrete Beams and Frames C Addendum*, (supplement to bibliographies published in ACI Journal in Nov. 1961 and Dec. 1962), J. ACI, Vol. 60, Oct. 1963, pp. 1471-1473.

10.9 Cohn, M.Z., *Limit-Design Solutions for Concrete Structures*, ASCE J. Struct. Div., Vol. 83, ST1, Feb. 1967, pp. 37-57.

CHAPTER 11   Serviceability Limits

11.1   INTRODUCTION

The two classes of limit states, namely *ultimate limit states* (concerned with safety) and *serviceability limit states* (concerned with satisfactory performance under intended use and occupancy) was explained in Chapter 3. Reinforced concrete structures and structural components are designed to have sufficient strength and stability to withstand the effects of factored loads, thereby satisfying the safety requirements. The design for serviceability limits is made at specified (service) loads. Generally, the design is first made for strength and the serviceability limits are checked. The two major serviceability requirements of reinforced concrete structures relate to limiting deflections and crack control (CSA A23.3-94(8.1)).

11.2   DEFLECTIONS

Accurate prediction of deflection of reinforced concrete members is difficult because of several factors, such as: (1) uncertainties in the flexural stiffness *EI* of the member, as influenced by the varying degrees of tensile cracking of the concrete, varying amounts of flexural reinforcement, and variations in the modulus of elasticity and modulus of rupture of concrete; (2) curvature and deflection induced by differential shrinkage and creep, and the uncertainties about the time-dependent and environmental effects that influence these; (3) inelastic flexural behaviour of members; and (4) the inherently high degree of variability of measured deflections, even under controlled laboratory conditions. Furthermore, the deflection that may be considered as allowable in a given situation is dependent on several aspects (such as aesthetic and sensory acceptability, effect on attached structural and non-structural elements, possible interference with expected performance or function, etc.) particular to that situation, and may often be somewhat arbitrary (Ref. 11.1). Thus, approximations and simplifications are essential in deflection computations. At the same time, because of the random nature of deflection and its high variability, the apparent accuracy of elaborate and sophisticated calculation procedures is often illusory. Results of deflection computations must be regarded as representative values used for the purpose of comparison with the empirically set limiting values.

Deflection of reinforced concrete members is considered in two parts; namely, the *immediate* (or short-time) deflection occurring on application of the load, and the additional *long-time deflection* resulting mostly from shrinkage and creep under sustained loading. The immediate deflection due to applied dead load is often not a controlling factor, because it can be compensated for by cambering the members. The long-time deflection is generally of more concern to the designer. The discussion that follows deals with deflection under static loading only.

11.3 DEFLECTIONS BY ELASTIC THEORY

Structural response at service load levels is nearly elastic. Therefore, short-time deflection calculations for serviceability limits are generally based on elastic theory. Expressions for the elastic deflections of homogeneous beams for any loading and support conditions can be derived using standard methods of structural analysis; formulas for several standard cases are presented in handbooks (Refs. 11.2-11.4). These are usually in the form:

$$\Delta = k \frac{WL^3}{EI} = k_1 \frac{ML^2}{EI} \qquad (11.1)$$

where Δ is typically the midspan deflection; W is the total load on the span; L is the span; EI is the flexural rigidity of a reference section; M is the maximum moment; and k and k_1 are constants, which depend on the load distribution, conditions of end restraint, and variation of EI (if any). For example, $k = 5/384$ for a uniformly loaded simple beam of uniform section. When a beam or loading is nonsymmetrical about the midspan, the maximum deflection is not at midspan. However, the midspan deflection of beams supported at both ends is very close to the maximum value (within about 3.5 percent for a span fixed at one end and simply supported at the other with distributed loading) so that for serviceability checks, the former value is generally used. A standard case frequently encountered in design is that of a continuous beam of uniform section with uniform loading w per unit length (Fig. 11.1). Since the support and midspan moments in such a case are generally known from the analysis made for design of the section, it would be convenient to express the midspan deflection of such a beam in terms of these moments. As shown in Fig. 11.1, this may be done by superimposing the deflections at midspan due to the three component loadings (computed in this case conveniently using the conjugate beam method). It may be recalled that (Ref. 11.2) the conjugate beam is subjected to a loading equal to the M/EI diagram of the real beam. The moment and shear in the conjugate beam at a section correspond to the deflection and slope of the real beam at the same section.

Assuming the flexural rigidity EI to be uniform along the span, and with the notations in Fig. 11.1, the midspan deflection, Δ_m, is obtained as:

CHAPTER 11 SERVICEABILITY LIMITS

Fig. 11.1 Midspan deflection - uniformly loaded continuous beam

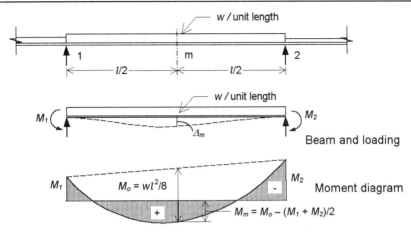

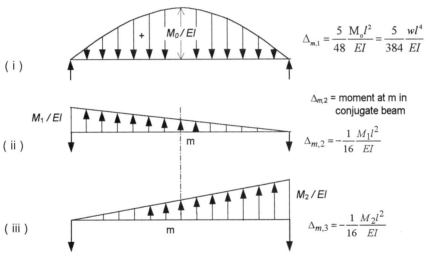

$$\Delta_m = \frac{5}{48}\frac{M_o L^2}{EI} - \frac{M_1 L^2}{16EI} - \frac{M_2 L^2}{16EI}$$

$$= \frac{5L^2}{48EI}[M_o - (3/5)(M_1 + M_2)]$$

Substituting $M_m = M_o - (1/2)(M_1 + M_2)$, and eliminating M_o,

$$\Delta_m = \frac{5L^2}{48EI}[M_m - (1/10)(M_1 + M_2)] \qquad (11.2)$$

Alternatively, eliminating $(M_1 + M_2)$,

$$\Delta_m = K\frac{5}{48}\frac{M_m L^2}{EI} \qquad (11.3)$$

where

$$K = 1.20 - 0.20\, M_o/M_m \qquad (11.3a)$$

Values of K for several cases are tabulated in Ref. 11.4. Similar expressions can also be worked out for concentrated loadings (Ref. 11.5). For the moment patterns associated with the Code moment coefficients, the values of K can be computed using Eq. 11.3a. For a few typical cases, K is given in Table 11.1.

11.4 IMMEDIATE DEFLECTIONS

The principal member property that influences short-time deflection is the flexural rigidity, EI. The modulus of elasticity of concrete, $E = E_c$, depends on factors, such as concrete quality, age, stress level, and rate or duration of applied stress. However, for short-time loading to service load levels, the approximate relation in Eq. 1.1 gives a satisfactory value for E_c.

The moment of inertia, I, of a section of a beam is influenced by the steel percentages and the extent of flexural cracking. Therefore, it is not a constant along the span even for a beam of uniform cross-section. Flexural cracking will depend on the applied bending moment at a section and the modulus of rupture of the concrete. During the first time loading of a reinforced concrete beam, the portions of a beam where the applied moment is less than the cracking moment ($M < M_{cr}$, see Example 4.1 for the computation of M_{cr}) remain uncracked and, hence, have the moment of inertia of the gross transformed section (including reinforcement). Where the moment exceeds M_{cr}, the concrete in tension fails at the outer fibres and develops flexural cracks at random spacings. However, the concrete close to the neutral axis and in between the cracks still carries some tension and contributes to the effective stiffness of the beam. The tensile stress in the steel in between cracks may be as low as 60 percent of the stress at the cracks due to the tension contribution of concrete — the so-called *tension stiffening* effect. Thus, even in the portion of the beam with flexural cracking, the effective stiffness will be greater than that corresponding to a fully cracked transformed section.

11.4.1 Effective Moment of Inertia

The typical load-deflection behaviour of a reinforced concrete beam under short-time loading is shown in Fig. 11.2. The initial slope of the curve corresponds closely to the gross transformed section moment of inertia. As the load increases and cracking spreads, the slope of the curve decreases. In this region, the effective moment of inertia, I_e, within the service load range, will have a value in between the gross section moment of inertia, I_g, and the cracked transformed section moment of inertia, I_{cr}, and that the value of I_e will depend on the load level.

Table 11.1 Coefficient K in Eq. 11.3

Loading case	K
simply supported	1.00
both ends fixed	0.60
propped cantilever (fixed-pinned)	0.80
propped cantilever	0.925*
Col. / beam	0.85*
cantilever	0.80*

* Using Code moment coefficients

For use in the computation of immediate deflection of simple spans, CSA A23.3-94(9.8.2.3) recommends Eq. 11.4, developed by Branson in 1973 (Ref. 11.5), giving an average effective moment of inertia I_e as:

$$I_e = I_{cr} + (I_g - I_{cr})\left(\frac{M_{cr}}{M_a}\right)^3 \leq I_g \qquad (11.4)$$

where M_{cr} = $f_r I_g / y_t$ = cracking moment (11.4a)
          M_a = maximum moment in the member at the load stage at which deflection is being computed, or at any previous stage
          I_g = moment of inertia of the gross concrete section about centroidal axis, neglecting the reinforcement

***Fig. 11.2** Typical load-deflection curve for reinforced concrete beams*

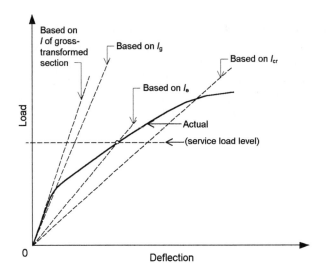

I_{cr} = transformed moment of inertia of the cracked section expressed as the moment of inertia of the equivalent concrete section

f_r = modulus of rupture of concrete, $= 0.6\lambda\sqrt{f'_c}$

y_t = distance from centroidal axis of the gross section, neglecting the reinforcement, to the extreme fibre in tension

l = factor to account for concrete density

Equation 11.4 is relatively simple and gives the correct limiting values of $I_e = I_g$ for $M_a \leq M_{cr}$ and $I_e \approx I_{cr}$ for $M_a >> M_{cr}$, where the beam is severely cracked. When $M_a / M_{cr} \geq$ about 3, $I_e \approx I_{cr}$. Although the Code recommends neglecting the reinforcement in computing I_g, the gross transformed section (including reinforcement) moment of inertia may be more appropriate in place of I_g in Eq. 11.4, particularly for beams with high reinforcement ratios and structural low density concrete (high modular ratio) (Refs. 11.5, 11.6).

Comparison with test results indicates reasonable agreement (within the range of about ±20 percent) between the deflections measured and those computed using Eq. 11.4 (Ref. 11.6). There are a number of other methods (such as that proposed by Yu and Winter (Ref. 11.7)) to estimate the effective I, which result in deflection predictions with the same degree of accuracy as obtained by the Branson formula (Ref. 11.6). A review and comparison of several methods proposed for computing short-time deflections is presented in Ref. 11.5.

Based on the effective stiffness, I_e, given by Eq. 11.4, the moment (or load)-deflection curve of a beam under short-time loading has the form shown in Fig. 11.3.

Fig. 11.3 Moment-deflection curve for short-term loading with I_e (Eq. 11.4)

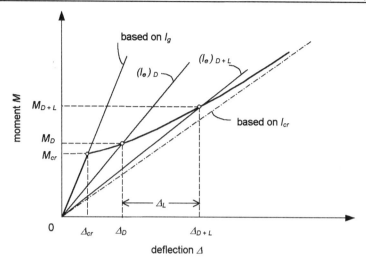

The value of I_e depends on the magnitude of the moment. Thus, for different load (and hence moment) levels, such as dead load alone or dead load plus live load, the respective deflections (Δ_D and Δ_{D+L}) should be computed with the I_e values given by Eq. 11.4 for the corresponding total moment levels (M_D and M_{D+L}) as indicated in Fig. 11.3. The incremental deflection, such as the part Δ_L due to live load, is obtained as the difference between the deflections computed with and without the live load, that is, $\Delta_L = \Delta_{D+L} - \Delta_D$.

11.4.2 Average Effective Moment of Inertia for Continuous Spans

Equation 11.4 was developed for use as a single average I_e for simply supported beams or for the region between inflection points of continuous beams. The negative moment regions of a continuous beam may have an effective stiffness, I_e, different from that of the positive moment region. The variation in I_e may be due to changes in cross-sectional dimensions, changes in reinforcement ratios, or differences in the influence of cracking (for example, in continuous T-beams, the value of I_{cr} may be substantially different under positive and negative moments). To account approximately for these variations, for continuous spans an average of I_e is obtained from Eq. 11.4 for the critical positive and negative moment sections. On this basis, for a beam continuous at one end,

$$\text{average } I_e = 0.75 \times (I_{e,m}) + 0.25 \times (I_{e,\text{cont.end}}) \qquad (11.5a)$$

and, for a beam continuous at both ends,

$$\text{average } I_e = \tfrac{1}{2}[I_{e,m} + \tfrac{1}{2}(I_{e,1} + I_{e,2})] \tag{11.5b}$$

where $I_{e,m}$ is the value of I_e at the critical positive moment section (taken as the midspan section), and $I_{e,1}$ and $I_{e,2}$ are values of I_e at the critical negative moment sections (continuous ends).

Typically, in a continuous (interior) span, the loading pattern producing the maximum deflection (and maximum positive moment) is different from the loading pattern that produces the maximum negative moment at supports. However, the degree of cracking in the negative moment regions of the beam will depend on the maximum negative moments there, as obtained from the moment envelope. To account for the influence of more widespread cracking in the negative moment region, which could have occurred under possibly larger negative moments, Branson recommends (Ref. 11.4, 11.5) the use of the moment envelope (for M_a in Eq. 11.4) to evaluate the negative and positive moment values for I_e. However, once the average I_e for the beam is so determined, the deflection is calculated using the loading pattern and corresponding moment diagram which produces the maximum deflection. In general, in continuous spans, the negative moment I_e values ($I_{e,1}$ and $I_{e,2}$) have a much smaller effect on the deflections than the positive moment I_e value ($I_{e,m}$). This can be seen from the conjugate beam method, where the span moment in the conjugate beam (which gives the deflection of the real beam) is more influenced by the loading on it ($= M/EI$) in the midspan regions than the loading near the ends. In fact, for continuous spans with (Neg. I_e)/(Pos.I_e) < 2, and with moment patterns in between those for a simple span and for a fixed-ended span, the use of midspan I_e alone ($= I_{e,m}$) gives reasonable accuracy. In a limited comparison with measured deflections of continuous prismatic members, this procedure was seen to give generally better results than the simple averaging procedure of Eqns. 11.5a and 11.5b (Refs. 11.8, 11.9). For spans with larger variation in flexural rigidity, or when the negative moment at either end is relatively larger, use of a weighted average I_e (such as in Eq. 11.6) is preferable (Refs. 11.5, 11.8, 11.9). For cantilever beams, I_e is appropriately based on the support section. For continuous prismatic members, CSA A23.3-94(9.8.2.4) gives the following equations to calculate the effective moment of inertia, based on the weighted average of the values obtained from Eq. 11.4.

For beams with both ends continuous,

$$\text{Weighted average } I_e = 0.7 I_{e,m} + 0.15(I_{e,1} + I_{e,2}) \tag{11.6a}$$

For beams with one end continuous,

$$\text{Weighted average } I_e = 0.85\, I_{e,m} + 0.15\, I_{e,cont.end} \tag{11.6b}$$

Fig. 11.4 Cracked transformed section properties

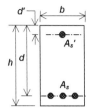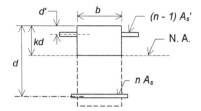

11.4.3 Cracked Section Properties

For the general case of a doubly reinforced rectangular section, the cracked transformed section is shown in Fig. 11.4. Under elastic behaviour, the neutral axis passes through the centroid of the section. Equating moment of areas about the centroid,

$$b\frac{(kd)^2}{2} + A_s'(n-1)(kd-d') = nA_s(d-kd)$$

Substituting $\rho = A_s/bd$ and $\rho' = A_s'/bd$, and solving for k,

$$k = \sqrt{[n\rho + (n-1)\rho']^2 + 2\left[n\rho + (n-1)\rho'\frac{d'}{d}\right]} - [n\rho + (n-1)\rho'] \qquad (11.7)$$

$$I_{cr} = \frac{b(kd)^3}{3} + (n-1)A_s'(kd-d')^2 + nA_s(d-kd)^2 \qquad (11.8)$$

When there is no compression reinforcement,

$$k = \sqrt{(n\rho)^2 + 2n\rho} - n\rho \qquad (11.9)$$

$$I_{cr} = \frac{b(kd)^3}{3} + nA_s(d-kd)^2 \qquad (11.10)$$

EXAMPLE 11.1

For the T-beam designed in Example 5.7, compute the immediate deflection due to dead load and live load. Use normal density concrete, $f_c' = 35$ MPa, and $f_y = 400$ MPa.

Fig. 11.5 Example 11.1

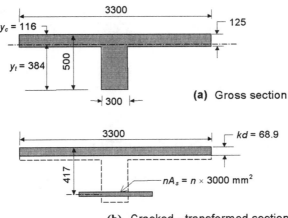

(a) Gross section

(b) Cracked - transformed section

SOLUTION

1. Loads and moments
   From Example 5.7,
   Dead load = 13 + 2.65 = 15.65 kN/m
   Specified live load = 21.5 kN/m
   Dead load moment, $M_D = 15.65 \times (8-0.24)^2/8 = 118$ kN·m
   Moment under combined dead plus live load,
   $$M_{D+L} = (15.65 + 21.5)(8 - 0.24)^2/8 = 280 \text{ kN·m}$$
   $$E_c = 4500\sqrt{f'_c} = 4500\sqrt{35} = 26622 \text{ MPa}$$
   The effective T-beam section is shown in Fig. 11.5. The cross-section properties are given in the following steps.

2. Gross section, I_g

   Depth to centroid, $y_c = \dfrac{(3300-300) \times 125 \times 62.5 + 300 \times 500 \times 250}{(3300-300) \times 125 + 300 \times 500}$

   $= 116$ mm

   $$I_g = \frac{3300 \times 116^3}{3} + \frac{(3300-300) \times 9^3}{3} + \frac{300 \times (500-116)^3}{3}$$

   $= 7.38 \times 10^9$ mm^4

3. Cracked transformed section, I_{cr}

The modular ratio, $n = \dfrac{E_s}{E_c} = \dfrac{200\,000}{26\,622} = 7.5$

The depth of neutral axis, kd, assuming it to be within the flange, is obtained by equating moments of area,

$$3300\dfrac{(kd)^2}{2} = 7.5 \times 3000(417 - kd)$$

Solving, $kd = 68.9$ mm ($< h_f$, OK)

$$I_{cr} = \dfrac{3300 \times 68.9^3}{3} + 7.5 \times 3000(417 - 68.9)^2 = 3.09 \times 10^9 \text{ mm}^4$$

4. The cracking moment, M_{cr}

$f_r = 0.6\lambda\sqrt{f_c'} = 3.55$ MPa

$$M_{cr} = f_r I_g / y_t = \dfrac{3.55 \times 7.38 \times 10^9}{(500 - 116)} \times 10^{-6} = 68.2 \text{ kN·m}$$

5. Effective stiffness, I_e

For dead load only, $(I_e)_D$, using Eq. 11.4, and with $M_a = M_D = 118$ N·m $> M_{cr}$

$(I_e)_D = 3.09 \times 10^9 + (7.38 - 3.09) \times 10^9 \times (68.2/118)^3 = 3.92 \times 10^9 \text{ mm}^4$

For dead plus live load, $(I_e)_{D+L}$ is

$(I_e)_{D+L} = 3.09 \times 10^9 + (7.38 - 3.09) \times 10^9 \times (68.2/280)^3 = 3.15 \times 10^9 \text{ mm}^4$

6. Immediate deflections

Midspan deflection for simply supported uniformly loaded beam is (Eq. 11.3 and Table 11.1),

$$\Delta = \dfrac{5}{48}\dfrac{M_m L^2}{EI}$$

For dead load only,

$$\Delta_D = \dfrac{5}{48} \times \dfrac{(118 \times 10^6) \times (8000 - 240)^2}{26\,622 \times 3.92 \times 10^9} = 7.1 \text{ mm}$$

For dead plus live load,

$$\Delta_{D+L} = \frac{5}{48} \times \frac{(280 \times 10^6) \times (8000 - 240)^2}{26\,622 \times 3.15 \times 10^9} = 20.9 \text{ mm}$$

Live load deflection is,
$$\Delta_L = 20.9 - 7.1 = 13.8 \text{ mm}$$

7. Allowable immediate deflection due to live load by CSA A23.3-94(9.8.2.6) (Section 11.11) is, for floors not supporting or attached to clements likely to be damaged by large deflection,
$$\frac{l_n}{360} = \frac{(8000 - 480)}{360} = 20.9 \text{ mm}$$

$\Delta_L < \Delta_{L,\, allow}$ OK

EXAMPLE 11.2

Compute the immediate deflection due to live load of the exterior panel slab of the one-way floor slab designed in Example 5.5.

SOLUTION

The Code moment coefficients will be used. Considering a slab strip of 1 m width, the specified loads are (Example 5.5):

Dead load = 3.94 kN/m; live load = 6.5 kN/m
$E_c = 4500 \sqrt{f'_c} = 26\,622$ MPa, and $n = 7.5$
$I_g = 1000 \times 125^3/12 = 163 \times 10^6$ mm^4
$f_r = 0.6 \sqrt{35} = 3.55$ MPa
$$M_{cr} = \frac{3.55 \times 163 \times 10^6}{62.5} \times 10^{-6} = 9.26 \text{ kN} \cdot \text{m}$$

1. Effective moments of inertia
   I_e for midspan section (I_m) and the two support sections ($I_{e,1}$ and $I_{e,2}$) will be computed for the respective maximum moments.

   (a) Midspan section (Fig. 11.6b)
   From Example 5.5 (Table 5.6),
   $d = 99.4$ mm, $A_s = 294$ mm^2, $\rho = A_s/bd = 0.00296$

Fig. 11.6 Example 11.2

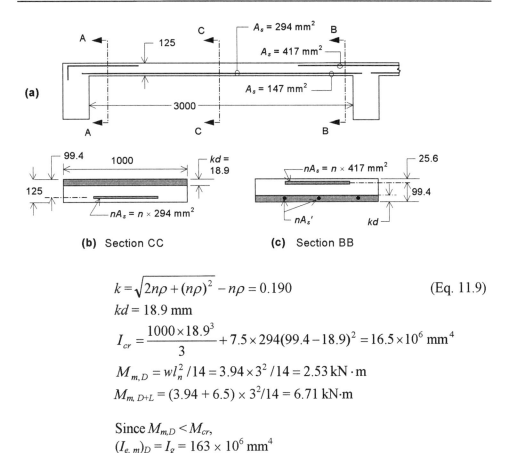

$$k = \sqrt{2n\rho + (n\rho)^2} - n\rho = 0.190 \quad \text{(Eq. 11.9)}$$
$$kd = 18.9 \text{ mm}$$
$$I_{cr} = \frac{1000 \times 18.9^3}{3} + 7.5 \times 294(99.4 - 18.9)^2 = 16.5 \times 10^6 \text{ mm}^4$$
$$M_{m,D} = wl_n^2/14 = 3.94 \times 3^2/14 = 2.53 \text{ kN} \cdot \text{m}$$
$$M_{m,\,D+L} = (3.94 + 6.5) \times 3^2/14 = 6.71 \text{ kN} \cdot \text{m}$$

Since $M_{m,D} < M_{cr}$,
$(I_{e,\,m})_D = I_g = 163 \times 10^6 \text{ mm}^4$

Since under both loadings, D and D + L, the moment is less than the cracking moment, M_{cr}, the section is uncracked, and there is no need to compute I_{cr} at this section.

$I_{e,m} = I_g = 163 \times 10^6 \text{ mm}^4$, for both dead load only and for dead plus live load.

(b) Exterior support
$$M_D = wl_n^2/24 = 3.94 \times 3^2/24 = 1.48 \text{ kN} \cdot \text{m}$$
$$M_{D+L} = (3.94 + 6.5) \times 3^2/24 = 3.92 \text{ kN} \cdot \text{m}$$

Since under both loadings, the moment is less than the cracking

moment, M_{cr}, the section is uncracked and there is no need to compute I_{cr} at this section.
$I_{e,1} = I_g = 163 \times 10^6$ mm^4, for both dead load only and for dead plus live load.

(c) Interior support (Fig. 11.6c)
$A_s = 417$ mm^2, $\rho = 0.00420$

The bottom reinforcement from midspan, continued into the support, acts as compression reinforcement and has area $A_s' = 147$ mm^2 ($\rho' = 0.00148$). The neutral axis depth may be computed using Eq. 11.7 as: $kd = 22.1$ mm, but this means that the bottom reinforcement is in tension. Therefore,

$$\frac{b(kd)^2}{2} = nA_s\left(d - \frac{kd}{2}\right) + nA_s'\left(d' - \frac{kd}{2}\right)$$

which gives $kd = 24$ mm at the interior support.

$$I_{cr} = \frac{1000 \times 24.0^3}{3} + 7.5 \times 417 \times (99.4 - 24.0)^2 + 7.5 \times 147$$
$$\times (25.6 - 24.0)^2 = 22.4 \times 10^6 \text{ mm}^4$$
$$M_D = wl_n^2/10 = 3.94 \times 3^2/10 = 3.55 \text{ kN} \cdot \text{m}$$
$$M_{D+L} = (3.94 + 6.5) \times 3^2/10 = 9.40 \text{ kN} \cdot \text{m}$$

Since $M_D < M_{cr}$, for dead load condition,
$(I_{e,2}) = I_g = 163 \times 10^6$ mm^4

For dead load plus live load, using Eq. 11.4,
$$(I_{e,2})_{D+L} = 22.4 \times 10^6 + (163 - 22.4) \times 10^6 \times (9.26/9.40)^3$$
$$= 156.8 \times 10^6 \text{ mm}^4$$

2. Average I_e for span (Eq. 11.6a)
Since the slab is integral with supporting beams, for I average, it will be considered as continuous at both ends. For dead load only, all critical sections have $I_e = I_g$ so that,
Average $(I_e)_D = I_g = 163 \times 10^6$ mm^4

For dead load plus live load,

Weighted average $(I_e)_{D+L} = [0.7 \times 163 + 0.15(163 + 156.8)] \times 10^6$
$$= 162 \times 10^6 \text{ mm}^4$$

Calculation of deflection:

$$\Delta = K \frac{5}{48} \frac{M_m L^2}{EI_e}$$
$$K = 1.20 - 0.20 M_o / M_m \quad \text{(Eq. 11.3a)}$$

Loading for maximum deflection is same as for maximum positive moment, for which $M_m = wL_n^2/14$

$$K = 1.20 - 0.20(wL^2/8)/(wL^2/14) = 0.85$$

$$\Delta_D = 0.85 \times \frac{5}{48} \times \frac{M_m L^2}{E(I_e)_D}$$
$$= 0.85 \times \frac{5}{48} \times \frac{2.53 \times 10^6 \times 3000^2}{26\,622 \times 163 \times 10^6} = 0.465 \text{ mm}$$

$$\Delta_{D+L} = 0.85 \times \frac{5}{48} \times \frac{6.71 \times 10^6 \times 3000^2}{26\,622 \times 162 \times 10^6} = 1.240 \text{ mm}$$

$$\Delta_L = 1.240 - 0.465 = 0.775 \text{ mm}$$

The maximum allowable immediate deflection due to live load for floors not supporting or attached to nonstructural elements likely to be damaged by large deflection is (CSA A23.3-94, Table 9-2):
$$\Delta_{all} = l_n / 360 = 3000/360 = 8.33 \text{ mm}$$
$$\Delta_L < \Delta_{all} \qquad \qquad \text{OK}$$

[The reinforcement ratio of 0.0029 at midspan is only $0.105\rho_b$ (Table 5.3). Therefore, it could have been anticipated that deflection would not be critical.]

11.5 TIME-DEPENDENT DEFLECTION

Under sustained loads, the deflection increases with time, due principally to the effects of creep and shrinkage, although additional factors such as formation of new cracks, widening of earlier cracks, and effects of repeated load cycles also contribute to these increases. The additional long-time deflection may be as large as two to three times the immediate deflection. The factors that influence creep and shrinkage were described in

356 REINFORCED CONTRETE DESIGN

Sections 1.11 and 1.12. Thus, the major parameters that influence the long-time deflection are the stress in concrete, amount of tensile and compressive reinforcement, size of member, curing conditions, temperature, relative humidity, age of concrete at time of loading, duration of loading, and conditions of restraint.

11.6 DEFLECTION DUE TO SHRINKAGE

In an unrestrained reinforced concrete member, shrinkage causes shortening of the member which is resisted by the reinforcement, thereby developing compressive stress in the steel and tensile stress in the concrete, which balance each other. When the reinforcement is symmetrically placed in the cross section, the shrinkage does not cause any curvature of the member. However, when the reinforcement is not symmetric, as is usually the case in flexural members, the shortening of the reinforced face is far less than that of the unreinforced (or lesser reinforced) face (Fig. 11.7a,b). This differential shrinkage causes a curvature of the member. The location of flexural reinforcement in beams is such that the shrinkage curvature has the same sign as the curvature due to flexure and, thereby, magnifies the deflections due to applied loads. Also, the addition of compression reinforcement decreases the differential shrinkage strains and resulting deflections.

Shrinkage stresses are sustained stresses, and shrinkage and creep are interdependent. The shrinkage curvature is also affected by the flexural cracking and, hence, the transverse load. However, as an approximation, the effects of shrinkage may be considered separate from those of creep and transverse loads.

Fig. 11.7 Shrinkage curvature

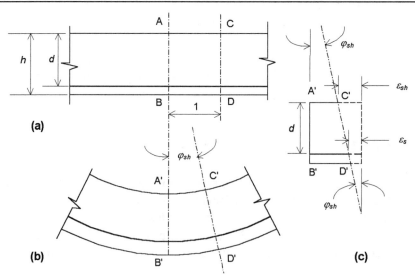

11.7 SHRINKAGE CURVATURE

The two broad approaches used for the computation of shrinkage curvature are (1) the so called "equivalent tensile force" method, and (2) empirical methods. Detailed reviews of these methods are contained in Refs. 11.5 and 11.6. Because of its better agreement with test results, and its recommendation by ACI Committee 435 (Ref. 11.6), only the empirical method proposed by Miller as modified by Branson is briefly discussed below.

For a singly reinforced section, Miller (Ref. 11.10) assumed a shrinkage strain distribution as shown in Fig. 11.7c, which gives the curvature as:

$$\varphi_{sh} = \frac{\varepsilon_{sh} - \varepsilon_s}{d} = \frac{\varepsilon_{sh}}{d}\left(1 - \frac{\varepsilon_s}{\varepsilon_{sh}}\right) \tag{11.11}$$

The fibres on the unreinforced face are assumed to have shrinkage strains equal to the free shrinkage strain of the concrete, ε_{sh}, and ε_s in Eq. 11.11 is the strain in the steel resulting from shrinkage. The value of $\varepsilon_s/\varepsilon_{sh}$ depends on the amount of reinforcement, and empirical values for this were suggested by Miller. As a modification and extension of Miller's method, Branson in 1963 suggested the following equations which are applicable as well to doubly reinforced beams.

For $(\rho - \rho') \leq 3$,

$$\varphi_{sh} = 0.7 \frac{\varepsilon_{sh}}{h}(\rho - \rho')^{1/3} \left(\frac{\rho - \rho'}{\rho}\right)^{1/2} \tag{11.12}$$

For $(\rho - \rho') > 3$,

$$\varphi_{sh} = \varepsilon_{sh}/h \tag{11.13}$$

where ρ and ρ' are tension and compression reinforcement ratios expressed as percentages (= $100 A_s/bd$ and $100 A_s'/bd$), and h is the thickness of the member. In a comparison of several methods of computing shrinkage deflection, Eqs. 11.12 and 11.13 were seen to give the closest agreement with the test results (Refs. 11.5, 11.6). For use in these equations, the free shrinkage strain, ε_{sh}, obtained from test data incorporating effects of local aggregates is preferred. Methods for estimating the free shrinkage, both at time t, $(\varepsilon_{sh})_t$, and at ultimate, $(\varepsilon_{sh})_u$, were explained in Section 1.12. For situations where $(\varepsilon_{sh})_u$ under local conditions is not known, ACI Committee 435 and Branson have suggested the use of $(\varepsilon_{sh})_u = 400 \times 10^{-6}$ mm/mm for routine deflection calculations (Refs. 11.5, 11.9).

11.8 COMPUTATION OF SHRINKAGE DEFLECTIONS

If the distribution of the shrinkage curvature, φ_{sh}, along the span is known, the resulting deflection may be computed by any of the classical procedures, such as integration, moment-area, or conjugate beam methods (with φ_{sh} replacing the curvature due to moment, M/EI). In restrained beams, the warping due to shrinkage may induce secondary moments, producing a curvature which should be accounted for in the deflection calculations. In general, the shrinkage deflections, Δ_{sh}, can be expressed as:

$$\Delta_{sh} = k_{sh}\varphi_{sh}L^2 \qquad (11.14)$$

where φ_{sh} in the shrinkage curvature, k_{sh} is a coefficient depending on the support conditions and the distribution of φ_{sh} along the span, and L is the span length. Assuming that $\rho - \rho'$, and hence φ_{sh}, has the same value along the span (including the positive and negative curvature regions in continuous beams), together with simplifying assumptions about the location of inflection points in continuous spans, the values given in Table 11.2 can be derived for k_{sh} (see Ref. 11.5, pp. 173-174 for derivations) for beams of uniform section.

In actual beams, the ratio $(\rho - \rho')$ may not be uniform. However, as in the case of deflections due to transverse loads, the positive curvature, φ_{sh}, in the midspan region has the dominant effect on deflection. Therefore, it is satisfactory to use in Eq. 11.14 the φ_{sh}, based on $(\rho - \rho')$ for the midspan region for simple and continuous spans, and based on $\rho - \rho'$ for the support section for cantilever spans. For T-beams, Ref. 11.9 recommends the use of an average of ρ and ρ_w in Eqs. 11.12 and 11.13, where $\rho = 100\, A_s/bd$ and $\rho_w = 100\, A_s/b_w d$.

Table 11.2 Values of k_{sh}

k_{sh}	Types of beam
0.5	cantilever beam
0.125	simply supported beam
0.0625	fixed ended beams
0.084	beams continuous at one end (in continuous beams with 2 spans only)
0.090	beams continuous at one end (in continuous beams with 3 or more spans)
0.065	beams continuous at both ends

11.9 DEFLECTION DUE TO CREEP

As a result of creep under sustained stresses, the compressive strain in the concrete at the extreme fibre increases with time, as shown in Fig. 11.8, while the strain in the tension steel increases only marginally. With the shifting of the neutral axis and consequent decrease in internal lever arm, steel force and, hence, strain increase slightly to maintain equilibrium with applied loads (Fig. 11.8c). The creep strain causes an increase in curvature, φ_{cp}, which, like the shrinkage curvature, causes additional deflection of the member.

Since $\varphi_i = \varepsilon_i/x_i$, and $\varphi_{cp} = \varepsilon_{cp}/x_{cp}$; and with $x_{cp} > x_i$, (Fig. 11.8),

$$\frac{\varphi_{cp}}{\varphi_i} = k_r \frac{\varepsilon_{cp}}{\varepsilon_i} = k_r C_t \tag{11.15}$$

where φ_i is the initial elastic curvature, ε_i is the initial elastic strain, C_t is the creep coefficient (Section 1.11), and the coefficient k_r has a value less than unity. The increase in deflection may be taken in proportion to the increase in curvature, so that,

$$\frac{\Delta_{cp}}{\Delta_i} = k_r C_t$$

$$\Delta_{cp} = k_r C_t \Delta_i \tag{11.16}$$

where Δ_{cp} is the additional deflection due to creep, and Δ_i is the immediate deflection due to the sustained part of the load. As in the case of shrinkage deflection, compression reinforcement will restrain creep strains, and reduce curvature and deflection due to creep. For k_r in Eq. 11.16, ACI Committee 435 recommends Eq. 11.17, which includes the effect of compression reinforcement (Ref. 11.9).

$$k_r = 0.85/(1+50\rho') \tag{11.17}$$

where $\rho' = A_s'/bd$

Expressions for computing the creep coefficient C_t at age t were presented in Section 1.11. For average conditions, in the absence of more precise data, the ACI

Fig. 11.8 Creep strain and curvature

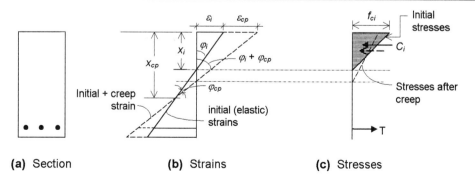

(a) Section   (b) Strains   (c) Stresses

360 *REINFORCED CONTRETE DESIGN*

Committee recommended a value of 1.6 for the ultimate creep coefficient, C_u, for use in Eqs. 1.6 and 11.16.

11.10 LONG-TIME DEFLECTION BY CODE

Because of the large number of interacting parameters affecting long-time deflections, the uncertainty regarding these parameters, and the lower degree of reliability of deflection computations in general, under normal situations it is more practical as well as satisfactory to estimate the combined additional deflection due to all time-dependent effects with a single time-dependent factor applied to the initial deflection. Such a procedure is recommended in CSA A23.3-94(9.8.2.5). According to this, the additional long-time deflection $\Delta_t = (\Delta_{cp} + \Delta_{sh})$ may be obtained by multiplying the immediate deflection, Δ_i, due to the sustained load considered, by a factor $[S/(1 + 50\,\rho')]$.

$$\Delta_t = \Delta_{cp} + \Delta_{sh} = \Delta_i \times \frac{S}{1+50\rho'} \qquad (11.18)$$

where $\rho' = A_s'/(bd)$ is the compression reinforcement ratio, taken as the value at midspan for simple and continuous spans and at the support for cantilevers, and S is a factor for creep deflections under sustained loads, taken as:

        2.0 for 5 years or more
        1.4 for 12 months
        1.2 for 6 months
        1.0 for 3 months

Also, Δ_i includes the immediate deflection due to dead load and due to the portion of live load that is sustained. The multiplication factor in Eq. 11.18 was developed by Branson, and Eq. 11.18 has been found to give good agreement with results of tests (Ref. 11.5, 11.9, 11.11). The factor accounts for the effect of compression reinforcement in reducing long-time deflection as well as for the duration of sustained loading.

The Code does permit the use of more detailed procedures such as computing Δ_{cp} and Δ_{sh} individually using Eqs. 11.14 and 11.16.

EXAMPLE 11.3

For the slab in Example 11.2, compute the long-time deflection and compare with the allowable values specified by the Code. Assume 20 percent of the live load to be sustained for over five years.

SOLUTION

1. By Code procedure
   From Example 11.2, the immediate deflections are:
   $(\Delta_i)_D = 0.465$ mm, $(\Delta_i)_L = 0.775$ mm
   Using Eq. 11.18, with $\rho' = 0$; 20% of live load sustained; and $S = 2.0$ for duration of sustained load of over five years,

   $$(\Delta_t) = \Delta_{cp} + \Delta_{sh} = \Delta_i \times \frac{S}{1+50\rho'} = (0.465 + 0.2 \times 0.775) \times \frac{2.0}{1+0} = 1.24 \text{ mm}$$

   Sum of long time deflection due to sustained load and immediate deflection due to live load, of which 20% is sustained, gives the incremental deflection occurring after the portion is installed.

   $\Delta_t + (\Delta_i)_L = 1.24 + 0.8 \times 0.775 = 1.86$ mm

   Allowable limits in CSA A23.3-94(9.8.2.6) are:

   (a) Floor attached to nonstructural elements not likely to be damaged by large deflections,

   $$\frac{l_n}{240} = \frac{3000}{240} = 12.5 \text{ mm} > 2.02 \text{ mm} \qquad \text{OK}$$

   (b) If attached to nonstructural elements likely to be damaged by large deflection,

   $$\frac{l_n}{480} = \frac{3000}{480} = 6.25 \text{ mm} > 2.02 \text{ mm} \qquad \text{OK}$$

   In both cases, the computed deflections are less than the maximum allowable deflection specified. Therefore, the design is satisfactory.

2. By detailed computation, using Eqs. 11.14 and 11.16,
   $$\Delta_{sh} = k_{sh}\varphi_{sh}L^2 \qquad (11.14)$$
   $$\Delta = k_r C_t \Delta_i \qquad (11.16)$$

   (a) Shrinkage deflection Δ_{sh}
   φ_{sh} is given by Eqs. 11.12 and 11.13. For the slab, based on midspan section, $\rho' = 0$, $\rho = 100 A_s/bd = 0.295 < 3$, and Eq. 11.12 controls.

Assuming average conditions and $(\varepsilon_{sh})_u = 400 \times 10^{-6}$,

$$\varphi_{sh} = 0.7 \frac{\varepsilon_{sh}}{h}(\rho-\rho')^{1/3}\left(\frac{\rho-\rho'}{\rho}\right)^{1/2}$$

$$= 0.7 \times \frac{400 \times 10^{-6}}{125}(0.295)^{1/3} = 1.49 \times 10^{-6} \text{ mm}^{-1}$$

For slab continuous at both ends, from Table 11.2 (since slab is integral with beams at both ends, it is considered continuous at both ends), $k_{sh} = 0.065$

$$\Delta_{sh} = 0.065 \times 1.49 \times 10^{-6} \times 3000^2 = 0.872 \text{ mm}$$

(b) Creep deflection Δ_{cp}
k_r, using Eq. 11.17, with $\rho' = 0$

$$k_r = \frac{0.85}{1+50\rho'} = 0.85$$

Ultimate C_t, for average conditions, = 1.6
$\Delta_{cp} = k_r C_t \Delta_i = 0.85 \times 1.6 \times (0.47 + 0.2 \times 0.78) = 0.851$ mm

(c) Total long-time deflection
$\Delta_t = \Delta_{sh} + \Delta_{cp} = 0.872 + 0.851 = 1.73$ mm, compared with 1.24 mm computed at 1.

Including immediate deflection due to live load,
$(\Delta_i)_L + \Delta_t = 0.78 + 1.73 = 2.51$ mm.

This is within the allowable limits.

EXAMPLE 11.4

The dimensions of the exterior span of a continuous interior beam of a slab-beam girder floor are shown in Fig. 11.9a. The slab is 110 mm thick, and the beams are spaced at 3.5 m centres. The specified loads consist of a dead load (including self weight) of 19.8 kN/m and a live load of 16.8 kN/m, of which 50 percent is sustained. Compute the immediate and long-term deflections of the beam and compare with the Code requirements. Use $f_c' = 30$ MPa and $f_y = 300$ MPa.

Fig. 11.9 Example 11 4

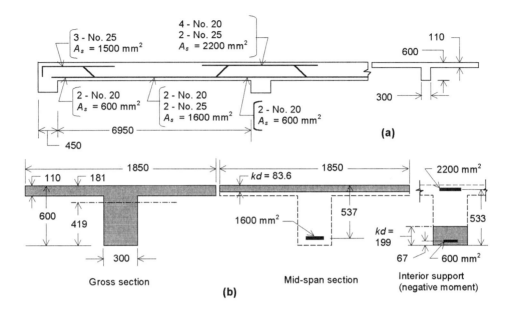

SOLUTION

$E_c = 4500\sqrt{30} = 24\,678$ MPa

$n = E_s/E_c = 8$

Effective flange width of T-section is

$0.2 \times \text{span} + b_w = 0.2 \times 7400 + 300 = 1780$ mm   (controls)

$24h_f + b_w = 2940$ mm

spacing $= 3500$ mm

The gross and cracked transformed sections of the beam at various critical sections are shown in Fig. 11.9b. The Code moment coefficients are used to compute maximum moments at critical sections. The computations for deflections are set up in Table 11.3. In computing the average moment of inertia for the continuous span, the beam is assumed (conservatively) as discontinuous at the exterior support in view of the relatively low negative moment at this section. Therefore, Eq. 11.6b is used in averaging I_e rather than Eq. 11.6a. Also note that at the interior support section, I_e, is computed for the maximum negative moment at this section ($wL^2/10$), although the moment here corresponding to the load causing maximum deflection will be less than this maximum value.

364 REINFORCED CONTRETE DESIGN

Table 11.3 Design Example 11.4

Mom. Coeff.	1/24	1/14	1/10	Comments
I_g, 10^9 mm^4	10.7	10.7	10.7	
I_{cr}, 10^9 mm^4	2.16	2.92	2.89	
M_{cr}, kN·m	-196	84.5	-196	Eq. 11.4a
Dead load only				
M_{Dmax}, kN·m	39.8	68.3	95.6	
M_{cr}/M_{Dmax}	>1	>1	>1	
$I_{e,D}$, 10^9 mm^4	10.7	10.7	10.7	
$(I_{e,D})_{av}$, 10^9 mm^4	-	10.7	-	Eq. 11.6b
Dead load + Live load				
$M_{D+L, max}$, kN·m	73.7	126	177	
$M_{cr}/M_{D+L, max}$	>1	0.671	>1	
$I_{e,D+L}$, 10^9 mm^4	10.7	5.27	10.7	
$(I_{e,D+L})_{av}$, 10^4 mm^4	-	6.08	-	Eq. 11.6b
Deflections				
$K = 1.2 - 0.2(wL^2/8)\,(wL^2/14) = 0.85$				Eq. 11.3a
$(\Delta_i)_D = 0.85 \times \dfrac{5}{48} \times \dfrac{68.31 \times 10^6 \times (6950)^2}{24678 \times 10.7 \times 10^9} = 1.11\text{ mm}$				
$(\Delta_i)_{D+l} = 0.85 \times \dfrac{5}{48} \times \dfrac{126 \times 10^6 \times (6950)^2}{24678 \times 6.08 \times 10^9} = 3.59\text{ mm}$				
$(\Delta_i)_L = 3.59 - 1.11 = 2.48$ mm				
$s/(1 + 50\,\rho') = 2.0$				
$\Delta_{cp+sh} = 2(1.11 + 0.5 \times 2.48) = 4.70$ mm				
$\Delta_{cp+sh} = (\Delta_i)_L = 4.70 + 0.5 \times 2.48 = 5.41$ mm				
Check allowable limits:				
$(\Delta_i)_L = 2.48$ mm $< l_n/360 = 19.3$ mm				OK
$\Delta_{cp+sh} + (\Delta_i)_L = 7.18 < l_n/480 = 14.5$ mm				OK
At both conditions, deflections are within limits.				

EXAMPLE 11.5

Compute the deflection due to creep and shrinkage, over a period of three years for the beam in Example 11.4. Assume that the ultimate creep coefficient, $C_u = 2.35$, ultimate shrinkage strain, $(\varepsilon_{sh})_u = 780 \times 10^{-6}$ mm/mm, and that the relative humidity is 60%.

SOLUTION

1. Creep deflection, using Eq. 11.16 is:

$$\Delta_{cp} = k_r C_t \Delta_i \quad \text{(Eq. 11.16)}$$

Considering the midspan section properties only, which has the dominant effect on deflections,

$$k_r = 0.85/(1+50\rho') \quad \text{(Eq. 11.17)}$$
$$= 0.85, \text{ as } \rho' = 0$$

$$C_t = \left(\frac{t^{0.6}}{10+t^{0.6}}\right) C_u \times (\text{Correction Factors } CF) \quad \text{(Eq. 1.6)}$$

CF for humidity is (Eq. 1.8):
$1.27 - 0.0067 \times (H - 60) = 0.868$

$$C_t = \frac{(3\times 365)^{0.6}}{10+(3\times 365)^{0.6}} \times 2.35 \times 0.868 = 1.77$$

From Example 11.4, immediate deflection due to sustained part of load is:
$\Delta = 1.11 + 0.5 \times 2.48 = 2.35$ mm
$\Delta_{cp} = k_r C_t \Delta_i = 0.85 \times 1.77 \times 2.35 = 3.54$ mm

2. Shrinkage deflection
$$\Delta_{sh} = k_{sh}\varphi_{sh}L^2 \quad \text{Eq. (11.14)}$$

Shrinkage strain is given by (Eq. 1.9):

$$(\varepsilon_{sh})_t = \frac{t}{35+t}(\varepsilon_{sh})_u \times CF$$

CF for humidity is (Eq. 1.11):
$(CF)_h = 1.40 - 0.01 H = 1.40 - 0.01 \times 60 = 0.8$

$$(\varepsilon_{sh})_{3\,years} = \frac{3\times 365}{35+3\times 365} \times 780\times 10^{-6} \times 0.8 = 605 \times 10^{-6}$$

Shrinkage curvature is:

$$\varphi_{sh} = 0.7\frac{\varepsilon_{sh}}{h}(\rho-\rho')^{1/3}\left(\frac{\rho-\rho'}{\rho}\right)^{1/2} = 0.7\frac{\varepsilon_{sh}}{h}\rho^{1/3} \quad \text{(Eq. 11.12)}$$

For T-beams
$\rho = 100 A_s/bd = 100 \times 1600/(1780 \times 533) = 0.168$
$\rho_w = 100 A_s/b_w d = 100 \times 1600/(300 \times 533) = 1.00$

Average $\rho = 0.584$ percent

$$\varphi_{sh} = 0.7 \times \frac{605 \times 10^{-6}}{600} \times (0.584)^{\frac{1}{3}} = 0.590 \times 10^{-6}$$

For beam continuous at one end (more than two spans):
$k_{sh} = 0.09$ (Table 11.2)
$\Delta_{sh} = 0.09 \times 0.590 \times 10^{-6} \times 6950^2 = 2.56$ mm
$\Delta_{cp} + \Delta_{sh} = 3.54 + 2.56 = 6.10$ mm

11.11 DEFLECTION CONTROL IN CODE

Two methods are used in CSA A23.3-94 for control of deflections of reinforced concrete members. The first method is to provide sufficient thickness for the member to have adequate flexural stiffness. Thus, for beams and one-way slabs (solid or ribbed), not supporting or attached to other construction likely to be damaged by deflection, recommended minimum thicknesses are given in the Code (Cl. 9.8.2.1) (also Table 5.2). For two-way construction, minimum thickness requirements are given in Clause 9.8.3. Only for these types of members, the recommended minimum thickness values may be used in lieu of deflection calculations. (A more extensive table including minimum thicknesses for beams and one-way slabs supporting or attached to nonstructural elements likely to be damaged by large deflections is presented in Ref. 11.9). The specified minimum thicknesses are established primarily from past experience with usual structures, and are not intended to be applied to special cases such as for members supporting heavy concentrated loads.

For members not meeting the minimum thickness requirements, or those supporting or attached to nonstructural elements likely to be damaged by large deflections, the deflections must be calculated and limited to the allowable values in the Code (Cl. 9.8.2.6).

11.12 CONTROL OF CRACKING

Cracks in reinforced concrete structures may result from (1) volume changes (due to shrinkage, creep, thermal and chemical effects, etc.), (2) imposed loads and/or displacements (such as continuity effects, settlement of supports, etc.), and (3) flexural stress due to bending (Ref. 11.15). Cracks due to the first two causes could mostly be controlled by ensuring good quality concrete and correct construction (control joints, temperature and shrinkage reinforcement, etc.). The discussion that follows deals with cracking due to flexural stresses.

Because of the low strength and failure strain of concrete in tension, some minute cracking of concrete in the tension side of flexural members is inevitable. This is particularly so when relatively higher strength reinforcing steels are used, in which case the steel stresses and strains at service load levels will also be high. From considerations of corrosion protection for reinforcement, as well as aesthetics, it is preferable to have several well distributed fine hairline cracks than a few wide cracks. This is the basic objective of crack control procedures.

The cracking of concrete is of a random nature, and the width of cracks is subject to relatively wide scatter. The methods for crack width calculation only attempt to predict the most probable maximum crack width as observed in laboratory tests. These are further simplified to arrive at satisfactory reinforcing details meeting the objectives of crack control.

Experimental investigations of both tensile and flexural members have shown that crack width is proportional to tensile steel stress, f_s. The other important parameters that influence crack width are (1) the thickness of concrete cover, (2) the effective tension area of concrete surrounding the flexural tension reinforcement and having the same centroid as the reinforcement (A_e in Fig. 11.10), and (3) the number of bars, n (Ref. 11.12). Based on statistical analysis of extensive test data reported by several investigators, Gergely and Lutz developed Eq. 11.19 for the probable maximum crack width in a flexural member (Ref. 11.12).

$$w = C\beta f_s \sqrt[3]{d_c A} \qquad (11.19)$$

where, w = the maximum crack width; β = ratio of distances to the neutral axis from the extreme tension fibre to that from the centroid of the main reinforcement (= h_2/h_1 in Fig. 11.10); f_s = stress in steel at specified load calculated by elastic cracked section theory; d_c = thickness of concrete cover measured from the extreme tension fibre to the centre of the bar located closest thereto; A = effective tension area of concrete surrounding the main tension reinforcement and having the same centroid as that reinforcement, divided by the number of bars, = A_e/n; n = number of bars, taken as A_s divided by area of largest bar used when bars are of different sizes; and C = a numerical constant determined from statistical analysis of experimental data. C has the inverse unit of stress and was determined in Ref. 11.12 as 76×10^{-6} in^2/kip with f_s in ksi. In SI units, the corresponding value of C is 11×10^{-6} mm^2/N with stress, f_s, in MPa (= N/mm^2). Eq. 11.19 suggests that the maximum crack width will be less with several moderately spaced small size bars well distributed over the zone of maximum tension than with a few bars of a larger diameter giving the same steel area. With a view to

Fig. 11.10 Parameters for crack width computation

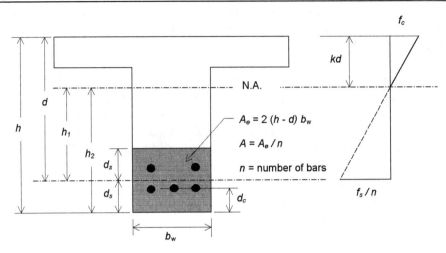

emphasising the selection of reinforcing details for crack control, Eq. 11.19 can be rearranged as:

$$z = f_s \sqrt[3]{d_c A} \times 10^{-3} = \frac{w}{C\beta} \times 10^{-3} \quad (11.20)$$

For general design use, β is assumed to have an approximate value of 1.2. The Code further assumes, on the basis of experience with existing structures, limiting crack widths of 0.40 mm and 0.33 mm for interior and exterior exposure, respectively. With these approximations, the right hand side of Eq. 11.20 reduces to:

for interior exposure, $\dfrac{0.40}{11 \times 10^{-6} \times 1.2} \times 10^{-3} = 30 \text{ kN/mm}$

for exterior exposure, $\dfrac{0.33}{11 \times 10^{-6} \times 1.2} \times 10^{-3} = 25 \text{ kN/mm}$

On this basis, CSA A23.3-94(10.6.1) limits the value of $z = f_s \sqrt[3]{d_c A} \times 10^{-3}$ to a maximum of 30 kN/mm for interior exposure and to 25 kN/mm for exterior exposure. In computing z, f_s is the tensile steel stress at specified loads computed by elastic cracked section theory (Section 4.6). However, the Code permits the use of $f_s = 0.6 f_y$, (f_y = specified yield stress) in lieu of a detailed calculation. In most practical situations, this is a conservative assumption as the actual steel stress at specified load levels will be below this value.

Equation 11.19 has been shown to be applicable also to one-way slabs (Ref. 11.13). However, the average value of β for slabs will be larger than for beams, and Ref. 11.14 suggests $\beta = 1.35$ for slabs. On this basis the maximum values of z for one-way slabs are:

for interior exposure, $z = \dfrac{1.2}{1.35} \times 30 = 26.7$ kN/mm

for exterior exposure, $z = \dfrac{1.2}{1.35} \times 25 = 22.2$ kN/mm

The limiting values for z given above are for normal conditions and do not apply to structures subject to very aggressive exposure or designed to be watertight (Code Cl. 10.6.1). Recommended maximum permissible crack widths for such conditions are given in Ref. 11.15.

In T-beams where the flange is in tension, too wide a spacing of the reinforcement may result in formation of wide cracks in the slab near the web. To safeguard against this the Code (Cl. 10.5.3) limits the distribution of negative moment reinforcement to a width of an overhanging flange equal to the lesser of 1/20 of the beam span, or the width defined in Clause 10.3 (also Section 4.9.3). The area of this reinforcement must be at least 0.004 times the gross area of the overhanging flange. Where necessary, the outer portions of the flange must be protected with additional reinforcement.

In relatively deep flexural members, controlling the crack width at the level of tension reinforcement by limiting the value of z may still leave a deep zone of tension in the web between the neutral axis and leave the reinforcement level unprotected and susceptible to wide cracking. Therefore, for flexural members with depth of web in excess of 750 mm, the Code (Cl. 10.6.2) requires the placement of longitudinal skin reinforcement near the vertical faces of the web in the tension zone.

EXAMPLE 11.6

The actual placement details of flexural reinforcement at midspan of the beam in Example 11.4 are shown in Fig. 11.11. Assume the steel as Grade 400. Check the section against the crack control requirements of the Code.

SOLUTION

Distance of centroid of steel area from bottom face is:
$$d_s = \dfrac{2 \times 500 \times (51.3 + 25.2/2) + 2 \times 300 \times (51.3 + 19.5/2)}{1600} = 63 \text{ mm}$$
$d = 600 - 63 = 537$ mm

Fig. 11.11 Example 11.6

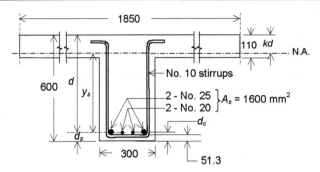

$n = 8$, and $\rho = A_s/bd = 0.0016$
Using Eqs. 11.9 and 11.10,
$kd = 81$ mm, $I_{cr} = 2.96 \times 10^9$ mm^4
Service load moment (from Example 11.4) = 126 kN·m

Steel stress, $f_s = n\dfrac{My_s}{I_{cr}} = \dfrac{8 \times 126 \times 10^6 \times (537-81)}{2.96 \times 10^9} = 155$ N/mm^2

From Fig. 11.11,

$d_c = 51.3 + 25.2/2 = 63.9$ mm
$A_e = 2\, d_s b_w = 2 \times 63 \times 300 = 37\,800$ mm^2

Number of bars, $n = \dfrac{A_s}{\text{area of largest bar}} = \dfrac{1600}{500} = 3.2$

$A = \dfrac{A_e}{n} = \dfrac{37\,800}{3.2} = 11\,800$ mm^2

$z = f_s \sqrt[3]{d_c A} \times 10^{-3} = 155 \times \sqrt[3]{63.9 \times 11\,800} \times 10^{-3} = 14.1$ kN/mm < 25 OK

PROBLEMS

1. A simply supported beam of rectangular section is subjected to the service loads shown in Fig. 11.12. For this beam,
   (a) Compute the long-term deflection due to the sustained dead load only.

   (b) Under full service loads, check whether the cracking is under control for "interior exposure" conditions. Take $f_c' = 25$ MPa, $f_y = 350$ MPa.

Fig. 11.12 Problem 1

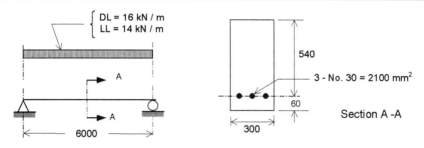

Fig. 11.13 Problem 2

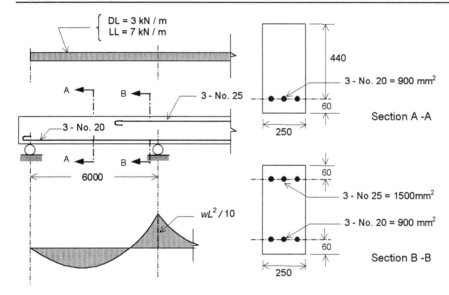

2. The end span of a continuous rectangular beam is shown in Fig. 11.13. Assuming that the negative moment at the first interior support is $wl^2/10$, compute the immediate deflection at midspan due to live load.
Dead load = 3 kN/m, live load = 7 kN/m, $f_c' = 30$ MPa, and $f_y = 300$ MPa.

3. The dimensions, reinforcement details and service dead load moments using Code coefficients for a one-way slab are shown in Fig. 11.14. Compute the long-time deflection due to sustained dead load only for the exterior span. Assume: Dead load, including weight of slab = 6 kN/m² and use $f_c' = 30$ MPa, $f_y = 400$ MPa.

4. For the beam in Problem 1 above (Fig. 11.12), check whether the crack width is within acceptable limits.

Fig. 11.14 Problem 3

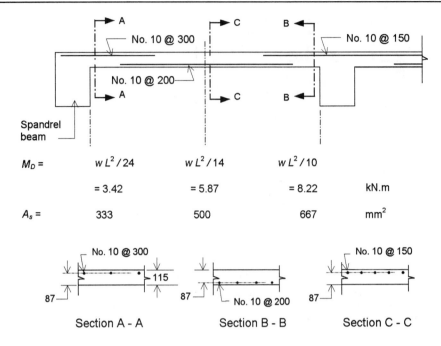

REFERENCES

11.1 Subcommittee 1, ACI Committee 435, *Allowable Deflections*, J. of ACI, Vol. 65, No. 6, June 1968, pp. 433-444.

11.2 Hibbeler, R.C., *Structural Analysis*, Prentice Hall, Inc., New Jersey, 1999, 600 pp.

11.3 Roark, R.J., and Young, W.C., *Formulas for Stress and Strain*, 5th Ed., McGraw Hill Book Co., New York, 1975.

11.4 *Concrete Design Handbook*, Canadian Portland Cement Association, Ottawa, Ontario, Canada, 1995.

11.5 Branson, D.E., *Deformation of Concrete Structures*, McGraw-Hill, Inc., New York, 1977.

11.6 ACI Committee 435, *Deflections of Reinforced Concrete Flexural Members*, J. of ACI, Vol. 63, No. 6, June 1966, pp. 637-674.

11.7 Yu, W.W., and Winter, G., *Instantaneous and Long-time Deflections of Reinforced Concrete Beams under Working Loads*, J. of ACI, Vol. 57, No. 1, July 1960, pp. 29-50.

11.8 ACI Committee 435, *Deflections of Continuous Concrete Beams*, J. of ACI, Vol. 70, No. 12, Dec. 1973, pp. 781-787.

11.9 ACI Committee 435, *Proposed Revisions by Committee 435 to ACI Building Code and Commentary Provisions on Deflections*, J. of ACI, Vol. 75, No. 6, June 1978, pp. 229-238.

11.10 Miller, A.L., *Warping of Reinforced Concrete Due to Shrinkage*, J. of ACI, Vol. 54, No. 11, May 1958, pp. 939-950.

11.11 Branson, D.E., *Compression Steel Effect on Long-time Deflections*, J. of ACI, Vol. 68, No. 8, Aug. 1971, pp. 555-559.

11.12 Gergely, P. and Lutz, L.A., *Maximum Crack Width in Reinforced Concrete Flexural Members*, Causes, Mechanism, and Control of Cracking in Concrete, (SP-20), American Concrete Institute, Detroit, 1968, pp. 87-117.

11.13 Lloyd, J.P., Rejali, H.M., and Kesler, C.E., *Crack Control in One-Way Slabs Reinforced with Deformed Wire Fabric*, J. of ACI, Vol. 66, No. 5, May 1969, pp. 366-376.

11.14 ACI Standard 318-95, *Building Code Requirements for Structural Concrete*, and *Commentary* (ACI 318R-95), American Concrete Institute, Detroit, Michigan, 1995, 369 pp.

11.15 ACI Committee 224, *Control of Cracking in Concrete Structures*, J. of ACI, Vol. 69, No. 12, Dec. 1972, pp. 717-753.

CHAPTER 12 One-Way Floor Systems

12.1 ONE-WAY SLAB - BEAM AND GIRDER FLOOR

Slabs are highly statically indeterminate. The need to control deflections of slabs usually dictates the selection of slab thickness, and results in thicknesses far greater than needed from flexural strength requirements. As a result, reinforced concrete slabs are usually very lightly reinforced (*under-reinforced* sections) and, hence, posses considerable ductility and moment redistribution capability. This facilitates the use of simplified analysis and design procedures. The structural behaviour and the analysis and design procedure to be adopted for reinforced concrete slabs depend on their geometric proportions and the manner in which they are supported (also Section 13.1). Slabs supported such that they deform into a nearly cylindrical surface, with curvature and primary moment only in one direction, are referred to as one-way slabs and are the subject of this Chapter. Apart from deflection control, crack control is also an important consideration in slab design.

A plan of a typical one-way slab-beam girder floor is shown in Fig. 12.1a. The beam and girder layout subdivides the floor slab into long rectangular panels, the length of the panel being greater than twice its width. Such slabs transfer the loads on them primarily along the short span onto the supporting beams on their longer sides (Section 2.3.2 and Fig. 2.4b). Therefore, the slab can be designed as a continuous one-way slab supported on the beams. A typical design strip of 1 m width is shown in Fig. 12.1b. Moments in continuous one-way slabs are usually determined by means of approximate coefficients, such as those presented in Section 10.3.3 (Code Cl. 9.3.3). More extensive tables of moment coefficients, covering a large range of parameters, are presented in the Appendix to Ref. 12.1. These include the cases of a single concentrated load at midspan, two equal loads at third-points, and three equal loads at quarter-points, as well as illustrative examples.

The load on the floor slab is in turn carried by the transverse beams. Each beam supports the load from approximately one-half the width of the slab panel on either side of it. Thus, an interior beam such as B1-B2-B1 carries the load from a tributary slab area of width, *b*, as shown in Fig. 12.1a, while the exterior beam B3-B4-B3 carries the load from a slab area of width, *b*/2. These transverse beams are continuous T-beams (L-beams for the exterior beam), with those along column lines framing into the columns, and the others framing into the longitudinal girders (Fig. 12.1 c, d). Two alternative procedures for analysing these beams for gravity loads were discussed in Section 10.5 (Fig. 10.7b). In one-way slab systems, usually each transverse beam, with its tributary floor slab, is analysed individually as a continuous T-beam (rather than lumping together all the transverse beams within a panel width, *B*, as an equivalent single slab-beam member framing into the columns in the transverse

CHAPTER 12 ONE-WAY FLOOR SYSTEMS 375

Fig. 12.1 One-way slab-beam and girder floor

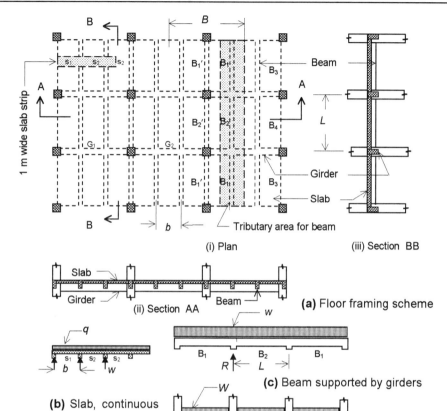

bent, and analysing the resulting partial frame). This procedure will be followed in the design example to follow in this Chapter. The transverse beams framing into the columns (column line beams) and those framing into the girders will have different degrees of restraint at their supports. In usual situations involving continuous beams of approximately equal spans with uniform loads, the moments may be determined using approximate coefficients as in the case of continuous slabs (Section 10.3.3 and Code Cl. 9.3.3). Additional tables giving coefficients for three different concentrated loadings are contained in Ref. 12.1. For regular spans and loadings in particular where moment coefficients are used to determine moments, there is little difference between the design of beams supported by columns and those supported by girders.

The longitudinal girders in Fig. 12.1e, such as G1-G2-G1, are continuous over the columns and are loaded mostly through the transverse beams. Interior girders have beams framing in from both sides, whereas exterior girders have beams on only one side. The floor loads transmitted through the beams act as concentrated loads on the girders. The Code moment coefficients, which are for uniform loads, are not applicable to the design of girders. Girders are usually analysed as a partial frame, following the procedure illustrated in Example 10.1. However, where appropriate, the approximate moment coefficients in Ref. 12.1 may be used also for determining the moments in girders.

12.2 DESIGN EXAMPLE

The design of a one-way slab-beam-girder floor is illustrated in this Section, with the design of a typical floor shown in Fig. 12.2. The specified live load is 4.8 kN/m^2. In addition to the weight of the floor itself, allowance is to be made for a dead load of 2.0 kN/m^2 obtained as follows: suspended ceiling = 0.25 kN/m^2, mechanical and electrical fixtures = 0.50 kN/m^2, floor finish = 0.25 kN/m^2, and partitions = 1.0 kN/m^2. Concrete of strength f_c' = 25 MPa (normal weight) and steel reinforcement of Grade 400 (f_y = 400 MPa) are specified.

12.2.1 Design of Slabs

The slab panels have length greater than twice the width. Therefore, they can be designed as one-way slabs. Typical slab panels are identified in Fig. 12.2 as S1 (exterior spans) and S2 (interior spans).

Clear Spans and Minimum Thickness of Slab
Assuming a width of stem for the transverse beams of 300 mm, clear span for
S1 = 3075 - 300 = 2775 mm, and for S2, = 3500 - 300 = 3200 mm.

Suggested minimum thickness of slab for deflection control (CSA A23.3-94 Table 9-1; also see Chapter 5, Table 5.2) is:

For exterior spans, $\dfrac{l_n}{24} = \dfrac{2775}{24} = 116$ mm

For interior spans, $\dfrac{l_n}{28} = \dfrac{3200}{28} = 114$ mm

A slab thickness of 120 mm will be used.

Fig. 12.2 Floor framing scheme

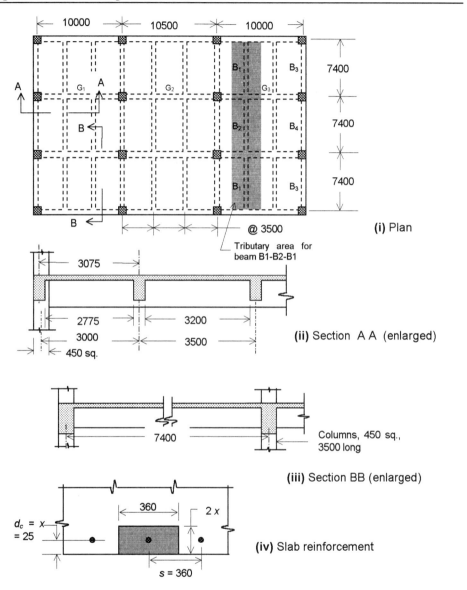

Factored Loads
Live load = $1.5 \times 4.8 = 7.20$ kN/m^2
Dead load, including slab weight = $1.25 \times (2 + 0.12 \times 24) = 6.10$ kN/m^2
Total load, $w_f = (7.20 + 6.10)$ kN/m$^2 = 13.3$ kN/m^2

The Code moment coefficients (Chapter 10, Table 10.1) are applicable for the continuous slab, and will be used to compute the moments. Considering a continuous strip of slab 1 m wide, the moment coefficients and design moments for critical sections are set forth in lines 1 and 3 of Table 12.1.

Design of Sections

The absolute maximum moment in the slab is at an interior support and equal to 12.4 kN·m. The required slab depth will be based on this moment. Selecting a reinforcement ratio $\rho = 0.5\rho_{max}$, from Table 4.1, in Chapter 4, $\rho = 0.01035$
For this ρ, $K_r = 3.01$ MPa.

Required $$d = \sqrt{M_f / K_r b} = \sqrt{\frac{12.4 \times 10^6}{3.01 \times 1000}} = 64.2 \text{ mm}$$

An overall thickness of 120 mm is adequate. Assuming No. 10 bars and a clear cover of 25 mm, the depth actually provided is:

$$d = 120 - 25 - 11.3/2 = 89.4 \text{ mm}$$

Actual $$K_r = \frac{12.4 \times 10^6}{1000 \times 89.4} = 1.55 \text{ MPa}$$

For $K_r = 1.55$, using Table 5.4 in Chapter 5, or solving Eq. 5.3, $\rho = 0.0049$

Required $A_s = 0.0049 \times 1000 \times 89.4 = 438$ mm^2

The minimum reinforcement requirement for slabs (Cls. 10.5.1.2 & 7.8) is $0.002A_g = 240$ mm^2 < 438 mm^2. Similar calculations for areas of reinforcement required at other critical sections are set forth in Table 12.1.

The top reinforcement (for negative moment) in the slab will also function as the transverse reinforcement in the flange of the T-beam forming the secondary beams, such as B1-B2-B1. The minimum area of such (flange) reinforcement is given in the Code Cl.10.5.3.2. (Although Cl.10.5.3.2 is for cases where principal slab reinforcement is *parallel* to the T-beam, it is prudent to assume that, where the principal slab reinforcement is also *perpendicular* to the T-beam, the area provided shall not be less than this minimum). Accordingly the minimum area of top reinforcement is also controlled in this case by the requirement:

$$A_{s,min} = \frac{0.2\sqrt{f'_c}}{f_y} b_t h$$

$$A_{s,\min} = 0.2 \times \frac{\sqrt{25}}{400} \times 1000 \times 120 = 300 \text{ mm}^2 / \text{m width}$$

From Table 12.1, the area of top reinforcement required for flexure is greater than this and, hence, controls in all locations *except* over the exterior support. The top reinforcement at exterior support is increased to meet this requirement, and extended past the web face of the exterior beam a distance of $0.3 l_n \approx 840$ mm. The bar curtailment and extension are done as per Fig. 5.6.

Check Crack Control (Cl. 10.6.1, see section 11.12)
At locations of widest spacing, where this is most critical, spacing = 360 mm.
From Fig. 12.2 (iv), $d_c = 25 + 11.3/2 = 30.7$ mm
$A = 360 \times 2 \times 30.7 = 22\,100$ mm^2; $f_s \approx 0.6 f_y = 0.6 \times 400 = 240$ MPa
$z = f_s (d_c A)^{1/3} = 240 \times (30.7 \times 22\,100)^{1/3} = 21\,086 < 30\,000$ N/mm,   OK

Check Shear
The maximum factored shear forces are:

in span, S1: $1.15 \dfrac{w_f l_n}{2} = 1.15 \times \dfrac{13.3 \times 2.775}{2} = 21.2$ kN

and in span S2: $\dfrac{w_f l_n}{2} = 13.3 \times 3.2/2 = 21.3$ kN

Factored shear resistance provided by concrete, V_c, is:
$V_c = 0.2 \lambda \phi_c \sqrt{f'_c} b_w d = 0.2 \times 1.0 \times 0.6 \times \sqrt{25} \times 1000 \times 89.4 \times 10^{-3} = 53.6$ kN

$V_f = 21.3$ kN $< V_c$ OK

Shrinkage and Temperature Reinforcement
Since the principal reinforcement is provided only along the short direction (one-way slab), secondary reinforcement normal to the principal reinforcement must be provided to resist shrinkage and temperature stresses. The minimum area for such reinforcement is (CSA A23.3-94(7.8)): $A_s = 0.002 \times 1000 \times 120 = 240$ mm^2

Maximum spacing of such reinforcement is: $5h = 550$ mm or 500 mm

No. 10 bars at 400 mm will be adequate to provide this area.
These secondary reinforcements are in a direction transverse to the girders (such as G1-G2-G1). Hence, over the girder, they can function as the required transverse reinforcement in the flanges of the T-beam girder. However, the minimum required area for this is 300 mm^2/m, as has been calculated earlier. Therefore, extra reinforcement is needed.

380 REINFORCED CONCRETE DESIGN

Table 12.1 Design of slabs

	SPAN S1			SPAN S2		
	Exterior support	Midspan	First interior support	First interior support	Midspan	All other supports
1. Moment of coefficients C	$-\dfrac{1}{24}$	$\dfrac{1}{14}$	$-\dfrac{1}{10}$	$-\dfrac{1}{11}$	$\dfrac{1}{16}$	$-\dfrac{1}{11}$
2. Length l_n, m	2.775	2.775	2.988	2.988	3.20	3.20
Factored moment $M_f = C w_f l_n^2$, kN·m	-4.27	7.32	-11.9	-10.8	8.51	-12.4
4. $K_r = \dfrac{M_f \times 10^6}{1000 d^2}$, MPa	0.534	0.916	1.49	1.35	1.06	1.55
5. 100ρ, (Eq. 5.3, or Table 5.4)	0.160	0.285	0.467	0.425	0.328	0.49
6. $A_{s,min} = \rho_{min} A_g$, mm²	←	240	→	←	240	→
7. $A_s = \rho \times 1000 \times 89.4$ mm²	143	255	417	380	293	438
8. Maximum, spacing, mm (Cl. 7.4.1.2)	←	360	→	←	360	→
9. Bar selection (No. and spacing in mm)	No. 10 @330	No. 10 @300	No. 10 @220		No. 10 @300	No. 10 @220
10. A_s provided, mm²	303	333	455		333	455
11. Sketch						

12.2.2 Design of Beams B1-B2-B1

The Code moment coefficients will be used for the design of transverse beams. All interior beams will be provided with the same cross section, although there is some difference between the restraints at supports for beams supported by girders and for beams framing into columns. The girders will be assumed to have a width of 450 mm. The clear span, $l_n = 7.4 - 0.45 = 6.95$ m.

Beam depth: the minimum depth for deflection control (Code, Table 9.1), for beams continuous at both ends is:

$$\frac{l_n}{21} = \frac{6950}{21} = 331 \text{ mm}$$

For the dead load calculation, an overall depth of 550 mm will be assumed. The beams are spaced at 3.5 m so that each interior transverse beam has a tributary floor area of width 3.5 m.

Live load = 7.20 × 3.5 = 25.2 kN/m
Dead load from floor = 6.1 × 3.5 = 21.4 kN/m
Weight of beam stem = 0.3 (0.55 – 0.12) × 24 × 1.25 = 3.87 kN/m
Total beam load = 50.5 kN/m

The beam spans and loads satisfy the limitations in CSA A23.3-94(9.3.3), and the approximate coefficients may be used to compute the design moments and shears at all critical sections in the beam. The coefficients, moments and shears for beams B1 and B2 are given in Fig. 12.3. The negative moment at the interior face of an exterior support is taken as $w_f l_n^2/24$, applicable for a spandrel beam support, and this is also used for all beams supported on girders, so that the same design can be used for all interior beams.

Design of Sections

The overall beam size is controlled by the maximum moment, which is at the interior support of B1 and equal to -244 kN·m. The beam effectively acts as a rectangular section under the negative moment. Assuming a reinforcement ratio of $\rho = 0.6 \, \rho_{max} = 0.0125$, from Table 4.1 in Chapter 4, $K_r = 3.5$

$$\text{Required } d = \sqrt{\frac{244 \times 10^6}{3.5 \times 300}} = 482 \text{ mm}$$

Keeping the overall depth at 550 mm, and assuming No. 25 bars, No. 10 stirrups and 40 mm clear cover, $d = 550 - (40 + 11.3 + 25.2/2) = 486$ mm

The adequacy of this depth is now checked for shear.

Shear design will be done with the Simplified Method in Clause 11.3 of the Code. Assuming that at least the minimum amount of transverse reinforcement is provided, the factored shear resistance provided by concrete is:

$$V_c = 0.2\lambda\phi_c\sqrt{f'_c}b_w d = 0.2 \times 1.0 \times 0.6 \times \sqrt{25} \times 300 \times 486 \times 10^{-3} = 87.5 \text{ kN}$$

Required shear resistance, V_s, to be provided by shear reinforcement is:

$$V_s = V_f - V_c = 202 - 87.5 = 115 \text{ kN}$$

Upper limit on V_s, (Cl. 11.3.4) is:

$$V_{s,\lim} = 0.8\lambda\phi_c\sqrt{f'_c}b_w d = 4 \times 87.5 = 350 \text{ kN} > V_s \text{ required} \qquad \text{OK}$$

The depth of 550 mm is adequate to carry shear with web reinforcement.

To compute flexural reinforcement at the exterior face of the first interior support, where the bending moment is maximum,

$$K_r = \frac{244 \times 10^6}{300 \times 486} = 3.44 \text{ MPa}$$
$$\rho = 0.0122 \text{ (Table 5.4)}$$
$$A_s = 0.0122 \times 300 \times 486 = 1779 \text{ mm}^2$$

This reinforcement ratio is approximately $0.59\rho_{max}$ and is satisfactory.

Checks for minimum reinforcement and crack control will be made for the exterior support, where the moment and flexural steel are lowest, and, hence, is the critical section for these.
Check for minimum bar spacing: (CSA A23.3 Cl.7.4.1.1)
The reinforcement selected is shown in Fig. 12.3. Providing the 3 – No. 20 straight bars in the outer layer and the 2 –No. 25 bent bars in the inner layer, and with 40 mm clear cover and No. 10 stirrups, the spacing between bars is:

$$\begin{aligned}s &= [300 - 2 \times 40 - 2 \times 11.3 - 3 \times 19.5]/2 = 69.5 \text{ mm} \\ &> 1.4\, d_b = 1.4 \times 19.5 = 27.3 \text{ mm} \\ &> 1.4\, a_{max} = 1.4 \times 25 = 35 \text{mm} \\ &> 30 \text{ mm}, \qquad\qquad\qquad\qquad\qquad\qquad\qquad\qquad \text{OK}\end{aligned}$$

CHAPTER 12 ONE WAY FLOOR SYSTEMS

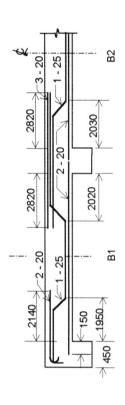

Fig. 12.3 Beam B1 - B2 - B1

1. Moment coefficient, C	-1/16	1/14	-1/10	1/16	-1/11
2. Moment = $C \times 50.5 \times 6.95^2$, (kN·m)	-152	174	-244	152	-222
3. Shear = $\dfrac{50.5 \times 6.95}{2}, 1.15 \times \dfrac{w_f l_n}{2}$ (kN)	175		202		175
4. A_s, required, (mm²)	1021	1073	1779	933	1578
5. Reinforcement selected and	1-No. 25 bt.	2-No. 20 st.	3-No. 20 st		2-No. 20 st
	2-No. 20 st.	1-No. 25 bt.	2-No. 25 bt.		1-No. 25 bt.
area, A_s (mm²)	1100	1100	1900		1100
6. Factored resistance, M_r, (kN·m)	-164	178	-261	179	-267

At exterior support of B1,

$$K_r = \frac{152 \times 10^6}{300 \times 486} = 2.15 \text{ MPa}$$

$$\rho = 0.0070$$

$$A_s = 1021 \text{ mm}^2$$

Check minimum reinforcement requirement (Cl. 10.5.1.1)
The actual reinforcement provided (Fig. 12.3) has an area of 1100 mm², compared to 1021 mm² required.

Hence, the actual $M_r \approx \dfrac{1100}{1021} \times 152 = 164 \text{ kN} \cdot \text{m}$

$$M_{cr} = 0.6\lambda\sqrt{f_c'}\,\frac{I}{c_t} = 0.6 \times 1 \times \sqrt{25} \times 300 \times \frac{550^3}{12} \times \frac{2}{550} \times 10^{-6} = 45.4 \text{ kN} \cdot \text{m}$$

$M_r = 164 \text{ kN} \cdot \text{m} \gg 1.2 M_{cr} = 1.2 \times 45.4 = 54.5 \text{ kN} \cdot \text{m}$ 　　　　　OK

Alternatively, using Cl. 10.5.1.2 (Eq. 10.4 of Code)

$$A_{s,\min} = \frac{0.2\sqrt{f_c'}}{f_y} b_t h = \frac{0.2\sqrt{25}}{400} \times 300 \times 550 = 413 \text{ mm}^2 \ll 1100 \text{ mm}^2 \quad \text{OK}$$

Check crack control (Cl. 10.6.1)
$d_c = 40 + 11.3 + 19.5/2 = 61.1 \text{ mm}, \ \bar{y} \approx 61.1 \text{ mm}$
$A = (300 \times 2 \times 61.1^2)/3 = 12\,220 \text{ mm}^2$
$z = f_s(d_c A)^{1/3} = 0.6 \times 400 \times (61.1 \times 12\,220)^{1/3} = 21773 \text{ N/mm} < 30\,000 \text{ N/mm}$　OK

At support of B2,

$$K_r = \frac{222 \times 10^6}{300 \times 486} = 3.13 \text{ MPa}$$

$$\rho = 0.0108$$

$$A_s = 1578 \text{ mm}^2$$

At midspan of B1 and B2, the beams act as T-beams. Effective overhang of flange is (Cl. 10.3.3),

　　　　1/10 of span = 740 mm　　(least)
　　　　12 h_f = 12 × 120 = 1440 mm
　　　　½ clear distance to web = ½ (3500 − 300) = 1600 mm

Hence, effective width, $b = 2 \times 740 + 300 = 1780$ mm

Since the same beam acting as a rectangular section at the support under a larger moment of 246 kN·m is under-reinforced, the T-section with the lower span moment will surely be under-reinforced and, hence, there is no need to check this.

For beam B1 at midspan:
The reinforcement area is determined using the alternative procedure explained in Example 5.7,

$$d - \frac{h_f}{2} = 486 - \frac{120}{2} = 426 \text{ mm}$$
$$0.9d = 0.9 \times 486 = 437 \text{ mm}$$
$$z \approx 437 \text{ mm}$$
$$\text{Trial } A_s = \frac{174 \times 10^6}{0.85 \times 400 \times 437} = 1171 \text{ mm}^2$$

Factor α_1, of compressive stress block = $0.85 - 0.0015 f_c' = 0.813$

$$a = \frac{1171 \times 0.85 \times 400}{0.813 \times 0.6 \times 25 \times 1780} = 18.3 \text{ mm} < h_f$$
$$z = d - \frac{a}{2} = 486 - \frac{18.3}{2} = 477 \text{ mm}$$
$$\text{Revised } A_s = \frac{174 \times 10^6}{0.85 \times 400 \times 477} = 1073 \text{ mm}^2$$

(Detailed calculation gives $A_s = 1171$ mm^2.)

For beam B2 at midspan:
$$M_f = 152 \text{ kN} \cdot \text{m}$$
$$A_s \approx 1073 \times \frac{152}{174} = 937 \text{ mm}^2$$
$$a = \frac{937 \times 0.85 \times 400}{0.813 \times 0.6 \times 25 \times 1780} = 14.7 \text{ mm}$$
$$z = 486 - 14.7/2 = 479 \text{ mm}$$
$$A_s = \frac{152 \times 10^6}{0.85 \times 400 \times 479} = 933 \text{ mm}^2$$

Reinforcement is provided as shown in Fig. 12.3. Since depth $h = 550$ mm is less than 750 mm, no skin reinforcement is needed. At sections under negative moment, the T-beam flange is in tension. The Code (Cl. 10.5.3.1) requires part of the flexural tension reinforcement at such locations to be placed in the flange. The locations where individual bars are no longer required for flexure may be determined using Fig. 10.6. This, together with the Code requirements in Fig. 5.14, determines the bar bending details. In arriving at the bend locations in Fig. 12.3, it has been assumed that for satisfying bar extension requirements, the bent bar is effective up to the section where it crosses the mid-depth of the beam. The bar cut-off and bending locations in Fig. 12.3 meet all the requirements for development (see Example 9.4). Note that with the No. 25 bars the effective depth will be 536 mm compared to the 533 mm used in the computations. This difference will hardly change the results of the calculations and, furthermore, the use of 533 mm is on the safe side. Part of the tension reinforcement (two No. 20 bars) at the top over the support section will be placed in the flanges of the T-section. The factored flexural strengths, M_r of all critical sections are also given in line 6 of Fig. 12.3 (see Example 9.4 for calculations).

Design of Shear Reinforcement
For Beam B1 the variation of shear force based on the Code coefficients is as shown in Fig. 12.4.

At distance d from face of first interior support,

$$V_f = \frac{202(3996 - 486)}{3996} = 177 \text{ kN}$$

Factored shear resistance provided by concrete, V_c, is 87.5 kN. Shear resistance to be provided by shear reinforcement is:

$$V_s = V_f - V_c = 177 - 87.5 = 89.5 \text{ kN}$$

To determine the minimum spacing (Cl. 11.2.11),

$$0.1\lambda\phi_c f'_c b_w d = 0.1 \times 1 \times .06 \times 25 \times 300 \times 486 \times 10^{-3} = 219 \text{ kN} > V_f = 202 \text{ kN}$$

Since $V_f < 0.1 \; \lambda \; \phi_c f'_c \; b_w d$, the maximum spacing is 600 mm or $0.7 \, d = 340$ mm.

Using No. 10 vertical U stirrups, the required spacing is:

$$s = \frac{\phi_s A_v f_y d}{V_s} = \frac{0.85 \times 200 \times 400 \times 486}{89.5 \times 10^3} = 369 \text{ mm}$$

A minimum area of shear reinforcement is required (Cl. 11.2.8) where V_f exceeds $0.5 \, V_c = 87.5/2 = 43.8$ kN. Since the area where $V_f < 0.5 \, V_c$ is only a short region near mid-span (mainly in the interior span), minimum shear reinforcement should be provided throughout the span.
Spacing required for providing the minimum shear reinforcement is (Cl. 11.2.8.4):

Fig. 12.4 Shear force in beams

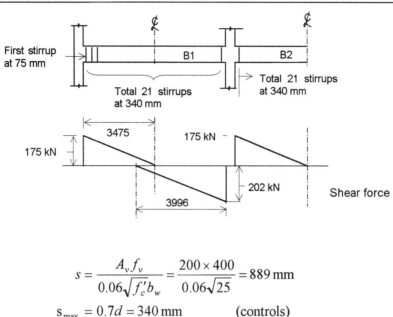

$$s = \frac{A_v f_y}{0.06\sqrt{f'_c}b_w} = \frac{200 \times 400}{0.06\sqrt{25}} = 889 \text{ mm}$$

$$s_{max} = 0.7d = 340 \text{ mm} \qquad \text{(controls)}$$

For both beams B1 and B2, the required spacing, being greater than the maximum permitted, the latter controls. For each span, provide a total of 21 stirrups at a spacing of 340 mm, starting the first stirrup at 75 mm from the face of the support.

Note that the edge beam B3-B4-B3 must be designed also for the torsion corresponding to the negative moment in the slab at its exterior support.

12.2.3 Design of Girders G1-G2-G1

The interior girders G1-G2-G1 have beams framing in from both sides. The girder moments and shears will be determined by an elastic analysis of the partial frame consisting of the girders and the columns immediately above and below. The girder cross-sectional dimensions are needed for the elastic analysis and this is first estimated using a preliminary design.

Preliminary Design of Girder Section
The loads on the girder are the dead and live loads from the floor slab and floor beams transmitted by the latter as point loads, and the distributed loading acting directly on the girder including its own weight.

The concentrated live load transmitted by the interior beams from a tributary

area of 3.5 m × 7.4 m is:
$$7.20 \times 3.5 \times 7.4 = 186 \text{ kN}$$

(This includes the live load on the part of the floor directly over the girder. This part of the live load is in fact distributed over the length of the girder; however, there is little error in combining it with the load from the rest of the floor transmitted through the beams as concentrated loads.)

Concentrated dead load from the beams is:
$$(21.4 + 3.87) \times 6.95 = 176 \text{ kN}$$

Distributed dead load due to partition, floor finish, etc., acting directly on the girder (width = 450 mm) is:
$$1.25 \times 2 \times 0.45 = 1.13 \text{ kN/m length}$$

Weight of girder, assuming a total depth of 1 m,
$$1.25 \times 0.45 \times 1 \times 24 = 13.5 \text{ kN/m}$$

Total distributed dead load = 1.13 + 13.5 = 14.6 kN/m

The loads and span dimensions are shown in Fig. 12.5a. For the preliminary design, the maximum negative moment in the girder may be taken as 80 percent of the simple span moment and the maximum shear as 1.15 times the simple span shear. These factors give the same ratios as those used in the Code coefficients for distributed loading ($w_f l_n^2 / 10 = 0.8 \times w_f l_n^2 / 8$). On this basis, for the longer span G2,

$$M_f \approx 0.8 \ [(186 + 176) \times 3.275 + 14.6 \times 10.05^2/8] = 1096 \text{ kN·m}$$
$$V_f \approx 1.15 \ (186 + 176 + 14.6 \times 10.05/2) = 501 \text{ kN}$$

With a reinforcement ratio of about $0.60\rho_{max}$, the depth required is:
$$d = \sqrt{\frac{1096 \times 10^6}{3.5 \times 450}} = 834 \text{ mm}$$

With this depth, shear resistance provided by concrete, V_c, is:
$$V_c = 0.2 \times 1.0 \times 0.6 \times \sqrt{25} \times 450 \times 834 \times 10^{-3} = 225 \text{ kN}$$

Shear resistance required to be provided by reinforcement is:
$$V_s = 501 - 225 = 276 \text{ kN}$$

Upper limit of V_s for section dimensions is:

Fig. 12.5 Analysis of girder

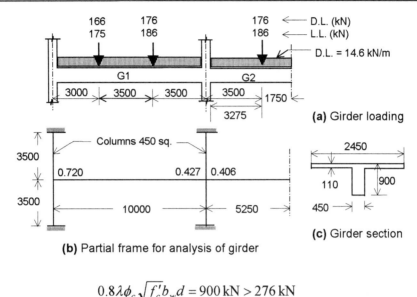

(a) Girder loading

(b) Partial frame for analysis of girder

(c) Girder section

$$0.8\lambda\phi_c\sqrt{f'_c}b_wd = 900 \text{ kN} > 276 \text{ kN}$$

The dimensions are adequate for shear with shear reinforcement.

Since the reinforcement ratio assumed is still relatively low, and the bending moment over most of the girder length will be lower than the value considered above, an overall girder depth of 900 mm will be assumed for use in detailed analysis.

Detailed Elastic Analysis

The partial frame dimensions are shown in Fig. 12.5b. The analysis will be made using the moment distribution method. The girder is assumed to have a T-section with an effective flange width of $0.2 \times$ span $+ b_w = 2450$ mm. The member stiffnesses are:

1. columns: $\qquad K_c = \dfrac{4E \times 450^4}{12} \times \dfrac{1}{3500} = 3.91E \times 10^6$

2. for girder (T-section): $I = 5.02 \times 10^{10}$ mm^4

$$\text{for G1}, K_{G1} = \frac{4E \times 5.02 \times 10^{10}}{10\,000} = 20.1E \times 10^6$$

$$\text{for G1}, K_{G2} = \frac{4E \times 5.02 \times 10^6}{10\,500} = 19.1E \times 10^6$$

Fig. 12.6 Load patterns and simple span moments

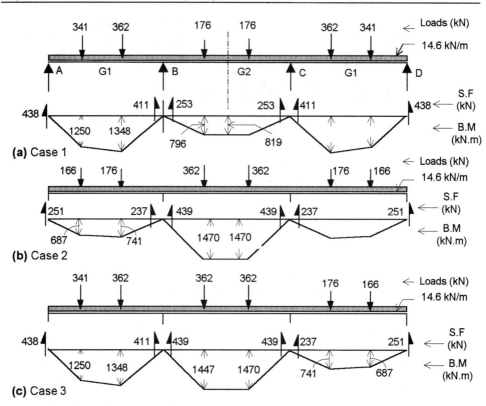

The distribution factors to the girders are (Fig. 12.5b):
at exterior support,

$$\text{to } G1 = \frac{20.1}{2 \times 3.91 + 20.1} = 0.720$$

at interior support,

$$\text{to } G1 = \frac{20.1}{2 \times 3.91 + 20.1 + 19.1} = 0.427$$

$$\text{to } G2 = \frac{19.1}{2 \times 3.91 + 20.1 + 19.1} = 0.406$$

The critical loading conditions together with the corresponding simple span moment diagrams are shown in Fig. 12.6. The results of the moment distribution are presented in Fig. 12.7 (moments) and Fig. 12.8 (shears).

Some economy may be achieved by redistributing the moments obtained by the elastic analysis, as permitted by CSA A23.3-94(9.2.4); however this is not attempted

Fig. 12.7 Theoretical moment diagrams

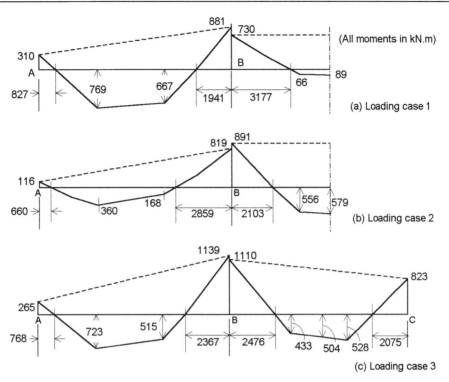

(a) Loading case 1

(b) Loading case 2

(c) Loading case 3

in this example. The complete moment envelope for girders G1-G2 is presented in Fig. 12.9.

Design of Sections
1. Section B (Fig. 12.9)

The maximum design moment for the girders is at the exterior face of the first interior column. The moment here being negative, the girder acts as a beam of rectangular section. The moment at the column face is (see Section 10.6) given by

(M_{CL} - $Vb/3$), and is:

$$1139 - \frac{498 \times 0.45}{3} = 1064 \text{ kN·m}$$

Since the moment is much lower at all other sections, a slightly higher reinforcement ratio of about 0.75 ρ_{max} will be used at this section.

For ρ = 0.75 ρ_{max} ≈ 0.0153, K_r = 4.1

Fig. 12.8 Shear force distribution

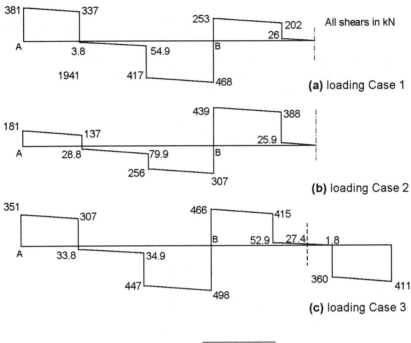

(a) loading Case 1

(b) loading Case 2

(c) loading Case 3

$$\text{Required } d = \sqrt{\frac{1064 \times 10^6}{4.1 \times 450}} = 759 \text{ mm}$$

Provide an overall depth of 900 mm. The negative reinforcement in the girder will be placed with a clear cover of 80 mm so that it will clear the reinforcement of the beams framing into it. Assuming No. 35 bars,

$$d = 900 - 80 - \frac{35.7}{2} = 802 \text{ mm}$$

$$K_r = \frac{1064 \times 10^6}{450 \times 802^2} = 3.68 \text{ MPa}$$

$$\rho = 0.0133$$

$$A_s = 0.0133 \times 450 \times 802 = 4800 \text{ mm}^2$$

2. Exterior support, A:

   Design moment = $310 - \frac{381 \times 0.45}{3} = 253 \text{ kN} \cdot \text{m}$

Fig. 12.9 Girder design

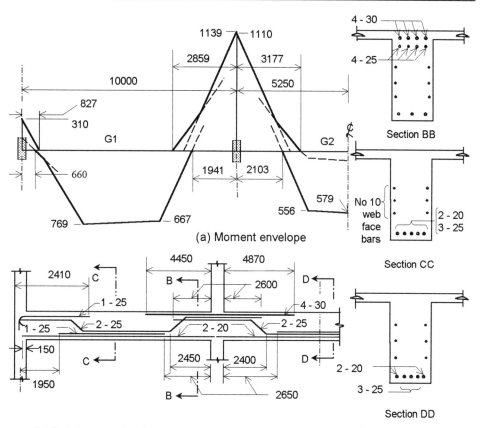

(a) Moment envelope

(b) Reinforcement details

(c) Cross sections

$$K_r = \frac{253 \times 10^6}{450 \times 802^2} = 0.874 \text{ MPa}$$

$$\rho = 0.0027, A_s = 974 \text{ mm}^2$$

Assuming the width of tension zone $b_t = b_w$, (Cl.10.5.1.2),

$$A_{s\min} = \frac{0.2\sqrt{f'_c}}{f_y} b_c h$$

$$= \frac{0.2\sqrt{25}}{400} \times 450 \times 900 = 1013 \text{ mm}^2$$

3. Span AB:
$M = 769$ kN·m
Girder has a T-Section, with $b = 2 \times 12 \times 120 + 450 = 3330$ mm. With 40 mm clear cover, No. 15 stirrups and No. 30 bars, $d = 900 - (40 + 16 + 29.9/2) = 829$ mm.

$$\text{Approximate lever arm} \left(d - \frac{h_f}{2}, \text{ or } 0.9d\right) = 769 \text{ mm}$$

$$A_s \approx \frac{769 \times 10^6}{0.85 \times 400 \times 769} = 2941 \text{ mm}^2$$

With this A_s,

$$a = \frac{2941 \times 0.85 \times 400}{0.813 \times 0.6 \times 25 \times 3330} = 24.6 \text{ mm}$$

$$d - \frac{a}{2} = 829 - \frac{24.6}{2} = 817 \text{ mm}$$

$$\text{Revised } A_s = \frac{769 \times 10^6}{0.85 \times 400 \times 817} = 2768 \text{ mm}^2$$

4. Span BC:
As in the case of span AB, the reinforcement area is obtained as 2077 mm^2. The flexural reinforcement details are shown in Fig. 12.9. The bar cut-off and bend details satisfy the Code requirements presented in Fig. 5.14. The bar cut-off and bending locations, anchorage and development requirements can be checked as demonstrated in Example 9.4.

Check minimum bar spacing
The most critical location is at mid-span of span AB. With 3-No. 25 and 2-No. 30 bars, clear spacing = $[450 - 2 \times 56 - 3 \times 25.2 - 2 \times 29.9]/4 = 50.7$ mm,

> $1.4\, d_b = 42$ mm
> $1.4\, a_{max} = 35$ mm (assuming 25 mm aggregates)
> 30 mm   OK

Check crack control (Cl. 10.6.1)
In the negative moment regions, the beam has heavy reinforcement. Furthermore, the slab forming the flange of the T-beam is in the tension side and has well distributed reinforcement. Hence crack control is more critical in the positive moment regions.

Considering span BC, which has the lightest reinforcement, for the arrangement shown:
$$d_c = 40 + 16 + 19.5/2 = 65.8 \text{ mm}$$

Distance from centroid of steel to bottom fibres, =:
$X = [3 \times 500 \times (56 + 25.2/2) + 2 \times 300 \times 65.8]/2100 = 67.8$ mm
$A = 2 \times 67.8 \times 450/5 = 12\,204$ mm^2
$Z = f_s (d_c A)^{1/3} = 0.6 \times 400 \times (65.8 \times 12204)^{1/3}$
$ = 22\,308 < 30\,000$ N/mm, OK

Shear Reinforcement
Maximum shear force at distance d from the face of support $(802 + 450/2 = 1027$ mm from centre of support) is:
$$V_f = 498 - 14.6 \times 1.027 = 483 \text{ kN}$$
$$V_c = 0.2 \times 1.0 \times 0.6 \times \sqrt{25} \times 450 \times 802 \times 10^{-3} = 217 \text{ kN}$$
$$V_s = 483 - 217 = 266 \text{ kN}, \quad < 0.1 \lambda \phi_c f_c' b_w d$$
$$= 0.1 \times 1 \times 0.6 \times 25 \times 450 \times 802 \times 10^{-3}$$
$$= 541 \text{ kN}$$

Hence, maximum spacing (Cl. 11.2.11) is 600 mm or $0.7d = 561$ mm. V_s is also $< 0.8 \lambda \phi_c \sqrt{f_c'} \, b_w d$ as checked initially.

Using No. 15-U stirrups, the required spacing is,
$$s = \frac{0.85 \times 400 \times 400 \times 802}{266 \times 10^3} = 410 \text{ mm}$$

Provide No. 15 stirrups at 400 mm spacing.

In the outer thirds of both girder spans, the shear is relatively high. Hence in this region a uniform spacing of 400 mm may be provided. The first stirrup is placed at 200 mm from the face of the support, followed by 8 spaces (ie, 8 stirrups) at 400 mm. In the remaining portion of the middle third of each span, the shear force is very low, and shear reinforcement to meet the minimum requirement (Cl. 11.2.8) may be provided. Five numbers of No. 10-U stirrups in this region will result in a spacing of about 542 mm, and will satisfy both the minimum shear reinforcement and maximum spacing requirements. (In this region $V_f < V_c/2$ and, hence, no shear reinforcement is required as per Code.)

Additional Considerations
Since the height of beam is in excess of 750 mm, skin reinforcement conforming to CSA A23.3-94(10.6.2) must be placed near the faces of the member over the half of

396 *REINFORCED CONCRETE DESIGN*

the height nearest the principal reinforcement. Three No. 10 bars on each face will be adequate for this. The Code (Cl. 10.5.3.1) requires some tension reinforcement to be distributed over the effective flange width when the flanges are in tension. In this case, the main reinforcement in the slab is placed parallel to the girder, and this will be adequate to meet the above requirement. Furthermore, as required in Clause 10.5.3.2, transverse reinforcement is needed at the top of the girder flange. The area of reinforcement so required ($= 0.2\sqrt{f'_c}\ b_t h_f / f_y$) has been calculated earlier as 300 mm²/m width. No. 10 bars at 320 mm will provide the required area. These may be extended past the web face a distance equal to 0.3 times the clear distance between webs of the T-beams, that is $0.3 \times (7400 - 450) \approx 2000$ mm. This reinforcement will also function as the needed shrinkage and temperature reinforcement in the slab in the region.

12.3 ONE-WAY JOIST FLOOR

12.3.1 General

A one-way concrete joist floor (also known as ribbed slab) consists of a series of closely spaced parallel joists with an integrally cast thin concrete slab on top, and framing into girders at both ends (Fig. 12.10). Each joist together with its share of the top slab acts as a continuous T-beam spanning in one direction between the girders. Joist floors are usually formed using standard removable steel forms (*pans*) placed between the joist ribs. Because much of the ineffective concrete in the tension zone is eliminated, joist floor construction can provide increased effective depth and stiffness for the floor member, without increasing its dead weight. Standard size reusable forms make this type of construction economical for large spans (in excess of about 5 m), where a solid one-way slab becomes uneconomical on account of its large self-weight.

Standard pan form dimensions include widths of 500 mm and 750 mm and depths from 150 mm to 500 mm (Ref. 12.2). Pans with tapered ends are also available for use where larger joist width is desired near the end supports due to higher shear or flexural (negative moment) requirements.

12.3.2 Code Requirements

The Code limitations on cross-sectional dimensions of standard concrete joist floor construction are given in CSA A23.3-94(10.4) and summarised in Fig. 12.11. The top slab spanning the joists must have flexural reinforcement at right angles to the joists, sufficient to carry applied slab loads. However, under normal joist spacings, the

Fig. 12.10 One-way concrete joist floor construction

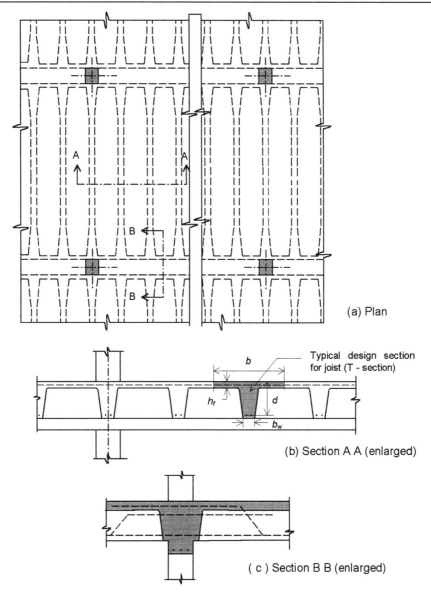

(a) Plan

(b) Section A A (enlarged)

(c) Section B B (enlarged)

minimum reinforcement specified for shrinkage and temperature will provide adequate flexural strength for the slab. With closely spaced and long span joists, the joist floor is a highly redundant structural system that allows substantial redistribution of loads and has adequate ductility. Hence, there is no minimum shear reinforcement requirement for joists (CSA A23.3-94 (11.2.8.1(b)).

Fig. 12.11 Code limitations for one-way joist construction

- A_s as required for flexure, but ≥ shrinkage & temp. reinf. (ie., minimum requirement for slabs)
- $h_f \geq s/12$, 50 mm
- No minimum web reinforcement
- $h \leq 3.5\, b_w$
- $b_w \geq 100$ mm
- $s \leq 800$ mm

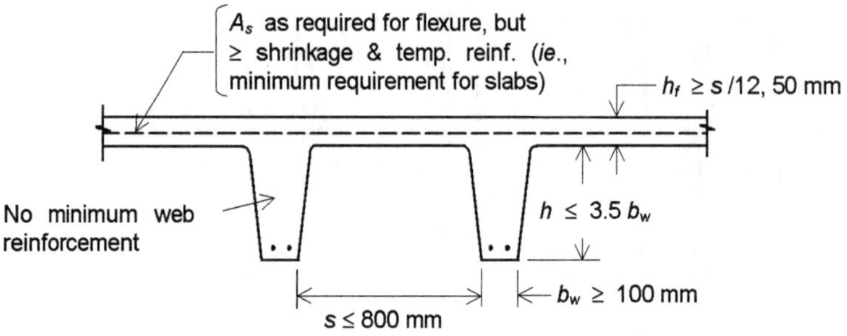

12.3.3 Design

Just as a one-way slab is designed considering a typical 1 m wide strip as a continuous beam, a one-way joist-floor is designed by considering a typical single joist together with its flange (of width equal to the centre to centre spacing of the joists) as a continuous T-beam (Fig. 12.10b). In the design of this T-beam, the width of stem, b_w, is taken as the least width of the joist rib, at the bottom of the joist, neglecting the slight increase in width at higher levels due to the taper of the sides of the forms. As for regular T-beams, for joists the minimum flexural reinforcement ratio, (Cl. 10.5.1.2), is based on the width of stem, b_w. Although web reinforcement may be used to increase the shear capacity of joists, it is usually avoided (limit $V_f \leq V_c$). The shear strength can be increased by using deeper joists or by widening the joists near the support using tapered forms.

In the design of the girder, the load transmitted by the joists is considered as a distributed loading. The girder spanning the columns is also designed as a continuous T-beam whose section has the shape shown in Fig. 12.10c.

Design examples and load tables for standard one- and two-way joist construction are presented (in imperial units) in Ref. 12.3.

PROBLEMS

1. For the floor system shown in Fig. 12.2, design the exterior beam B3-B4-B3.

2. For the floor system shown in Fig. 12.2, design the exterior girder.

REFERENCES

12.1 ACI, *Reinforced Concrete Design Handbook*, 3rd ed., ACI Publication SP-3, American Concrete Institute, Detroit, Michigan, 1965, pp. 255-271.

12.2 CRSI, *Code of Standard Practice - Joist Construction*, CRSI Manual of Standard Practice, Concrete Reinforcing Steel Institute, Chicago, Illinois, 1977.

12.3 CRSI, *CRSI Handbook 1980*, 4th ed., Concrete Reinforcing Steel Institute, Chicago, Illinois, 1980.

CHAPTER 13 Two-Way Slabs on Stiff Supports

13.1 ONE- AND TWO-WAY ACTION OF SLABS

Slabs that deform essentially into a cylindrical surface can be designed as one-way slabs or beam-strips spanning in the direction of the curvature. Such behaviour truly exists in rectangular slabs supported only on the two opposite sides by unyielding supports and carrying a load that is uniformly distributed along the direction parallel to the supports (Fig. 2.4a, d). When an elastic, homogeneous and isotropic slab is bent into a cylindrical surface, secondary moments in the direction parallel to the supports are developed due to the Poisson effect (Ref. 13.1). However, in reinforced concrete slabs, the magnitude of this secondary (longitudinal) moment is relatively small, and it is taken care of by the temperature and shrinkage reinforcement provided along this direction at right angles to the principal reinforcement.

When the support and/or loading conditions are different from those mentioned above, the slab bends into a surface with curvature in both the longitudinal and transverse directions, and the slab develops two-way action, with primary moments in both directions. Principal flexural reinforcement is then needed in both directions. Typical elastic theory distributions of longitudinal and transverse moments in uniformly loaded square and rectangular plates, simply supported on all four sides, are shown in Fig. 13.1. For the square panel, the maximum moments in both directions are equal. In addition to the bending moments, the slab elements are also subject to twisting moments that have their maximum values at the corners of the panel. These twisting moments can be resolved into principal moments (bending), M_1 and M_2, at the corners. At slab corners, the moment M_1 is along the diagonal and negative, causing tension on the top fibres of the slab along the diagonal; and the moment M_2 is positive and at right angles to the diagonal, causing tension in this direction on the bottom fibres of the slab. The potential crack patterns resulting from the twisting moments are shown in Fig. 13.1c.

As the ratio long span/short span increases, the curvature and moments along the long direction progressively decrease, and more and more of the slab load is transferred to the two longer supports by flexure in the short direction (Fig. 13.1b). Thus, in general, edge-supported rectangular reinforced concrete slabs, with ratios of long to short span of two or more, can also be designed as a one-way slab spanning in the short direction, with the principal reinforcement placed at right angles to the long support. The small moment in the long direction and the lengthwise distribution of normal non-uniform loads can usually be taken care of by the shrinkage and temperature reinforcement in the long direction. In addition, a minimum amount of negative reinforcement must be provided perpendicular to the short edge, and some

CHAPTER 13 TWO-WAY SLABS ON STIFF SUPPORTS 401

Fig. 13.1 *Moment variation in elastic homogenous slab supported on all four sides*

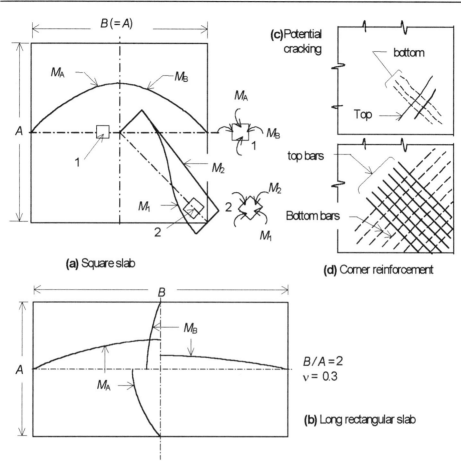

special reinforcement is also needed at the corners of panels to carry the twisting moment effects (Fig. 13.1d).

Slab panels that deform with significant curvature in both directions must be designed as two-way slabs with principal reinforcement placed in both directions. This category includes rectangular slabs (generally with long to short span ratios of less than two) supported by walls or by beams framing into columns (Figs. 2.4c, 2.9b); all slabs supported directly on columns (flat plates and flat slabs - Figs. 2.5 and 2.6); and waffle slabs (Fig. 2.8).

13.2 BACKGROUND OF TWO-WAY SLAB DESIGN

Prior to the 1973 edition of CSA A23.3, rectangular two-way floor systems supported on all sides by walls or relatively stiff beams (that is, essentially non-deflecting edge supports) were designed using moment coefficients based on semi-elastic analysis of homogeneous isotropic slabs. This procedure assumed that all loads on the slab panel is transmitted to the surrounding walls or beams, rather than directly to the columns. On the other hand, flat plates and flat slabs, which transmit the load directly to the columns, develop significant frame action and were designed on the basis of elastic analysis (frame analysis) considering the interaction between the slab and the columns. The 1973 and 1977 editions of CSA A23.3 provided for the design of all two-way slab systems (with or without beams along the edges of the panel) on the basis of the same general principles, considering it as part of a frame consisting of a row of columns or supports and the supported slab-beam members. These procedures, namely, the "Direct Design Method" and the "Equivalent Frame Method", are more complex and are described in Chapter 14. They are essentially meant for two-way slab systems, with or without beams, and supported by columns at the corners of the panels (Figs. 2.5, 2.6, 2.8, 2.9b).

The 1984 edition of CSA A23.3 retained as standard design procedures both the Direct Design Method and the Equivalent Frame Method for the design of two-way slab systems in general. In addition, the 1984 Code permitted the use of a method set out in Appendix E of the Code for rectangular slabs supported on four sides by walls or stiff beams. This Appendix E, though not a mandatory part of the Standard, reintroduced one of the earlier methods of using moment coefficients for determining slab thickness, moments and shears in slabs and loads on supporting beams or walls. CSA A23.3-94 has retained this method in its Appendix B. This method of using moment (and shear) coefficients for edge-supported slabs is simpler but slightly less economical. It is still widely used for such cases, and is particularly useful in situations where there is little or no interaction between the floor systems and supporting columns, that is, no frame action is developed. An example of this is a two-way slab system supported entirely on masonry walls. Similarly, in simple cases, such as the design of a one or two-panel slab supported on walls or beams, the elaborate and complex procedure of Chapter 14 may not often be justified, and the simpler pre-1973 procedure may be more appropriate. The use of this simpler design procedure for edge-supported slabs is detailed in this chapter.

As a general design procedure for flexure of two-way slab systems, CSA A23.3-94 permits any procedure satisfying conditions of equilibrium and compatibility with supports, provided all strength and serviceability requirements are met. In particular, the Code recommends the use of Elastic Plate Theory, Plastic Theory, Elastic Frame Analysis (Equivalent Frame Method), or the Direct Design Method. The general case of two-way slab systems is presented in Chapter 14.

13.3 DESIGN OF TWO-WAY SLABS SUPPORTED ON WALLS AND STIFF BEAMS

Only rectangular uniformly loaded slabs supported on all four sides by walls or relatively stiff beams (in general, slabs with essentially non-deflecting edge-supports) are considered in this category. In such cases, the design of the slab part and that of the supporting elements (walls and beams) can be separated. The former is done using moment coefficients for edge-supported slabs based on elastic analysis, considerations of inelastic redistribution of moments, and simplifying approximations. In the case of wall supports, this completes the floor design part. When the edge support consists of a stiff beam between columns, the beam is designed for the share of slab load that is transmitted directly to it, considering frame action, if any.

Three alternative methods for computing design moments in two-way slabs at critical sections are presented in Ref. 13.2. All three methods allow for the effects of continuity at supports and the ratio of short span to long span. Subject to their respective specified limitations, all three methods lead to satisfactory designs. One of these is based on classical plate theory (Ref. 13.2) and is the method recommended in CSA A23.3-94 (Appendix B) described in this Section. To adopt this method, a supporting beam may be considered stiff if the ratio $b_w h_b^3/(l_n h_s^3)$ is not less than 2.0, where b_w, h_b and l_n, are, respectively, the width of web, overall depth and clear span of the beam, and h_s, is the overall depth of the slab. (This ratio is different from the ratios α and α_m defined in Clause 13 of the Code)

The two-way slab panel is divided in both directions into a middle strip, one-half panel in width and symmetrical about the panel centreline, and two column strips, each one-quarter panel in width and occupying the two quarter-panel areas outside the middle strip, as shown in Fig. 13.2. The bending moments (per unit width of slab) in the middle strips are given by:

$$M_A = C_A w A^2 \tag{13.1}$$

and

$$M_B = C_B w B^2 \tag{13.2}$$

where M_A, M_B = moment in the middle strip along the short span (*A*) and long span (*B*), respectively. These are moments per unit width of slab.

C_A, C_B = moment coefficients given in Tables 13.1-13.3

w = applicable factored uniform load per unit area; total (dead + live) load for negative moments and shears, but separated into dead load and live load for positive moments

A, B = length of clear span in short and long directions, respectively

Fig. 13.2 *Column and middle strips, and variation of moments for edge-supported two-way slab*

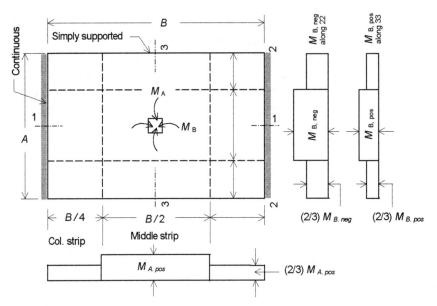

Values of C_A and C_B for positive and negative moments for various edge conditions and A/B ratios are given in Tables 13.1-13.3. CSA A23.3-94 (Appendix B) also gives identical coefficients in Tables B-1 and B-2. The support conditions considered in these tables include only combinations of essentially fixed (or continuous) and simply supported (supports with negligible torsional resistance) edge conditions. Similarly, the tables cover a range of A/B ratios from 1.0 to 0.5. When this ratio is less than 0.5, the slab should be designed as a one-way slab spanning the short span; however, negative moment reinforcement as required for a ratio of 0.5 must be provided along the short edge (reinforcement near the top face and perpendicular to the short edge).

The negative moment coefficients are given in Table 13.1 and are applicable to the respective continuous (or fixed) edges, indicated by cross-hatching. To allow for the slight restraint that may be provided by the torsional rigidity of the supporting beam or wall, all discontinuous (simply supported) edges must be designed for a negative moment equal to three-quarters of the positive moment in the same direction (CSA A23.3-94-App. B). For negative moment calculations with Table 13.1, w is the total factored uniform load (dead plus live). The critical sections for negative moment

Table 13.1 Coefficients for Negative Moments in Slabs *

$$M_{A\,neg} = C_{A\,neg} \times w \times A^2$$
$$M_{B\,neg} = C_{B\,neg} \times w \times B^2$$

where w = total uniform dead plus live load

Ratio $m = \dfrac{A}{B}$		Case 1	Case 2	Case 3	Case 4	Case 5	Case 6	Case 7	Case 8	Case 9
1.00	$C_{A\,neg}$	0.045		0.050	0.075	0.071		0.033	0.061
	$C_{B\,neg}$	0.045	0.076	0.050			0.071	0.061	0.033
0.95	$C_{A\,neg}$	0.050		0.055	0.079	0.075		0.038	0.065
	$C_{B\,neg}$	0.041	0.072	0.045			0.067	0.056	0.029
0.90	$C_{A\,neg}$	0.055		0.060	0.080	0.079		0.043	0.068
	$C_{B\,neg}$	0.037	0.070	0.040			0.062	0.052	0.025
0.85	$C_{A\,neg}$	0.060		0.066	0.082	0.083		0.049	0.072
	$C_{B\,neg}$	0.031	0.065	0.034			0.057	0.046	0.021
0.80	$C_{A\,neg}$	0.065		0.071	0.083	0.086		0.055	0.075
	$C_{B\,neg}$	0.027	0.061	0.029			0.051	0.041	0.017
0.75	$C_{A\,neg}$	0.069		0.076	0.085	0.088		0.061	0.078
	$C_{B\,neg}$	0.022	0.056	0.024			0.044	0.036	0.014
0.70	$C_{A\,neg}$	0.074		0.081	0.086	0.091		0.068	0.081
	$C_{B\,neg}$	0.017	0.050	0.019			0.038	0.029	0.011
0.65	$C_{A\,neg}$	0.077		0.085	0.087	0.093		0.074	0.083
	$C_{B\,neg}$	0.014	0.043	0.015			0.031	0.024	0.008
0.60	$C_{A\,neg}$	0.081		0.089	0.088	0.095		0.080	0.085
	$C_{B\,neg}$	0.010	0.035	0.011			0.024	0.018	0.006
0.55	$C_{A\,neg}$	0.084		0.092	0.089	0.096		0.085	0.086
	$C_{B\,neg}$	0.007	0.028	0.008			0.019	0.014	0.005
0.50	$C_{A\,neg}$	0.086		0.094	0.090	0.097		0.089	0.088
	$C_{B\,neg}$	0.006	0.022	0.006			0.014	0.010	0.003

* A shaded edge indicates that the slab continues across or is fixed at the support; an unmarked edge indicates a support at which torsional resistance is negligible.

406 REINFORCED CONCRETE DESIGN

Table 13.2 Coefficients for Dead Load Positive Moments in Slabs *

$$M_{A\,posDL} = C_{A\,DL} \times w_{DL} \times A^2$$
$$M_{B\,posDL} = C_{B\,DL} \times w_{DL} \times B^2$$

where w_{DL} = total uniform dead

Ratio $m = \dfrac{A}{B}$		Case 1	Case 2	Case 3	Case 4	Case 5	Case 6	Case 7	Case 8	Case 9
1.00	$C_{A\,DL}$	0.036	0.018	0.018	0.027	0.027	0.033	0.027	0.020	0.023
	$C_{B\,DL}$	0.036	0.018	0.027	0.027	0.018	0.027	0.033	0.023	0.020
0.95	$C_{A\,DL}$	0.040	0.020	0.021	0.030	0.028	0.036	0.031	0.022	0.024
	$C_{B\,DL}$	0.033	0.016	0.025	0.024	0.015	0.024	0.031	0.021	0.017
0.90	$C_{A\,DL}$	0.045	0.022	0.025	0.033	0.029	0.039	0.035	0.025	0.026
	$C_{B\,DL}$	0.029	0.014	0.024	0.022	0.013	0.021	0.028	0.019	0.015
0.85	$C_{A\,DL}$	0.050	0.024	0.029	0.036	0.031	0.042	0.040	0.029	0.028
	$C_{B\,DL}$	0.026	0.012	0.022	0.019	0.011	0.017	0.025	0.017	0.013
0.80	$C_{A\,DL}$	0.056	0.026	0.034	0.039	0.032	0.045	0.045	0.032	0.029
	$C_{B\,DL}$	0.023	0.011	0.020	0.016	0.009	0.015	0.022	0.015	0.010
0.75	$C_{A\,DL}$	0.061	0.028	0.040	0.043	0.033	0.048	0.051	0.036	0.031
	$C_{B\,DL}$	0.019	0.009	0.018	0.013	0.007	0.012	0.020	0.013	0.007
0.70	$C_{A\,DL}$	0.068	0.030	0.046	0.046	0.035	0.051	0.058	0.040	0.033
	$C_{B\,DL}$	0.016	0.007	0.016	0.011	0.005	0.009	0.017	0.011	0.006
0.65	$C_{A\,DL}$	0.074	0.032	0.054	0.050	0.036	0.054	0.065	0.044	0.034
	$C_{B\,DL}$	0.013	0.006	0.014	0.009	0.004	0.007	0.014	0.009	0.005
0.60	$C_{A\,DL}$	0.081	0.034	0.062	0.053	0.037	0.056	0.073	0.048	0.036
	$C_{B\,DL}$	0.010	0.004	0.011	0.007	0.003	0.006	0.012	0.007	0.004
0.55	$C_{A\,DL}$	0.088	0.035	0.071	0.056	0.038	0.058	0.081	0.052	0.037
	$C_{B\,DL}$	0.008	0.003	0.009	0.005	0.002	0.004	0.009	0.005	0.003
0.50	$C_{A\,DL}$	0.095	0.037	0.080	0.059	0.039	0.061	0.089	0.056	0.038
	$C_{B\,DL}$	0.006	0.002	0.007	0.004	0.001	0.003	0.007	0.004	0.002

* A shaded edge indicates that the slab continues across or is fixed at the support; an unmarked edge indicates a support at which torsional resistance is negligible.

Table 13.3 *Coefficients for Live Load Positive Moments in Slabs* *

$$M_{A\,posLL} = C_{A\,LL} \times w_{LL} \times A^2$$
$$M_{B\,posLL} = C_{B\,LL} \times w_{LL} \times B^2$$

where w_{LL} = total uniform live load

Ratio $m = \dfrac{A}{B}$		Case 1	Case 2	Case 3	Case 4	Case 5	Case 6	Case 7	Case 8	Case 9
1.00	$C_{A\,LL}$	0.036	0.027	0.027	0.032	0.032	0.035	0.032	0.028	0.030
	$C_{B\,LL}$	0.036	0.027	0.032	0.032	0.027	0.032	0.035	0.030	0.028
0.95	$C_{A\,LL}$	0.040	0.030	0.031	0.035	0.034	0.038	0.036	0.031	0.032
	$C_{B\,LL}$	0.033	0.025	0.029	0.029	0.024	0.029	0.032	0.027	0.025
0.90	$C_{A\,LL}$	0.045	0.034	0.035	0.039	0.037	0.042	0.040	0.035	0.036
	$C_{B\,LL}$	0.029	0.022	0.027	0.026	0.021	0.025	0.029	0.024	0.022
0.85	$C_{A\,LL}$	0.050	0.037	0.040	0.043	0.041	0.046	0.045	0.040	0.039
	$C_{B\,LL}$	0.026	0.019	0.024	0.023	0.019	0.022	0.026	0.022	0.020
0.80	$C_{A\,LL}$	0.056	0.041	0.045	0.048	0.044	0.051	0.051	0.044	0.042
	$C_{B\,LL}$	0.023	0.017	0.022	0.020	0.016	0.019	0.023	0.019	0.017
0.75	$C_{A\,LL}$	0.061	0.045	0.051	0.052	0.047	0.055	0.056	0.049	0.046
	$C_{B\,LL}$	0.019	0.014	0.019	0.016	0.013	0.016	0.020	0.016	0.013
0.70	$C_{A\,LL}$	0.068	0.049	0.057	0.057	0.051	0.060	0.063	0.054	0.050
	$C_{B\,LL}$	0.016	0.012	0.016	0.014	0.011	0.013	0.017	0.014	0.011
0.65	$C_{A\,LL}$	0.074	0.053	0.064	0.062	0.055	0.064	0.070	0.059	0.054
	$C_{B\,LL}$	0.013	0.010	0.014	0.011	0.009	0.010	0.014	0.011	0.009
0.60	$C_{A\,LL}$	0.081	0.058	0.071	0.067	0.059	0.068	0.077	0.065	0.059
	$C_{B\,LL}$	0.010	0.007	0.011	0.009	0.007	0.008	0.011	0.009	0.007
0.55	$C_{A\,LL}$	0.088	0.062	0.080	0.072	0.063	0.073	0.085	0.070	0.063
	$C_{B\,LL}$	0.008	0.006	0.009	0.007	0.005	0.006	0.009	0.007	0.006
0.50	$C_{A\,LL}$	0.095	0.066	0.088	0.077	0.067	0.078	0.092	0.076	0.067
	$C_{B\,LL}$	0.006	0.004	0.007	0.005	0.004	0.005	0.007	0.005	0.004

* A shaded edge indicates that the slab continues across or is fixed at the support; an unmarked edge indicates a support at which torsional resistance is negligible.

408 REINFORCED CONCRETE DESIGN

Table 13.4 Ratio of Load w in A and B Directions for Shear in Slab and Load on Supports *

Ratio $m = \dfrac{A}{B}$		Case 1	Case 2	Case 3	Case 4	Case 5	Case 6	Case 7	Case 8	Case 9
1.00	W_A	0.50	0.50	0.17	0.50	0.83	0.71	0.29	0.33	0.67
	W_B	0.50	0.50	0.83	0.50	0.17	0.29	0.71	0.67	0.33
0.95	W_A	0.55	0.55	0.20	0.55	0.86	0.75	0.33	0.38	0.71
	W_B	0.45	0.45	0.80	0.45	0.14	0.25	0.67	0.62	0.29
0.90	W_A	0.60	0.60	0.23	0.60	0.88	0.79	0.38	0.43	0.75
	W_B	0.40	0.40	0.77	0.40	0.12	0.21	0.62	0.57	0.25
0.85	W_A	0.66	0.66	0.28	0.66	0.90	0.83	0.43	0.49	0.79
	W_B	0.34	0.34	0.72	0.34	0.10	0.17	0.57	0.51	0.21
0.80	W_A	0.71	0.71	0.33	0.71	0.92	0.86	0.49	0.55	0.83
	W_B	0.29	0.29	0.67	0.29	0.08	0.14	0.51	0.45	0.17
0.75	W_A	0.76	0.76	0.39	0.76	0.94	0.88	0.56	0.61	0.86
	W_B	0.24	0.24	0.61	0.24	0.06	0.12	0.44	0.39	0.14
0.70	W_A	0.81	0.81	0.45	0.81	0.95	0.91	0.62	0.68	0.89
	W_B	0.19	0.19	0.55	0.19	0.05	0.09	0.38	0.32	0.11
0.65	W_A	0.85	0.85	0.53	0.85	0.96	0.93	0.69	0.74	0.92
	W_B	0.15	0.15	0.47	0.15	0.04	0.07	0.31	0.26	0.08
0.60	W_A	0.89	0.89	0.61	0.89	0.97	0.95	0.76	0.80	0.94
	W_B	0.11	0.11	0.39	0.11	0.03	0.05	0.24	0.20	0.006
0.55	W_A	0.92	0.92	0.69	0.92	0.98	0.96	0.81	0.85	0.95
	W_B	0.08	0.08	0.31	0.08	0.02	0.04	0.19	0.15	0.05
0.50	W_A	0.94	0.94	0.76	0.94	0.99	0.97	0.86	0.89	0.97
	W_B	0.06	0.06	0.24	0.06	0.01	0.03	0.14	0.11	0.03

* A shaded edge indicates that the slab continues across or is fixed at the support; an unmarked edge indicates a support at which torsional resistance is negligible.

calculations are along the edges of the panel at the faces of the supports. At supports over which the slab is continuous, if the negative moment on one side is less than 80 percent of that on the other side, the difference is distributed between the two sides in proportion to the relative stiffnesses of the slabs.

The coefficients for positive moments are given in Table 13.2 for dead load and in Table 13.3 for live load. Dead and live loads are treated separately here to allow for their different patterns of placement to produce the maximum positive moment. The critical section for positive moment is along the centreline of the panel. The middle strips are designed for the full intensity of negative and positive moments computed using the Tables 13.1 to 13.3, assuming the moment variation to be uniform across the width of the strips. The moments per unit width (both positive and negative) in the column strip are taken as two-thirds of the corresponding bending moments in the middle strip, as shown in Fig. 13.2.

For computing the shear stresses in the slab, Ref. 13.2 recommends that the load, w, on the slab be assumed distributed in the A and B directions in the proportions given in Table 13.4. The same proportional distribution is assumed to compute the load on the supporting beams (or walls), provided that the load on the beam along the short direction is taken as not less than that on a slab area bounded by 45° lines from the corners. However, CSA A23.3-94(App. B) does not contain the equivalent of Table 13.4. Instead, the Code specifies that the loads on the supporting beams may be assumed as the load within the tributary areas of the panel bounded by the intersection of 45° lines from the corners and the median line of the panel parallel to the long sides. The slab shears may be computed on the assumption of the above distribution of the load to the supports. Furthermore, the bending moments in the supporting beams may be determined using an equivalent uniform load given by Eqs. 13.3 and 13.4. Equivalent uniform loads on beams are:

(i) for beam along short span: $\dfrac{w_f A}{3}$ (13.3)

(ii) for beam along long span: $\dfrac{w_f A}{3} \times \dfrac{(3-m^2)}{2}$ (13.4)

where $m = A/B$

The effects of twisting moments described in Section 13.1 and illustrated in Fig. 13.1c are significant at exterior corners. Hence, special reinforcement should be provided at exterior corners in both the bottom and top of the slab, details for which are presented in CSA A23.3-94(13.13.5). Accordingly, this reinforcement should be provided for a distance in each direction from the corner equal to one-fifth the shorter span. Both the bottom and top reinforcement should be sufficient to resist a moment

410 REINFORCED CONCRETE DESIGN

equal to the maximum positive moment per metre width in the slab. The top reinforcement should be placed in a single band in the direction of the diagonal from the corner, and the bottom reinforcement in a direction at right angles to the diagonal (see Fig. 13.1d). Alternatively, at both the top and bottom of the slab, the reinforcement may be placed in two bands parallel to the sides of the slab, with each band having an area sufficient to resist the moment mentioned above.

Additional requirements to be met by two-way slab reinforcement in general are given in CSA A23.3-94(13.11). The maximum spacing of bars at critical sections in solid two-way slabs are given in Cl. 13.11.3 (also Example 13.1). All positive reinforcement perpendicular to a discontinuous edge should be extended to the edge of the slab and embedded for at least 150 mm into spandrel beams, walls, or columns. Similarly, all negative reinforcement perpendicular to a discontinuous edge must be anchored into spandrel beams, walls, or columns, to be developed at the face of the support. The area of reinforcement must be adequate to resist the bending moments at all critical sections, but should not be less than the minimum specified as shrinkage and temperature reinforcement (Code Cl.7.8).

CSA A23.3-94(App. B and Cls. 9.8.3 and 13.3) requires that the slab have adequate thickness for deflection control (Section 14.3); however, for slabs designed by this method, the thickness should in no case be less than (a) 100 mm, (b) the perimeter of the slab divided by 140, in the case of slabs discontinuous on one or more edges, or (c) the perimeter of the slab divided by 160, in the case of slabs continuous on all edges.

Finally, the Code permits the Appendix B procedure to determine slab thickness, moments and shears in slabs and loads on supporting beams or walls. All other requirements such as area of minimum reinforcement, spacing, crack control, etc., should be in accordance with the appropriate provisions of the Code.

EXAMPLE 13.1

The plan of a floor consisting of two-way slabs supported on and cast integrally with stiff beams is shown in Fig. 13.3. The specified dead load, in addition to the slab weight, is 1.5 kN/m^2 and the specified live load is 4.8 kN/m^2. Design the corner panel. Assume $f_c' = 20$ MPa, $f_y = 400$ MPa, and the beams to be 300 mm wide.

SOLUTION

1. Minimum Thickness for Deflection Control
   Clear spans are, 7 - 0.3 = 6.7 m; and 6 - 0.3 = 5.7 m. For design using the provisions of CSA A23.3-94(App. B), the minimum thickness should be

Fig. 13.3 Example 13.1

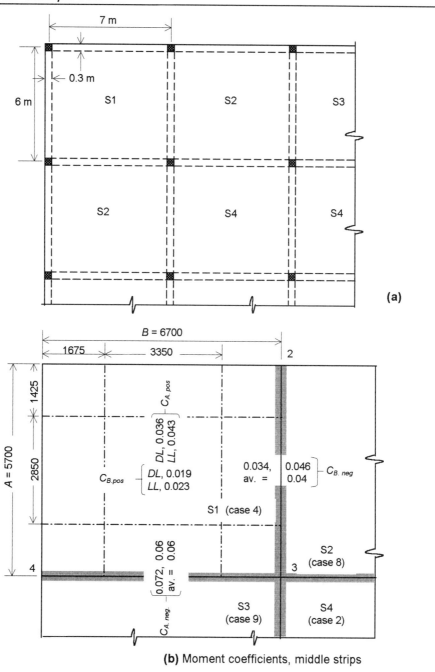

(b) Moment coefficients, middle strips

determined by Cl. 13.3, but in no case should it be less than (this being a corner panel with two discontinuous edges):

$$h = \frac{\text{perimeter of slab}}{140} = \frac{2(6700 + 5700)}{140} = 177 \text{ mm}$$

Minimum thickness using CSA A23.3-94 Equation 13.3 (also Section 14.3) is given by:

α_m, assumed > 2 for stiff beams, and hence limited to 2 in Eq. 13.3
β = ratio of long to short clear spans
= 6.7/5.7 = 1.18
l_n = the longer clear span = 6700 mm

$$h \geq \frac{l_n(0.6 + f_y/1000)}{30 + 4\beta\alpha_m} = \frac{6700 \times (0.6 + 400/1000)}{30 + 4 \times 1.18 \times 2} = 170 \text{ mm}$$

The controlling minimum thickness is perimeter/140 = 177 mm. In this, case, a thickness of 180 mm is selected. (A smaller thickness may be used, provided deflections are computed and shown to be within allowable limits).

$$\text{Weight of slab} = 0.18 \times \frac{2400}{1000} \times 9.81 = 4.24 \text{ kN/m}^2$$

Factored loads are:
Dead load = 1.25 (1.5 + 4.24) = 7.17 kN/m^2
Live load = 1.5 × 4.8 = 7.20 kN/m^2
Total = 14.4 kN/m^2

Ratio, short to long span = m = 5.7/6.7 = 0.85

2. Design Moments in Middle Strip

The moment coefficients are given in Table 13.1 for negative moments at continuous ends and in Tables 13.2 and 13.3 for positive moments at midspan (for dead and live load, respectively). These coefficients are shown in Fig. 13.3b at the appropriate locations. At continuous edges, the negative moment coefficients for the neighbouring panels are also shown to facilitate redistribution of the unbalanced value if greater than 20 percent. Here, such a situation exists at the continuous edge 2-3, with the coefficients being

0.034 and 0.046. Since the span lengths and loading are the same, the coefficients instead of the moment may be distributed. The slab thickness being the same for both slabs, the difference is distributed equally between the two sides, to yield a moment coefficient of $(0.034 + 0.046)/2 = 0.04$. The difference in the restraint conditions at the far edges has little effect on the relative stiffnesses of the slabs at the near edges, and the average is used.

At the continuous edge 3-4 (Fig. 13.3b), the negative moment coefficients on the two sides do not differ by more than 20 percent. The procedure described in Section 13.3 does not specify a distribution of the difference in the moments for such cases. However, normally it is the same negative reinforcement that resists the moment on either side of the beam along edge 3-4. Therefore, the slab has to be designed for the larger of the moments from either side (see also Code Cl. 13.10.3.4), that is in this case for a moment with a coefficient $C_{A, neg} = 0.072$.

Alternatively, the reinforcement may be designed for the moment obtained by distributing the difference in proportion to the relative stiffnesses (i.e., $C_{A, neg} = (0.066 + 0.072)/2 = 0.069$) as for the edge 2-3; and this latter procedure is followed in this example. Slabs are highly indeterminate structures and are normally very much under-reinforced so that some amount of inelastic moment redistribution can be expected. CSA A23.3-94(13.10.3.3) also permits a modification of the moment by up to 15 percent, provided there is a reserve capacity elsewhere to accommodate the total static moment. Since the positive and negative reinforcements are designed for different loading patterns, for a given loading condition generally there is a reserve strength elsewhere to accommodate the total static design moment.

The factored moments at various critical sections are computed below. Subscripts A and B refer to short and long spans, respectively.

(i)   Negative moments at continuous edges:

$M_{A, neg} = 0.069 \times 14.4 \times 5.7^2$ $= 32.3$ kN·m

$M_{B, neg} = 0.04 \times 14.4 \times 6.7^2$ $= 25.9$ kN·m

(ii)   Positive moments at midspan:

$M_{A, pos, DL} = 0.036 \times 7.17 \times 5.7^2$ $= 8.39$ kN·m

$M_{A, pos, LL} = 0.043 \times 7.2 \times 5.7^2$ $= \underline{10.1 \text{ kN·m}}$

$M_{A, pos, total}$ $= 18.5$ kN·m

$M_{B, pos, DL} = 0.019 \times 7.17 \times 6.7^2$ $= 6.12$ kN·m

$M_{B, pos, LL} = 0.023 \times 7.2 \times 6.7^2$ $= \underline{7.43 \text{ kN·m}}$

$M_{B, pos, total}$ $= 13.6$ kN·m

414 REINFORCED CONCRETE DESIGN

(iii) Negative moment at discontinuous edges, equal to 3/4 of positive moment at midspan:
$M_{A,\,neg} = 0.75 \times 18.5$ = 13.9 kN·m
$M_{B,\,neg} = 0.75 \times 13.6$ = 10.2 kN·m

3. **Design of Sections**

The largest moment anywhere in the slab is the negative moment at the continuous edge = 32.3 kN·m. Selecting 20 mm clear cover, and No. 10 bars, and placing the reinforcement parallel to the short side as the outer layer (to provide the greater effective depth in this direction to resist the relatively larger moment along the short span),

$d_A = 180 - 20 - 11.3/2 = 154$ mm (Fig. 13.4)
and $d_B = 180 - 20 - 11.3 - 11.3/2 = 143$ mm

(i) At continuous end of short span, middle strip

$M_{A,\,neg} = 32.3 \text{ kN} \cdot \text{m}$

Required $K = \dfrac{M_f}{bd^2} = \dfrac{32.3 \times 10^6}{1000 \times 154^2} = 1.36$ MPa

Referring to Table 5.4, this gives a reinforcement ratio of $\rho = 0.00434$, which is within limits and in the desirable range for slabs.

Area of steel $A_s = 0.00434 \times 1000 \times 154 = 667$ mm^2
No. 10 bars at spacing of 150 mm gives $A_s = 667$ mm^2, as required

Alternatively, in the entire middle strip, which has a width of 6700/2 = 3350 mm, the number of No. 10 bars required is = 667 ×

Fig. 13.4 *Effective depths for short and long spans*

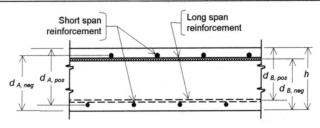

Section of slab parallel to long side

3.35/100 = 22.3. A total of 23 bars can be distributed giving an area per metre width of 2300/3.35 = 687 mm^2 (compared to 667 mm^2 required) and resulting in a spacing of 3350/23 = 146 mm. While no longer common practice, these bars may be made up of, to the extent available, bent up bars to a maximum of one-half the positive reinforcement from the midspan regions, on both sides of the support. The balance may be provided by additional straight bars.

(ii) The computations for the reinforcement at all other critical sections in middle strips, following the same procedure as illustrated above, are set forth in Table 13.5.

(iii) Reinforcement in column strip.
Across the width of each column strip, the moment per unit width is taken as two-thirds of the corresponding moment in the middle strip (Fig. 13.2). The reinforcement required to resist this moment can be computed as for the middle strip. However, it is sufficiently accurate and slightly conservative to determine the reinforcement in proportion to the moments from the values already determined for the middle strip. Thus, each column strip has one-half the width of the middle strip and two-thirds its moment. Hence, the total number of bars required will be roughly one-third the number of bars required in the middle strip. For example, each column strip in the short direction will need, at the continuous end, one-third of 23 or approximately 8 bars. On this basis, the number of bars required for moment in each column strip at all critical sections is given in row 9 of Table 13.5. These numbers may have to be increased to meet maximum spacing requirements.

4. Bar spacing
Spacing of bars at critical sections shall not exceed the limits set in Code Clause 13.11.3. Accordingly, for all positive moment reinforcements (in both middle and column strips) and the negative moment reinforcements in the middle strip, the maximum spacing shall not exceed $3h_s$ nor 500 mm, where h_s is the overall thickness of slab. The number and spacing of bars shown in rows 9 and 10 of Table 13.5 satisfy these requirements.
For the negative moment reinforcements in the column strips, at interior columns, at least 1/3 of the reinforcement for total factored negative moment shall be located in a band width extending a distance $1.5h_s$ from the side of the

416 REINFORCED CONCRETE DESIGN

Table 13.5 Example 13.1

		Short Span			Long Span		
		Cont. end Neg. mom.	Midspan Pos. mom.	Discont. End Neg. mom.	Cont. end Neg. mom.	Midspan Pos. mom.	Discont. End Pos. mom.
Middle Strip:							
1. Factored moment, M_f	kN·m	32.3	18.5	13.9	25.9	13.6	10.2
2. Effective depth, d	mm	154	154	154	143	143	143
3. $K_r = M_f/(1000\,d^2)$	MPa	1.36	0.780	0.586	1.27	0.665	0.499
4. 100ρ, (Table 5.1)		0.434	0.242	0.176	0.401	0.200	—
5. $100\rho_{\min}$		—	—	0.200	—	0.200	0.200
6. $A_s = \rho \times 1000\,d$, per m	mm²	667	373	—	573	286	—
7. $A_{s\min} = 0.002 \times 1000 \times h$	mm²	—	—	360	—	360	360
8. Width of middle strip	mm	—	3350	—	—	2850	—
9. No. 10 bars required for full middle strip	Nos.	23	12	12	17	11	11
10. Resulting spacing, ($\le 3h$, ≤ 500)	mm	146	279	279	168	259	259

CHAPTER 13 TWO-WAY SLABS ON STIFF SUPPORTS

Column Strip:					
11. Full width of each column strip	mm	—	1675	—	1425
12. No. of bars required in each column strip = 1/3 (number in mid-strip)		8	4	6	4
13. Selected number and spacing for pos. moment ($s \leq 35$ and $s \leq 500$)	mm		4 @ 420		4 @ 356
14. Width of band for neg. mom. rein. concentration	mm	270	270	270	270
15. No. of bars in band, $\geq 1/3$ (Cl.13.12.2)		3	2	2	2
16. Spacing in band ($\leq 1.5h_s$, 250 mm)		90	135	135	135
17. Width of col. strip outside band	mm	1405	1405	1155	1155
18. No. of bars in above width		5	2	4	2
19. No. of bars selected in strip at (17), (spacing $s \leq h$ and $s \leq 500$ mm)		5 @ 281	3 @ 468	4 @ 289	3 @ 385

1.5h_s nor 250 mm (Cl.13.11.3(a)). In this example, the width of this band is 1.5h_s = 270 mm. Similarly, in the column strip at the exterior column, reinforcement for the total factored negative moment shall be placed within a band with a width extending a distance 1.5h_s from the sides of the column (Cl. 13.12.2.2), and again the spacing of such negative moment reinforcement within the band shall not exceed 1.5h_s, nor 250 mm (Cl. 13.11.3(a)). The column strip negative moment reinforcements outside the above bands shall be spaced not farther than 3h_s, nor 500 mm. Based on these, the negative moment reinforcement numbers and spacing within the band in the column strip and in the remaining portion of the column strip are worked out in rows 11 to 19 of Table 13.5.

5. Reinforcement Detailing

To detail bar cut-off (and bend points, if used) the inflection points may be assumed to be at one-sixth of the span length from the face of supports over which the slab is continuous. At exterior spans, all positive reinforcement perpendicular to the discontinuous edge should extend to the edge of the slab and have embedment (straight or hooked) of at least 150 mm into spandrel beams or walls. The bar development requirements are the same as for beams. The negative reinforcement must be well anchored at a discontinuous end into the supporting beam or wall, so as to develop the bar stress at the critical section at the face of the support. Typical bar bending details are shown in Fig. 13.5a.

6. Shear in Slab

Using the coefficients in Table 13.4, the larger fraction of load transferred, in the short direction, to the beams on the longer side is 0.66.

On a 1 m-wide slab strip in the short direction, the shear force is:

$$V_f = \frac{w_a A}{2} = \frac{0.66 \times 14.4 \times 5.7}{2} = 27.1 \text{ kN}$$

$$V_c = 0.2\lambda\phi_c\sqrt{f_c'}b_w d = 0.2 \times 1 \times 0.6 \times \sqrt{20} \times 1000 \times 154 \times 10^{-3}$$
$$= 82.7 \text{ kN}$$

$V_f < V_c$, no shear reinforcement needed.   OK

Alternatively, if the load distribution to the support is assumed as in CSA A23.3-94(App. B 3.1), the maximum shear force per metre on a slab strip in short direction at midspan is,

$$V_f = 0.5 w_f A = 0.5 \times 14.4 \times 5.7 = 41 \text{ kN}$$

Again, $V_f < V_c$ OK

Fig. 13.5 Reinforcement details, Example 13.1

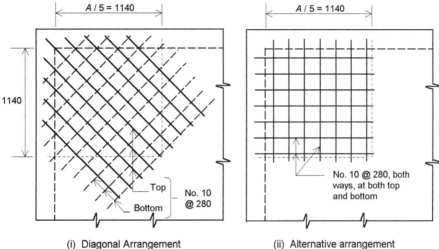

(a) Suggested reinforcement details

(i) Diagonal Arrangement (ii) Alternative arrangement

(b) Details of corner reinforcement, exterior corners

7. Load Transmitted to Beams

If the coefficients in Table 13.4 are used, the load per unit length on a long beam from the slab panel is:

$$\frac{0.66 \times 14.4 \times 5.7}{2} = 27.1 \, \text{kN/m}$$

Load on short beam = $\dfrac{0.34 \times 14.4 \times 6.7}{2} = 16.4 \, \text{kN/m}$

Minimum = $\dfrac{wA}{3} = \dfrac{14.4 \times 5.7}{3} = 27.4 \, \text{kN/m} > 16.4 \, \text{kN/m}$

Loading on shorter beam from slab panel = 27.4 kN/m

Alternatively, using Eqs. 13.3 and 13.4 specified in CSA A23.3-94, load on the longer beam is:

420 REINFORCED CONCRETE DESIGN

$$\frac{w_f A}{3} = \frac{(3-m^2)}{2} = \frac{14.4 \times 5.7}{3} = \frac{(3-0.85^2)}{2} = 31.2 \text{ kN/m}$$

and load on shorter beam is:

$$\frac{w_f A}{3} = \frac{14.4 \times 5.7}{3} = 27.4 \text{ kN/m}$$

These loads may be used as equivalent uniform load for computing bending moments on the respective beams.

8. Corner Reinforcement

The corner reinforcement required at exterior corners is equal to the maximum positive moment reinforcement, that is No. 10 bars at 280 mm (≈ 279 mm) (Table 13.5). These bars are provided for a length of 5700/5 = 1140 mm, as shown in Fig. 13.5b.

13.4 DIFFERENCE BETWEEN WALL SUPPORTS AND COLUMN SUPPORTS

As explained in Section 13.3, in general, a load w placed on a two-way slab is transmitted partly (w_A) along the short span to the longer edge supports, and partly (w_B) along the long span to the shorter edge supports. In wall supported panels, the portions w_A and w_B of the load are transmitted by the respective walls vertically, directly to their foundations (or other supports) (Fig. 13.6a). In contrast, when the edge supports consist of beams between columns, (Fig. 13.6b), the portion, w_A, of the load transmitted in the North-South (N-S) direction by the slab is in turn transmitted in the E-W direction to the supporting columns by the beams along the E-W direction. Thus, considering the combined action of both slab and E-W direction beams, the full load (part w_B by slab and part w_A through the beams) is transmitted in the E-W direction, as is evident also from the equilibrium considerations of the section of floor, as shown in Fig. 13.6b (ii). The same considerations also show that the slab together with the beams in the N-S direction (Fig. 13.6b(iii)) transmit the full slab load along the N-S direction, to the supporting columns.

In general, in column-supported slabs, with or without beams along column lines, 100 percent of the slab load has to be transmitted by the floor system in both directions towards the columns. For the case of the stiff beams considered in this chapter, the analysis and design of the slab part and the beam part are considered separately. When the beams are relatively flexible (or shallow), and when there are no beams, the analysis of the floor system (including beams, if any) is integrated, and the frame action developed between the floor system and supporting columns is taken into consideration. This procedure is described in the next chapter.

Fig. 13.6 *Load transfer in wall-supported and column-supported slabs*

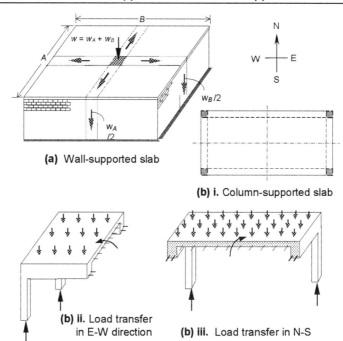

(a) Wall-supported slab

(b) i. Column-supported slab

(b) ii. Load transfer in E-W direction

(b) iii. Load transfer in N-S

PROBLEMS

1. A garage, 6 m × 4 m, is to be provided with a single panel slab roof. The slab is to be supported on walls on three sides and on a monolithically cast beam on the entrance side (4 m). The service live load on the roof is 2 kN/m^2. Use $f_c' = 30$ MPa and $f_y = 300$ MPa. Design the roof.

2. For the floor in Example 13.1, (Fig. 13.3) design the reinforcement details for an interior panel. Assume the slab thickness to be 180 mm.

REFERENCES

13.1 Timoshenko, S., and Woinosky-Krieger, S., *Theory of Plates and Shells*, McGraw-Hill Book Co., New York, 1959, 580 pp.

13.2 ACI Committee 318, *Building Code Requirements for Structural Concrete* (ACI 318R-95) and *Commentary* (ACI 318R-95), American Concrete Institute, Detroit, 1995, 369 pp.

CHAPTER 14 Design of Two-Way Slab Systems

14.1 INTRODUCTION

This chapter deals with the general case of slab systems supported on columns, with or without beams along column lines, and includes flat plates, flat slabs, waffle slabs, and solid two-way slabs with beams along column lines. The essential difference between the slab discussed in Chapter 13 and the slab here is that while in the former the deflection of the slab is (near) zero all along the supported edges of the panel, in this case the deflection is zero only at the column supports, and the slab may have significant deflections at all other points, including along the line between columns (ie, panel edges). The moment coefficients given in Chapter 13 were based on classical *elastic plate theory*. Elastic plate theory solutions are possible for this case also, especially by resorting to numerical techniques such as finite element analysis. The Code permits such procedures to determine of design moments in the slab.

Yet another method that can be used is the *plastic theory*. This is particularly suited for slabs with irregular geometry and support points. This theory is based on the assumption of an elastic-plastic moment-curvature relationship for the section. Therefore, care should be taken to ensure a ductile response (see also Section 10.7.1). Furthermore, the plastic theory solution corresponds to a failure mechanism and, hence, takes care of the ultimate limit state under factored loads only. Separate checks have to be made for serviceability limit states requirements under service loads.

Design moments and shear forces in regular two-way slab systems can also be determined by an *elastic analysis* of plane frames along column lines taken longitudinally and transversely, of which the horizontal line members connecting the columns are made up of the slab-beam elements. This being the most frequently used method, it will be explained in detail below.

It was shown in Section 13.4 that in two-way slabs supported on columns, 100 percent of the floor loads have to be carried in both the longitudinal and transverse directions to the supporting columns. The mechanisms of load transfer from the slab to the columns include flexure, shear, and torsion. To obtain the load effects on the elements of the floor system and its supporting members using an elastic analysis, the structure may be considered as a series of bents, each consisting of a row of columns (or other supports) and the portion of the floor system tributary to it, which can be taken as the part of the floor member bounded by the panel centrelines on either side of the columns. Such bents must be taken both longitudinally and transversely in the building, to assure

Fig. 14.1 Equivalent frame

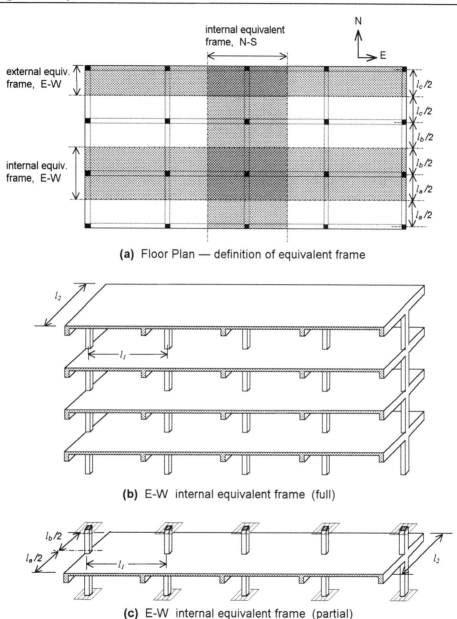

(a) Floor Plan — definition of equivalent frame

(b) E-W internal equivalent frame (full)

(c) E-W internal equivalent frame (partial)

load transfer in both directions, as indicated in Fig. 14.1a. One such isolated bent is shown in Fig. 14.1b. The horizontal member consists of the slab, together with drop panels or beams along column lines, if any.

Part of the structure in Fig. 14.1b is very similar to the plane frame considered in Fig. 10.3a and b, the primary difference being that the horizontal member in the former is a relatively wide slab-beam member. In the elastic analysis of the plane frame in Fig. 10.3, it was shown in Section 10.2 that several approximations can be made subject to certain limitations. Similar approximations can also be made in the present case. For example, in the analysis for gravity load effects, instead of considering the entire frame, a partial frame consisting of each floor (or roof) may be analysed separately together with the columns immediately above and below, the columns being assumed fixed at their far ends (Fig. 14.1c). Such a procedure is described in the *Elastic Frame Method (Analogy)* presented in Section 14.9 (compare with the simplification in Fig. 10.3c and d). When frame geometry and loading meet certain limitations, the positive and negative factored moments at critical sections of the slab-beam member may be calculated using moment coefficients applied to the simple span moments, M_o. This is analogous to the use of the CSA moment coefficients in the analysis of continuous beams described in Section 10.3.3. A somewhat similar procedure for two-way slabs, termed the *Direct Design Method* is given in Section 14.5.

In the plane frames considered in Chapters 10 and 12, the horizontal member is a beam, which can be proportioned for the full design moment obtained from the analysis. In contrast, the horizontal member in the frame in Fig. 14.1c is a very wide beam (slab with or without beam along the column line), which is supported generally on a limited width. Hence, the outer portions of this member are less restrained than the part along the support lines, and the distribution of the moment along the width of the member is not uniform. Considering one panel of the frame, the variation of the moment in the floor member along the span is as shown in Fig. 14.2b. Here M_{ab} represents the moment in the slab-beam member over its full width AB.

This moment is distributed over the width AB non-uniformly, as shown (qualitatively) by the solid curve in Fig. 14.2c. The actual variation along AB depends on several factors, such as the span ratio, relative stiffness of beams along column lines (if any), presence of drop panels, torsional stiffness of transverse beams (if any), etc., and is highly indeterminate. Therefore, approximations are desired. The procedure generally adopted is to divide the slab panel into column strips (along the column lines) and middle strips (along the panel centrelines), and then to apportion the moment between these strips, the distribution of the moment within the width of each strip being assumed uniform. This is shown by the broken lines in Fig. 14.2c and also in Fig. 14.2d. When beams are provided along the column line, the beam portion is relatively stiffer than the slab and resists a significant share of the moment at the section. In this case, the moment has to be apportioned between the beam part and the slab part of the slab-beam member (Fig. 14.2e).

Fig. 14.2 Moment variations in two-way slab panel

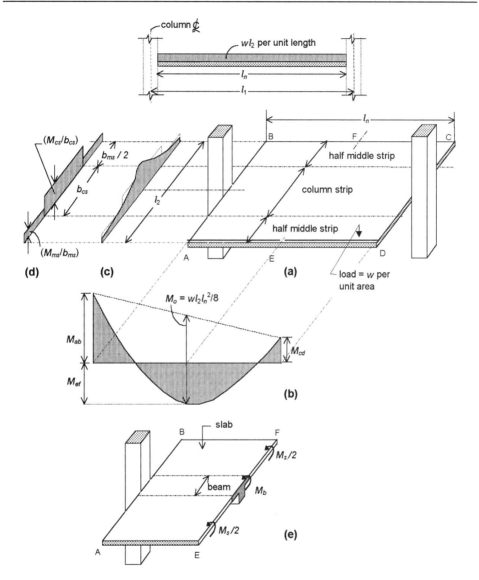

14.2 DESIGN PROCEDURES IN CSA CODE

CSA A23.3-94(13.6) permits the design of two-way slab systems by any procedure that satisfies the conditions of equilibrium and compatibility with the supports and the requirements of strength and serviceability. This general provision enables the designer to use any rational method that satisfies principles of structural mechanics and all safety and serviceability criteria. Apart from the specific method for design of rectangular slabs supported on four sides by walls or stiff beams described in Chapter 13, the Code identifies four specific methods, namely, *Elastic Plate Theory*, *Plastic Analysis*, *Elastic Frame Analogy* and *Direct Design Method*. General requirements and limitations for these procedures are also given. Two of these methods, namely the procedure of analysing slab systems as part of elastic frames (Code Cl. 13.9 – formerly known as the *Equivalent Frame Method* and herein after termed *Elastic Frame Method*) and the *Direct Design Method* (Cl. 13.10) are described in detail in the Code for design of regular two-way slab systems under gravity loads

The two methods detailed in the Code are applicable for *regular two-way slab systems* (see definition below) reinforced for flexure in more than one direction, such as flat plates, flat slabs, waffle slabs, and two-way slabs with beams along column lines. These methods are based on the analyses of results of extensive series of tests, comparison with theoretical results obtained using the theory of flexure for plates, and design practices used successfully in the past for two-way slabs deforming with nearly equal curvature in the two directions. Some of the background material used in the development of the design procedures recommended in the Code is presented in Refs. 14.1-14.3. Although the fundamental principles behind these design methods (viz. elastic frame analysis) are quite general, the detailed design rules for the two methods in Section 13 of the Code are applicable, in the case of slabs with beams, only when the beams are located along the edges of the panel and when the beams are supported by columns (or other essentially non-deflecting supports) at the corners of the panel. The methods are not applicable to slabs reinforced for flexure in one direction only (one-way slabs).

A *regular two-way slab system* is defined as one consisting of approximately rectangular panels, supporting primarily uniform gravity loading and meeting the following geometric limitations (Cl. 13.1):
    (a) in a panel the ratio of longer to shorter span, centre-to-centre of supports, shall not be greater than 2.0;
    (b) for slab systems with beams between supports, the relative stiffness of beams in the two directions $(\alpha_1 l_2^2)/(\alpha_2 l_1^2)$ shall not be less than 0.2 nor greater than 5.0;
    (c) column offsets shall not be greater than 20% of the span (in the direction of the offset) from either axis between centre lines of successive columns; and
    (d) the reinforcement is placed in an orthogonal grid.
These limitations are to ensure that the curvatures in the two directions will be approximately equal ensuring two-way action.

CHAPTER 14 DESIGN OF TWO-WAY SLAB SYSTEMS

In both the Direct Design Method and the Elastic Frame Method, the portion of the slab system including beams and supports along a column line and bounded by the centre lines of the panels on each side is termed a *design strip*, which is again divided into *column strips* and *middle strips* (Fig. 14.3a). A column strip is defined as the portion of the design strip, along the column line and having a width equal to the lesser of $0.25l_1$ or $0.25l_1$ on each side of the column centreline, and includes within this width any drop panel or beam along the column line. Here, l_1 is the length of span in the direction moments are to be determined and l_2 (l_{2a}, the average of l_2 if different for panels on either side of the column line) is the length of span transverse to l_1, both measured centre to centre

Fig. 14.3 Column strip and middle strip

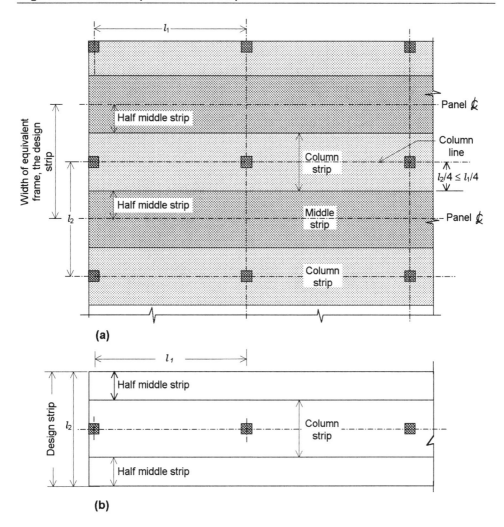

of the supports. (In general, subscript 1 identifies parameters in the direction moments are being determined, and subscript 2 those in the transverse direction.) The middle strip is the portion of the design strip outside the column strips. Within a panel, the full middle strip is the part of the slab bounded by the two column strips in it. In considering an elastic frame along a column line, the slab width, l_2, consists of two half middle strips flanking the column strip (Fig. 14.3b). The definition of *column strip* and *middle strip* used in this chapter are different from those used in Chapter 13 with respect to slabs with stiff supports on all the four sides. When monolithic (or fully composite) beams are used along column lines, the effective beam sections are usually T-shaped, and are considered to include a portion of the slab on each side of the beam extending a distance equal to the projection of the beam above or below the slab, whichever is greater, but not greater than four times the slab thickness, as shown in Fig. 14.4. In cases where the beam stem is short, the T-beam may be assumed to have a width equal to that of the support (Fig. 14.4c). In the matter of specific provisions and details, the Code differentiates between *two-way slabs without beams* and *two-way slabs with beams between all supports*. Slabs without beams include slabs with beams only along discontinuous edges. To classify as a *slab with beams*, there must be beams between all supports (columns).

The two methods differ only in the manner of determining the longitudinal distribution of bending moments in the slab-beam member between

Fig. 14.4 Definition of beam section

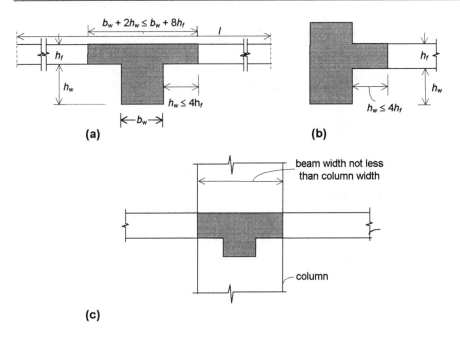

the negative and positive moment sections (Fig. 14.2b). While the Direct Design Method uses moment coefficients to determine the longitudinal distribution, the Elastic Frame Method uses an elastic partial frame analysis. The procedure for the lateral distribution of the factored moments between column strip and middle strip for regular slabs, without beams, and between the beam part and the slab part when beams are provided between all columns, is the same for both design methods. So also is the procedure for shear design. Both methods require the values of a few relative stiffness parameters to obtain the longitudinal and transverse distributions of factored moments in the design strips. Hence, it is necessary to assume, initially, the gross section dimensions of the floor system and the columns (or at least the relative stiffness values). If, on the basis of the analysis and subsequent design of sections, these dimensions have to be modified so as to affect the relative stiffness values significantly, the analysis and design have to be revised. The two design procedures are described in detail in Sections 14.5 and 14.9.

In the design for flexure, the *critical section* for negative moment at an interior support shall be taken at the face of rectilinear supports, but not at a distance greater than $0.175l_1$ from the centre of the column. At exterior supports provided with brackets or capitals, the critical section for negative moment in the span perpendicular to the edge shall be taken at a distance from the face of the supporting element not greater than one-half of the projection of the bracket or capital beyond the face of the supporting element. In locating these sections, circular or regular polygonal supports are to be treated as square supports having the same area.

14.3 MINIMUM THICKNESS OF TWO-WAY SLABS FOR DEFLECTION CONTROL

In two-way slabs, minimum required slab thickness is generally controlled by serviceability requirements, particularly the need to control deflections and crack width. In two-way slabs, without beams between column supports, the shear strength requirement of the slab at column supports is another critical factor, which may control the required slab thickness. In particular, in flat plates and at exterior column supports, which have to transfer the unbalanced moment between slab and column, the shear stresses are likely to be high (Section 14.4). Maximum permissible deflection limits are given in Table 9.2 of the Code.

Computation of deflections of two-way slab systems is quite complex. For regular two-way slab systems, CSA A23.3-94(13.3) recommends empirical equations (Eqs. 14.1 to 14.3 below) for the minimum overall thickness of slabs necessary for the control of deflections. If these minimum thickness requirements are satisfied, deflections need not be computed. These equations are also applicable for two-way slabs supported on stiff beams, considered in Chapter 13. Equations 14.1 to 14.3 were established based on results of extensive tests on floor slabs, and have been supported by past experience with this type of construction and normal values of uniform gravity loading. However, these thicknesses may not be the most economical thicknesses in all cases, and may even be inadequate for slabs with large live to dead load ratios. In any case, the slab thickness is not less than 120 mm, as required by the Code.

For flat plates and slabs with column capitals, the minimum overall thickness of slab, h_s, is:

$$h_s \geq \frac{l_n(0.6 + f_y/1000)}{30} \tag{14.1}$$

However, at discontinuous edges, an edge beam is provided with a stiffness ratio, α, of not less than 0.80, failing which the thickness required by Eq.14.1 is increased by 10 per cent. Even in slabs without beams between supports, it is good practice to provide an edge beam, which apart from the favourable effect on minimum thickness requirement, helps in stiffening the discontinuous edge, increasing the shear capacity at the critical exterior column support, and supporting exterior walls, cladding, etc. With such an edge beam, Eq. 14.1 leads to a minimum thickness of $l_n/30$ for reinforcement having $f_y = 400$ MPa. For slabs with drop panels, the minimum thickness of slab is:

$$h_s \geq \frac{l_n(0.6 + f_y/1000)}{30\left[1 + \left(\dfrac{2 \times d}{l_n}\right)\left(\dfrac{h_d - h_s}{h_s}\right)\right]} \tag{14.2}$$

Where $x_d/(l_n/2)$ is the smaller of the values determined in the two directions, and x_d is not greater than $l_n/4$, and $(h_d - h_s)$ is not larger than h_s.

For slabs with beams between all supports, the minimum thickness of is:

$$h_s \geq \frac{l_n(0.6 + f_y/1000)}{30 + 4\beta\alpha_m} \tag{14.3}$$

where α_m is not greater than 2.0. The limit on α_m is imposed to ensure that with heavy beams all around the panel, the slab thickness does not become too thin. In the above equations,

l_1 = length of span in the direction that moments are being determined, centre-to-centre of supports;
l_2 = length of span transverse to l_1, centre-to-centre of supports;
l_n = length of clear span in long direction;

f_y = specified yield strength of reinforcement, in MPa;
h_s = overall thickness of slab, mm;
h_d = overall thickness of drop panel, mm;
x_d = dimension from face of column to edge of drop panel, mm;
α_m = average value of α for beams on the four sides of the panel;
α = the beam stiffness ratio defined as the ratio of moment of inertia of beam section to moment of inertia of a width of slab bounded laterally by the centrelines of the adjacent panels, if any, on each side of the beam, $= I_b/I_s$;
α_1, α_2 = α in directions of l_1 and l_2, respectively; and
β = ratio of clear span in long to short direction.

14.4 TRANSFER OF SHEAR AND MOMENTS TO COLUMNS

Lateral loads and unbalanced gravity loads cause transfer of moments between the slab system, supporting columns, and walls built integrally with the slab. This moment is computed by the frame analysis when the Elastic Frame Method is used. When the Direct Design Method is used, the unbalanced moment at an interior support due to unbalanced gravity loads (partial loadings in adjacent spans) is obtained using an empirical equation, and the moment at the exterior support is taken as equal to the exterior negative factored moment in the exterior span. The total unbalanced moment at a support must be resisted by the columns above and below the slab in proportion to their stiffnesses (Fig. 14.5a).

In slabs without beams along the column line, the transfer of the moment from the slab to the columns requires special consideration. Based on data in Ref. 14.4, CSA A23.3-94(13.11.2) specifies that a portion, M_{fb}, of the total unbalanced moment, M_f, be considered as transferred to the columns by flexure; and the balance, M_{fv}, through shear and torsion, as given in Eqs. 14.4-14.6, and shown in Fig. 14.5.

$$M_{fv} = \gamma_v M_f \qquad (14.4)$$

$$M_{fb} = (1-\gamma_v)M_f \qquad (14.5)$$

$$\gamma_v = 1 - \frac{1}{1+\frac{2}{3}\sqrt{\frac{b_1}{b_2}}} \qquad (14.6)$$

432 REINFORCED CONCRETE DESIGN

Fig. 14.5 Transfer of unbalanced moments to columns

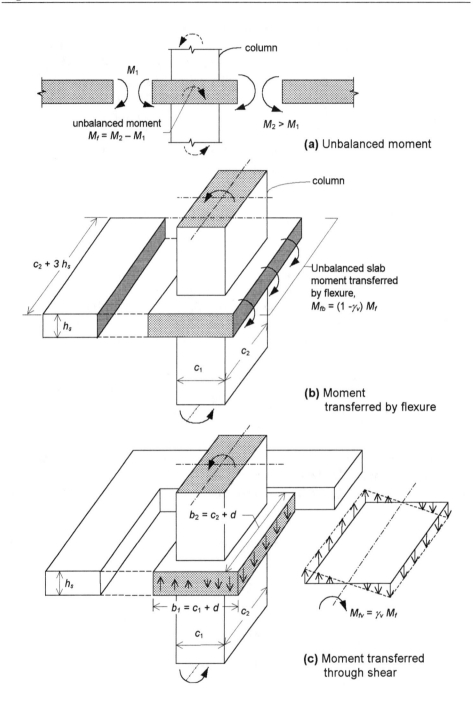

(a) Unbalanced moment

(b) Moment transferred by flexure

(c) Moment transferred through shear

where M_{fb} = portion of moment transferred by flexure;
M_{fv} = portion of moment transferred by shear;
M_f = total unbalanced moment transferred to columns;
b_1, b_2 = width of critical section for shear measured in the direction moments are determined, and transverse to it, respectively ($= c_1+d, c_2+d$);
c_1, c_2 = the size of equivalent rectangular column, capital or bracket, measured in the direction moments are being determined, and transverse to it, respectively; and
γ_v = fraction of unbalanced moment transferred by eccentricity of shear.

For square and round columns, $\gamma_v = 0.4$. Based on test results and experience, the width of slab considered effective in resisting the moment M_{fb} is taken as the width between lines a distance $1.5h_s$ on either side of the column or column capital (Fig. 14.5b). Therefore this strip needs adequate reinforcement to resist this moment. The section considered for moment transfer by eccentricity of shear stress is at a distance $d/2$ from the periphery of the column or column capital (Fig. 14.5c). The shear stresses introduced because of the moment transfer must be added to the shear stresses due to the vertical support reaction. Shear stresses are considered in Section 14.6. The detailing of the reinforcement for the transfer of moments from slab to the columns, particularly at the exterior column where the unbalanced moment is usually the largest, is critical to both the performance and the safety of flat slabs without edge beams.

14.5 DIRECT DESIGN METHOD

14.5.1 Limitations

The Direct Design Method is a simplified elastic frame analogy procedure, where the negative and positive design moments at critical sections in the slab-beam member and the unbalanced moments transferred to columns are computed using empirical moment coefficients. To ensure that the moments obtained by this procedure are not significantly different from those obtained by an elastic frame analysis, the Direct Design Method is restricted to *regular two-way slab systems* meeting the following additional limitations (Code (Cl. 13.10.1)):

1. There must be at least three continuous spans in each direction.
2. The successive span lengths, centre-to-centre of supports, in each direction must not differ by more than one-third of the longer span.

434 REINFORCED CONCRETE DESIGN

3. The factored live load must not exceed two times the factored dead load (the loads are assumed to be gravity loads, uniformly distributed over the entire panel).

14.5.2 Negative and Positive Factored Moments in Slab-Beam Member

(a) Total Factored Static Moment

In any given span l_1, the *total factored static moment* for the span considering the full width of design strip, M_o, (same as the simple span moment) is given by Eq. 14.7. From considerations of equilibrium and safety (see Fig. 14.2b), the absolute sum of the positive and average negative design moments must not be less than M_o.

$$M_o = w_f l_{2a} l_n^2 / 8 \qquad (14.7)$$

where
- w_f = factored load per unit area of the slab;
- l_1 = length of span in the direction moments are being determined;
- l_{2a} = the average of transverse spans on either side of the column line, same as the full width of the slab-beam member of the equivalent plane frame; and
- l_n = clear span in direction moments are determined, measured face-to-face of supports (columns, capitals, brackets, or walls), but not less than $0.65\, l_1$. Circular or regular polygonal supports are treated as equivalent square supports having the same area.

(b) Longitudinal Distribution of Factored Moments

The factored static moment is distributed between critical negative moment sections (taken at the face of rectangular supports) and positive moment sections (at or near midspan), as shown in Fig. 14.6. For interior spans, the distribution is as follows:

1. negative factored moment, $M_{neg} = 0.65 M_o$ (14.8)
2. positive factored moment, $M_{pos} = 0.35 M_o$ (14.9)

In the end span, the total factored static moment, M_o, is distributed according to the factors given in the Table in Fig. 14.6c. The Code also permits the modification of the negative and positive factored moments as determined above by a maximum of 15%, provided that the total static moment for the span

Fig. 14.6 *Longitudinal distribution of factored static moments – Direct Design Method*

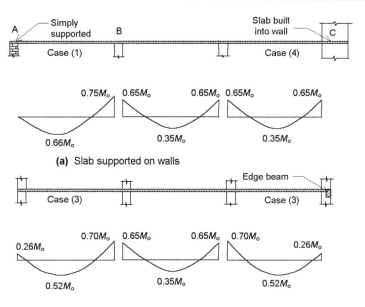

(a) Slab supported on walls

(b) Slab without beams between interior supports

Case	(1) Exterior edge unrestrained	(2) Slab with beams between all supports	(3) Slab without beams between interior supports	(4) Exterior edge fully restrained
Interior negative factored moment	0.75	0.70	0.70	0.65
Positive factored moment	0.66	0.59	0.52	0.35
Exterior negative factored moment	0	0.16	0.26	0.65

(c) Moment factors for end span

after such modification is not less than that given by Eq. 14.7. This is in recognition of the moment redistribution that can occur in slabs, which are highly indeterminate systems.

At interior supports, such as *B* in Fig. 14.6, the negative moment sections must be designed to resist the larger of the two negative factored moments determined for the spans on either side of the support. However, the moments

Fig. 14.7 Torsional member and restraint at exterior edge

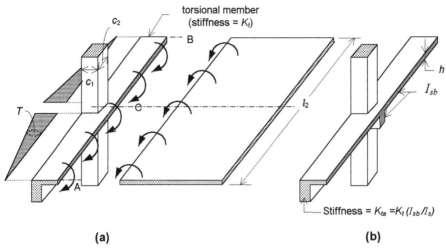

(a)   (b)

may first be modified as given above. At exterior supports, edge beams or edges of slabs acting as torsional members (see below) must be designed to resist, in torsion, their share of the exterior negative factored moments. For determining the torsion in the edge beam, it may be assumed conservatively that the negative factored moment at the exterior support is uniformly distributed over the width of the stab-beam member, l_2, with the torsion being zero at the panel centrelines, as shown in Fig. 14.7a. Alternatively, the distribution of torsion on the edge beam may be taken to be the same as the transverse distribution of the negative moment in the slab-beam member (Section 14.5.3).

If the maximum factored torque is less than 0.25 times the cracking torque, T_{cr}, of the torsional member, then torsion can be neglected (Cl.11.2.9.1). Torsional cracking can reduce the torsional stiffness, and this being an indeterminate system, may result in a reduction in the induced torsional moment. Hence the Code permits the reduction of the torque to $0.67T_{cr}$ if the calculated torque exceeds $0.67T_{cr}$, provided corresponding adjustment is made in the slab moments, maintaining the total factored static moment in the span as given by Eq. 14.7 (Fig. 14.8)

(c) Torsional Member and Stiffness
The torsional member (Figs. 14.7 and 14.24) is assumed to have a constant cross section throughout its length, consisting of the larger of (Code (Cl. 13.9.2.7)):
1. a portion of the slab having width equal to that of the column, bracket, or capital measured in the direction of l_1;
2. the portion of the slab as in 1 above, plus that part of the transverse beam above and below the slab; and
3. the transverse beam section as defined in Fig. 14.4.

Fig. 14.8 *Torsion in spandrel beam and adjusted span moments, due to torsional cracking*

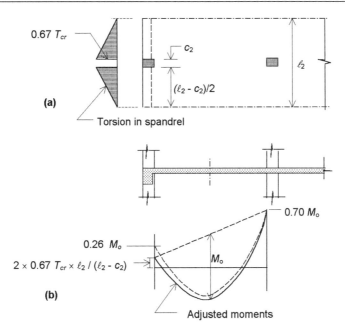

These three definitions for the section of the torsional member are illustrated in Fig. 14.9. The torsional stiffness, K_t, of the torsional member, needed for the Elastic Frame Method (for computing the equivalent column stiffness, K_{ec} - see Section 14.9), may be computed using the approximate expression:

$$K_t = \sum \frac{9E_c C}{l_t \left(1 - \dfrac{c_2}{l_t}\right)} \qquad (14.10)$$

where E_c = modulus of elasticity of concrete;
c_2 = size of rectangular or equivalent rectangular column, capital, or bracket, measured transverse to the direction in which moments are being determined;
C = cross-sectional property of the torsional member having the same relationship to the torsional rigidity as does the polar moment of inertia for circular cross sections; and
l_t = length of attached torsional member, equal to the smaller of l_{2a} or l_{1a} of spans adjacent to the joint, where l_{1a} is the average of adjacent spans in the l_1 direction, and l_{2a} is the average of adjacent spans in the l_2 direction.

Fig. 14.9 Definition of torsional member

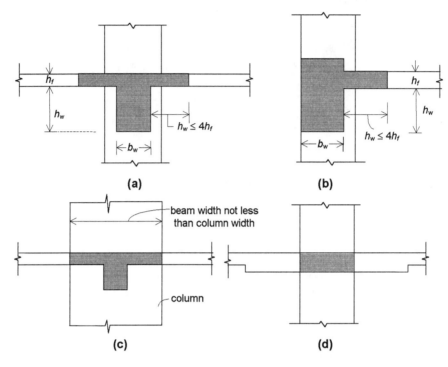

The summation in Eq. 14.10 applies to the case where there is a torsional member on each side of the column extending to the centrelines of the adjacent panels. For an exterior equivalent frame, the attached torsional member exists on one side of the column only. For the torsional member cross section, which may be T- or L-shaped, the computation of an exact value for C is very difficult. As an approximation, C is computed by subdividing the cross section into separate rectangular parts and summing up the C-values for each of the component rectangles using the approximate expression for C for a rectangular section. On this basis,

$$C = \sum \left(1 - 0.63\frac{x}{y}\right)\frac{x^3 y}{3} \qquad (14.11)$$

where x and y are the shorter and longer dimensions of a rectangular part (Fig. 14.10). Since this procedure will always give a value for C less than the correct value, the subdivision of the section into rectangles must be such as to maximise C. This is done by subdividing so as to minimise the length of the common boundaries.

When beams frame into the column in the direction moments are being determined, as shown in Fig. 14.7b, the value of K_t computed using Eq. 14.10

Fig. 14.10 Computation of torsional property C

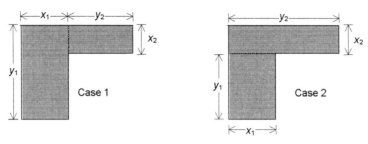

Use the larger C computed for Cases 1 and 2

leads to K_{ec} values which are too low. In such cases, the value of K_t given by Eq. 14.10 must be multiplied by the ratio of the moment of inertia of the slab with such beam to the moment of inertia of the slab without such beam, as follows:

$$K_{ta} = K_t \frac{I_{sb}}{I_s}$$

where
- K_{ta} = increased torsional stiffness due to beam in l_1 direction;
- I_s = moment of inertia of the slab having the full width l_2, excluding the portion of the beam stem extending above and below the slab, $= l_2 h_s^3 /12$ (Fig. 14.7b); and
- I_{sb} = moment of inertia of the slab-beam member of width l_2, including the portion of beam stem extending above and below the slab (full T-section in Fig. 14.7b).

14.5.3 Transverse Distribution of Moments at Critical Sections

Having obtained the factored negative and positive moments at all critical sections in the slab-beam member, the next step is to distribute these laterally across the design strip to the column strips and middle strips (or beam part and the slab parts) at each critical section.

Two-way slabs are highly statically indeterminate systems. Added to this, they are usually greatly under-reinforced. This makes slab sections capable of undergoing substantial increases in curvature at constant moment after yielding of the reinforcement at sections of peak moment. This amounts to the formation of a *plastic hinge* (Section 10.7.1) which in the case of slabs extends along its width, forming *yield lines*. The effect of this is a considerable degree of moment redistribution to sections of lesser moment. This inherent ability of the slab gives the designer considerable leeway in adjusting the moment field and designing the reinforcement accordingly, as long as static equilibrium conditions and strength and serviceability requirements are met. This flexibility is reflected in the Code provisions for apportioning the moment at critical sections between the column strip and the middle strip in regular slabs without beams between supports. In regular slabs with beams between all supports, the beam itself, because of its higher stiffness absorbs a substantial part of the moments at critical sections. Thus, the portion of the moment assigned to the beam is dependent on its stiffness ratio, α_1, and the span ratio l_2/l_1. The rules for transverse distribution of moments at critical sections are given by Code Clauses 13.12 and 13.13 and are summarised below. These are applicable to both the Direct Design Method (DDM) and Elastic Frame Method (EFM).

(a) Regular slabs without beams
This includes flat plates and slabs with drop panels and/or column capitals, which may or may not have beams along the discontinuous edges.

(i) Factored moments in column strips

The *column strips* are designed to resist the total negative or positive factored moments at the critical sections (given in Fig. 14.6 for the DDM or determined by analysis in EFM) multiplied by an appropriate factor within the following ranges:

    a) negative moment at an interior column -   0.6 to 1.00
    b) negative moment at an exterior column -   1.00
    c) positive moment at all spans -   0.50 to 0.70

Furthermore, at interior columns, at least one-third of the reinforcement for the total factored negative moment shall be located in a band with a width extending a distance of $1.5h_s$ from the sides of the column. Similarly, reinforcement for the total factored negative moment at exterior columns is placed within such a band width. This is to facilitate the transfer of the unbalanced moment to the column by flexure.

    By specifying a range of values for the portion of moments to be assigned to the column strip (except for the negative moment section at the exterior column), the Code gives more flexibility and freedom to the designer in selecting the slab reinforcement placing.

(ii) Factored Moments in Middle Strips

At all critical sections, the portion of the negative and positive factored moments not resisted by the column strip is assigned proportionately to the two half middle strips on either side of the column strip. Each full middle strip in a panel has moments assigned to its two halves from the

CHAPTER 14 DESIGN OF TWO-WAY SLAB SYSTEMS 441

Fig. 14.11 Factored moments in middle strips

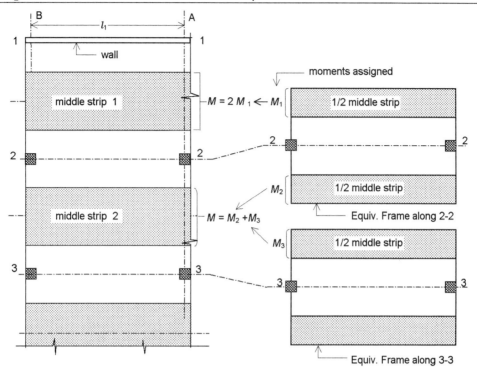

equivalent frames on either side, and is designed for the total moment, as shown in Fig. 14.11 for middle strip 2. The middle strip adjacent to and parallel with an edge supported by a *wall* must be designed for twice the moment assigned to its interior half portion forming part of the plane frame along the first row of interior supports (see middle strip 1 in Fig. 14.11).

(iii) Torsion in Spandrel Beam
From Fig. 14.7, the exterior negative moment in a slab is transferred to the columns, in part, through the edge beam (or spandrel beam) acting as a torsional member. The spandrel beam must be designed for this torsion if it exceeds 25% of the cracking torque, T_{cr}, of the beam (Cl.11.2.9.1). The distribution of torsion in the spandrel beam shown in Fig. 14.7, which assumes that the negative moment at exterior support is uniformly distributed over the width of the design strip, is conservative. An alternative is to calculate the torsion from the transverse distribution of the negative moment given at (i) and (ii) above. If the calculated torque exceeds $0.67T_{cr}$, the possibility of torsional cracking, reduction in torsional stiffness, and relaxation in the induced torque to a value equal

to $0.67T_{cr}$ can be allowed for (Cl.11.2.9.2, also Section 14.5.2b and Fig. 14.8). Similar action, although to a much lesser degree, develops in the transverse torsional member along interior supports.

(b) Regular slabs with beams between all supports
For regular slabs with beams between all supports, the design strip is divided into the beam part (as defined in Cl. 13.1 – see also Section 14.2 and Fig. 14.4) and the slab part which is the portion of the design strip outside the beam part.

(i) Factored moments in beams
The beam shall be designed to resist the following fractions of the total negative and positive factored moments at the critical sections, determined by analysis or determined as in Fig. 14.6:

(a) negative moment at interior columns
and positive moment in all spans $\dfrac{\alpha_1}{[1+(l_2/l_1)^2]}$ (14.12)

(b) negative moment at an exterior column   100 percent

In Eq. 14.12, α_1 is not larger than 1.0

In addition, the beam must also resist moments caused by concentrated or linear loads applied directly to the beams, including the weight of the beam stem. The reinforcement for factored negative moment in the beam at exterior supports is placed within a band with a width extending a distance $1.5h_s$ past the sides of the column or the edge of the beam web, whichever is larger.

(ii) Factored moments in slabs
At all critical sections in the design strip, the portion of the negative and positive factored moments not resisted by the beam is assigned to the slab parts outside the beam. For the negative moment at interior supports, the required slab reinforcement is uniformly distributed over the width of the slab. Slab reinforcement for positive moment may also be distributed uniformly.

14.5.4 Moments in Supporting Member and Effects of Pattern Loadings

As described in Section 14.4, columns and walls built monolithically with the slab and their joints must be designed to resist the unbalanced moments transferred from the slab. The moment transferred to the exterior column is the same as the negative factored moment at the exterior support, computed using the factors given in Fig. 14.6c in the DDM and obtained by the frame analysis in the EFM. The negative factored moments at faces of interior supports computed by Eq. 14.8 and the factors in Fig. 14.6c correspond to the action of full factored live load plus dead load, whereas the maximum unbalanced moment at the support would occur under pattern loadings. The Code specifies that in the DDM, the factored unbalanced gravity load moment at an interior support may be computed using Eq. 14.13. The total unbalanced moment must be resisted by the joint and the supporting elements above and below the slab in proportion to their stiffnesses.

$$M_f = 0.07\left[(w_{df} + 0.5\ w_{lf})\ l_{2a}\ l_n^2 - w_{df}'\ l_{2a}'\ (l_n')^2\right] \times 10^{-3} \qquad (14.13)$$

where w_{df} is the factored dead load per unit area; w_{lf} is the factored live load per unit area; w_{df}', l_{2a}' and l_n' refer to the shorter of the two spans on either side of the support, assumed to carry factored dead load only; and w_{df}, w_{lf}, l_{2a} and l_n refer to the longer of the two spans, assumed to carry factored dead load plus one-half the factored live load.

At the exterior support, the torsional member (edge beam or edge portion of the slab (see Fig. 14.8)) must be designed to resist, in torsion, its share of the exterior negative factored moment.

14.6 SHEAR IN TWO-WAY SLABS

14.6.1 One-Way Shear (Beam Shear)

In two-way slabs supported on walls or relatively stiff beams along column lines (considered in Chapter 13), a portion of the slab load is transmitted in each of the two directions. The shear strength of the slab can be checked by considering typical slab strips, one metre wide, taken in each of the two directions and loaded with the appropriate share of the load transmitted in the respective directions. In this computation, the slab strip in each direction is considered as a beam one metre wide, and the procedure for shear design is identical to that for beams discussed in Chapter 6. This process was illustrated in Example 13.1. This kind of shear design in slabs, considering beam-type action, may be termed one-way shear or beam shear. The critical section for shear in one-way or beam shear is taken at a distance equal to the effective depth, d, away from the face of the support. In beam-supported slabs, in addition to designing the slab to resist the shear force in it, the beam must also be designed to resist the shear force caused by the loads in the area tributary to it (Fig. 14.12).

444 REINFORCED CONCRETE DESIGN

Fig. 14.12 Tributary areas for beam shear

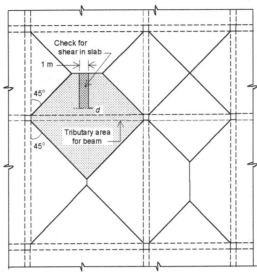

(a) Shear in slab systems with beams

In two-way slab systems with beams between supports, shear design considerations are mostly limited to one-way shear. The Code (Cl.13.5) recommends a procedure similar to the one discussed in the preceding paragraph. Thus, when the beams have stiffness ratios such that $\alpha_1 l_{2a}/l_1 \geq 1.0$, the Code specifies that the beam be designed for the shear caused by factored loads on tributary areas bounded by $45°$ lines drawn from the corners of the panels and the centrelines of the adjacent panels parallel to the long sides, (Fig. 14.12). When $\alpha_1 l_{2a}/l_1 < 1.0$, the beam shears are obtained by linear interpolation, assuming that beams carry no load at $\alpha_1 = 0$. This means that in such cases the beams framing into the columns transmit only part of shear from the panels to the column, and the balance of the shear is transmitted by the slab directly to the columns. In such cases, it is necessary to check the total shear strength of the slab-beam-column connection, to ensure that resistance to total shear occurring on a panel is provided. This involves checking the shear strength of the slab-beam part around the column perimeter as in the case of flat slabs (this is the limited instance where the concepts of two-way shear discussed in Section 14.6.2 has to be used in slabs with beams). Beams must also resist shears due to factored loads applied directly on the beams, in addition to the shear due to slab loads. Design procedures for shear in beams have been discussed in Chapter 6.

Apart from ensuring adequate shear strength for the beams, as described above, the slab must also be checked for one-way shear. For this, the slab shears

CHAPTER 14 DESIGN OF TWO-WAY SLAB SYSTEMS 445

may be computed based on the assumption of load distribution from the slab to the beam discussed above (Cl. 13.5.4). Calculations may be made for a one metre wide strip near mid-span, the critical section for shear being at a distance d from the face of the beam. Such a strip is identified in Fig. 14.12.

(b) One-way shear in two-way slabs without beams (Code Cl. 14.4.6)
Such slabs in the vicinity of columns are designed for both one-way shear and two-way shear. One-way shear design in such cases is discussed below, while two-way shear is considered later in Section 14.6.2. One-way shear design in this

Fig. 14.13 *Critical sections for one-way shear (wide-beam action)*

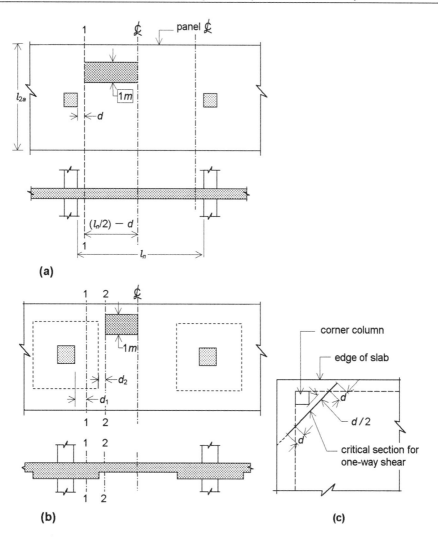

446 REINFORCED CONCRETE DESIGN

case is done by considering the slab as wide beam supported on and spanning between the columns. The critical section for one-way shear is at a distance, d, from the face of support and extending in a plane across the entire width of the design strip, (section 1-1 in Fig. 14.13a). The shear stress may be computed for the full slab width, l_{2a}, or for a typical strip of slab, one metre wide, shown shaded in Fig. 14.13a. Considering the one metre wide strip and assuming the shear to be zero at mid-span, the factored shear, V_f is given by

$$V_f = w_f \times 1 \times \left(\frac{l_n}{2} - d\right)$$

In the case of a slab with drop panels (Fig. 14.13b), a second section where one-way shear stress may be critical is at section 2-2, a distance d_2 from the edge of the drop panel, where d_2 is the effective depth of slab outside the drop.

In considering one-way shear strength of the slab in the vicinity of the corner column, the critical section is taken along a straight line having a minimum length and located not farther than $d/2$ from the corner column. In the case the slab cantilevers beyond the face of the corner column, the critical section may be extended into the cantilevered portion by a length not exceeding d, (Fig. 14.3c).

Since shear reinforcement is not generally used in slabs (except, in unusual cases, at the column head to take care of two-way shear), the factored shear force, V_f, in one-way action must not exceed the factored shear resistance of concrete, V_c. For one-way action, the value for V_c is given by Eq. 6.11. Slabs are generally safe in one-way shear, and, where necessary, the shear resistance is controlled by increasing the slab thickness.

14.6.2 Shear in Two-Way Action (Punching Shear or Perimeter Shear) - Two-Way Slabs without Beams

(a) Two-way shear and factored shear resistance

In a relatively thin flat slab supported on columns and *without beams* between columns (or the case of a large concentrated load applied on a small slab area, such as in footings supporting a column) and subjected to bending in two directions, shear failure may occur by punching through of the loaded area along a truncated cone or pyramid, with the surface sloping out in all directions from the perimeter of the loaded area, as shown in Fig. 14.14a. This type of shear is termed *two-way, perimeter* or *punching* shear (The term punching shear is no longer used widely, although it conveys the type of failure clearly). The shear

Fig. 14.14 Shear stresses in slabs

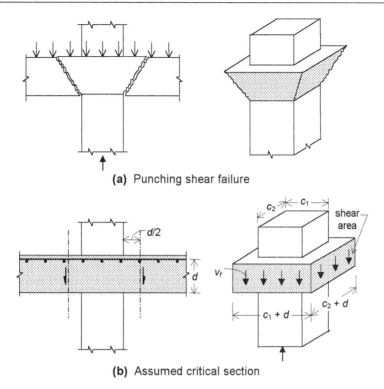

(a) Punching shear failure

(b) Assumed critical section

strength in the vicinity of the slab-to-column connections, especially when moments are transferred, is of primary importance in the satisfactory performance of flat plate and flat slab construction. The strength of two-way slabs in two-way shear is evaluated by comparing the factored shear stress, v_f, computed at a critical section to a factored shear stress resistance, v_r.

Based on an evaluation of extensive research studies, Ref. 14.5 has shown that the critical section governing the ultimate shear strength in two-way action of slabs and footings is along the perimeter of the loaded area. Furthermore, for square columns or loaded areas and for practical design situations, the ultimate shear stress at this section is a function of $\sqrt{f'_c}$ and the ratio of the side dimensions of the loaded area to the effective depth of slab. The effect of the latter parameter can be accounted for by assuming the shear area to be that at a vertical section located at a distance $d/2$ beyond the edge of the loaded area (Fig. 14.14b), in which case the ultimate shear stress will depend only on $\sqrt{f'_c}$. On this basis, the factored shear stress resistance in slabs without shear reinforcement for two-way action, computed at an assumed critical section located at $d/2$ from the periphery of the loaded area (for square or nearly square loaded areas, see Fig. 14.14b), was determined as $0.33\sqrt{f'_c}$ MPa. The shear stress resistance is also dependent on the shape and size of the loaded area (Ref. 14.6). With the introduction of $\phi_c = 0.6$ as the material resistance factor for concrete, CSA A23.3-94 (13.4.4) recommends that the factored shear stress resistance, v_r, of *slabs without shear reinforcement* for *two-way shear* be taken as the least of:

(a) $$v_r = v_c = 0.4\lambda\phi_c\sqrt{f'_c} \qquad (14.14)$$

(b) $$v_r = v_c = \left(1 + \frac{2}{\beta_c}\right)0.2\lambda\phi_c\sqrt{f'_c} \qquad (14.15)$$

(c) $$v_r = v_c = \left(\frac{\alpha_s d}{b_o} + 0.2\right)\lambda\phi_c\sqrt{f'_c} \qquad (14.16)$$

where α_s = 4 for interior columns and 3 for edge columns;
β_c = ratio of the long side to the short side of the concentrated load or reaction area; and
b_o = perimeter of critical section assumed located at distance $d/2$ from the periphery of loaded area, = $2(c_1 + c_2 + 2d)$ for the column in Fig. 14.14b.

Tests have shown (Ref. 14.6) that the shear resistance decreases with increasing rectangularity of the loaded area. The factor β_c in Eq. 14.15 takes into account this reduction. As β_c becomes very large (ie. $\beta_c \to \infty$, as in the case of a wall support), $v_c = 0.2\lambda\phi_c\sqrt{f'_c}$, which is the shear resistance for one-way shear applicable in such cases. Yet another factor that influences the shear resistance is the size of the loaded area. This is accounted for by the factor $\alpha_s d/b_o$ in Eq. 14.16. For example, for a square interior column having a size greater than $4d$, Eq. 14.16 gives a lower value for v_c than Eqs. 14.14 and 14.15.

In general, the factored shear force, V_f, causing punching shear may be computed as the net upward column reaction (or other concentrated load) less the downward load within the area of slab enclosed by the perimeter of the critical section. When a general frame analysis is made, as for the EFM, the column reaction is obtained from the analysis. In the DDM, and for preliminary design purposes, V_f may be computed as the total design load acting on the shaded area shown in Fig. 14.15.

Fig. 14.15 *Critical sections and loading for punching shear*

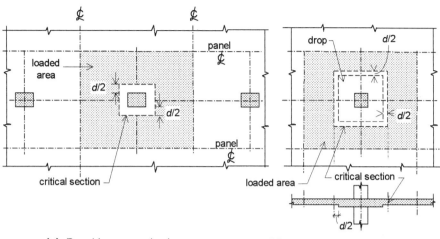

(a) Punching around column (b) Punching around drop panel

(b) Shear stress due to factored shear force

The shear stress due to factored shear force, V_f, acting at the centroid of the critical shear section is assumed to be uniformly distributed over the section. Thus, the factored shear stress, v_f, due to shear force (shear due to moment transfer is considered later in Section 14.6.3) is given by:

$$v_f = \frac{V_f}{b_o d} \qquad (14.17)$$

where b_o is the perimeter of the critical section for two-way shear.

For computing v_f, the critical section is considered as the section perpendicular to the plane of the slab and located so that its perimeter, b_o, is a minimum and approaches no closer than $d/2$ to the perimeter of the concentrated load or reaction area. An illustration of this for a non-rectangular loaded area and the associated value of β_c is shown in Fig. 14.16 (Ref. 14.7). Where there is any change in slab thickness, a section in the thinner slab located at $d/2$ from the face of the thicker slab must also be investigated (Fig. 14.15b). For a square or rectangular loaded area, the critical section is assumed to have four straight sides (Fig. 14.14b, 14.15).

Where openings in slabs are located at a distance of less than ten times the slab thickness from a concentrated load or reaction area, or within column strips in slabs without beams, the portion of the perimeter of the critical section, which is enclosed by straight lines projecting from the centre of the loaded area and tangent to the boundaries of the opening, must be considered ineffective in

Fig. 14.16 Critical sections and β_c for non-rectangular loaded area (ref. 14.7)

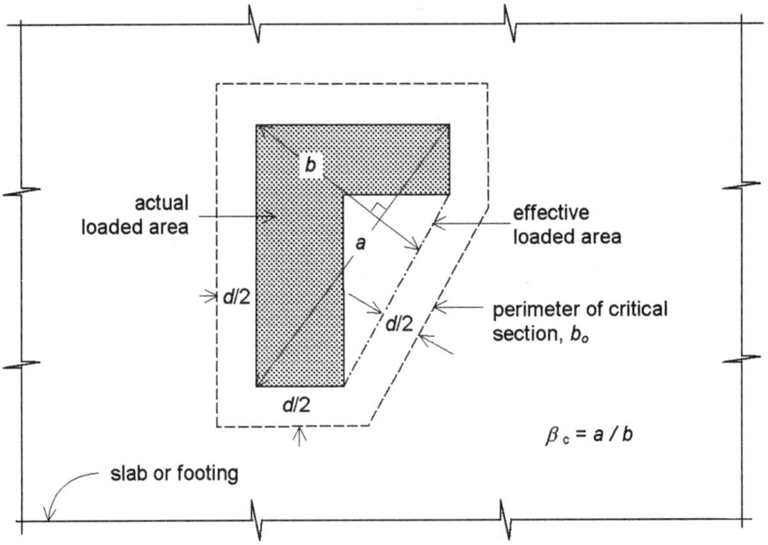

[Figure: diagram showing actual loaded area, effective loaded area, perimeter of critical section b_o, with dimensions $d/2$, and $\beta_c = a/b$, in a slab or footing]

computing the shear stress. Some examples of the effective portions of critical sections for slabs with openings are illustrated in Fig. 14.17.

The factored shear stress resistance, v_r, must be equal to or greater than the maximum factored shear stress, v_f, due to the factored shear force and the unbalanced moments (Section 14.6.3 below). The stress v_f is determined for full load on all spans and any other patterns of loading which might result in larger stresses.

$$v_r \geq v_f \qquad (14.18)$$

Shear reinforcement design for two-way shear is presented in (d) below.

(c) Shear Stress Including Transfer of Moments

The vertical shear stress, v, in the slab at the critical section around a column (Figure 14.14b) transferring a design shear force, V_f, is given by Eq. 14.17 as:

$$v = \frac{V_f}{A_c} = \frac{V_f}{b_o d} \qquad (14.17)$$

where $A_c = b_o d$ is the area of concrete resisting shear at the critical section. As described in Section 14.4, a portion, M_{fv}, of the total unbalanced moment transferred between slab and column is considered transferred by the eccentricity of the shear about the centroid of the critical section (Fig. 14.5c). Combining the

Fig. 14.17 Effective perimeter for slabs with openings (ref. 14.7)

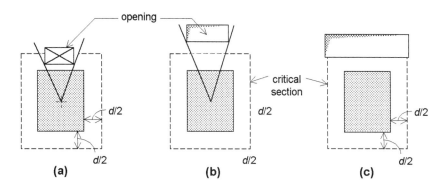

effects of both V_f and M_{fv}, and assuming the shear stress to vary linearly about the centroid of the critical section, the shear stress distribution at the critical section is as illustrated in Fig. 14.18. The maximum shear stress is given by:

$$v_{AB} = \frac{V_f}{A_c} + \frac{M_{fv}}{J_c} c_{AB} \qquad (14.19)$$

where A_c = area of concrete at the critical section = $b_o d$
J_c = property of the critical section analogous to the polar moment of inertia; and
c_{AB} = distance from the centroidal axis, yy, of critical section to the face AB of the critical section, respectively.

Expressions for J_c are also given in Fig. 14.18 for two cases. Equations for J_c for other cases are presented in Ref. 14.8. Including the shear stress due to moment transfer also, the requirement in Eq. 14.18 reduces to:

$$v_f = \frac{V_f}{b_o d} + \frac{M_{fv}}{J_c} c_{AB} \leq v_r \qquad (14.20)$$

When there are unbalanced moments about both principal axes, as in the case of a corner column, Eq. 14.20 takes the form:

$$v_f = \frac{V_f}{b_o d} + \left[\frac{M_{fv}}{J} e\right]_{xx} + \left[\frac{M_{fv}}{J} e\right]_{yy} \leq v_r \qquad (14.21)$$

where e is the distance from centroid of the critical section to the extreme point where the stress is being calculated. For *members without shear reinforcement,*

Fig. 14.18 Combined shear due to punching and transfer of moments

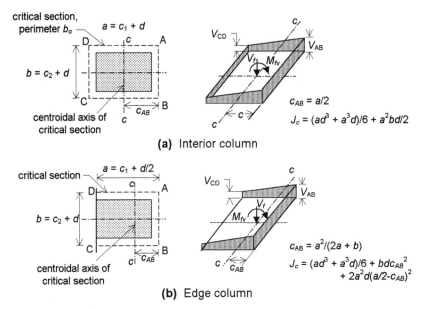

the value of v_r is given by Eqs. 14.14 – 14.16. The loading condition for individual maximum values of V_f and M_{fv} might not be the same in all cases. Thus, the combination of V_f, $(M_{fv})_{xx}$ and $(M_{fv})_{yy}$ to be used in Eq. 14.21 should be for a consistent loading that would produce the worst combined effect.

(d) Shear Reinforcement for Slabs Without Beams
If the factored shear stress, v_f, at the critical section given by Eq. 14.21 is greater than the factored shear stress resistance, v_r, for concrete without shear reinforcement given by Eqs. 14.14 to 14.16, shear reinforcement has to be provided. However, in slabs it is advisable to avoid shear reinforcements, for which the factored shear stress has to be reduced by increasing the slab thickness, providing column capitals and/or providing drop panels.

When required, shear reinforcement for two-way shear in slabs without beams may consist of headed shear reinforcement, stirrups (in slabs with depth ≥ 300 mm) or shear heads (Fig. 14.19). With shear reinforcement, the factored shear stress, v_f, cannot exceed the factored shear stress resistance, v_r, which is taken as $v_c + v_s$

$$v_f \leq v_r = v_c + v_s \qquad (14.22)$$

The shearing resistance should be investigated at the critical section and at successive sections more distant from the support. The reinforcement is extended to the section where $V_f / (b_o d)$ is not greater than $0.2\lambda\sqrt{f'_c}$.

Fig. 14.19 *Shear reinforcement for slabs at columns heads*

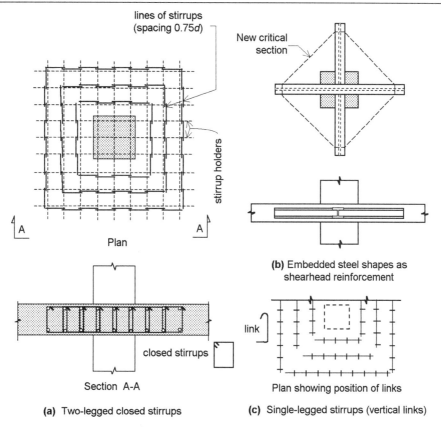

(a) Two-legged closed stirrups

(b) Embedded steel shapes as shearhead reinforcement

(c) Single-legged stirrups (vertical links)

Headed shear reinforcement (Code Cl.13.4.8) are vertical bars, anchored at each end by a plate or head bearing against the concrete and having an area of at least ten times the cross sectional area of the bar. These are located along concentric lines parallel to the perimeter of the column cross section, the first line of such reinforcement being at a distance $s/2$ from the column face, s being the spacing of such reinforcement. Spacing s is determined based on the shear stress resistance requirement as given below:

For headed shear reinforcement, the factored shear stress resistance of the concrete is $0.3\lambda\phi_c\sqrt{f'_c}$ and the factored shear stress, v_f, is less than or equal to $0.3\lambda\phi_c\sqrt{f'_c}$. Thus:

$$v_f \leq v_c + v_s \quad (14.22)$$

but

$$v_f \leq 0.8\lambda\phi_c\sqrt{f'_c} \quad (14.23)$$

$$v_c = 0.3\lambda\phi_c\sqrt{f'_c} \quad (14.24)$$

and
$$v_s = \frac{\phi_s A_{vs} f_{yv}}{b_o s} \tag{14.25}$$

where, A_{vs} = the area of headed shear reinforcement on a concentric line parallel to the perimeter of the column;
f_{yv} = specified yield strength of shear reinforcement; and
s = spacing of shear reinforcement measured perpendicular to perimeter b_o and shall be limited to:

$$s \leq 0.75\,d, \text{ when } v_f \leq 0.6\lambda\phi_c\sqrt{f'_c}$$
$$\leq 0.5\,d, \text{ when } 0.6\lambda\phi_c\sqrt{f'_c} < v_f \leq 0.8\lambda\phi_c\sqrt{f'_c}$$

When *stirrups* are used, in Eq. 14.22,

$$v_f \leq 0.6\lambda\phi_c\sqrt{f'_c}$$
$$v_c = 0.2\lambda\phi_c\sqrt{f'_c}$$

and v_s is as given by Eq. 14.25, with A_{vs} as the cross sectional area of stirrups.

Shear-head reinforcement consists of steel I-sections or channel shapes embedded within the slab and designed in accordance with Ref. 14.9. Such shearhead reinforcement effectively moves the critical section farther out from the column perimeter, as shown in Fig. 14.19c.

14.7 DESIGN EXAMPLE FOR DIRECT DESIGN METHOD

EXAMPLE 14.1

The plan and cross section of a two-way slab floor, with beams along column lines, are shown in Fig. 14.20. The service live load is 5 kN/m^2, and there is a superimposed dead load of 1.0 kN/m^2. Material properties are f'_c = 30 MPa and f_y = 400 MPa. The member sizes shown are all based on preliminary estimates. Determine the design moments in the various strips in the E-W direction for an edge and an interior panel. Use the Direct Design Method.

Fig. 14.20 Example 14.1

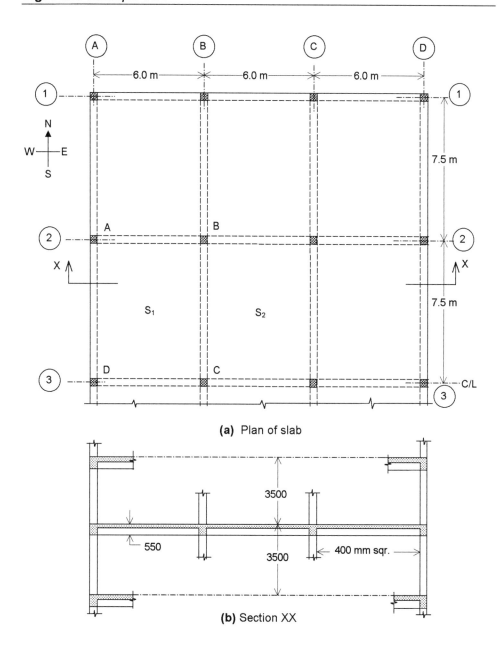

(a) Plan of slab

(b) Section XX

SOLUTION

The moments in panels $S1$ and $S2$ in Fig. 14.20 will be determined by considering the design strip along column line 2-2, isolated and shown in Fig. 14.21a.

1. Check whether the limitations for the use of the DDM (CSA A23.3-94(13.10)) are applicable. To qualify as a "regular" two-way slab system:
   - (i) Panels to be rectangular   OK
   - (ii) Primarily uniform gravity loading   OK
   - (iii) Ratio of longer to shorter span within a panel $= 7.5/6.0 = 1.25 < 2.0$,   OK
   - (iv) No offset columns.
   - (v) Reinforcement will be placed in an orthogonal grid.
   - (vi) Relative effective stiffness of the beams:
     $$\alpha_1 = I_{b1}/I_{s1}$$
     $$\alpha_2 = I_{b2}/I_{s2}$$
     Since the beam stem dimensions are the same along all edges of the panel, to check this limitation, it may be assumed $I_{b1} \approx I_{b2}$ (the effective beam section along exterior edge has flange only on one side, but this difference is neglected), so that,
     $$\alpha_1/\alpha_2 = I_{s2}/I_{s1} = l_1/l_{2a}$$

     ratio, $\dfrac{\alpha_1 l_2^2}{\alpha_2 l_1^2} \approx \dfrac{l_1}{l_{2a}} \times \dfrac{l_{2a}^2}{l_1^2} = \dfrac{l_{2a}}{l_1} = \dfrac{7.5}{6} = 1.25 < 5.0$

     Hence, qualifies as regular two-way slab. Furthermore, to qualify for DDM:
   - (vii) There are three or more spans in each direction.
   - (viii) Difference between successive spans $= 0 < 1/3 \times$ longer span
   - (ix) Dead load, assuming a slab thickness of 180 mm, is:
     $1 + 0.18 \times 2400 \times 9.8 \times 10^{-3} = 5.23$ kN/m^2
     Live load $= 5$ kN/m^2
     Ratio, factored live load/factored dead load
     $= (1.5 \times 5)/(1.25 \times 5.23) = 1.15 < 2$
     Also, all loads are gravity loads and uniformly distributed over the entire panels.

     All limitations are satisfied, and the Direct Design Method is applicable.

Fig. 14.21 Slab-beam system, example 14.1

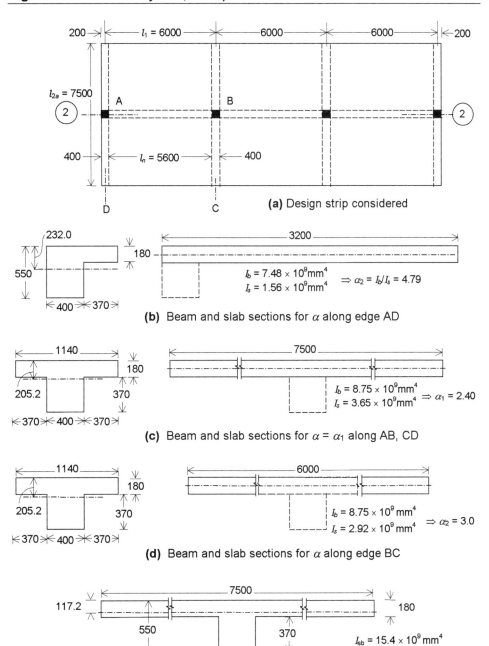

(a) Design strip considered

(b) Beam and slab sections for α along edge AD

(c) Beam and slab sections for $\alpha = \alpha_1$ along AB, CD

(d) Beam and slab sections for α along edge BC

(e) Slab–beam member along AB

2. Determine slab thickness for deflection control.
The equation for minimum thickness, Eq. 14.3, applicable for two-way slabs having beams between all supports, is in terms of the beam stiffness parameter, $\alpha = I_b/I_s$. The thickness requirement will be evaluated for the edge panel $S1$.

The beam and corresponding slab section for computing α for the edges of the panel are shown in Fig. 14.21 b-d, wherein the centroidal moments of inertia of the gross sections are also given.

For edge AD, $I_b = 7.48 \times 10^9 \text{ mm}^4$
$I_s = 3200 \times 180^3/12 = 1.56 \times 10^9 \text{ mm}^4$
$\alpha_2 = 7.48/1.56 = 4.79$

For edges AB, CD $I_b = 8.75 \times 10^9 \text{ mm}^4$
$I_s = 7500 \times 180^3/12 = 3.65 \times 10^9 \text{ mm}^4$
$\alpha_1 = 8.75/3.65 = 2.40$

For edge BC, $I_b = 8.75 \times 10^9 \text{ mm}^4$
$I_s = 6000 \times 180^3/12 = 2.92 \times 10^9 \text{ mm}^4$
$\alpha_2 = 8.75/2.92 = 3.00$

Average $\alpha = \alpha_m = (4.79 + 2 \times 2.4 + 3.0)/4 = 3.15$
But α_m is not greater than 2.0.

$$\beta = \frac{\text{larger clear span}}{\text{shorter clear span}} = \frac{7100}{5600} = 1.27$$

Minimum thickness required is (Eq. 14.3):
$$h_s \geq \frac{7100(0.6 + 400/1000)}{30 + 4 \times 1.27 \times 2.0} = 177 \text{ mm}$$

A slab thickness of 180 mm is adequate and will be used in the design.

3. Factored static moment
Factored loads are:
Live load on stab = $1.5 \times 5 = 7.50 \text{ kN/m}^2$
Dead load on slab, including slab weight, = $1.25 \times 5.23 = 6.54 \text{ kN/m}^2$
Weight of beam stem = $1.25 (0.4 \times 0.37 \times 2400 \times 9.8 \times 10^{-3}) = 4.35 \text{ kN/m}$

Factored static moment (ie, sum of the absolute values of the positive and average negative factored moments) in the E-W direction in a span is:

$$M_o = \frac{1}{8} [7.5 (7.50 + 6.54) + 4.35] \times 5.6^2 = 430 \text{ kN·m}$$

The weight of the beam stem has been included in the calculation of the static design moment. The Code (Cl. 13.13.2.3) specifies that moments caused by loads applied on the beam, and not considered in the slab design, be determined directly and the beam proportioned to resist such moments also. Thus, the weight of the beam stem could be considered separately as a direct loading on the beam. However, when such loading is not significant compared to the floor loading, as is usually the case with the self-weight of the stem, it is convenient and satisfactory to consider the former along with the slab loading, as is done in this example.

4. Longitudinal distribution of factored moments
In the interior span, the moment is distributed as follows (Eqs. 14.8-14.9 and Fig. 14.6):

Negative factored moment at face of support = $0.65 \times 430 = 280$ kN

Positive factored moment at midspan = $0.35 \times 430 = 151$ kN·m

In the exterior span, the distribution is as given in Fig. 14.6c for Case (2) - slab with beams between all supports.

Negative factored moment at exterior support,

$0.16 \times 430 = 68.8$ kN·m

Negative factored moment at exterior face of first interior support,

$0.70 \times 430 = 301$ kN·m

Positive factored moment

$0.59 \times 430 = 254$ kN·m

CSA A23.3-94(13.10.3.3) permits a 15 percent modification of the factored moments, but this is not applied in this example.
The longitudinal distribution of moments is shown in Fig. 14.22a.

5. Transverse distribution of moments within design strips.

At each critical section, part of the design moment is assigned to the beam and the balance to the slab portion. The fraction of the positive and interior negative factored moment to be resisted by the beam is given by Eq. 14.12, with α_1 taken not larger than 1.0 (here actual $\alpha_1 = 2.40$), as:

$$\frac{1.0}{1+(7.5/6)^2} = 0.39 = 39 \text{ percent}$$

Hence, the slab moment in these locations is 61 per cent.

At the exterior support, the beam is to be designed for 100 per cent of the exterior negative moment.

The actual factored moments in the beam part and the slab part are shown in Fig. 14.22b. The beam width is 1140 mm (Fig.14.21c), and each slab on either side of beam is 3180 mm wide.

6. Slab reinforcement

The design of flexural reinforcement in the E-W direction for the slab panels S1 and S2 is set up in Table 14.1. The effective depth of the beam, assuming 40 mm cover, No. 10 bar stirrups and No. 20 bars as

Fig. 14.22 Design moments in various elements in E–W direction

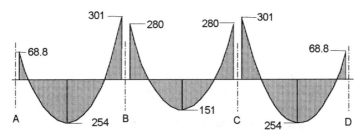

(a) Positive and negative design moments

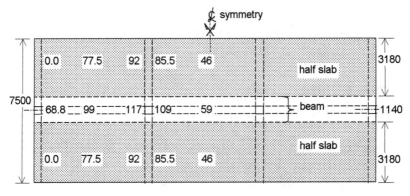

(b) Beam and slab moments (kN.m)

reinforcement, is 489 mm. For the larger moment at the interior support, $K_r = 1.22$, which corresponds to a reinforcement ratio of close to $0.16\,\rho_{max}$, as seen from Table 4.1, thus ensuring that the beam is under-reinforced throughout.

For the slab panel, the E-W direction being the shorter span, the reinforcement in this direction is placed as the outer layer of slab reinforcement. Assuming 20 mm cover and No. 10 bar reinforcement, the effective depth of the slab in the E-W direction is 154 mm. The largest moment per metre width in the slab is 28.9 kN·m at the first interior support. If a reinforcement ratio of $0.4\rho_{max}$ is selected (Table 4.1), $K_r = 2.80$ MPa, and the required effective depth is:

$$d = \sqrt{\frac{28.9 \times 10^6}{2.80 \times 1000}} = 102\,\text{mm}$$

The effective depth provided, 154 mm, is more than this, and the slab will be under-reinforced throughout. The slab thickness in this case is dictated by the minimum required for deflection control. A lower thickness can be used, if it can be shown by computations that deflections will not exceed the stipulated limits (Cl. 13.3.2).

The minimum reinforcement in the slab, for shrinkage and temperature stresses, is $0.002bh = 0.002 \times 1000 \times 180 = 360\,\text{mm}^2$ per metre width. Where the area of reinforcement computed for flexure is less than this minimum, the latter value is provided. The maximum bar spacing for slabs with beams is (Cl. 13.11.3) limited to $3h_s \leq 500$ mm. Where necessary, the number of bars must be increased to meet this maximum spacing requirement (not critical in the design in Table 14.1). The negative moment section at the interior support is designed to resist the larger of the two interior negative design moments determined for the spans on either side. There is a portion of slab of width 1140 mm over the beam that forms the flange of the beam. This part will have the beam reinforcement distributed over most of this width where the beam moment is negative. However, where beam moment is positive, this part of slab will have no reinforcement at all. It is recommended that the minimum reinforcement of No. 10 bars at 500 mm be provided in the slab in this region.

Details of flexural reinforcement for slabs are described in Section 14.10.

7. Transfer of moments to the columns
In this example, since beams are provided along column lines, transfer of unbalanced moment to the supporting columns is not critical. For

instance, the largest unbalanced moment is at the exterior support and equal to 68.8 kN·m. The beam is designed to resist 100% of this. The requirement for moment transfer through flexure (Eq. 14.5) is only 60%. (Note that the Code requirement in Clause 13.11.2 is concerned primarily with slab systems without beams.)

There is no negative moment assigned to the slab part at the exterior support. Hence, the slab at this section is provided with the minimum required reinforcement. The torsion transmitted to the edge beam by the slab is also zero as per the above moment assignment. In actual behaviour, some torque will be introduced, but this will be low. The torque being very low, the torsional stresses in the edge beam may be neglected. Similar stresses at interior supports will be even less.

Table 14.1 *Design of Slab Reinforcement in E-W Direction for Panels S1 and S2*

			Exterior Panel $S1$			Int. Panel $S2$	
			Ext. support	Mid-span	Int. support	Support	Mid-Span
(1) Beam design							
Moment, M_f		kN·m	68.8	99	117	109	59
$K_r = M_f \times 10^6/(400 \times 489^2)$		MPa	0.719	1.04	1.22	‡	0.617
ρ, from Table 5.4			0.00216	0.00322	0.00377	–	0.00185
$A_s = \rho \times 400 \times 489$		mm^2	422	630	737	–	362
$A_{s,min} = \dfrac{0.2\sqrt{f_c'}}{f_y} b_E h$			602*	–	–	–	602*
Number of No. 20 bars			2+ 1-No. 10	3	3	3	2+ 1-No. 10
(2) Slab design, 2 strips, each 3.18 m, full width							
Moment in 6.36 m		kN·m	nil	155	184	171	92
Moment/m width, M_f		kN·m/m	–	24.4	28.9	–	14.5
$K_r = M_f \times 10^6/(1000 \times 154^2)$		MPa	–	1.03	1.22	–	0.611
ρ, from Table 5.4			–	0.00319	–	0.00377	0.00183
$A_s = \rho \times 1000 \times 154$		mm^2	–	491	581	–	282
$A_s = 0.002 \times 1000 \times 180$		mm^2	360*	–	–	–	360*
A_s for full 6.36 m width		mm^2	2290	3123	3695	3695	2290
Number of No. 10 bars required			23	32	37	37	23
Spacing (≤ 500)		mm	270	190	170	170	270

* Minimum reinforcement requirement controls.
‡ Provide same reinforcement as at section to the left of support.

8. Shear in slab and beams
   For the beam, the parameter:
   $\alpha_1 l_{2a}/l_1 = 2.40 \times 7.5/6.0 = 3 > 1.0$

   For the N-S direction also, this parameter is greater than 1.0. Since all beams in this example have $\alpha_1 l_2/l_1$ values greater than 1.0, they are proportioned to resist the full shear from the loads on the respective tributary areas, and no part of this need be assigned to the slab to be resisted in two-way action (Code Cl. 13.5).

   For one-way shear, considering a one metre wide strip of slab, the factored shear force at the critical section at a distance d from the face of the beam is:

   $$V_f = (7.5 + 6.54) \times \left(\frac{5.6}{2} - 0.154\right) = 37.2 \text{ kN}$$

   $$V_c = 0.2\lambda\phi_c\sqrt{f'_c}\, b_w d$$
   $$= 0.2 \times 1.0 \times 0.6 \times \sqrt{30} \times 1000 \times 154 \times 10^{-3} = 101 \text{ kN}$$

   $V_f < V_c$ \hfill OK

The design of the beam for shear follows the procedures described and illustrated in Chapter 6.

Fig. 14.23 *Shear in slab and beams*

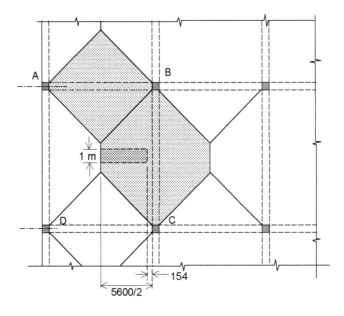

14.8 DESIGN AIDS

The discussions in preceding sections, and the previous design example show that the design of two-way slabs involves the computation of stiffness parameters and section properties. This is also true for the Elastic Frame Method, discussed in the next section. To assist in these computations several design aids in the form of charts and tables have been prepared, and these are presented in Ref. 14.8.

14.9 ELASTIC FRAME METHOD

This is the method used most frequently to determine the design moments in *regular* two-way slab systems. A brief description of the background of the method, with references to detailed descriptions, is presented in Ref. 14.2. The method (Code (Cl. 13.9)) simplifies the analysis of a three-dimensional building by subdividing it into a series of two-dimensional frame bents centred on column lines in both longitudinal and transverse directions, as shown in Fig. 14.1. However, unlike in the Direct Design Method, the longitudinal distribution of the moments in the slab-beam member is determined using an elastic analysis of an equivalent plane frame composed of equivalent *line members*, intersecting at member centrelines, such as in Fig. 10.1b. The analysis may be done considering an entire plane frame (Fig. 14.1b) or, for gravity loading, each floor may be analysed separately, considering the corresponding partial frame shown in Figs. 14.1c and 10.3c. Consistent with the usual approximations made in the analysis of continuous frames (Chapter 10) it may further be assumed in determining the moment at a given support or span that the slab-beam member is fixed at any support two panels distant, provided the slab continues beyond that point.

It was stated above that the elastic analysis is done for a plane frame composed of line members, like the one shown in Fig. 10.3c. There is a major difference between the behaviour of the actual partial frame composed of three-dimensional slab-beam-column system, shown in Fig. 14.1c, and of the corresponding partial skeletal plane frame consisting of line members (Fig. 10.3c). In the analysis of the skeletal plane frame, it is justifiably assumed that at a beam-column joint, all members framing into the joint undergo the same rotation. In contrast, in the frame shown in Fig. 14.1c, the column restrains the slab only locally. Thus, in a transverse section of the slab along a column support, such as section *ACB* in Fig. 14.7a, under gravity loading, the slab rotation at junction *C* will be equal to the column rotation; however, the slab rotation at points farther away, like at *A* and *B*, will be much larger. This reduced restraint offered by the column to the slab-beam member should be accounted for in the analysis. For gravity load analysis, this is usually done by reducing the effective stiffness of the column.

The accuracy in extending the elastic plane frame analysis to the three-dimensional slab-beam-column system will depend on the appropriate assigning of effective stiffnesses to all the members. The Code recommends two methods for modelling of member stiffnesses, namely, *prismatic* and *non-prismatic*. *Prismatic modelling* (Cl. 13.9.3) is especially suited when the analysis of the elastic plane frame is done using standard frame analysis programs based on the direct stiffness method. In this model, the moment of inertia of the slab-beam element is based on the gross area of the concrete outside the joint or column capital. If the moment of inertia varies outside the joint, as with the use of drop panels, the slab-beam element is modelled as a series of prismatic elements. The reduced rotational restraint offered by the column is accounted for by reducing the effective moment of inertia of the column by a factor given by:

$$\psi = \left[0.25 + 0.75\left(\frac{\alpha l_{2a}}{l_1}\right)\right]\left(\frac{l_{2a}}{l_1}\right)^2$$

but $0.3 \leq \psi \leq 1.0$, and $(\alpha l_{2a}/l_1)$ is not be taken greater than 1.0. With the stiffness of all members so determined, the analysis of the plane frame is fairly straightforward, and hence will not be detailed here.

In the *non-prismatic modelling* (Cl. 13.9.2), the frame is modelled using member centrelines and the actual variation of the gross moment of inertia along this line from centre of joint to centre of joint for all members. The reduction in the effective column stiffness (reduced rotational restraint) is achieved by using the concept of an attached torsional member, as discussed in Section 14.9.1. This method can be used for hand computation using 'moment distribution' procedure.

14.9.1 Elastic Frame Method with Non-Prismatic Modelling of Member Stiffness

It can be assumed that the load transfer between the slab system and the supporting columns takes place through three interconnected elements at each joint, namely, (1) the horizontal slab-beam member, (2) the vertical supporting elements (columns or walls), and (3) a transverse torsional member at the column supports. These three distinct parts are identified in Fig. 14.24a. To account for the increased flexibility of the slab-to-column connection, in the gravity load analysis of the frame, at all joints, the real columns above and below and the transverse torsional members are replaced by equivalent columns of stiffness, K_{ec}. With this modification at all supports, the frame that is analysed elastically is

Fig. 14.24 Elastic frame method

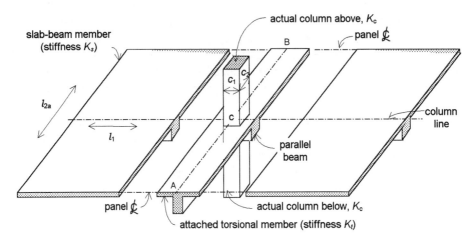

(a) Elements of equivalent frame at a connection

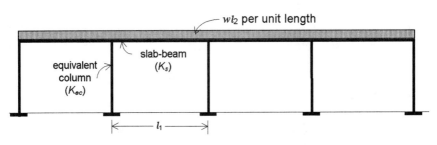

(b) Equivalent frame for analysis

a plane single storey multiple bay portal frame, shown in Fig. 14.24b. The procedure is described below in detail.

(a) Slab-Beam Member

The slab-beam member is bounded laterally by the centreline of the panel on each side of the column line. For an exterior frame, the slab-beam extends laterally from the edge to the centreline of the adjacent panel. Thus, the slab-beam includes drop panels for flat slabs, and beams along column lines for two-way slabs, if used. The moment of inertia of the slab-beam between the faces of supports (columns or column capitals) is computed on the basis of the gross cross-sectional area of concrete. From the face of the support to the centreline of the support, the moment of inertia of the slab-beam is taken as the moment of inertia at the face of the support divided by $(1 - c_2/l_{2a})^2$, where c_2 and l_{2a} are the column width and slab width, respectively, measured transverse to the direction

CHAPTER 14 DESIGN OF TWO-WAY SLAB SYSTEMS 467

in which moments are being determined (Fig. 14.24a). The magnification factor $1/(1 - c_2/l_{2a})^2$ allows for the increased cross-section of the slab-beam within the bounds of the supporting column cast monolithically with the slab. Variation of the moment of inertia along the span of the slab-beam for a typical case is shown in Fig. 14.25b. The stiffness factors, carry-over factors and fixed-end moments (required when the analysis is done by the moment distribution method) of the slab-beam are dependent on the variation of the moment of inertia along the span. These factors for slab-beams with common geometric and loading configurations are given in many standard Handbooks (such as the 1985 edition of Ref. 14.8). Factors for two typical cases are reproduced in Tables 14.2 and 14.3.

Fig. 14.25 Variations of moments of inertia along member axis

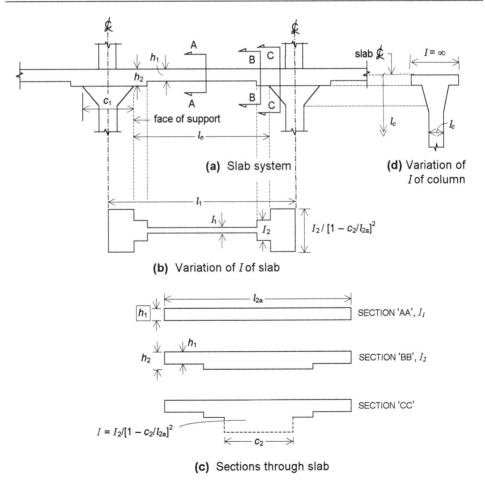

Table 14.2 Moment Distribution Constants for Slab-Beam Elements

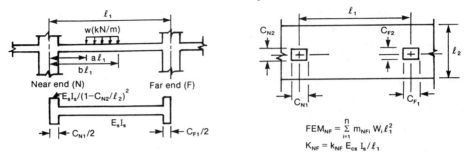

$$FEM_{NF} = \sum_{i=1}^{n} m_{NFi} W_i \ell_1^2$$

$$K_{NF} = k_{NF} E_{cs} I_s / \ell_1$$

C_{N1}/ℓ_1	C_{N2}/ℓ_2	Stiffness Factors k_{NF}	Carry Over Factors C_{NF}	Unif. Load Fixed end M. Coeff. (m_{NF})	Fixed end moment Coeff. (m_{NF}) for (b−a) = 0.2				
					a = 0.0	a = 0.2	a = 0.4	a = 0.6	a = 0.8
\multicolumn{10}{c}{$C_{F1} = C_{N1}$; $C_{F2} = C_{N2}$}									
0.00	—	4.00	0.50	0.0833	0.0151	0.0287	0.0247	0.0127	0.00226
0.10	0.00	4.00	0.50	0.0833	0.0151	0.0287	0.0247	0.0127	0.00226
	0.10	4.18	0.51	0.0847	0.0154	0.0293	0.0251	0.0126	0.00214
	0.20	4.36	0.52	0.0860	0.0158	0.0300	0.0255	0.0126	0.00201
	0.30	4.53	0.54	0.0872	0.0161	0.0301	0.0259	0.0125	0.00188
	0.40	4.70	0.55	0.0882	0.0165	0.0314	0.0262	0.0124	0.00174
0.20	0.00	4.00	0.50	0.0833	0.0151	0.0287	0.0247	0.0127	0.00226
	0.10	4.35	0.52	0.0857	0.0155	0.0299	0.0254	0.0127	0.00213
	0.20	4.72	0.54	0.0880	0.0161	0.0311	0.0262	0.0126	0.00197
	0.30	5.11	0.56	0.0901	0.0166	0.0324	0.0269	0.0125	0.00178
	0.40	5.51	0.58	0.0921	0.0171	0.0336	0.0276	0.0123	0.00156
0.30	0.00	4.00	0.50	0.0833	0.0151	0.0287	0.0247	0.0127	0.00226
	0.10	4.49	0.53	0.0863	0.0155	0.0301	0.0257	0.0128	0.00219
	0.20	5.05	0.56	0.0893	0.0160	0.0317	0.0267	0.0128	0.00207
	0.30	5.69	0.59	0.0923	0.0165	0.0334	0.0278	0.0127	0.00190
	0.40	6.41	0.61	0.0951	0.0171	0.0352	0.0287	0.0124	0.00167
0.40	0.00	4.00	0.50	0.0833	0.0151	0.0287	0.0247	0.0127	0.00226
	0.10	4.61	0.53	0.0866	0.0154	0.0302	0.0259	0.0129	0.00225
	0.20	5.35	0.56	0.0901	0.0158	0.0318	0.0271	0.0131	0.00221
	0.30	6.25	0.60	0.0936	0.0162	0.0337	0.0284	0.0131	0.00211
	0.40	7.37	0.64	0.0971	0.0168	0.0359	0.0297	0.0128	0.00195
\multicolumn{10}{c}{$C_{F1} = 0.5 C_{N1}$; $C_{F2} = 0.5 C_{N2}$}									
0.00	—	4.00	0.50	0.0833	0.0151	0.0287	0.0247	0.0127	0.0023
0.10	0.00	4.00	0.50	0.0833	0.0151	0.0287	0.0247	0.0127	0.0023
	0.10	4.16	0.51	0.0857	0.0155	0.0296	0.0254	0.0130	0.0023
	0.20	4.31	0.52	0.0879	0.0158	0.0304	0.0261	0.0133	0.0023
	0.30	4.45	0.54	0.0900	0.0162	0.0312	0.0267	0.0135	0.0023
	0.40	4.58	0.54	0.0918	0.0165	0.0319	0.0273	0.0138	0.0023
0.20	0.00	4.00	0.50	0.0833	0.0151	0.0287	0.0247	0.0127	0.0023
	0.10	4.30	0.52	0.0872	0.0156	0.0301	0.0259	0.0132	0.0023
	0.20	4.61	0.55	0.0912	0.0161	0.0317	0.0272	0.0138	0.0023
	0.30	4.92	0.57	0.0951	0.0167	0.0332	0.0285	0.0143	0.0024
	0.40	5.23	0.58	0.0989	0.0172	0.0347	0.0298	0.0148	0.0024
0.30	0.00	4.00	0.50	0.0833	0.0151	0.0287	0.0247	0.0127	0.0023
	0.10	4.43	0.53	0.0881	0.0156	0.0305	0.0263	0.0134	0.0023
	0.20	4.89	0.56	0.0932	0.0161	0.0324	0.0281	0.0142	0.0024
	0.30	5.40	0.59	0.0986	0.0167	0.0345	0.0300	0.0150	0.0024
	0.40	5.93	0.62	0.1042	0.0173	0.0367	0.0320	0.0158	0.0025
0.40	0.00	4.00	0.50	0.0833	0.0151	0.0287	0.0247	0.0127	0.0023
	0.10	4.54	0.54	0.0884	0.0155	0.0305	0.0265	0.0135	0.0024
	0.20	5.16	0.57	0.0941	0.0159	0.0326	0.0286	0.0145	0.0025
	0.30	5.87	0.61	0.1005	0.0165	0.0350	0.0310	0.0155	0.0025
	0.40	6.67	0.64	0.1076	0.0170	0.0377	0.0336	0.0166	0.0026
\multicolumn{10}{c}{$C_{F1} = 2 C_{N1}$; $C_{F2} = 2 C_{N2}$}									
0.00	—	4.00	0.50	0.0833	0.0151	0.0287	0.0247	0.0127	0.0023
0.10	0.00	4.00	0.50	0.0833	0.0150	0.0287	0.0247	0.0127	0.0023
	0.10	4.27	0.51	0.0817	0.0153	0.0289	0.0241	0.0116	0.0018
	0.20	4.56	0.52	0.0798	0.0156	0.0290	0.0234	0.0103	0.0013
0.20	0.00	4.00	0.50	0.0833	0.0151	0.0287	0.0247	0.0127	0.0023
	0.10	4.49	0.51	0.0819	0.0154	0.0291	0.0240	0.0114	0.0019
	0.20	5.11	0.53	0.0789	0.0158	0.0293	0.0228	0.0096	0.0014

Reproduced, with permission, from Ref. 14.9

Table 14.3 Moment Distribution Constants for Slab-Beam Elements, Drop Thickness = 0.50h

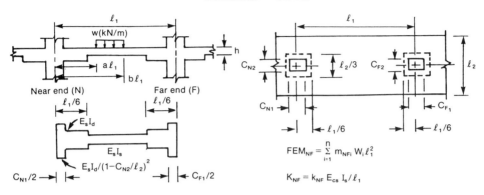

C_{N1}/ℓ_1	C_{N2}/ℓ_2	Stiffness Factors k_{NF}	Carry Over Factors C_{NF}	Unif. Load Fixed end M. Coeff. (m_{NF})	Fixed end moment Coeff. (m_{NF}) for (b−a) = 0.2				
					a = 0.0	a = 0.2	a = 0.4	a = 0.6	a = 0.8
$C_{F1} = C_{N1}$; $C_{F2} = C_{N2}$									
0.00	—	5.84	0.59	0.0926	0.0164	0.0335	0.0279	0.0128	0.0020
0.10	0.00	5.84	0.59	0.0926	0.0164	0.0335	0.0279	0.0128	0.0020
	0.10	6.04	0.60	0.0936	0.0167	0.0341	0.0282	0.0126	0.0018
	0.20	6.24	0.61	0.0940	0.0170	0.0347	0.0285	0.0125	0.0017
	0.30	6.43	0.61	0.0952	0.0173	0.0353	0.0287	0.0123	0.0016
0.20	0.00	5.84	0.59	0.0926	0.0164	0.0335	0.0279	0.0128	0.0020
	0.10	6.22	0.61	0.0942	0.0168	0.0346	0.0285	0.0126	0.0018
	0.20	6.62	0.62	0.0957	0.0172	0.0356	0.0290	0.0123	0.0016
	0.30	7.01	0.64	0.0971	0.0177	0.0366	0.0294	0.0120	0.0014
0.30	0.00	5.84	0.59	0.0926	0.0164	0.0335	0.0279	0.0128	0.0020
	0.10	6.37	0.61	0.0947	0.0168	0.0348	0.0287	0.0126	0.0018
	0.20	6.95	0.63	0.0967	0.0172	0.0362	0.0294	0.0123	0.0016
	0.30	7.57	0.65	0.0986	0.0177	0.0375	0.0300	0.0119	0.0014
$C_{F1} = 0.5C_{N1}$; $C_{F2} = 0.5C_{N2}$									
0.00	—	5.84	0.59	0.0926	0.0164	0.0335	0.0279	0.0128	0.0020
0.10	0.00	5.84	0.59	0.0926	0.0164	0.0335	0.0279	0.0128	0.0020
	0.10	6.00	0.60	0.0945	0.0167	0.0343	0.0285	0.0130	0.0020
	0.20	6.16	0.60	0.0962	0.0170	0.0350	0.0291	0.0132	0.0020
0.20	0.00	5.84	0.59	0.0926	0.0164	0.0335	0.0279	0.0128	0.0020
	0.10	6.15	0.60	0.0957	0.0169	0.0348	0.0290	0.0131	0.0020
	0.20	6.47	0.62	0.0987	0.0173	0.0360	0.0300	0.0134	0.0020
$C_{F1} = 2C_{N1}$; $C_{F2} = 2C_{N2}$									
0.00	—	5.84	0.59	0.0926	0.0164	0.0335	0.0279	0.0128	0.0020
0.10	0.00	5.84	0.59	0.0926	0.0164	0.0335	0.0279	0.0128	0.0020
	0.10	6.17	0.60	0.0907	0.0166	0.0337	0.0273	0.0116	0.0015

Reproduced, with permission, from Ref. 14.9

(b) Equivalent Column
The restraint to the slab-beam member at the support sections is offered by the flexural stiffness of the column and the torsional stiffness of the transverse torsional member (Figs. 14.7 and 14.24a). The actual column stiffness has to be modified to account for the torsional flexibility of the transverse torsional member. Referring to Fig. 14.24a, the angle of twist of the torsional member, and hence the rotation of the slab at support section, progressively increases from a minimum (equal to the flexural rotation of the column) at the column support, C, to a maximum at the outer ends A and B. To allow for this increased flexibility of the connection, the actual columns above and below the slab and the transverse beam (or torsional member) are replaced by an *equivalent column* whose flexibility (inverse of the stiffness) is equal to the sum of the flexibilities of the actual columns and the torsional member. The stiffness, K_{ec}, of this equivalent column is thus obtained from:

$$\frac{1}{K_{ec}} = \frac{1}{\Sigma K_c} + \frac{1}{K_t} \tag{14.26}$$

or
$$K_{ec} = \frac{(\Sigma K_c) K_t}{\Sigma K_c + K_t} \tag{14.27}$$

where ΣK_c = sum of flexural stiffnesses of the columns above and below the slab
K_t = torsional stiffness of the torsional member

Eq. 14.26 can be seen to approach the correct limits. When the torsional member is infinitely stiff, ($K_t = \infty$), the slab-beam member along the entire length AB undergoes the same rotation as that of the column, and in this case Eq. 14.26 rightly gives $K_{ec} = \Sigma K_c$. On the other hand, if $\Sigma K_c = \infty$, $K_{ec} = K_t$. In this case, the column, being infinitely stiff, does not rotate, but the slab still undergoes the same rotation as that of the torsional member (except for the width c_2 directly over the column). For the case $K_c = 0$, as for instance with a simple masonry support along the length AB of the slab, the transverse beam becomes torsionally unrestrained by the supporting member and can rotate freely as a rigid body. Hence, the slab-section AB is flexurally unrestrained and in this case Eq. 14.27 appropriately gives $K_{ec} = 0$. For the hypothetical case of $K_t = 0$ (transverse member with zero torsional stiffness), Eq. 14.27 again gives $K_{ec} = 0$. In this case also, the slab is flexurally unrestrained along the length AB (except for the short stretch of width c_2 directly over the column). When the slab is supported on a monolithic reinforced concrete wall, the slab rotation will be nearly uniform along the length AB and equal to the rotation of the wall. Here, the flexural stiffness of the wall replaces ΣK_c, and the torsional flexibility $1/K_t$ of the wall may be assumed zero, so that $K_{ec} = \Sigma K_{wall}$.

The concept of the equivalent column stiffness, K_{ec}, as defined in Eq. 14.26 and explained in the preceding paragraph is applicable for gravity load analysis only. For lateral load analysis, the effect of the flexibility of the connection from column-to-slab will be to reduce the effective slab stiffness.

In computing the equivalent column stiffness, K_{ec}, at a support using Eq. 14.26, K_t and K_c are the stiffnesses of the torsional member and of the column. The definition of the torsional member and the calculation of its torsional stiffness, K_t, were detailed in Section 14.5.2(c).

The length of an actual column, l_c, is measured centre-to-centre of floors. For computing the column stiffness, K_c, the moment of inertia of the column section outside the joint is based on the gross concrete section. The column section moment of inertia is assumed to be infinite for the portion of the column integral with the slab-beam, which is from the top to the bottom of the slab-beam at a joint. Variation of the column moment of inertia for the top half of a column with a capital is shown in Fig. 14.25d. Stiffness and carry-over factors of columns for the common configurations are presented in Table 14.4 page 472.

(c) Analysis Procedure

Having computed the stiffness properties of the individual members of the equivalent frame shown in Fig. 14.24b, the elastic analysis of this frame for determining the bending moments and shear forces in the members may be made using any of the conventional methods, such as moment distribution, slope-deflection, and matrix analysis. In the analysis illustration given in Example 14.2, the moment distribution procedure is used.

(d) Loading Pattern

When specific loading patterns are known, the frame is analysed for those patterns. When the live load is uniformly distributed, and the specified live load does not exceed three-quarters of the specified (unfactored) dead load, or the nature of live load is such that all panels will be loaded simultaneously, the Code recommends a single loading case of full factored loads (dead plus live) on all spans, for determining the maximum positive and maximum negative factored moments at all critical sections in all spans (Fig. 14.26a). For loading conditions other than the above, the maximum positive factored moment in a span is determined with three-quarters of full factored live load on that span and on alternate spans (plus factored dead load on all spans, Fig. 14.26b). For maximum negative moment at a support, three-quarters of full factored live load is placed on the two adjacent spans only with factored dead loads on all spans (Fig. 14.26c). However, in no case are the design moments to be taken as less than those occurring with full factored loads (dead plus live) on all spans. Use of

Table 14.4 Stiffness and Carry-Over Factors for Columns

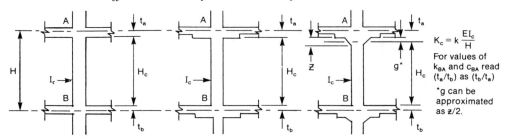

t_a/t_b	H/H_c	1.05	1.10	1.15	1.20	1.25	1.30	1.35	1.40	1.45	1.50
0.00	k_{AB}	4.20	4.40	4.60	4.80	5.00	5.20	5.40	5.60	5.80	6.00
	C_{AB}	0.57	0.65	0.73	0.80	0.87	0.95	1.03	1.10	1.17	1.25
0.2	k_{AB}	4.31	4.62	4.95	5.30	5.65	6.02	6.40	6.79	7.20	7.62
	C_{AB}	0.56	0.62	0.68	0.74	0.80	0.85	0.91	0.96	1.01	1.07
0.4	k_{AB}	4.38	4.79	5.22	5.67	6.15	6.65	7.18	7.74	8.32	8.94
	C_{AB}	0.55	0.60	0.65	0.70	0.74	0.79	0.83	0.87	0.91	0.94
0.6	k_{AB}	4.44	4.91	5.42	5.96	6.54	7.15	7.81	8.50	9.23	10.01
	C_{AB}	0.55	0.59	0.63	0.67	0.70	0.74	0.77	0.80	0.83	0.85
0.8	k_{AB}	4.49	5.01	5.58	6.19	6.85	7.56	8.31	9.12	9.98	10.89
	C_{AB}	0.54	0.58	0.61	0.64	0.67	0.70	0.72	0.75	0.77	0.79
1.0	k_{AB}	4.52	5.09	5.71	6.38	7.11	7.89	8.73	9.63	10.60	11.62
	C_{AB}	0.54	0.57	0.60	0.62	0.65	0.67	0.69	0.71	0.73	0.74
1.2	k_{AB}	4.55	5.16	5.82	6.54	7.32	8.17	9.08	10.07	11.12	12.25
	C_{AB}	0.53	0.56	0.59	0.61	0.63	0.65	0.66	0.68	0.69	0.70
1.4	k_{AB}	4.58	5.21	5.91	6.68	7.51	8.41	9.38	10.43	11.57	12.78
	C_{AB}	0.53	0.55	0.58	0.60	0.61	0.63	0.64	0.65	0.66	0.67
1.6	k_{AB}	4.60	5.26	5.99	6.79	7.66	8.61	9.64	10.75	11.95	13.24
	C_{AB}	0.53	0.55	0.57	0.59	0.60	0.61	0.62	0.63	0.64	0.65
1.8	k_{AB}	4.62	5.30	6.06	6.89	7.80	8.79	9.87	11.03	12.29	13.65
	C_{AB}	0.52	0.55	0.56	0.58	0.59	0.60	0.61	0.61	0.62	0.63
2.0	k_{AB}	4.63	5.34	6.12	6.98	7.92	8.94	10.06	11.27	12.59	14.00
	C_{AB}	0.52	0.54	0.56	0.57	0.58	0.59	0.59	0.60	0.60	0.61
2.2	k_{AB}	4.65	5.37	6.17	7.05	8.02	9.08	10.24	11.49	12.85	14.31
	C_{AB}	0.52	0.54	0.55	0.56	0.57	0.58	0.58	0.59	0.59	0.59
2.4	k_{AB}	4.66	5.40	6.22	7.12	8.11	9.20	10.39	11.68	13.08	14.60
	C_{AB}	0.52	0.53	0.55	0.56	0.56	0.57	0.57	0.58	0.58	0.58
2.6	k_{AB}	4.67	5.42	6.26	7.18	8.20	9.31	10.53	11.86	13.29	14.85
	C_{AB}	0.52	0.53	0.54	0.55	0.56	0.56	0.56	0.57	0.57	0.57
2.8	k_{AB}	4.68	5.44	6.29	7.23	8.27	9.41	10.66	12.01	13.48	15.07
	C_{AB}	0.52	0.53	0.54	0.55	0.55	0.55	0.56	0.56	0.56	0.56
3.0	k_{AB}	4.69	5.46	6.33	7.28	8.34	9.50	10.77	12.15	13.65	15.28
	C_{AB}	0.52	0.53	0.54	0.54	0.55	0.55	0.55	0.55	0.55	0.55
3.2	k_{AB}	4.70	5.48	6.36	7.33	8.40	9.58	10.87	12.28	13.81	15.47
	C_{AB}	0.52	0.53	0.53	0.54	0.54	0.54	0.54	0.54	0.54	0.54
3.4	k_{AB}	4.71	5.50	6.38	7.37	8.46	9.65	10.97	12.40	13.95	15.64
	C_{AB}	0.51	0.52	0.53	0.53	0.54	0.54	0.54	0.53	0.53	0.53
3.6	k_{AB}	4.71	5.51	6.41	7.41	8.51	9.72	11.05	12.51	14.09	15.80
	C_{AB}	0.51	0.52	0.53	0.53	0.53	0.53	0.53	0.53	0.53	0.52
3.8	k_{AB}	4.72	5.53	6.43	7.44	8.56	9.78	11.13	12.60	14.21	15.95
	C_{AB}	0.51	0.52	0.53	0.53	0.53	0.53	0.53	0.52	0.52	0.52
4.0	k_{AB}	4.72	5.54	6.45	7.47	8.60	9.84	11.21	12.70	14.32	16.08
	C_{AB}	0.51	0.52	0.52	0.53	0.53	0.52	0.52	0.52	0.52	0.51
4.2	k_{AB}	4.73	5.55	6.47	7.50	8.64	9.90	11.27	12.78	14.42	16.20
	C_{AB}	0.51	0.52	0.52	0.52	0.52	0.52	0.52	0.51	0.51	0.51
4.4	k_{AB}	4.73	5.56	6.49	7.53	8.68	9.95	11.34	12.86	14.52	16.32
	C_{AB}	0.51	0.52	0.52	0.52	0.52	0.52	0.51	0.51	0.51	0.50
4.6	k_{AB}	4.74	5.57	6.51	7.55	8.71	9.99	11.40	12.93	14.61	16.43
	C_{AB}	0.51	0.52	0.52	0.52	0.52	0.52	0.51	0.51	0.50	0.50
4.8	k_{AB}	4.74	5.58	6.53	7.58	8.75	10.03	11.45	13.00	14.69	16.53
	C_{AB}	0.51	0.52	0.52	0.52	0.52	0.51	0.51	0.50	0.50	0.49
5.0	k_{AB}	4.75	5.59	6.54	7.60	8.78	10.07	11.50	13.07	14.77	16.62
	C_{AB}	0.51	0.51	0.52	0.52	0.52	0.51	0.51	0.50	0.49	0.49
6.0	k_{AB}	4.76	5.63	6.60	7.69	8.90	10.24	11.72	13.33	15.10	17.02
	C_{AB}	0.51	0.51	0.51	0.51	0.50	0.50	0.49	0.49	0.48	0.47
7.0	k_{AB}	4.78	5.66	6.65	7.76	9.00	10.37	11.88	13.54	15.35	17.32
	C_{AB}	0.51	0.51	0.51	0.50	0.50	0.49	0.48	0.48	0.47	0.46
8.0	k_{AB}	4.78	5.68	6.69	7.82	9.07	10.47	12.01	13.70	15.54	17.56
	C_{AB}	0.51	0.51	0.50	0.50	0.49	0.49	0.48	0.47	0.46	0.45
9.0	k_{AB}	4.79	5.69	6.71	7.86	9.13	10.55	12.11	13.83	15.70	17.74
	C_{AB}	0.50	0.50	0.50	0.50	0.49	0.48	0.47	0.46	0.45	0.45
10.0	k_{AB}	4.80	5.71	6.74	7.89	9.18	10.61	12.19	13.93	15.83	17.90
	C_{AB}	0.50	0.50	0.50	0.49	0.48	0.48	0.47	0.46	0.45	0.44

Reproduced, with permission, from Ref. 14.9

Fig. 14.26 Arrangements of live loads for elastic frame analysis

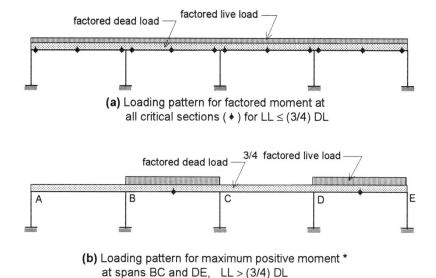

(a) Loading pattern for factored moment at all critical sections (♦) for LL ≤ (3/4) DL

(b) Loading pattern for maximum positive moment *
at spans BC and DE, LL > (3/4) DL

(c) Loading pattern for maximum negative moment *
at support B, LL ≥ (3/4) DL

* Note: factored moments shall not be taken less than those from loading pattern at (a)

three-quarters of full factored live load in the two pattern loadings given above is in recognition of the fact that the maximum negative and maximum positive live load moments in a span do not occur simultaneously, and that some inelastic redistribution of moments does occur in slabs which are usually very much under-reinforced.

The results of the frame analysis using centreline dimensions gives negative moments at the centrelines of the supports, from which the factored negative moments at critical sections must be determined, as shown in Fig. 14.27 (Cl. 13.9.5). The critical section of the slab-beam member for negative moment at interior supports is taken at the face of the support (column, capital, or bracket), but in no case at a distance greater than $0.175l_1$ from the centre of the column. At exterior supports with brackets or capitals, the critical section is taken at a distance from the face of the column not greater than one-half the projection of the bracket or capital beyond the face of the column (Fig. 14.27).

The moment in the equivalent column, obtained from the elastic analysis, must be distributed between the actual columns above and below the slab-beam, in proportion to their stiffness, and these moments used in their design (see Fig. 14.27c).

After determining the negative and positive factored moments at all critical sections for the slab-beam member, these factored moments must be distributed transversely at each critical section to the column strip and middle strip in the case of slabs without beams, and to the beam and slab in the case of slabs with beams between all supports. This transverse distribution procedure is the same as the one used in the Direct Design Method, and explained in Section 14.5.3. The design procedure for shear is as detailed in Section 14.6. A design example using the Elastic frame method is given in Section 14.11.

14.10 SLAB REINFORCEMENT DETAILS

14.10.1 General

The area of reinforcement must be that required for the factored moments at the critical sections, but not less than the minimum specified for shrinkage and temperature stresses (Code (Cl. 7.8)). The Code limits the spacing of bars at critical sections in solid slabs to a maximum of three times the slab thickness h_s, but not to exceed 500 mm except for the negative moment reinforcement within a limited band width directly over columns in slabs without beams (Cl. 13.11.3). Within such a band with a width extending a distance of $1.5h_s$ from the sides of the column (Section 14.5.3(a)(i)), the negative moment reinforcement spacing is limited to a maximum of 1.5 h_s, but ≤ 250 mm. This limitation is intended to ensure slab action, reduce cracking, and provide for the distribution of concentrated loads.

The flexural restraint along the exterior edge of the slab panel is subject to considerable variation and uncertainty. Depending on the torsional rigidity of the spandrel beam or slab edge, and its interaction with an exterior wall, if any, the flexural restraint at the slab edge may range from one of almost complete fixity to a condition of simple support. To provide for such uncertainties, special requirements are made in the Code for detailing flexural reinforcement in the exterior span. Thus, in these spans, all positive reinforcement perpendicular to the discontinuous edge is required to be extended to the slab edge and embedded for a length, straight or hooked, of at least 150 mm. Similarly all negative moment reinforcement perpendicular to the discontinuous edge must be anchored

Fig. 14.27 *Factored negative moments in slab-beam and column*

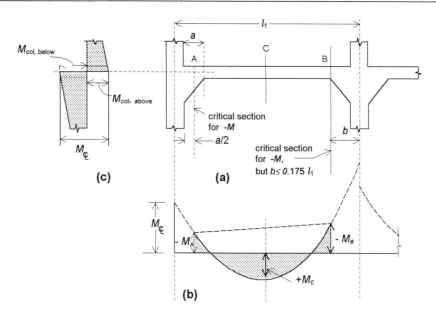

into the spandrel beam, wall, or column (or within the slab when no spandrel beam or wall is present), to provide development at the face of such support.

For slabs supported on beams along column lines having a value of α greater than 1.0, special corner reinforcement is required at exterior corners, as described in Example 13.1.

Location of bar cut-off (or bend points if used) must be based on the moment envelopes and the requirements for development length and bar extensions described in Section 5.10. When the Elastic Frame Method is used, the frame analysis gives the moment envelopes and locations of points of inflection.

14.10.2 Slabs Without Beams

In addition to the above requirements, for regular slabs without beams, and meeting the limitations given in Section 14.5.1, the Code prescribes bend point locations and minimum extensions for reinforcement, as shown in Fig. 14.28. However, for slab systems not meeting these limitations, the length of reinforcement must be determined by analysis, but must not be less than those prescribed in Fig. 14.28.

Fig. 14.28 *Minimum length of reinforcement-slabs without beams*

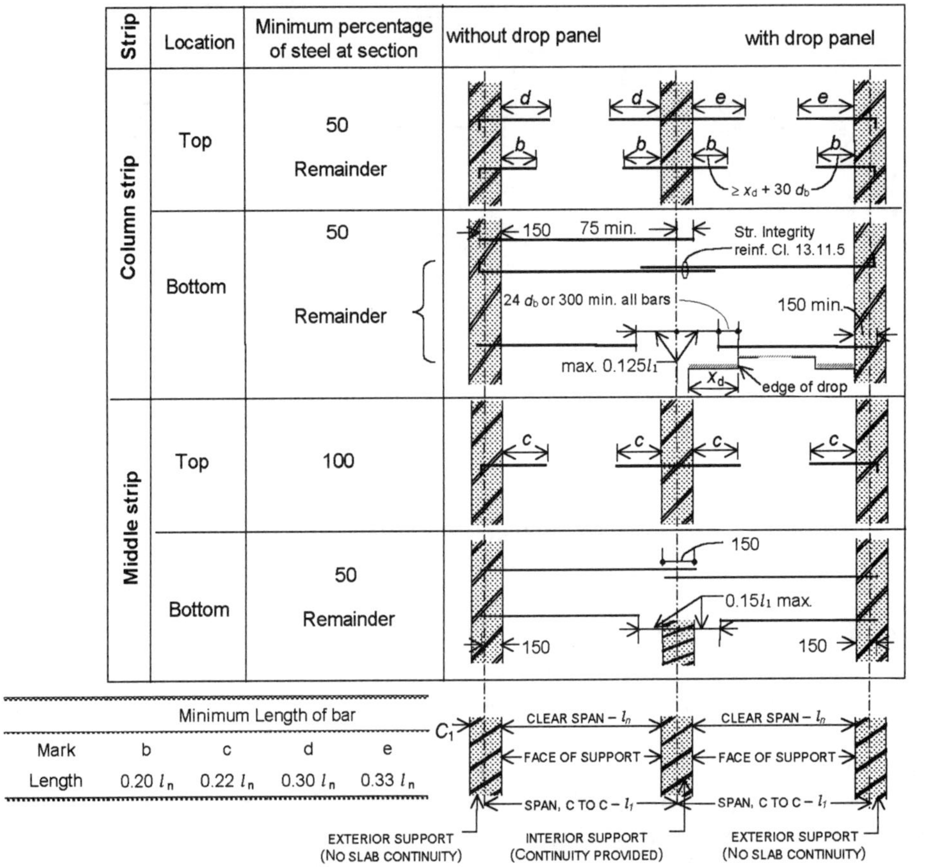

Strip	Location	Minimum percentage of steel at section	Minimum Length of bar Mark	Length
Column strip	Top	50		
		Remainder		
	Bottom	50		
		Remainder		
Middle strip	Top	100		
	Bottom	50		
		Remainder		

Mark	b	c	d	e
Length	$0.20\,l_n$	$0.22\,l_n$	$0.30\,l_n$	$0.33\,l_n$

Notes: With the permission of CSA International, this material is reproduced from CSA Standard A23.3-94, *Design of Concrete Structures*, which is copyrighted by CSA International, 178 Rexdale Boulevard, Rexdale, Ontario, M9W 1R3.

14.10.3 Minimum Bottom Reinforcement for Structural Integrity

Shear failure of flat slabs around a column type support may lead to the tearing out of the top reinforcement over the support section from the top surface of the slab (Fig. 14.29a) resulting in a punch through at the support. This, in turn, may lead to a progressive type of collapse. To provide adequate post-punching-failure resistance, thereby preventing progressive collapse of the structure after a local failure, a minimum amount of properly anchored bottom reinforcing bars, made effectively continuous over the span, must pass through the columns (or other

CHAPTER 14 DESIGN OF TWO-WAY SLAB SYSTEMS 477

support reaction areas) in each span direction (Fig. 14.29b). The Code (Cl. 13.11.5) prescribes Eq. 14.28 for the minimum area of reinforcement, A_{sb}, so required, and the area A_{sb} must be made up of at least two bars in each span framing into the column.

$$\Sigma A_{sb} = \frac{2V_{se}}{f_y} \quad (14.28)$$

where ΣA_{sb} gives the summation of the area of bottom reinforcement connecting the slab to the column (or column capital) on all faces of the periphery of the column (or column capital), and V_{se} is the shear transmitted to the column due to specified loads, but not less than the shear corresponding to twice the self-weight of the slab. This reinforcement enables the slab to hang from the support, even after local failure of the slab around the support (Fig. 14.29b). To make these bars effectively continuous, they may be lap spliced over a column with a Class A tension lap splice (Chapter 10), or spliced outside the reaction area with a lap length of $2l_d$. At discontinuous edges, this reinforcement must be anchored over the support by bends, hooks or otherwise, so as to develop yield stress at the face of the support.

If there are beams containing shear reinforcement in all spans framing into the column, the above reinforcement is not required.

14.11 DESIGN EXAMPLE FOR ELASTIC FRAME METHOD

EXAMPLE 14.2

Figure 14.30 shows the floor plan and cross section of a flat slab floor for an office building. In addition to the self weight of the slab, there is a superimposed dead load of 1.70 kN/m^2, made up of: ceiling = 0.20 kN/m^2, mechanical and electrical fittings = 0.50 kN/m^2, and partitions 1.00 kN/m^2. The specified live load is 2.4 kN/m^2. Assume concrete weighs 24 kN/m^3, f'_c = 30 MPa, and f_y = 400 MPa.

Analyse one typical interior bay in the E-W direction, shown shaded in Fig. 14.30a, using the Elastic Frame Method. Compute the reinforcement in the various design strips and check shear stresses.

478 REINFORCED CONCRETE DESIGN

Fig. 14.29 Minimum bottom reinforcement required to pass through column

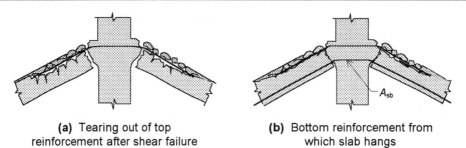

(a) Tearing out of top reinforcement after shear failure

(b) Bottom reinforcement from which slab hangs

Fig. 14.30 Example 14.2

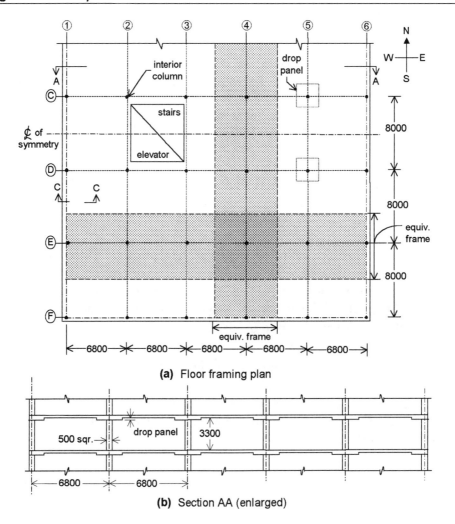

(a) Floor framing plan

(b) Section AA (enlarged)

SOLUTION

Drop panels are used over all columns to reduce the amount of negative reinforcement necessary and to decrease the shear stresses around the column head.

1. Slab thickness for deflection control
   The floor slab is provided the same thickness in all panels. At discontinuous edges, edge beams with a stiffness ratio $\alpha \geq 0.8$ will be provided so that no increase in slab thickness is required for the exterior panels. Equation 14.2 gives the minimum thickness specified by the Code for deflection control.

   Assume drop panels extending 1.5 m in each direction from column centre (3 m × 3 m panels) and projecting by half the slab thickness below the slab. With this, in Eq. 14.2,

   l_n = larger clear span = 8000 - 500 = 7500 mm
   x_d = distance from face of column to edge of drop panel

   $\qquad$ = 1500 – 250 = 1250 mm < $l_n/4$, $\qquad\qquad\qquad\qquad$ OK

   Of the two directions, the smaller value of
   $x_d/(l_n/2) = 2 \times 1250/7500 = 1/3$
   $(h_d - h_s)/h_s = 0.5$

   With these values, Eq. 14.2 gives,

   $$h_s \geq \frac{7500(0.6 + 400/1000)}{30[1 + (1/3)(1/2)]} = 214 \text{ mm}$$

   A slab of thickness of 200 mm is selected. This is marginally less than the calculated value above. This means a deflection computation is required to check that deflections will not exceed the limits stipulated in Table 9.2 of the Code.

2. Drop panels
   A drop panel extending 1.5 m in each direction from column centre (3 m × 3 m panel) has already been assumed. The projections of drop below the slab = half slab thickness (assumed) = 100 mm. The drop panels are shown in Fig. 14.31.

480 REINFORCED CONCRETE DESIGN

3. Edge beam
   As already indicated, an edge beam with $\alpha \geq 0.8$ is provided along discontinuous edges. Try a beam with rib dimensions of 450 mm depth and 250 mm width, as shown in Fig. 14.31.
   $I_b = 2.71 \times 10^9 \text{ mm}^4$

   For a beam along longer edge,
   $I_s = 2.43 \times 10^9 \text{ mm}^4$ and,
   $\alpha_a = I_b/I_s = 2.71/2.43 = 1.12 > 0.8$

   For a beam along shorter edge of 6.8 m,
   $I_s = 4250 \times 200^3/12 = 2.83 \times 10^9 \text{ mm}^4$
   $\alpha = 2.71/2.83 = 0.96 > 0.8$

4. Effective depths
   Minimum cover for slab by CSA A23.1-94 is 20 mm. However, minimum cover to give a two hour fire resistance rating (Ref. 14.10) is 25 mm. Reinforcement will be placed in the outer layer for the N-S direction bars to resist the larger moments in this direction, and in the inner layer for the E-W direction. Selecting a 25 mm cover and assuming

Fig. 14.31 Slab dimensions

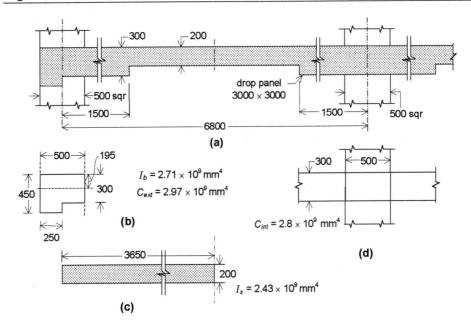

CHAPTER 14 DESIGN OF TWO-WAY SLAB SYSTEMS 481

No. 15 bars, the effective depths are (Fig. 14.32) worked out below. The Code (Cl. 13.11.6) restricts the thickness of drop panel below slab assumed for effective depth at drop panels for negative moment to $x_d/4$. Here actual $(h_d-h_s) < x_d/4$, and hence full depth of drop is effective.

$d_{E-W(slab)}$ = 200-25-16-16/2 = 151 mm
$d_{E-W(slab)}$ = 300-25-16-16/2 = 251 mm
$d_{N-S(slab)}$ = 200-25-16/2 = 167 mm
$d_{N-S(drop)}$ = 300-25-16/2 = 267 mm

5. Factored loads
   Dead loads are:
   Slab concrete 0.2×24 = 4.8 kN/m^2
   Additional DL = 1.7 kN/m^2
   Total DL on slab = 6.5 kN/m^2
   Concrete in drop panel 0.1×24 = 2.4 kN/m^2

   Live load = 2.4 kN/m^2
   Factored loads are:
   Dead load on slab 1.25×6.5 = 8.13 kN/m^2
   Live load on slab 1.5×2.4 = 3.60 kN/m^2
   Total on slab = 11.7 kN/m^2

   Dead load due to drop panel 1.25×2.4 = 3.00 kN/m^2

6. Preliminary checking of shear
   To ascertain the overall adequacy of slab thickness, the shear strength is checked now. However, for shear in two-way action, a more detailed check, including shear due to moment transfer, is deferred until after the

Fig. 14.32 Effective depths

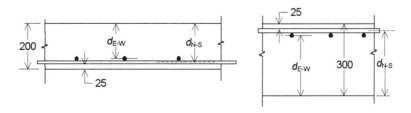

Slab section (N - S direction)   Drop panel section (N - S direction)

482 REINFORCED CONCRETE DESIGN

moments are determined from the frame analysis. To ensure safety when moment transfer effects are included, the punching shear stress due to vertical shear only, calculated below, must be kept well below the allowable value.

a. One-way (beam) shear
The critical sections are at distance d from column face, and also from the edge of the drop panel (Fig. 14.33).

At $d = 251$ mm from the column face, considering (conservatively) a 1 m wide strip outside the drop, shown shaded in Fig. 14.32 (to be more exact, the full width of slab may be considered as a single wide beam, allowing for the larger load and depth of slab within the drop panel):

$V_f = 11.7 \times 1 \times 2.899 = 33.9$ kN

$V_c = 0.2 \times 1.0 \times 0.6 \times \sqrt{30} \times 1000 \times 151 \times 10^{-3} = 99.3$ kN

$V_f \ll V_c$ \hfill OK

The shear stress will be also within limits at the critical section, a distance d from the edge of the drop. Note that one-way shear is more critical along the longer (8 m) span, than it is in the N-S direction. The corresponding shear stress also will be within limits in this example.

b. Two-way action (punching shear)
Critical sections are at $d/2$ from the face of the columns and from the edge of the drop panel. At an interior column, at distance $d/2 = 251/2$ from the face of the column (Fig. 14.33):

$V_f = 11.7 \times (6.8 \times 8 - 0.751^2) + 3.00 \times (3 \times 3 - 0.751^2)$
$\quad = 655$ kN
$b_o = 4 \times 0.751 = 3.004$ m

Average factored shear stress,

$$v_f = \frac{V_f}{A_c} = \frac{655 \times 10^3}{3004 \times 251} = 0.869 \text{ MPa}$$

For slab without shear reinforcement, the factored shear stress calculated above, together with the shear stress due to moment

Fig. 14.33 *Critical Sections for shear (one-way and two-way action)*

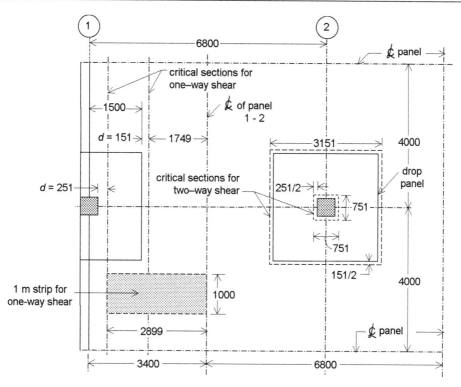

transfer (yet to be computed), is limited to the factored shear stress resistance, v_r, which is the least of:

(i) $\quad v_r = v_c = (1 + 2/\beta_c)\, 0.2\, \lambda\, \phi_c\, \sqrt{f_c'}$

$\quad\quad \beta_c = 1.0$ for square columns (hence this is greater than (iii) below)

(i) $\quad v_r = \left(\dfrac{\alpha_s d}{b_o} + 0.2\right) \lambda \phi_c \sqrt{f_c'}$,

where $\alpha_s = 4$ for interior columns

$\quad = \left(\dfrac{4 \times 251}{3004} + 0.2\right) \lambda \phi_c \sqrt{f_c'}$

$\quad = 0.53\, \lambda \phi_c \sqrt{f_c'}$ (which is also greater than (iii) below)

(iii) $\quad v_r = 0.4\, \lambda \phi_c \sqrt{f_c'} = 0.4 \times 1.0 \times 0.6 \times \sqrt{30}$

$\quad\quad = 1.31$ MPa $\rightarrow$ controls.

Factored shear = 0.869 MPa $\ll v_r$

484 REINFORCED CONCRETE DESIGN

At $d/2 = 151/2$ mm from edge of drop panel, (Fig. 14.33):
$V_f = 11.7 \times (6.8 \times 8 - 3.151^2) = 520$ kN

Factored shear stress $v_f = \dfrac{520 \times 10^3}{(4 \times 3151) \times 151} = 0.273$ MPa $<< v_r$

Similar calculations show that punching shear stress is less critical at the exterior column. Furthermore, at the exterior column, the edge beam transmits part of the shear by beam action.

7. Relative stiffness parameters of frame members
   a. Column stiffnesses, K_c
   Referring to Fig. 14.34,

$$\left. \begin{array}{l} H = 3.3 \text{ m} \\ H_c = 3.3 - (0.2 + 0.1) = 3.0 \text{ m} \end{array} \right\} H/H_c \ 3.3/3 = 1.1$$

$$\left. \begin{array}{l} t_a = 200/2 + 100 = 200 \text{ mm} \\ t_b = 200/2 = 100 \text{ mm} \end{array} \right\} \begin{array}{l} t_a/t_b = 200/100 = 2.0 \\ t_b/t_a = 100/200 = 0.5 \end{array}$$

From Table 14.4 giving stiffness and carry-over factors for columns:

for $\left. \begin{array}{l} H/H_c = 1.1 \\ t_a/t_b = 2 \end{array} \right\} \quad \begin{array}{l} K_{AB} = 5.34 \\ C_{AB} = 0.54 \end{array}$

Fig. 14.34 Column properties

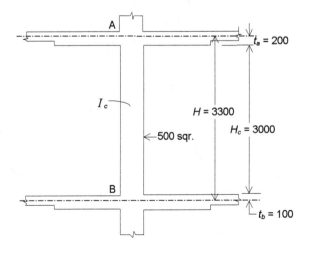

CHAPTER 14 DESIGN OF TWO-WAY SLAB SYSTEMS 485

$$\left.\begin{array}{l}H/H_c = 1.1\\t_b/t_a = 0.5\end{array}\right\} \quad \begin{array}{l}K_{AB} = 4.85\\C_{BA} = 0.60\end{array}$$

$K_c = K\,EI_c/H$, $I_c = 500^4/12 = 5.21 \times 10^9$ mm^4
$K_{C,\,below} = 5.34\,E_c \times 5.21 \times 10^9/3300 = 8.43\,E_c \times 10^6$
$K_{C,\,above} = 4.85\,E_c \times 5.21 \times 10^9/3300 = 7.66\,E_c \times 10^6$

b. Torsional member stiffness, K_t
Torsional member at the two outer columns (1) and (6) has the section shown in Fig. 14.31b.

$$C_{ext} = \left(1 - 0.63 \times \frac{150}{250}\right) \times \frac{150^3 \times 250}{3} + \left(1 - 0.63 \times \frac{300}{500}\right)$$

$$\times \frac{300^3 \times 500}{3}$$

$$= 2.97 \times 10^9 \text{ mm}^4$$

For interior columns, the torsional member is the portion of slab having a width equal to the column, that is, a section 500 mm wide and 300 mm deep (Fig. 14.31d).

$$C_{int} = \left(1 - 0.63 \times \frac{300}{500}\right) \times 300^3 \times \frac{500}{3} = 2.8 \times 10^9 \text{ mm}^4$$

For all columns, $c_2/l_2 = 500/8000 = 0.0625$
Using Eq. 14.10,

$$K_{t,ext} = \frac{2 \times 9E_c \times 2.97 \times 10^9}{8000(1-0.0625)^3} = 8.11E_c \times 10^6$$

$$K_{t,int} = \frac{2 \times 9E_c \times 2.8 \times 10^9}{8000(1-0.0625)^3} = 7.65E_c \times 10^6$$

c. Equivalent column stiffnesses, K_{ec}, (Eq. 14.27)

$$K_{ec,ext} = \frac{(8.43+7.66)\times 10^6 E_c \times 8.11 \times 10^6 E_c}{(8.43+7.66)\times 10^6 E_c + 8.11 \times 10^6 E_c} = 5.39 \times 10^6 E_c$$

$$K_{ec,int} = \frac{(8.43+7.66)\times 10^6 E_c \times 7.65 \times 10^6 E_c}{(8.43+7.66)\times 10^6 E_c + 7.65 \times 10^6 E_c} = 5.18 \times 10^6 E_c$$

486 REINFORCED CONCRETE DESIGN

d. Slab stiffnesses and fixed-end moment coefficients.
Although the Table 14.3 is for a drop extension of $l_1/6 = 1.13$ m, Table 14.3 will be used in this example with drop extensions of 1.5 m. The slab-beam geometry is given in Fig. 14.31a and the loading in Fig. 14.35a.

$C_{F1} = C_{N1}, C_{F2} = C_{N2}$
$C_{N1}/l_1 = 0.5/6.8 = 0.074$
$C_{N2}/l_2 = 0.5/8.0 = 0.063$

Fig. 14.35 Loading details for slab-beam

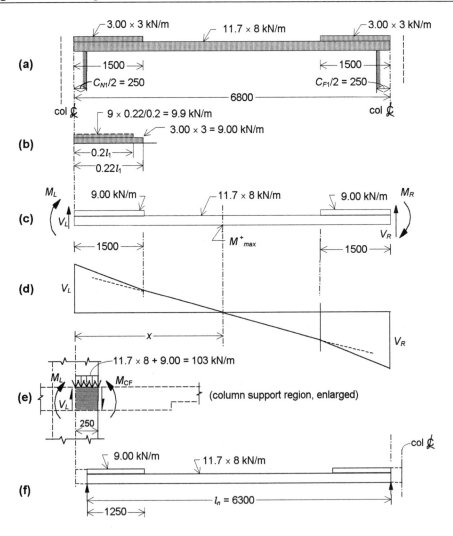

Interpolating from Table 14.3 for these values,
Stiffness factor, $K_{NF} = 5.93$
Carry-over factor, $C_{NF} = 0.60$
FEM coefficient, $m_{NF1} = 0.0931$, for uniform load

For additional load due to drop panel, Table 14.3 gives moment coefficients for partial loading over lengths of $0.2\, l_1$ ($b - a = 0.2$). Weight/m of span due to drop = $3.00 \times 3 = 9.00$ kN, acting for lengths of 1.5 m (= $0.22\, l_1$) from column centrelines. For using Table 14.3, this load may be considered as $9.00 \times 0.22/0.20 = 9.90$ kN/m distributed over a length of $0.2 l_1$, as indicated in Fig. 14.35b.

For weight of drop at near end, $a = 0$, $b - a = 0.2$, and for $C_{N1} / l_1 = 0.074$ and $C_{N2} / l_2 = 0.063$, interpolating from Table 14.3, fixed-end moment coefficient is:
$m_{NF,\,2} = 0.0165$

Similarly, for weight of drop at far end,
$a = 0.8,\ b - a = 0.2$

and $m_{NF,\,3} = 0.00191$

Moment of inertia of the slab section beyond drop, I_s is:
$I_s = 8000 \times 200^3/12 = 5.33 \times 10^9$ mm^4
$K_s = K_{NF} E_c I_s / l_1$
$= 5.93 \times E_c \times 5.33 \times 10^9 / 6800 = 4.65 \times 10^6\, E_c$

8. Elastic frame analysis
The analysis of the frame is done by the method of moment distribution in this example.
The moment distribution factors for slab-beam members are:
at exterior support,

$$DF_{1-2} = DF_{6-5} = \frac{K_s}{\Sigma K} = \frac{4.65 \times 10^6 E_c}{4.65 \times 10^6 E_c + 5.39 \times 10^6 E_c} = 0.463$$

at interior supports,

$$DF = \frac{4.65 \times 10^6}{2 \times 4.65 \times 10^6 + 5.18 \times 10^6} = 0.321$$

Since specified live load (= 2.4 kN/m^2) is less than three-quarters of dead load (= 6.5 × 3/4 = 4.88 kN/m^2), the factored moments at all critical sections may be obtained from the single loading pattern of full factored live load plus dead load on all spans. The factored loads on the slab-beam member are (Fig. 14.35):

uniformly distributed load = 11.7 × 8 = 93.6 kN/m

dead load due to weight of drop assumed redistributed over $0.2 l_1$ of span $\bigg\}$ = 9.90 kN/m

Slab fixed end moments are:
$$FEM_{NF} = \sum m_{NF} w l_1^2$$
$$= 0.0931 \times (11.7 \times 8) \times 6.8^2 + 0.0165 \times 9.90 \times 6.8^2$$
$$+ 0.00191 \times 9.90 \times 6.8^2$$
$$= 411 \, kN \cdot m$$

The moment distribution is set up in Table 14.5, wherein only the slab moments are given.

The freebody diagram for a typical span is shown in Fig. 14.35c. The shear forces at each end of the span are given by:
taking moments about right end,
$$V_L \times 6.8 + M_L + M_R = 9.00 \times 1.5 \times (6.8 - 0.75)$$
$$- (11.7 \times 8 \times 6.8^2)/2 - (9.00 \times 1.5^2)/2 = 0$$

$$V_L = 331.7 - \frac{(M_L + M_R)}{6.8}$$

Similarly, $V_R = 331.7 - \dfrac{(M_L + M_R)}{6.8}$

Location of zero shear (and maximum moment) is given by (Fig. 14.35d):

$$V_L = 9.00 \times 1.5 + 11.7 \times 8(x)$$
$$x = \frac{V_L - 13.5}{93.6}$$

Maximum positive moment at distance (x) is given by:
$$+M_{max} = M_L + V_L(x) = 9.00 \times 1.5 \times (x - 0.75) - 11.7 \times 8(x)^2/2$$

Values of V_L, V_R, x, and $+M_{max}$ are also tabulated in Table 14.5 for all spans.

The critical sections for negative design moment are at the faces of the columns. The moment at face of columns may be computed as (Fig. 14.35e):

$$M_{CF} = M_L + V_L \times 0.25 = 103 \times 0.25^2/2$$

and similarly for the left face of the support at right. These values are also computed and given in Table 14.5.

9. Transverse distribution of moments
   As there are no beams between supports, the distribution is as given in Section 14.5.3(a) (Code Cl. 13.12).

   a. Column strip share of moments
      Negative moment at an interior column may be between 60% and 100%. Here the share in column strip may be taken as 75%.
      Negative moment in the column strip at the exterior column is 100%.
      Positive moment at all spans in the column strip may be between 50% and 70%. Here the share in column strip may be taken as 60%.

   b. Middle strip share
      Interior negative factored moment = 100 - 75 = 25 percent
      Exterior negative factored moment = nil
      Positive factored moment   = 100 - 60 = 40 percent

      Both column strip and middle strip moments are shown in Fig. 14.36.

10. Column moments
    Total unbalanced slab moment at exterior support, from Table 14.5, is 234 kN·m.
    For column above,
    stiffness = $7.66 \times 10^6 E_c$, carry-over factor (COF) = 0.60
    For column below,

Table 14.5 *Equivalent Frame Analysis (Moment Distribution)*

	1	2	3	4	5	6	7	8	9	10
$K_s \times 10^6 E_c =$	5.39	4.65	5.18	4.65	4.65	5.18	4.65	4.65	5.18	5.39
$K_{ec} \times 10^6 E_c =$										
DF	0.463	0.321	0.32	0.321	0.321	0.32	0.32	0.321	0.321	0.463
COF		0.6			0.6			0.6		0.6
FEM	−411	411	−411	411	411	−411	−411	411	411	−411
Bal.	190									−190
C.O.		114		−37					37	
Bal.		−37							−114	
C.O.	−22					−22	22			22
Bal.	10			7	−7	7	−7			−10
C.O.		6	4	−4		−4	4	−6		
Bal.		−3	−3	1	1	−1	−1	3	3	
C.O.	−2				−1	2	−1			2
Bal.	1		1	1	1	−1	−1	−1		−1
C.O.		—	—	—				—	—	
M, Col. Centreline Σ (kN·m)	−234	491	−446	396	−407	407	−396	446	−491	234
V_L, V_R (kN)	294	369	339	324	332	332	324	339	369	294
x (m)	3.00		3.48		3.40	3.40		3.48		3.00
$+M_{max}$ (kN·m)	196		130		145	145		130		196
$-M_{CF}$ (kN·m)	−164	402	−364	318	−327	327	−318	364	−402	164
Design Negative moment, (kN·m)	164	402	364	318	327	327	318	364	402	164
Design Positive moment, (kN·m)	196		130		145	145		130		196

Fig. 14.36 *Factored moments in column and middle strips*

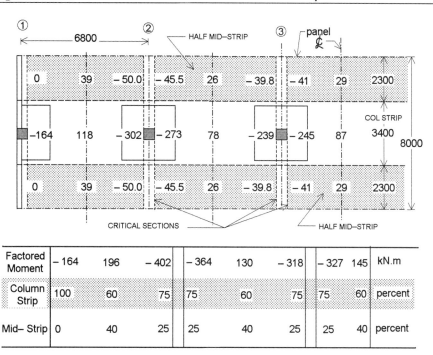

$$\text{stiffness} = 8.43 \times 10^6 E_c, \text{ COF} = 0.54$$

Proportion of moment in column above $= \dfrac{7.66}{7.66 \times 8.43} = 0.476$

Proportion of moment in column below $= 1 - 0.476 = 0.524$

At exterior support of slab,
    In column above: M at base $= 0.476 \times 234 = 111$ kN·m
                                    M at top $= 0.60 \times 111 = 66.8$ kN·m

    In column below: M at top $= 0.524 \times 234 = 123$ kN·m
                                      M at base $= 0.54 \times 123 = 66.2$ kN·m

At columns 2 and 5,

$$M = 491 - 446$$
$$= 45 \text{ kN·m}$$

In column above: M at base = 0.476 × 45 = 21.4 kN·m
M at top = 0.60 × 21.4 = 12.9 kN·m

In column below: M at top = 0.524 × 45 = 23.6 kN·m
M at base = 0.54 × 23.6 = 12.7 kN·m

At columns 3 and 4,

M, from Table 14.5 = 407 - 396 = 11 kN·m

Using Eq. 14.13,

$$l_1 = l_1' = 6.8 \text{ m}, l_n = 6.8 - 0.5 = 6.3 \text{ m}$$
$$w_d = w_d' = 8.13 \text{ kN/m}^2 \text{ (neglecting weight of drop)}$$
$$0.5 w_L = 3.60/2 = 1.80 \text{ kN/m}^2$$

$$M = 0.07[8.13 + 1.80)8 \times 6.3^2 - 8.13 \times 8 \times 6.3^2] = 40.0 \text{ kN} \cdot \text{m}$$
$$> 11 \text{ kN} \cdot \text{m}$$

In column above: M at base = 0.476 × 40 = 19.0 kN·m
M at top = 0.60 × 19.0 = 11.4 kN·m

In column below: M at top = 0.524 × 40 = 21 kN·m
M at base = 0.54 × 21.0 = 11.3 kN·m

All column moments are shown in Fig. 14.37.

11. Flexural reinforcement

The flexural reinforcement at all critical sections is computed in Table 14.6 for column strip and Table 14.7 for middle strips.

The Code (Cl. 13.12.2) requires a heavier concentration of negative moment reinforcement (100% at exterior column and not less than 33.3% at interior column) over the column within a band having a width of $c_2 + 3h_s$ = 500 + 3 × 200 = 1100 mm. The maximum spacing in this band is $1.5h_s \leq$ 250 mm, that is 250 mm in this case. This is also worked out in Table 14.6. The spacing limit at all other locations is $3h_s \leq 500$ mm, that is 500 mm in this case.

Table 14.6 Design of Reinforcement in Column Strip, E-W Direction

		1			2			3		
Moment in col. strip, M_f;	kN·m	164	118	302	273	78	239	245	78	87
Width of strip or drop, b;	mm	3000	3400	3000	3000	3400	3000	3000	3400	3000
Effective depth, d;	mm	251	151	251	251	151	251	251	151	251
$K_r = M_f \times 10^6/(bd^2)$;	MPa	0.867	1.52	1.60	—	1.01	—	1.30	—	1.12
ρ (Table 5.4);		0.0025	0.0046	0.00488	—	0.00307	—	0.0040	—	0.00342
			6					1		
$A_s = \rho bd$;	mm²	1883	2392	3675	—	1576	—	3020	—	1756
Number of No. 15 bars;	No.	10	12	19	—	9	—	16	—	9
Spacing of positive moment reinforcement.	mm		283			378			378	
No. of bars over column in a band 1100 mm wide, and spacing	mm	10 @110			7 @155			6 @180		
No. of bars in remaining width of 2300 mm and spacing	mm	nominal @500			12 @190			10 @230		

Table 14.7 Design of Reinforcement in Middle Strip, E-W Direction

		1			2			3		
Middle strip moment, M_f;	kN·m	0.0	78	100	91	52	79.6	82	58	—
Width of Strip b;	mm	5000	4600	5000	5000	4600	5000	5000	4600	—
Effective Depth, d;	mm	151	151	151	151	151	151	151	151	—
$K_r = M_f \times 10^6/(bd^2)$	MPa	(0.002)*	0.744	0.877	—	0.496	—	0.719	0.553	—
ρ		(0.002)*	0.00225	0.00263	—	(0.002)*	—	0.0021	(0.002)*	—
								7		
$A_s = \rho bd$;	mm²	1510	1560	1984	—	1390	—	1638	1390	—
Number of No. 15 bars;	No.	8	8	10	—	7	—	9	7	—
Spacing;	mm	625‡	575‡	500	—	657‡	—	556‡	657‡	—
Number of No. 15 bars to limit spacing $< 3h_s \leq 500$	No.	10	10	10	—	10	—	10	10	—

* Minimum reinforcement controls – (0.002) refers to $A_{s\,min} = 0.002A_g$
‡ Maximum allowable spacing controls.

Fig. 14.37 Column moments

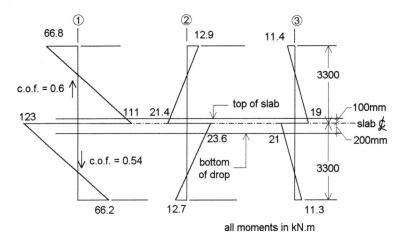

The areas and spacings of reinforcement in column strips are both within the specified limits. Note that No. 20 bars will give slightly more convenient spacings at interior supports.

In the middle strips, reinforcement at the exterior support and at midspan in interior spans is controlled by minimum area requirements. However, with No. 15 bars, the actual number of bars at all sections is controlled by the maximum allowable spacing limitation. Five bars are required in each of the two half middle strips.

12. Continuous bottom reinforcement required for structural integrity

self weight of slab + specified dead load $\quad = 6.5 \text{ kN/m}^2$

equivalent self weight of drop $= \dfrac{2.4 \times 3^2}{8 \times 6.8} \quad = 0.4 \text{ kN/m}^2$

specified live load $\quad = 2.4 \text{ kN/m}^2$

total specified load $\quad = 9.3 \text{ kN/m}^2$

$w_s \quad = 9.3 \text{ kN/m}^2$, but not less than twice dead load
$\quad = 2 \times (6.5 + 0.4) = 13.8 \text{ kN/m}^2$

Shear transmitted to column, $V_{se} = 13.8 \times (8 \times 6.8 - 0.5^2) = 747 \text{ kN}$

$\Sigma A_{sb} = \dfrac{2 \times 7.47 \times 10^3}{400} = 3736 \text{ mm}^2$

CHAPTER 14 DESIGN OF TWO-WAY SLAB SYSTEMS

Five No. 15 bottom bars each must be made effectively continuous and pass through the column within its width of 500 mm, in both E-W and N-S directions, giving a $\Sigma A_{sb} = 5 \times 4 \times 200 = 4000$ mm^2.

13. Checking shear stresses

    The shear stresses in two-way action around the column-slab connection, including the shear due to transfer of moments must now be checked.

    a. Interior column

    The maximum column reaction and unbalanced slab moment is at columns 2 and 5. From Table 14.5, the vertical column reaction is obtained as the sum of the shears on either side, and is:
    $R = 369 + 339 = 708$ kN

    For shear in two-way action, the average of the effective depths in E-W and N-S directions may be used for d.
    $d = (251 + 267)/2 = 259$ mm

    For the critical section at $d/2$ from the column face (Fig. 14.38), the shear is
    $V_f = 708 - (11.7 + 3.00) \times 0.759^2 = 700$ kN

    and perimeter $b_o = 4 \times 759 = 3036$ mm

    Moment transferred by shear is 40 percent of unbalanced moment, and equal to:
    $M_{fv} = 0.4 \times (491 - 446) = 18$ kN·m

    For an interior column, the term J_c/c in Eq. 14.19 is given by (Fig. 14.18):

    $c_{AB} = (c_1 + d)/2 = 759/2 = 380$ mm

    $$J_c = \frac{(c_1+d)d^3}{6} + \frac{(c_1+d)^3 d}{6} + \frac{(c_1+d)^2(c_2+d)d}{2}$$

    $$= \frac{759 \times 259^3}{6} + \frac{759^3 \times 259}{6} + \frac{759^2 \times 759 \times 259}{2}$$

    $$= 77.7 \times 10^9 \text{ mm}^4$$

    $$\frac{J_c}{c_{AB}} = \frac{77.7 \times 10^9}{380} = 204 \times 10^3 \text{ mm}^3$$

Maximum factored shear stress is (Eq. 14.19):

$$v_f = \frac{700 \times 10^3}{3036 \times 259} + \frac{18 \times 10^6}{2.04 \times 10^8} = 0.978 \text{ MPa} < v_r = 1.31 \text{ MPa}$$

b. Exterior column
Because of the edge beam, part of the shear will be transmitted by this beam to the column by beam action. However, to simplify the calculation, the shear stress is conservatively computed neglecting the contribution of the edge beam.

Referring to Table 14.5 and Fig. 14.38:
Column reaction = 294 kN
At critical section $d/2$ from column face,
$V_f = 294 - (11.7 + 3.00) \times 0.759 \times 0.630 = 287$ kN

Fig. 14.38 Critical sections for two-way shear

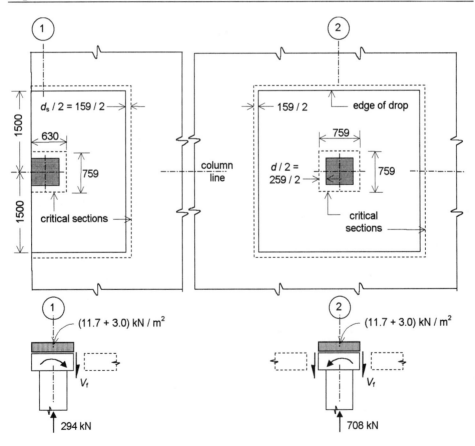

$M_{fv} = 0.4 \times 234 = 93.6$ kN·m
$b_o = 2 \times 630 + 759 = 2019$ mm

From Fig. 14.18,

$$C_{AB} = \frac{(c_1 + d/2)^2}{2c_1 + c_2 + 2d} = \frac{(500 + 259/2)^2}{2 \times 500 + 500 + 2 \times 259} = 196 \text{ mm}$$

$$J_c = \frac{(c_1 + d/2)d^3 + (c_1 + d/2)^3 d}{6} + (c_2 + d)d(c_{AB})^2$$

$$+ 2(c_1 + d/2)d[(c_1 + d/2)/2 - c_{AB}]^2$$

$$= \frac{630 \times 259^3 + 630^3 \times 259}{6} + 759 \times 259 \times 196^2 + 2$$

$$\times 630 \times 259(630/2 - 196)^2$$

$$= 24.8 \times 10^9 \text{ mm}^4$$

$$\frac{J_c}{c_{AB}} = \frac{24.8 \times 10^9}{196} = 126 \times 10^6 \text{ mm}^3 \qquad \text{OK}$$

$$v_f = \frac{287 \times 10^3}{2019 \times 259} + \frac{93.6 \times 10^6}{1.26 \times 10^8} = 1.29 \text{ MPa} < v_r = 1.31 \text{ MPa}$$

The two-way shear must also be checked at a critical section, a distance $d_s/2$ (half depth of slab) from the edge of the drop panels. The procedure is the same as shown above and the stresses at these sections are well within the limits in this example.

PROBLEMS

1. Redesign the interior equivalent frame in Example 14.2, assuming that beams 300 mm wide and 500 mm deep (overall) are provided along the column lines, and no drop panels are used.

2. For the floor designed in Example 14.2, design the exterior equivalent frame in the N-S direction, using the Elastic Frame Method.

3. For the floor system in Example 14.2, using the Direct Design Method, determine the design moments in the various design strips of a N-S interior equivalent frame.

REFERENCES

14.1 Sozen, M.A. and Siess, C.P., *Investigation of Multiple-Panel Reinforced Concrete Floor Slabs: Design Methods - Their Evolution and Comparison*, ACI Journal, Proc. V. 60, No. 8, Aug. 1963, pp. 999-1028.

14.2 Corley, W.G., and Jirsa, J.O., *Equivalent Frame Analysis for Slab Design*, ACI Journal, Proc. V. 67, No. 11, Nov. 1970, pp. 875-884.

14.3 Gamble, W.L., *Moments in Beam Supported Slabs*, ACI Journal, Proc. V. 69, No. 3, Mar. 1972, pp. 149-157.

14.4 Hanson, N.W., and Hanson, J.M., *Shear and Moment Transfer Between Concrete Slabs and Columns*, Journal, PCA Research and Development Laboratories, V. 10, No. 1, Jan. 1968, pp. 2-16.

14.5 ACI-ASCE Committee 326, *Shear and Diagonal Tension*, ACI Journal, Proc. V. 59, No. 3, Mar. 1962, pp. 352-396.

14.6 Explanatory notes on CSA Standard A23.3-94, *Concrete Design Handbook*, Canadian Portland Cement Association - 1995

14.7 *Commentary on Building Code Requirements for Reinforced Concrete* (ACI 318R-95), American Concrete Institute, Detroit, Michigan, 1995.

14.8 *Concrete Design Handbook*, Canadian Portland Cement Association, Ottawa, Canada, 1995.

14.9 ACI Standard 318-95, *Building Code Requirements for Structural Concrete* (ACI 318-95), American Concrete Institute, Detroit, Michigan, 1995.

14.10 *National Building Code of Canada* 1995 (Appendix D), Canadian Commission on Building and Fire Code, National Research Council of Canada, 1995.

CHAPTER 15 Design of Compression Members - Short Columns

15.1 COMPRESSION MEMBERS

15.1.1 Types of Columns

In a building, columns are the vertical members carrying primarily compressive forces. CSA A23.3-94 classifies a compression member having a ratio of height to least lateral dimension of three or greater as a column, and refers to less slender members as pedestals. Several other structural elements, such as inclined members, rigid frame members, truss members, and beams, may also be subjected to substantial axial compressive forces. The design considerations explained in this chapter with reference to *columns* are applicable to compression members in general.

There are few compression members that are subjected to purely axial compressive forces. In general, compressive forces in a member exist in combination with bending. The bending moments may result from applied end moments, lateral loading on the member and/or eccentricity of applied compressive forces.

The three basic types of reinforced concrete columns are shown in Fig. 15.1. Tied columns are reinforced with main longitudinal bars enclosed within lateral *ties*. The spiral column has its longitudinal bars enclosed within closely spaced and continuously wound spiral reinforcement. A composite column is reinforced longitudinally with structural steel shapes, pipe, or tubing, with or without longitudinal bars. This chapter is primarily concerned with the first two types, as these are most common in reinforced concrete construction. The most common cross-sectional shapes for tied columns are the square and rectangle, although T-, L-, and other shapes are occasionally used, particularly for edge and corner columns in buildings. Spiral columns are usually circular in cross-section, the alternative shapes being square and octagonal sections.

15.1.2 General Requirements on Reinforcement

(a) Longitudinal Bars
CSA A23.3-94(10.9) limits the longitudinal reinforcement in reinforced concrete compression members to between 1 and 8 percent of the gross area of a section. The lower limit of 1 percent is specified to ensure the use of a reinforced concrete (rather

Fig. 15.1 *Types of reinforced concrete columns*

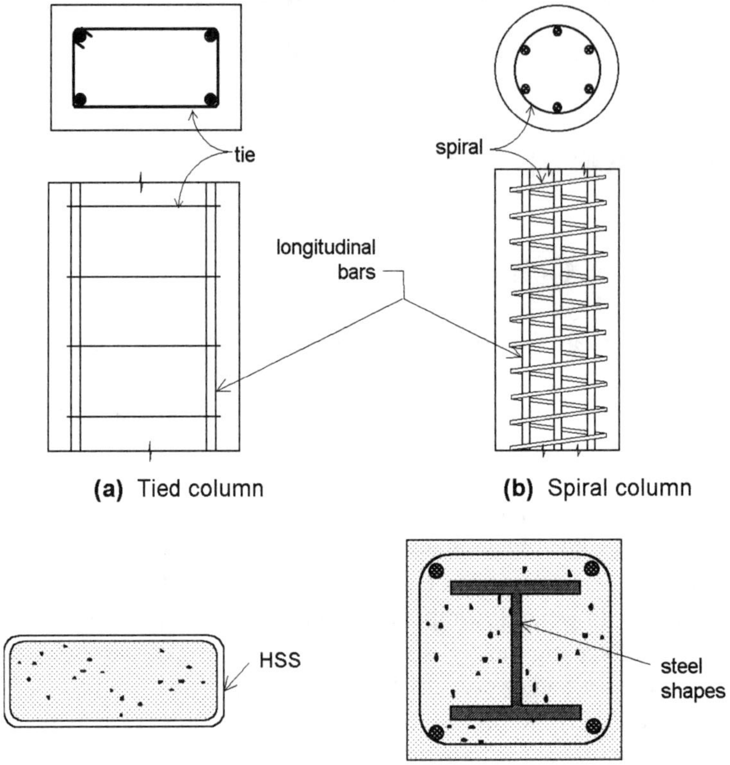

(a) Tied column

(b) Spiral column

(c) Composite columns

than plain concrete) column and to provide some flexural resistance even for members where computations show no applied moment. Furthermore, shrinkage and creep effects tend to transfer the load from the concrete to the longitudinal reinforcement. The minimum reinforcement ensures that these effects do not lead to the eventual yielding of the reinforcement under sustained service loads. Occasionally, considerations other than strength (architectural, for instance) may dictate the overall column size, resulting in an oversize column. In such a case, the minimum reinforcement percentage applied to the gross section may lead to excessive reinforcement. Therefore, CSA A23.3-94(10.10.5) permits reinforcement percentages less than 1 percent, but larger than 0.5 percent; however, the factored axial and moment resistances must be multiplied by the reinforcement percentage used.

The upper limit of 8 percent represents a practical maximum for longitudinal reinforcement from considerations of both economy and requirements for proper placing of concrete. The maximum steel ratio in the regions containing lap-splices must also not exceed 8 percent. Therefore, where column bars are required to be lap-spliced, the longitudinal reinforcement ratio should usually be limited to a maximum of 4 percent. Use of reinforcement in excess of 4 percent in regions outside lap splices may also result in difficulties in placing and compacting concrete. For spirally reinforced columns, a minimum of six bars and for columns with rectangular or circular ties a minimum of four bars (one at each corner) are required. For other shapes (L, T, etc.) one bar must be provided at each corner. Additional detailing requirements for longitudinal reinforcement in columns are given in CSA A23.3-94(7.5).

(b) Lateral Reinforcement
All longitudinal reinforcement in a compression member must be enclosed within lateral reinforcement. The latter may consist of closed ties or spirals. Lateral reinforcement helps to: (1) hold the longitudinal bars in position during construction; (2) provide lateral restraint to the individual longitudinal bars in compression and prevent their premature buckling; (3) at close spacings, provide lateral confinement to the concrete; and (4) carry shear and torsion when reinforcing for such stresses are required.

(c) Ties
The Code (Cl. 7.6.5) prescribes the requirements for ties. Reinforcement for ties must have a diameter of at least 30 percent of that of the largest longitudinal bar for bars up to No. 30, and must be at least No. 10 in size for larger size longitudinal bars and for bundled bars. The spacing of ties must not exceed the least of: (1) 16 times the diameter of the smallest longitudinal bar; (2) 48 times the tie diameter; (3) the least lateral dimension of the compression member; and (4) 300 mm when bundled bars are used. For specified concrete compressive strengths greater than 50 MPa, the tie spacing determined from the foregoing is multiplied by 0.75. Additional ties are usually provided at each end of lap spliced bars, above and below end bearing splices, and immediately below sloping regions of offset bent longitudinal bars.

The arrangement of ties must be such as to enclose and laterally support every corner bar and every alternate longitudinal bar by the corner of a tie having an included angle of not more than 135°. Furthermore, no bar is to be farther than 150 mm clear on either side from such a laterally supported bar. Several arrangements of ties to satisfy this requirement are presented in Refs. 15.1 and 15.2.

(d) Spirals
Requirements for spirals are presented in CSA A23.3-94(7.6.4). The minimum amount of spiral reinforcement prescribed and the basis for this will be described in Section 15.2 (Eq. 15.4). The diameter of the reinforcing steel used for the spiral must not be less than 6 mm. The pitch of the spiral must not exceed one-sixth of the diameter of the core (measured out-to-out of spirals), and the clear spacing between successive turns must be between 25 mm and 75 mm, inclusive.

15.1.3 Slenderness Effects

(a) Limiting Slenderness Ratio
Columns which are long and slender may fall due to instability at a load lower than the strength of a short column of identical cross-sectional dimensions. Furthermore, slender columns have larger lateral deflections under the applied primary moments, which create greater secondary moments due to the axial compression interacting with lateral deflection, the so-called P-Δ effect. These effects are discussed in more detail in Section 16.1. The ratio kl_u/r is taken as the measure of the slenderness, where l_u is the unsupported length of the member, k is the effective length factor, and r is the radius of gyration of the cross section. Relatively short and stocky columns with kl_u/r ratios such that the reduction of strength due to slenderness effects is small enough to be neglected will be termed *short* columns. The strength of such a column is dependent on the cross-sectional dimensions and material strength. It is reported, based on surveys of existing practice, that over 90 percent of the columns in braced frames and over 40 percent of the columns in unbraced frames fall into this category (Ref. 15.3).

The Code (Cl. 10.15.2) prescribes that for *compression members braced against lateral deflection*, the effects of slenderness may be neglected if:

$$\frac{kl_u}{r} \le \frac{25 - 10(M_1 - M_2)}{\sqrt{P_f / (f'_c A_g)}} \tag{15.1}$$

where M_1 and M_2 are the numerically smaller and larger factored end moments, respectively, but M_1/M_2 is not taken less than 0.5. These moments are obtained from a conventional (first order) frame analysis. M_1/M_2 is positive if the member is bent in single curvature, and negative if bent in double curvature. P_f is the factored axial load, and A_g is the gross area of the section. The radius of gyration, r, may be taken as 0.3 times the overall depth of a section, h, for rectangular sections, and 0.25 times the diameter for circular sections (Code (Cl. 10.14.2)).

For compression members *not* braced against lateral deflection, the Code requires the effects of slenderness, both in terms of member stability and lateral drift, to be taken into account irrespective of the value of the ratio kl_u/r of the member (see Chapter 16). Similarly, for all compression members with kl_u/r greater than 100, the design forces and moments must be determined from a "second order" analysis including the effects of the changing geometry caused by deflections, of the influence of axial loads and variable moment of inertia on member stiffnesses, and of the duration of loads.

(b) Effective Length Factor, k
The effective length, kl, of a column of length l is the distance between the inflection points in the deflected shape of the column. Columns in actual structures are usually

attached to beams or other framing members, which introduce some rotational restraint at the column ends. When relative transverse displacement between the upper and lower ends of columns is prevented, the frame is said to be braced against sidesway. A column in a braced frame is shown schematically in Fig. 15.2a. If the framing beam members have infinite flexural stiffness, the column ends cannot rotate and the column is effectively fixed at both ends, as shown in Fig. 15.2b. The effective length of the column in this case is $kl = l/2$, or $k = 0.5$. If, on the other hand, the beams have zero flexural stiffness, the column behaves as pinned at both ends (Fig. 15.2c) for which

Fig. 15.2 Effective length of columns

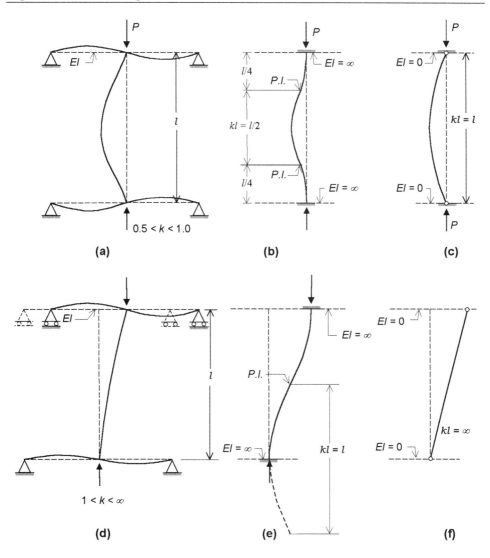

$kl = l$. Thus columns in braced frames will have $0.5 < k < 1.0$, and it is conservative to take $k = 1.0$.

Columns in an unbraced frame can sway sideways, as shown in Fig. 15.2d. In this case, beams of infinite flexural stiffness will prevent end rotation and the deflected column shape will be as shown in Fig. 15.2e for which $k = 1.0$. At the other theoretical limit of zero flexural stiffness for the beams, the column reduces to one that is pinned at both ends and permitted to sway. In this case the column is unstable and $k = \infty$ (that is, the critical load reduces to zero). Therefore, braced frames have $1.0 < k < \infty$. Approximate methods for computing k factors for columns in frames are given in Section 16.2.

15.2 DESIGN OF SHORT COLUMNS

Design of short columns is considered in the rest of this chapter. Slenderness effects and design of slender columns are described in Chapter 16.

15.3 STRENGTH AND BEHAVIOUR UNDER AXIAL COMPRESSION

As already mentioned, compression members are usually subjected also to flexure. It may be observed that CSA A23.3-94 also treats flexure and axial loads in the same section. Thus, the assumptions regarding stress and strain distributions across the cross section of an axially loaded member are the same as those used in flexure and given in Section 4.8 (Code (Cl. 10.1)). The theoretical factored resistance, P_{ro}, of the section under axial loading is a useful property and gives one of the limiting points in the strength behaviour of the column. Axial loading may be defined as loading that produces a uniform stress distribution across the cross section. For symmetrically reinforced sections (Fig. 15.3a), this would correspond to a load applied at the geometric centroid of the section. With the assumptions of stress and strain distributions given in Section 4.8, the ultimate strength of the section is reached when the axial strain reaches 0.0035. The average factored concrete stress now may be assumed as $\alpha_1 \phi_c f_c'$, and the steel at this strain would have yielded. The strain and stress distributions are as shown in Fig. 15.3a, and the corresponding factored axial load resistance, P_{ro}, is given by:

$$P_{ro} = \alpha_1 \phi_c f_c' (A_g - A_{st}) + \phi_s f_y A_{st} \qquad (15.2)$$

where P_{ro} = factored axial load resistance with zero moment
A_g = gross area of column cross section

Fig. 15.3 Axial loading

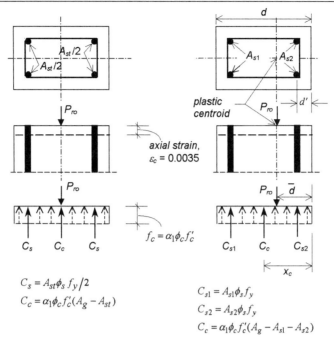

(a) Symmetrically reinforced   (b) Unsymmetrically reinforced

and A_{st} = total area of longitudinal reinforcement

When the section is not symmetrically reinforced, as with the section shown in Fig. 15.3b, axial load conditions exist when the load P_{ro}, once again given by Eq. 15.2, coincides with the resultant of the internal stresses shown. The point through which P_{ro} acts is termed the *plastic centroid*, whose location can be obtained from Eq. 15.3. In eccentrically loaded members, the eccentricity, e, is measured from the plastic centroid so that $e = 0$ corresponds to an axially loaded member. For the symmetrically reinforced section in Fig. 15.3a the plastic centroid coincides with the geometric centroid.

$$\bar{d} = \frac{C_c x_c + C_{s1} d + C_{s2} d'}{C_c + C_{s1} + C_{s2}}$$

For composite members, the value of P_{ro} can similarly be obtained as:

$$P_{ro} = \alpha_1 \phi_c f_c'(A_g - A_{st} - A_t) + \phi_s f_y A_{st} + \phi_a F_y A_t \qquad (15.3)$$

where A_t is the area of structural steel shape, pipe, or tubing in the composite section, F_y is the specified yield strength of the structural steel section, and $\phi_a = 0.90$ is the resistance factor for structural steel.

The Code (Cl. 10.3.5) limits the maximum factored axial load resistance, $P_{r(max)}$, of compression members to $0.85P_{ro}$ for members with spiral reinforcement and to $0.80P_{ro}$ for members with tie reinforcement. This limitation is to allow for unforeseen factors and accidental eccentricities that may result from misalignment and other construction errors, and to recognise that concrete strength may be less than f_c' under sustained high loads. With this limitation, the maximum factored axial load resistances are given by Eqs. 15.4 and 15.5.

For columns with spiral reinforcement,

$$P_{r(max)} = 0.85\left[\alpha_1 \phi_c f_c'\left(A_g - A_{st}\right) + \phi_s f_y A_{st}\right] \tag{15.4}$$

For columns with tie reinforcement,

$$P_{r(max)} = 0.80\left[\alpha_1 \phi_c f_c'\left(A_g - A_{st}\right) + \phi_s f_y A_{st}\right] \tag{15.5}$$

If no allowance is made for the material resistance factors (i.e., $\phi_c = \phi_s = 1.0$), the *nominal axial ultimate strength* with zero moment, P_{no}, of a column is given by:

$$P_{no} = \alpha_1 f_c'\left(A_g - A_{st}\right) + f_y A_{st} \tag{15.6}$$

A tied column falls at a nominal load, P_{no}, given by Eq. 15.6. The failure is by crushing and shearing of the concrete and by buckling of the longitudinal bars, and is sudden and brittle. A typical load versus axial shortening diagram has the shape shown in Fig. 15.4.

A spirally reinforced column behaves in a manner identical to a tied column of the same section, up to a nominal load, P_{no}, given by Eq. 15.6. At this load, the outer

Fig. 15.4 Behaviour of axially loaded tied and spiral columns

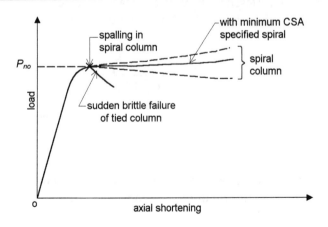

shell of concrete covering the spiral spalls off; however, the concrete in the core, laterally confined by the closely spaced spirals, continues to carry load. Failure ultimately takes place when the spiral reinforcement yields. The load capacity beyond spalling depends on the amount of spiral reinforcement, as indicated by the load-axial shortening curves shown in Fig. 15.4. Based on extensive experimental results, it has been found (Ref. 15.5) that the strength contributed by the spiral is approximately given by $2f_{sy}A_{sp}$, where f_{sy} is the yield strength of spiral reinforcement, and A_{sp} is the volume of spiral reinforcement per unit length of column. However, this additional strength is not realised until the column has been subjected to a load and deformation sufficient to cause the spalling of the concrete shell. If the amount of spiral reinforcement is such that the load capacity contributed by it compensates the loss of strength due to spalling of the concrete shell, there will be no sudden deterioration of strength and the column will continue to carry the load at spalling, given by Eq. 15.6. To achieve this behaviour, the amount of spiral reinforcement needed is obtained by equating the strength of the concrete in the shell to the strength contributed by spiral, as follows:

$$\alpha_1 f_c'(A_g - A_c) = 2f_y A_{sp}$$

or

$$\rho_s = \frac{A_{sp}}{A_c} = 0.425\left(\frac{A_g}{A_c} - 1\right)\frac{f_c'}{f_y} \quad (15.7)$$

where A_c = total area of concrete core, measured out-to-out of the spirals
ρ_s = ratio of volume of spiral reinforcement to volume of the core, per unit length of the column

CSA A23.3-94(10.9.4) requires a minimum spiral reinforcement of

$$\rho_s \geq 0.425\left(\frac{A_g}{A_c} - 1\right)\frac{f_c'}{f_y} \quad (15.8)$$

With an amount of spiral reinforcement as given by Eq. 15.8, the strength of a spiral column will be as shown by the continuous line in Fig. 15.4. The column strength is maintained even after spalling of the concrete cover, and failure is gradual, preceded by considerable yielding. Therefore, the spiral column has substantial ductility and toughness. Spirals in excess of that given by Eq. 15.8 will result in an increase in column strength after spalling; however, such increases are not of much practical value in normal situations, because of the accompanying large deformations and extensive cracking and spalling.

With the minimum amount of spiral reinforcement provided, the *factored* axial load resistance, P_{ro}, of both tied and spirally reinforced columns under axial loading is given by Eq. 15.2. Their differences in behaviour are with respect to their ductility, toughness and mode of failure. The ductile failure mode of the spiral column accounts for its larger $P_{r(\max)} = 0.85P_{ro}$, compared to $P_{r(\max)} = 0.80P_{ro}$ for tied columns.

15.4 COMPRESSION WITH BENDING

15.4.1 General Behaviour

The behaviour of members under combined axial compression and bending or, alternatively, under eccentric compressive force is discussed in this section. For simplicity, a rectangular section, with reinforcement placed near opposite faces, is used initially. Nonetheless, the basic principles involved are quite general and applicable to other shapes and distributions of reinforcement. The assumptions are the same as those described in Section 4.8 with reference to flexure. The two general principles used are the comparability of strains and equilibrium of forces. The compressive stress distribution in concrete will be assumed to be that represented by the equivalent rectangular stress block described in Section 4.8. Combined axial load, P, and bending moment, M, can be represented by a statically equivalent eccentric load, P, acting at an eccentricity $e = M/P$ (Fig. 15.5). The distribution of strain across the section at failure load, for various values of e, is shown in Fig. 15.6a. As seen in the previous section, axial compression alone ($e = 0$) causes a uniform strain in the cross section (line 1 in Fig. 15.6a). With eccentric loading, the strain variation is non-uniform. At low eccentricities the entire cross section may still be in compression. As eccentricity increases, part of the section may be subjected to tensile strains and stresses. In the limit, when e becomes infinity (that is, $P = 0$), the section is under pure flexure. The strain and stress distributions corresponding to this have already been discussed in Chapters 4 and 5.

Fig. 15.5 Compression plus bending (eccentric loading)

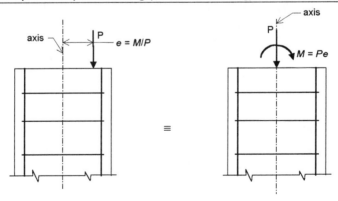

Fig. 15.6 Strength under combined compression and bending

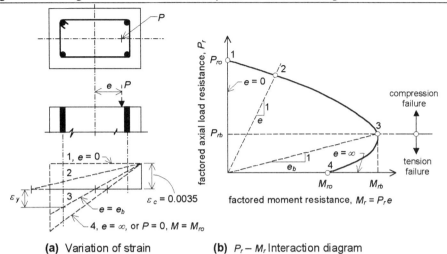

(a) Variation of strain   (b) $P_r - M_r$ Interaction diagram

In between the limiting cases of $e = 0$ and $e = \infty$, there exists a certain eccentricity, $e = e_b$, for which the strain distribution is such that as the failure load is reached, the yielding of the steel on the tension side and the attainment of a compressive strain $\varepsilon_c = 0.0035$ at the extreme fibre in the concrete occur simultaneously. This is the balanced strain condition. The factored axial load resistance corresponding to this condition is the balanced axial load resistance, P_{rb}. For eccentricity $e > e_b$, failure will be initiated by yielding of the steel on the tension side (tension failure). For $e < e_b$, failure occurs by crushing of the concrete (compression failure). The combinations of factored axial load resistance and factored moment resistance corresponding to these different values of e can be represented by a curve such as shown in Fig. 15.6b. For any given e, the value of the corresponding factored resistances, P_r and $M_r = P_r e$, may be computed using the equilibrium and comparability conditions as illustrated below for three different cases.

15.4.2 Balanced Strain Conditions

The cross-sectional dimensions, the strain distribution and the factored internal forces for this case are shown in Fig. 15.7. From the known maximum strains in concrete and steel, the depth to the neutral axis, x_b, and the depth of the equivalent rectangular concrete stress block, a_b, are directly calculated as:

$$x_b = \frac{0.0035}{f_y/E_s + 0.0035} d = \frac{700}{f_y + 700} d \qquad (15.9)$$

Fig. 15.7 Balanced strain condition

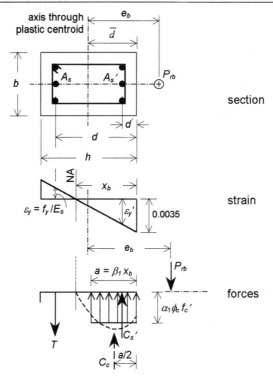

$$a = \beta_1 x_b$$

The strain and factored stress in the reinforcement on the compression side can then be computed as:

$$\varepsilon_s' = \frac{0.0035(x_b - d')}{x_b}$$

and
$$\phi_s f_s' = \phi_s E_s \varepsilon_s' \leq \phi_s f_y$$

Knowing the stress distribution, the internal forces in concrete, the compression reinforcement and tension reinforcement are given, respectively, by:

$$\begin{aligned}
C_c &= \alpha_1 \phi_c f_c' b a \\
C_s' &= A_s'(\phi_s f_s' - \alpha_1 \phi_c f_c') \\
&\cong A_s' \phi_s f_s' \\
T &= A_s \phi_s f_y
\end{aligned} \qquad (15.10)$$

Since the external load P_{rb} and internal forces are in equilibrium, summing up the forces vertically,

$$P_{rb} = C_c + C_s' - T \tag{15.11}$$

and equating moments about the plastic centroid

$$M_{rb} = P_{rb}e_b = C_c\left(\bar{d} - \frac{a}{2}\right) + C_s'(\bar{d} - d') + T(d - \bar{d}) \tag{15.12}$$

Equations 15.11 and 15.12 give the values of P_{rb}, and M_{rb} (or P_{rb} and e_b) corresponding to a balanced strain condition. A numerical example is given below.

EXAMPLE 15.1

For the column having the cross-sectional dimensions shown in Fig. 15.8 and loaded on the *x*-axis, compute the factored axial load resistance and factored moment resistance under balanced strain conditions. Take $f_c' = 30$ MPa and $f_y = 400$ MPa.

SOLUTION

Strain distribution for balanced conditions and corresponding internal forces are shown in Fig. 15.8. From these diagrams,

$\varepsilon_y = 400/200\,000 = 0.002$

Fig. 15.8 Example 5.1

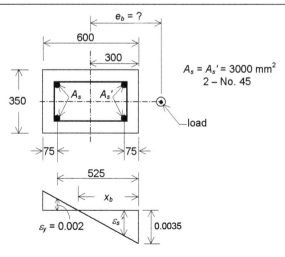

$$\frac{x_b}{525} = \frac{0.0035}{0.002 + 0.0035}, \quad x_b = 334 \text{ mm}$$

For $f_c' = 30$ MPa, $\alpha_1 = 0.805$ and $\beta_1 = 0.895$.
$a = \beta_1 x_b = 0.895 \times 334 = 299$ mm
$$\varepsilon_s' = \frac{0.0035(334 - 75)}{334} = 0.0027 > \varepsilon_y, \quad f_s' = f_y = 400 \text{ MPa}$$
$C_c = 0.805 \times 0.6 \times 30 \times 350 \times 299 \times 10^{-3} = 1516$ kN
$C_s' = 3\,000 \times (0.85 \times 400 - 0.805 \times 0.6 \times 30) \times 10^{-3} = 977$ kN
$T = 3000 \times 0.85 \times 400 \times 10^{-3} = 1020$ kN

The load, P_{rb}, and eccentricity, e_b, are such that the internal forces are in equilibrium with P_{rb}.
$P_{rb} = 1516 + 977 - 1020 = 1473$ kN

Taking moments about centroidal (plastic) axis
$P_{rb}\, e_b\, [1516 \times (300 - 299/2) + 977 \times (300 - 75) + 1020 \times (300 - 75)] \times 10^{-3} = 677$ kN·m
$e_b = 677/1473 = 0.460$ m

15.4.3 Compression Failure

When $e < e_b$, the concrete strain reaches 0.0035 before the steel on the tension side reaches yield strain. Depending on the value of e, the neutral axis depth, x, may even exceed h. If the neutral axis falls within the section, the strain distribution and the forces are as shown in Fig. 15.9a. Assume that the depth of neutral axis is x. All stresses and internal forces can be expressed in terms of x, and then the condition of equilibrium used to evaluate x and, hence, the load. The relevant equations are worked out below.

In terms of x,
$$\varepsilon_s' = \frac{0.0035(x - d')}{x}$$
and
$$f_s' = \varepsilon_s' E_s \leq f_y$$

Similarly for the steel in tension,
$$\varepsilon_s = \frac{0.0035}{x}(d - x)$$
and
$$f_s = E_s \varepsilon_s = E_s \left(\frac{0.0035(d - x)}{x} \right)$$

Fig. 15.9 Compression failure

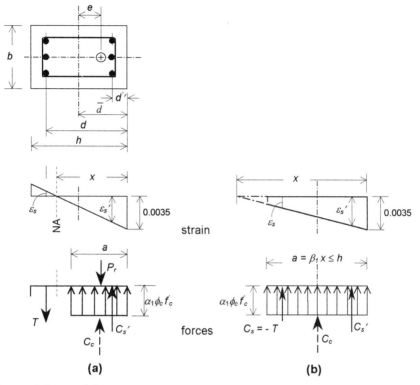

The internal factored forces are:

$$\left. \begin{array}{l} C_c = \alpha_1 \phi_c f'_c b \beta_1 x \\ C'_s = A'_s (\phi_s f'_s - \alpha_1 \phi_c f'_c) \\ T = A_s \phi_s f_s = A_s \phi_s E_s \dfrac{0.0035(d-x)}{x} \end{array} \right\} \quad (15.13)$$

C_c, C_s' and T can all be expressed in terms of x. These forces together with P_r form a system in equilibrium. Taking moments about P_r,

$$C_c \left(e - \overline{d} + \beta_1 x / 2 \right) - C_s{}' \left(\overline{d} - e - d' \right) - T \left(e - \overline{d} + d \right) = 0 \quad (15.14)$$

(Note that, in setting up the equilibrium equation, moments may be taken about any point. However, the point of application of P_r is a convenient point, because, in this case, P_r is eliminated from the equation).

Substituting in this equation, expressions for C_c, C_s' and T, a cubic equation in x will be obtained, which can be solved for x. Once x is determined, all internal forces can be computed using Eq. 15.13. The factored axial load resistance is given by:

$$P_r = C_c + C_s' - T$$

In substituting the expression for C_s' in Eq. 15.14, it is not known whether the compression steel has yielded or not. Therefore, a suitable assumption may be made, such as $f_s' = f_y$, and the assumption back-checked after determining x.

If $x > d$, the steel stress, f_s, is also compressive and, therefore, the force T is compressive (in this case, to be more exact, $T = A_s (\phi_s f_s' - \alpha_1 \phi_c f_c')$. If the value of x is such that the depth of stress block, $a = \beta_1 x$, exceeds h, then a in Eq. 15.13 has to be limited to h (see Fig. 15.9b).

Computation of the failure load for $e < e_b$ is illustrated with Example 15.2.

EXAMPLE 15.2

For the column analysed in Example 15.1, compute the factored axial load resistance at an eccentricity $e = 90$ mm.

SOLUTION

Example 15.1 showed that the steel near the compression face yielded even at balanced conditions. Since, for $e < e_b$ the strain e_s' will be even higher, it can be taken that $f_s' = f_y$. If the depth to neutral axis is x, referring to the assumed strain distribution in Fig. 15.9a,

$$f_s = (200\,000)\frac{0.0035}{x}(525 - x) = \frac{700(525 - x)}{x}$$

For $f_c' = 30$ MPa, $\alpha_1 = 0.805$ and $\beta_1 = 0.895$.
The factored forces are:
$C_c = 0.805 \times 0.6 \times 30 \times 350 \times 0.895x \times 10^{-3} = 4.54\,x$ kN
$C_s' = 3000\,[0.85 \times 400 - 0.805 \times 0.6 \times 30] \times 10^{-3} = 977$ kN
$$T = 3000(0.85)\left(\frac{700(525-x)}{x}\right)10^{-3} = 1785\left(\frac{525-x}{x}\right) \text{ kN}$$

Taking moments about the resultant along P_r,
$(4.54\,x)(210 - 0.895x/2) + 977(210 - 75) + 1785(525 - x)(525 - 210)/x = 0$
Simplifying,
$x^3 - 469x^2 - (37.8 \times 10^3)x - 14.24 \times 10^6 = 0$
Solving, $x = 577$ mm

Fig. 15.10 Example 15.2

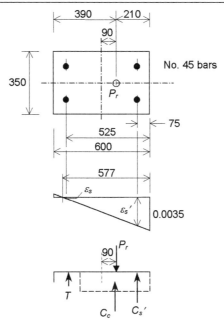

The actual strain and force distribution is as shown in Fig. 15.10. Note that $x > d$, the steel, A_s, will be in compression, and T will have a negative value (that is, compressive). Substituting for x, the factored forces are:
$C_c = 4.54x = 2620$ kN
$T = 1785 (525 - 577)/577 = -161$ kN (compression)
$P_r = C_s' + C_c - T = 977 + 2620 + 161 = 3758$ kN
The moment accompanying P_r is, $M_r = P_r e = 3758 (0.09) = 338$ kN·m

15.4.4 Tension Failure

For eccentricities greater than e_b, the steel on the tension side will yield first. This will be followed by a shifting of the neutral axis towards the compression side, an increase in the compressive strain in the extreme fibre, and ultimately failure of the concrete by crushing. At failure, the maximum compressive strain in the concrete is taken as 0.0035. Once again, the depth of the neutral axis at failure, x, may be taken as the unknown, and a solution obtained using strain comparability and equilibrium conditions as in the case of compression failure. The procedure is illustrated in Example 15.3.

EXAMPLE 15.3

If the column in Example 15.1 is loaded at an eccentricity of 900 mm, compute the nominal strength.

SOLUTION

Since $e > e_b$, the steel on the tension side yields and $f_s = f_y = 400$ MPa. It may be initially assumed that the compression steel also yields. This can be checked after the neutral axis is located. The depth of the neutral axis is assumed as x.

The factored forces are:

$C_c = \alpha_1 \phi_c f_c' b \, \beta_1 x = (0.805)(0.6)(30)(350)(0.895x) \, 10^{-3} = 4.54x$ kN
$C_s' = A_s' (\phi_s f_y - \alpha_1 \phi_c f_c') = 3000 \, (0.85(400) - 0.805(0.6)(30)) \, 10^{-3} = 977$ kN
$T = A_s \phi_s f_y = 3000(0.85)(400) 10^{-3} = 1020$ kN

Taking moments about the line of action of P_r (Fig. 15.11),
$4.54x(600 + 0.895x/2) + 977 \, (600 + 75) - 1020 \, (600 + 525) = 0$
Solving, $x = 160$ mm

Fig. 15.11 Example 15.3

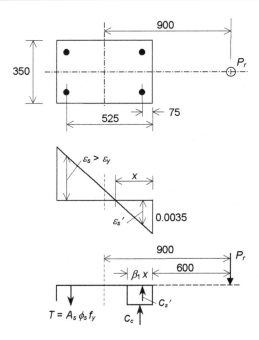

To verify whether the compression steel does in fact yield,
$$\varepsilon'_s = \frac{0.0035(160-75)}{160} = 0.00186 < \varepsilon_y$$
The compression steel does not yield. Hence, revise the calculation with,
$$f'_s = (200\,000)\frac{0.0035(x-75)}{x} = \frac{700(x-75)}{x}$$
$$C'_s = A_s'(\phi_s f_s' - \alpha_1 \phi_c f_c')$$
$$= 3000\left(\frac{0.85(700)(x-75)}{x} - 0.805(0.6)(30)\right)10^{-3} = \frac{1785(x-75)}{x} - 43.5$$

Using this correct expression for C_s' (in place of 977 kN) in the moment equation yields the solution:
$x = 167$ mm

Substituting for x,
$C_c = 4.54(167) = 758$ kN
$$C_s' = \frac{1785(167-75)}{167} - 43.5 = 940 \text{ kN}$$
$P_r = C_c + C_s' - T = 758 + 940 - 1020 = 678$ kN
$M_r = P_r e = 678(0.90) = 610$ kN·m

15.4.5 General Case

The general case of analysis involves the determination of the factored axial load resistance of a column of known cross section, loaded at a given eccentricity.

The procedures described in Section 15.3.3 and 15.3.4 are quite general and can be used for sections with reinforcement distributed around the perimeter of the section at varying distances from the compression face (Fig. 15.12). If x represents the depth to the neutral axis at failure, the strains, factored stresses, and factored forces in the reinforcement (C_s and T_s in Fig. 15.12c) and the factored compressive force in concrete, C_c can all be expressed in terms of x, using the strain diagram and cross-sectional dimensions. Considering equilibrium of internal forces and applied load,

$$P_r = \Sigma(\text{internal forces})$$

and, moment of $P_r = \Sigma(\text{moments of internal forces})$

These two equilibrium equations enable the determination of x and P_r. Verification of the yielding (or otherwise) of reinforcement and consequent revisions may be necessary.

Fig. 15.12 General case

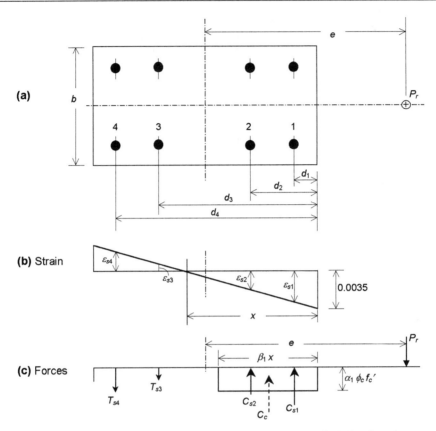

The procedure was illustrated with Examples 15.2 and 15.3, wherein an exact solution was obtained directly by setting up and solving algebraic equations. A solution may also be obtained by a trial and adjustment method. Here a neutral axis depth is initially assumed and the corresponding factored internal forces computed. The correct depth to neutral axis is such that the resultant of the internal forces will act at the given eccentricity, e, and the magnitude of this resultant will give the factored axial load resistance, P_r. The neutral axis depth is adjusted until these conditions are satisfied.

15.5 STRENGTH INTERACTION DIAGRAMS

The discussion in the preceding section has shown that the factored axial load resistance, P_r, of a column depends on the eccentricity of loading. For the particular column cross section shown in Fig. 15.8, the combinations of factored axial load resistance, P_r, and factored moment resistance, M_r, for three values of $e = M_r/P_r$, were

Fig. 15.13 *Limiting cases of ultimate strength*

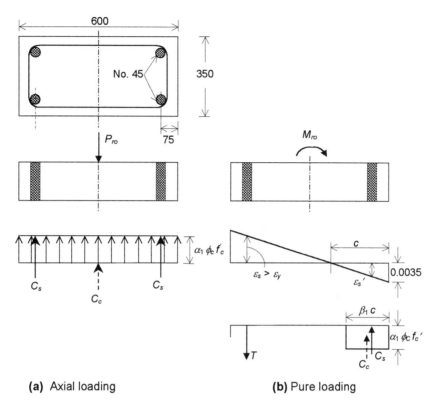

(a) Axial loading (b) Pure loading

computed in Examples 15.1-15.3. For the same section, the factored axial load resistance under purely axial loading (zero eccentricity), P_{ro}, is given by Eq. 15.3 (Fig. 15.13a).

$P_{ro} = [0.805 \times 0.6 \times 30\,(350 \times 600 - 6000) + 0.85 \times 400 \times 6000]10^{-3} = 4996$ kN

At the other extreme, under pure bending, the strains and forces are as shown in Fig. 15.13b. In this case,

$$T = C_c + C_s'$$

$300 \times 0.85 \times 400 = 0.805 \times 0.6 \times 30 \times 350 \times 0.895c$

$$+ 3000\left[0.85 \times 200000 \times \frac{0.0035(c-75)}{c} - 0.805 \times 0.6 \times 30\right]$$

Solving, $c = 110$ mm
Check: $\varepsilon_s' = 0.0035(c - 75)/c = 0.00111 < \varepsilon_y$ OK
Taking moments about tension steel,

$$M_{ro} = (0.805 \times 0.6 \times 30 \times 350 \times 0.895 \times 110(525 - 0.895 \times 110/2)$$
$$+ 3000[0.85 \times 200\,000 \times [0.0035(110 - 75)]/110$$
$$- 0.805 \times 0.6 \times 30](525 - 75)) \times 10^{-6}$$
$$= 474 \text{ kN} \cdot \text{m}$$

The combinations of factored axial load and moment resistances for this section under five different eccentricities (ranging from $e = 0$ to $e = \infty$) are presented graphically (points A to E) in Fig. 15.14. The curve $ABCDE$ drawn through these points represents the full range of combinations of P_r and M_r, and is termed a factored axial load and moment resistances interaction diagram. Combinations of P and M that are plotted on this curve represent factored resistances, whereas all points within the curve represent combinations that are safe.

As already explained in Section 15.3, to allow for unforeseen factors, accidental eccentricities, etc., the Code limits the maximum axial load resistance of compression members to 80 percent of P_{ro} for tied columns and to 85 percent of P_{ro} for columns with spiral reinforcement (Eqs. 15.4 and 15.5). For the column shown in Fig. 15.14, this limiting factored resistance is $P_{r\,(max)} = 0.8 \times 4996 = 3997$ kN. This limitation is indicated by the cut off line $A'F$ in Fig. 15.14. The axial load limitation is roughly equivalent to specifying that all columns be designed for a minimum eccentricity of load of $e = 0.1h$ for tied columns (point F in Fig. 15.14) and $e = 0.05h$ for spirally reinforced columns.

Fig. 15.14 *Factored resistance interaction diagram*

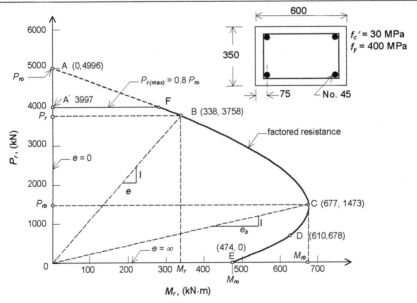

The complete factored resistance interaction diagram for the tied column shown in the inset in Fig. 15.14 is represented by the curve $A'FBCDE$ in Fig. 15.14 and is typical for common column sections. Points on the curve give factored resistances, P_r and M_r, corresponding to limiting strength conditions specified in CSA A23.3-94. The design criteria for strength require that the factored load effects do not exceed the factored resistances. It is apparent that at very low eccentricities (segment $A'F$), the factored axial load resistance is independent of the value of e, as the limiting maximum resistance controls. The region FC of the curve represents compression failure, where an increase in axial load is accompanied by a decrease in the moment the section can carry. Columns with relatively high axial loads, such as those in the lower storeys of a high rise building, fall into this category. In the region CDE of the curve, the failure is initiated by yielding of the reinforcement on the tension side. In this zone, as the axial load decreases the moment that can be resisted simultaneously also decreases. Therefore, the loading condition giving the maximum moment in the column and the minimum axial load that accompanies this moment is of greater significance in this region. It should be noted that for columns CSA A23.3-94(10.10.1) requires the consideration of the loading condition giving the maximum and minimum ratios of bending moment to axial load.

15.6 INTERACTION DIAGRAMS AS DESIGN AIDS

The factored resistance curve in Fig. 15.14, plotted in terms of P_r and M_r, is applicable for the section shown in the inset in Fig. 15.14. The factored resistance of this section, loaded at any given eccentricity, e, can be directly obtained from the graph by drawing the radial line OB (see Fig. 15.14) having a slope of $1/e$, and reading off the co-ordinates of the point B on the curve. However, by making the relationship independent of cross-sectional dimensions, as re-plotted in Fig. 15.15, the curve can be used for all sections geometrically similar to and having the same material properties as the one shown in the inset in Fig. 15.14. In Fig. 15.15, the curve is plotted in terms of parameters, P_r/A_g, and $M_r/(A_g h)$, both of which have the unit of MPa. The curve may also be plotted as fully non-dimensional, in terms of $P_r/(A_g f_c')$ and $M_r/(A_g h f_c')$.

Interaction relations such as that shown in Fig. 15.15, in tabular or graphical form, are presented in several publications (Refs. 15.6 to 15.8) as an aid to design (there are slight differences regarding the reference parameters chosen to represent the x and y axes). An extensive set of interaction diagrams, similar to that shown in Fig. 15.15, is presented in Ref. 15.6, from which the graphs shown in Figs. 15.16 to 15.22 have been reproduced. These graphs present the interaction between factored resistances, P_r and M_r, including the effect of material resistance factors ϕ_c and ϕ_s. The resistances of circular columns with ties and with spiral reinforcement (Figs. 15.20 to 15.22) differ only at very small eccentricities where the cut-off line (af in Fig. 15.15)

Fig. 15.15 Interaction diagram as a design aid

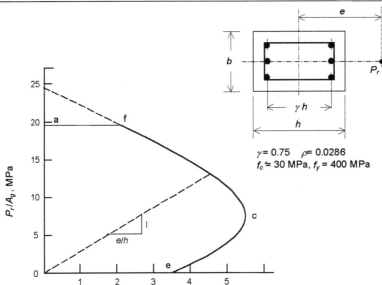

corresponding to the maximum limit for factored axial load resistance (Eqs. 15.4 and 15.5) controls. The broken lines in Figs. 15.20 to 15.22 indicate the cut-off line for a spirally reinforced column. For values of γ in between those for which interaction diagrams are available, the factored resistances may be obtained by interpolation; however, it is conservative and generally satisfactory to use an interaction diagram with γ smaller than the actual/expected value. The curves in Figs. 15.16 to 15.22 are for f_y = 400 MPa. These curves can be used for other values of f_y by multiplying the ρ_g obtained from these curves with the ratio $400/f_y$ to obtain the ρ_g corresponding to f_y. The radial lines in the interaction diagram give values of the tensile stress in the steel layer closest to the tensile face of the column. Figures 15.17 to 15.19 are applicable for columns with equal reinforcement on the two end faces, and these should be used for the common case of rectangular columns with four bars located at the corners of the column. Analysis and design procedures with the aid of these diagrams are illustrated in Examples 15.4 to 15.6.

The discussions in Section 15.4, have shown that the strength of a column section depends on several factors, such as cross-sectional dimensions, reinforcement ratio, disposition of reinforcement in the cross section, material strengths, and M/P ratio. It is apparent that for a design problem with given P_f, M_f, f_c' and f_y, there is no unique solution, and several alternative designs are possible. The design procedure involves an initial appropriate selection of some of these factors and then determining the remaining unknowns using the design aids.

Fig. 15.16 Interaction diagrams for axial load and moment resistance for rectangular columns with an equal number of bars on all four faces

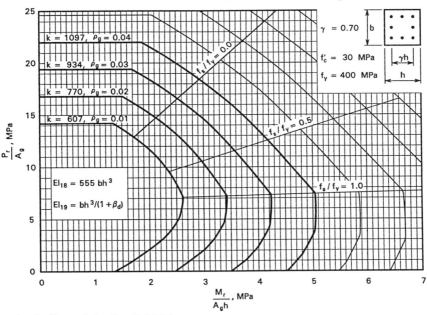

Reproduced, with permission, from Ref. 15.6

15.7 ANALYSIS OF SECTIONS USING INTERACTION DIAGRAMS

Analysis problems involve the determination of the design strength of a given section loaded at a given eccentricity. A method of solution of such problems using basic principles was explained in Section 15.4 and illustrated in Examples 15.1 to 15.3.

Solutions to analysis problems can be obtained more easily using the interaction diagrams. The procedure is illustrated in Example 15.4.

EXAMPLE 15.4

A column in a braced frame has an effective length of 3.5 m and the cross-sectional dimensions shown in Fig. 15.23. The service load effects are: dead load = 720 kN, live load = 1040 kN, dead load moment = 120 kN·m, and live load moment = 195 kN·m. All moments are about the major axis. Check the adequacy of the column for this loading condition. $f_c' = 30$ MPa and $f_y = 400$ MPa.

524 REINFORCED CONCRETE DESIGN

Fig. 15.17 Interaction diagrams for axial load and moment resistance for rectangular columns with bars in the two end faces only

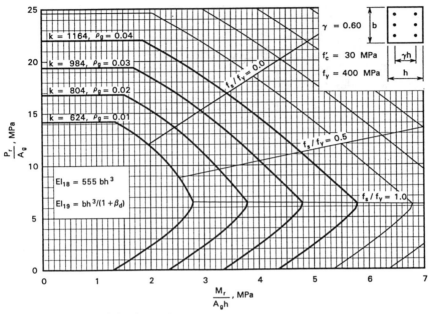

Reproduced, with permission, from Ref. 15.6

Fig. 15.18 Interaction diagrams for axial load and moment resistance for rectangular columns with bars in the two end faces only

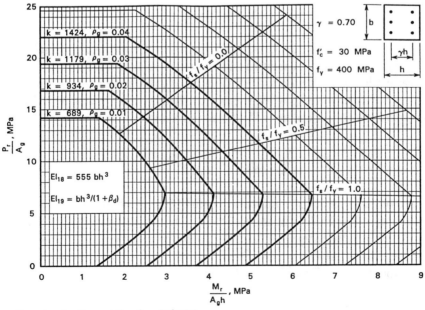

Reproduced, with permission, from Ref. 15.6

Fig. 15.19 *Interaction diagrams for axial load and moment resistance for rectangular columns with bars in the two end faces only*

Reproduced, with permission, from Ref. 15.6

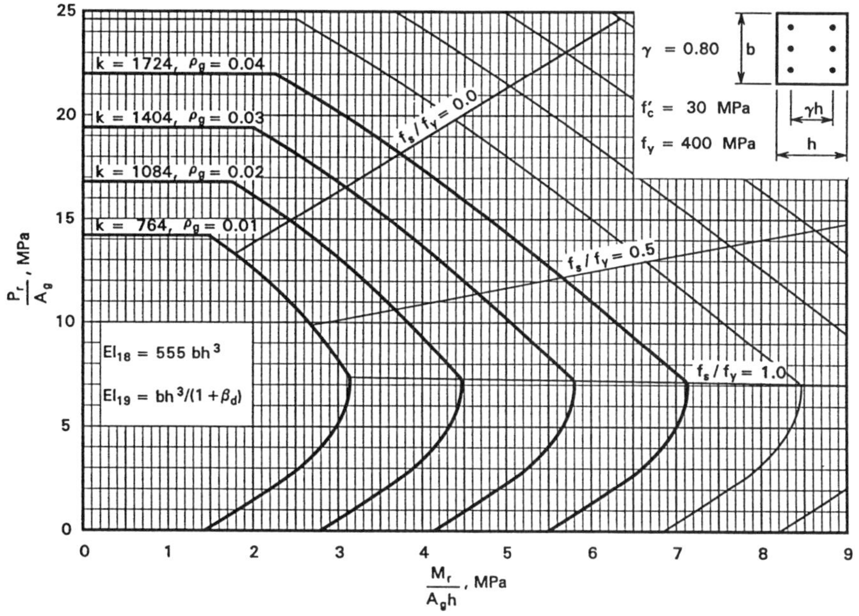

Fig. 15.20 *Interaction diagrams for axial load and moment resistance for circular tied or spiral columns*

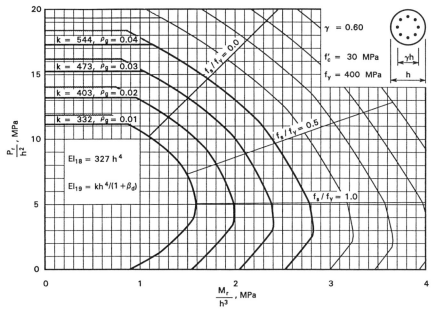

Reproduced, with permission, from Ref. 15.6

Fig. 15.21 Interaction diagrams for axial load and moment resistance for circular tied or spiral columns

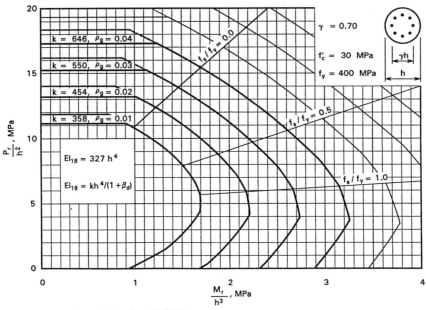

Reproduced, with permission, from Ref. 15.6

Fig. 15.22 Interaction diagrams for axial load and moment resistance for circular tied or spiral columns

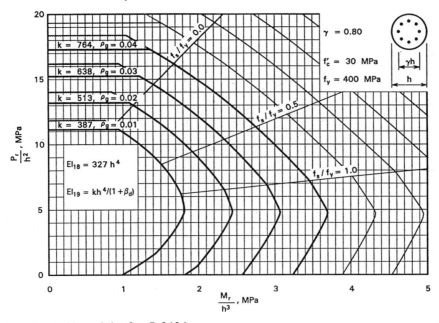

Reproduced, with permission, from Ref. 15.6

Fig. 15.23 Example 15.4

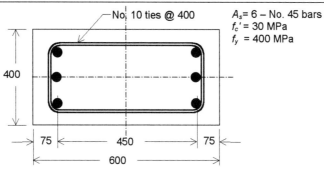

SOLUTION

1. Factored loads and moments
   Factored load = 1.25 × 720 + 1.5 × 1040 = 2460 kN
   Factored moment = 1.25 × 120 + 1.5 × 195 = 443 kN·m
   $e = M/P = 443/2460 = 0.180$ m

2. Check whether long or short column
   Assuming conservatively that the column bends in symmetrical single curvature bending (that is, $M_1/M_2 = 1$), the limiting slenderness ratio for a short column is:

   $$\frac{kl_u}{r} \leq \frac{25 - 10(M_1 - M_2)}{\sqrt{P_f/(f_c' A_g)}} = \frac{25 - 10 \times 1}{\sqrt{2460/(30 \times 400 \times 600)}} = 25.7$$

   The radius of gyration, r, may be taken as $0.3h$ for a rectangular section (Code (10.14.2)):
   $r = 0.3 \times 600 = 180$ mm
   Limiting $kl_u = 25.7 \times 180 = 4626$ mm > 3500 mm
   The slenderness effects can be neglected and the column treated as a short column.

3. Check reinforcement ratio,
   $$\rho = \frac{6 \times 1000}{400 \times 600} = 0.025 = 2.5 \text{ percent and is within the limits of 1 to 8 percent.}$$

4. Compute factored resistance
   For the section, $\gamma = 450/600 = 0.75$.
   For the material properties and reinforcement configuration, interaction

diagrams in Figs. 15.17-15.19 are applicable. Since interaction diagram for $\gamma = 0.75$ is not included here, the diagram in Fig. 15.18 for $\gamma = 0.7$ will be used. (Note that increasing γ will slightly increase the strength of a section. Therefore, using a graph for a γ value lower than the actual γ gives slightly conservative results).

Here, $e/h = 180/600 = 0.3$. In the graph in Fig. 15.19, drawing a radial line corresponding to $e/h = 0.3$, and interpolating between the curves for $\rho = 2$ percent and $\rho = 3$ percent gives:
For $\gamma = 0.7$, $e/h = 0.3$, and $\rho = 2.5$ percent
$P_r/A_g = 12.0$ MPa
$P_r = 12.0 A_g = 12.0 \times 400 \times 600 \times 10^{-3} = 2880$ kN > 2460 kN　　　　OK
and $M_r = P_r e = 2880 \times 0.180 = 518$ kN·m > 443 kN·m　　　　OK
The strength is adequate.

5. Check ties (Code (7.6.5)):
Minimum required tie size for No. 35 bars is No. 10 bars　　　　OK
For maximum spacing limits:
16 × bar diameter = 16 × 35.7 = 571 mm
48 × tie diameter = 48 × 11.3 = 542 mm
least section dimension = 400 mm
The 400 mm spacing is OK.

Clear distance between the middle longitudinal bar and the laterally supported bars on either side

$$\frac{400 - 2 \times 75}{2} - 35.7 = 89.3 \text{ mm} < 150 \text{ mm}$$

Therefore, the middle bars need not be enclosed by corners of additional ties. Use of No. 10 ties at 400 mm, as provided, is satisfactory.

15.8 DESIGN OF SECTIONS USING INTERACTION DIAGRAMS

Design of columns is usually done with the aid of interaction diagrams. It is possible to design sections using equilibrium equations following a procedure similar to that used in the analysis problems illustrated in Examples 15.1 to 15.3. However, because of the several factors involved, such a procedure requires many initial assumptions and a process of trial and adjustment. Finally, it is always possible to assume a trial section

CHAPTER 15 DESIGN OF COMPRESSION MEMBERS **529**

and then check whether it has the required strength by using the analysis procedures discussed before.

The design procedure using interaction diagrams is illustrated with Examples 15.5 and 15.6.

EXAMPLE 15.5

The analysis of a multi-storey braced building frame was done by considering one floor at a time, as suggested in CSA A23.3-94(9.3.2). From the analysis under specified loads, the loads and moments acting on an exterior column at the ground floor level is obtained as follows (see also Fig. 15.24a):

1. Column load and moment due to the loads on a single adjacent span,
   (a) Dead load: axial load on column = 144 kN
                   moment on column   = 96 kN·m
   (b) Live load: axial load on column = 78 kN
                   moment on column   = 52 kN·m

2. Column axial load accumulated from all remaining floors above:
   due to dead load = 1180 kN
   due to live load (allowing for the reduction for large tributary areas) = 370 kN
   The unsupported length of the column is 3.5 m. Design a tied column, taking $f_c' = 30$ MPa, and $f_y = 400$ MPa.

SOLUTION

1. Factored loads
   The weight of the column itself, assumed to act at its top, is calculated for an estimated column size of 500 mm square as:
   $0.5 \times 0.5 \times 3.5 \times 2.4 \times 9.81 = 20.6$ kN

   a. For loading condition giving maximum axial load and moment,
      axial load, $P_f = 1.25(144 + 1180 + 20.6) + 1.5(78 + 370) = 2353$ kN
      moment, $M_f = 1.25 \times 96 + 1.5 \times 52 = 198$ kN·m
      $e = M_f/P_f = 84.1$ mm
   b. For the loading condition giving the maximum ratio of bending moment to axial load (that is, live load only on the adjacent span),
      axial load = $1.25 (144 + 1180 + 20) + 1.5 \times 78 = 1797$ kN
      bending moment = 198 kN·m

530 REINFORCED CONCRETE DESIGN

Fig. 15.24 Example 15.5

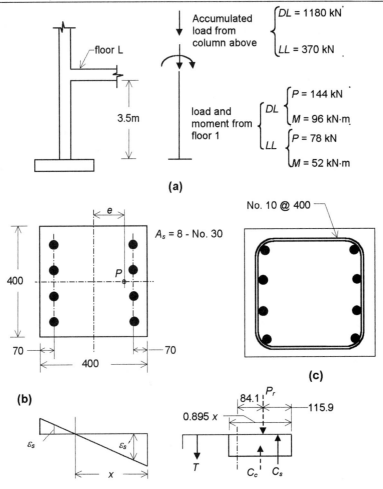

2. Approximate column size

A column with reinforcement placed on two opposite sides is selected, for which design charts in Figs. 15.17 to 15.19 are applicable. Since the column is in the ground floor, a relatively high reinforcement ratio of $\rho_g = 4$ percent may be selected, so that the same column size with lesser amounts of reinforcement may be used for the columns at the next few levels.

Figure 15.19 indicates that for $\rho_g = 0.04$, the maximum value of $P_r/A_g = 22.0$ MPa for a column loaded at low eccentricity. This gives:

Minimum $A_g = P_f/22.0 = 2353 \times 10^3/22.0 = 107 \times 10^3 \text{ mm}^2$
or, for a square column, $h_{min} = \sqrt{A_g} = 327$ mm

Since the column is loaded with $e = 84.1$ mm, which is greater than $0.1h$, a somewhat larger size is needed, and a size 400 mm × 400 mm is selected as a trial section.

3. Check whether slenderness effects can be neglected
For braced frames, limiting kl_u/r is:

$$\frac{kl_u}{r} \leq \frac{25 - 10(M_1 - M_2)}{\sqrt{P_f/(f'_c A_g)}}$$

Because of some fixity at the foundation, a column at the ground floor level in a braced frame will bend in reverse curvature. However, conservatively assuming $M_1/M_2 = 0$,

$$\frac{kl_u}{r} \leq \frac{25}{\sqrt{1797/(30 \times 400 \times 400)}} = 40.9 \; ;$$

and $(kl_u)_{\lim} = 40.9 \times 0.3 \times 400 = 4908$ mm

The actual unsupported length is 3.5 m, and since for braced frames $k \leq 1$, the actual $kl_u < (kl_u)_{\lim}$, the slenderness effects can be neglected. (Note that CSA A23.3-94 (10.15.1) stipulates that for braced columns, k be taken as 1.0 unless analysis shows that a lower value may be used.)

4. Select reinforcement
For 400 mm × 400 mm column, assuming 40 mm cover, No. 10 ties, and No. 35 longitudinal bars, $\gamma = 0.65$.
For loading condition (a):

$P_f/A_g = 2353 \times 10^3/(400)^2 = 14.7$ MPa and
$M_f/A_g h = 198 \times 10^6/(400)^3 = 3.09$ MPa
For these values, Fig. 15.18 gives: for $\gamma = 0.60$, $\rho_g = 3.0$ percent

For loading condition (b):
$P_f/A_g = 1797 \times 10^3/400^2 = 11.23$ MPa and $M_f/A_g h = 3.09$ MPa
The ρ required for this condition is less than that for loading case (a).
$A_s = \rho_g A_g = 0.030 \times 400^2 = 4800$ mm^3

Eight No. 30 bars give close to the required area. This will be provided and the actual strength for this area verified.

5. **Check strength**

   The strength of the section designed may now be checked by statics. Selecting the eight No. 30 bar arrangement shown in Fig. 15.24b, from the interaction diagram, it is apparent that the failure is in compression. The strain profile is shown in Fig. 15.24b. The forces are (assuming compression reinforcement to yield),

   $C_c = 0.805 \times 0.6 \times 30 \times (0.895x) \times 400 \times 10^{-3} = 5.19x$ kN
   $C_s' = 4 \times 700 (0.85 \times 400 - 0.805 \times 0.6 \times 30) \times 10^{-3} = 912$ kN
   $T = 4 \times 700 \times 0.85 \times 200\,000 \times 0.0035 (330 - x)/x \times 10^{-3}$
   $ = 1667 (330 - x)/x$ kN
   $e = M/P = 198 \times 10^6/2353 \times 10^3 = 84.1$ mm

   Taking moments about P_f,
   $912 \times (115.9 - 70) - (5.19x) \times (0.895x/2 - 115.9)$
   $+ 1667 (330 - x) (284.1-70)/x = 0$
   Solving, $x = 324$ mm

   Check,
   $\varepsilon_s' = \dfrac{0.0035(324 - 70)}{324} = 0.00274 > \varepsilon_y = 0.002$, and compression steel has yielded.

   The internal forces are:
   $C_c = 5.19 \times 324 = 1682$ kN
   $T = 1667 \times (330 - 324)/324 = 30.9$ kN
   $P_r = 1682 + 912 - 30.9 = 2563$ kN
   $M_r = [1682(200 - 0.895 \times 324/2) + 912(200 - 70) + 30.9(200 - 70)] \times 10^{-3}$
   $ = 218$ kN·m
   Check $M_r/P_r = 218/2563 = 0.0851$ m　　　　　　　　　　　　　　　　OK

   Factored resistances are:
   $P_r = 2563$ kN and $M_r = 218$ kN·m

   The actual self-weight of the 400 mm square column is 13 kN against the estimated weight of 20.6 kN. With this correction, the actual factored load is 2345 kN. Therefore, the factored resistances are adequate.

6. **Design of ties**
   No. 10 bar size is satisfactory.

Maximum spacing limits are:
    16 × bar diameter = 16 × 29.9 = 478 mm
    48 × tie diameter = 48 × 11.3 = 542 mm
    lateral dimension = 400 mm
Provide No. 10 ties at 400 mm spacing. The arrangement shown in Fig. 15.24c may be used.

15.9 COLUMNS OF CIRCULAR SECTION

In previous Sections, the computation of the strength of columns was illustrated with reference to rectangular sections. The procedure for calculating the strength of columns of circular cross section (or of columns of other shapes, bent in a plane containing an axis of symmetry) is identical in principle to that for a rectangular column. However, a circular column with spiral reinforcement has far greater ductility and toughness than a tied column, as shown in Fig. 15.4. Because of this, the maximum limiting axial load resistance of a spiral column is taken slightly larger than that of a similar tied column.

The cross section of a circular column is shown in Fig. 15.25a. The failure of the column may be tensile or compressive, depending on the relative eccentricity of the applied load. In either case, at failure conditions, the maximum compressive strain in the concrete is 0.0035. The variation of strain in the cross section is as shown in Fig. 15.25b, where x is the depth to neutral axis. From this diagram, the strains, and hence the stresses, in the various groups of reinforcement can be determined, in terms of x. The internal factored forces in the reinforcement are given by $A_{s1}\phi_s f_{s1}$, $A_{s2}\phi_s f_{s2}$, etc., as shown in Fig. 15.25c.

The concrete area in compression has the shape of the segment of a circle. However, the assumption of an average factored stress of $\alpha_1 \phi_c f_c'$ acting over a reduced area of depth $a = \beta_1 x$ is still applicable. On this basis, the magnitude and location of the resultant compressive force in the concrete, C_c, can be determined in terms of x. The internal forces must be in equilibrium with the factored resistance P_r applied at the eccentricity e. Hence, the value of x is such that the resultant of all internal forces acts along P_r. This condition determines x. The factored resistance, P_r, is then obtained as the resultant of all internal factored forces in the cross section. These computational procedures were illustrated earlier in Examples 15.1 to 15.3, with reference to a rectangular section. However, in the case of circular sections the distribution of the longitudinal reinforcement bars around the perimeter at varying distances from the extreme fibre in compression and the shape of the area of concrete in compression make the expressions somewhat more complex.

As for rectangular columns, in practice, analysis and design of columns of circular sections are done with the aid of interaction diagrams. The procedure is illustrated with Example 15.6.

534 REINFORCED CONCRETE DESIGN

Fig. 15.25 Circular section

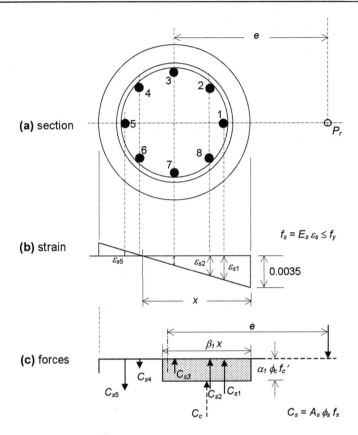

EXAMPLE 15.6

The column at the lowest level of a braced frame has an effective length $kl_u = 2900$ mm. The design loads (factored) on the column are: axial load $P_f = 3500$ kN, and moment $M_f = 250$ kN·m. Design a circular, spirally reinforced column, $f_c' = 30$ MPa, and $f_y = 400$ MPa.

SOLUTION

1. Eccentricity $e = M_f/P_f = 250 \times 10^6/(3500 \times 10^3) = 71.4$ mm

2. Approximate column size
   Since the column is at the lowest level, a large reinforcement ratio of

$\rho = 4$ percent may be selected so that the same column size with less reinforcement may be used for upper levels.

From the chart in Fig. 15.21, for $f_c' = 30$ MPa, $f_y = 400$ MPa, $\gamma = 0.7$, and $\rho = 0.04$, the maximum value of P_f/h^2 is 17.3 MPa.

minimum required $h^2 = P_f/17.3$
Hence, $= 3500 \times 10^3/17.3 = 202 \times 10^3$ mm^2
$h_{min} = 450$ mm

The actual $e = 71.4$ mm is greater than $0.05 h$. It is apparent that the actual e/h is in the neighbourhood of 0.16. Hence, the P_f/h^2 for the actual section will be somewhat less than the maximum, and a larger size column is necessary. A column 500 mm in diameter will be tried.

3. Check whether slenderness effects can be neglected.
$$\left(\frac{kl_u}{r}\right)_{lim} = \frac{25 - 10(M_1 - M_2)}{\sqrt{P_f/(f_c' A_g)}}$$

Taking, conservatively, $M_1/M_2 = 0$

$$(kl_u)_{lim} = \frac{25}{\sqrt{3500 \times 10^3/(30 \times 500 \times 500)}} \times r = 36.6 \times 0.25 \times 500 = 4575 \text{ mm}$$

Actual $kl_u = 2900$ mm $< (kl_u)_{lim}$, and slenderness effects may be neglected.

4. Reinforcement requirement
Assuming spirals with No. 10 bars, 40 mm clear cover, and No. 35 longitudinal bars, (Fig. 15.25):
$$\gamma = \frac{500 - 2 \times 40 - 2 \times 11.3 - 35.7}{500} = 0.72$$
The chart in Fig. 15.21 for $\gamma = 0.7$ can be used.

$P_f/h^2 = 3500 \times 10^3/500^2 = 14.0$ MPa
$M_f/h^3 = 250 \times 10^6/500^3 = 2$ MPa
For these coordinate values, Fig. 15.21 gives: for $\gamma = 0.7$, $\rho = 4.0$ percent.

Total area, $A_s = 0.040 \times \dfrac{\pi \times 500^2}{4} = 7850$ mm^2

Select eight No. 35 bars giving an area of 8000 mm^2, which is only slightly in excess of the requirement. For the reinforcement selected, it is now possible to carry out an exact analysis and determine the actual strengths, as was done for the rectangular section in Example 15.5. However, because of the lengthy computations involved, it is not included here.

5. Design of spirals
The minimum ratio of spiral reinforcement is (Eq. 15.8):

$$\rho_{s,min} = 0.45\left(\frac{A_g}{A_c} - 1\right)\frac{f_c'}{f_y}$$

The core diameter, D_c, measured to the outside of the spiral:

$D_c = 500 - 2 \times 40 = 420$ mm

$$\rho_{s,min} = 0.45\left(\frac{\pi \times 500^2/4}{\pi \times 420^2/4} - 1\right)\frac{30}{400} = 0.0141$$

ρ_s is the ratio of volume of spiral reinforcement to the total volume of the core.
Considering a length of column equal to the pitch, s, of the spiral (Fig. 15.26):

$$\rho_s = \frac{\text{Volume of spiral in one loop}}{\text{Volume of core in length } s} = \frac{a_s \pi (D_c - d_s)}{(\pi D_c^2/4)s} \geq \rho_{s,min}$$

Fig. 15.26 Example 15.6

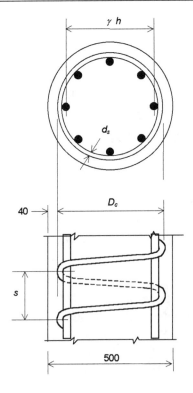

where a_s = area of spiral reinforcement
d_s = diameter of spiral reinforcement
D_c = diameter of core
and s = pitch of spiral

Rearranging this relation, in order to get the required spacing: $s \leq \dfrac{4a_s(D_c - d_s)}{D_c^2 \times \rho_{s,\min}}$

Substituting $D_c = 420$ mm, and $\rho_{s,\min} = 0.0141$, $s \leq \dfrac{a_s(420 - d_s)}{622}$

To design the spiral reinforcement, the spacings with three alternative choices for spiral bar size are tabulated below:

Bar/wire size	d_s mm	a_s mm^2	$s_{\max}$ mm	clear spacing mm
No. 10	11.3	100	65.7	54.4
9 mm	9	63.6	42.0	33.0
8 mm	8	50.3	33.3	25.3

CSA A23.3-94 (7.6.4.3) limits the pitch to one-sixth of core diameter = 420/6 = 70 mm; the clear spacing to not more than 75 mm nor less than 25 mm; and the bar size to not less than 6 mm. All three alternatives above meet these requirements; however, select No. 10 bar spiral at a pitch of 50 mm.

15.10 COMBINED AXIAL COMPRESSION AND BIAXIAL BENDING

15.10.1 General

In all the discussions so far, columns subjected to bending moment about only one principal axis (uniaxial bending) were considered. Frequently, columns such as those at corners of buildings are subjected to bending moments about both principal axes (biaxial bending). This creates no difficulty for circular columns, as the section can be designed as a column subject to uniaxial bending with the factored moment, M_f, taken as the resultant moment, $M_f = (M_{fx}^2 + M_{fy}^2)^{\frac{1}{2}}$, where M_{fx} and M_{fy} are the factored moments acting about the x and y axes (principal axes), respectively.

538 REINFORCED CONCRETE DESIGN

In the case of rectangular columns, biaxial bending results in a neutral axis inclined to the principal axes. Further, the neutral axis is not necessarily at right angles to the plane of the resultant applied moment. For a known neutral axis location, the strain distribution causing failure strain (0.0035) at the extreme compression fibre can be drawn (see Fig. 15.27). From this the factored stresses and internal forces in the steel and concrete can be computed using the procedures already used in Examples 15.1 and 15.2. The factored resistances, P_r, M_{rx} and M_{ry}, corresponding to these internal forces can readily be determined from statics. In most practical problems, the factored loads and moments are the quantities known, and it is required to design an adequate

Fig. 15.27 Biaxial bending and compression

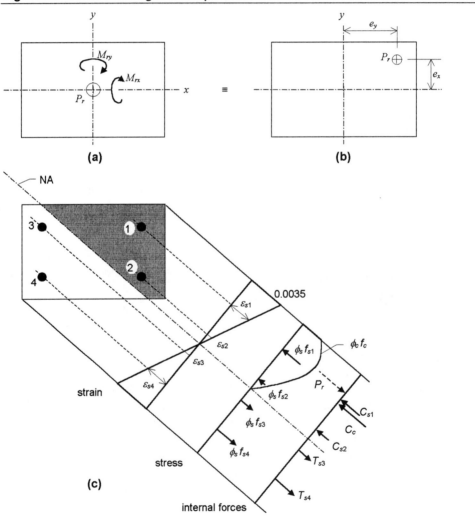

CHAPTER 15 DESIGN OF COMPRESSION MEMBERS 539

cross section to carry these loads. However, the complexities of the computation preclude a direct design procedure. Even with a trial section selected, the determination of the location of the actual neutral axis is a lengthy process of trial and error. Therefore, an analytical computation of its factored resistance is very difficult.

Several simplified and approximate procedures have been proposed (Refs. 15.9, 15.10) for the design of biaxially loaded columns, mostly based on the shape of the interaction (failure) surface; one of these is described in Section 15.10.2. Combinations of the factored axial load, P_r, and biaxial moments M_{rx} and M_{ry} (about x and y axes, respectively) corresponding to failure (that is, $e_{c,\,max} = 0.0035$) of a biaxially loaded column can be represented by a failure surface in a three-dimensional

Fig. 15.28 *Failure surface for biaxially loaded column*

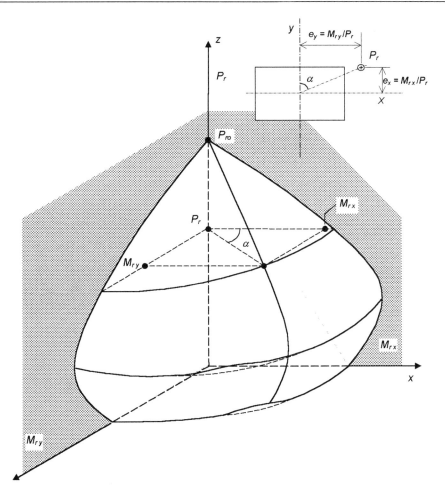

plot of $P_r = M_{rx} = M_{ry}$, as shown in Fig. 15.28. One set of values of P_r, M_{rx}, and M_{ry}, such as those corresponding to the neutral axis location shown in Fig. 15.27c, represents one point on the failure surface. The traces of the failure surface on the x-z and y-z planes correspond to the interaction curves for uniaxial bending about the x and y axes, respectively, and have the shape of the curve in Fig. 15.14.

15.10.2 Reciprocal Load Method for Biaxially Bent Columns

This approximate method to compute the axial load capacity of a section when the load is applied at eccentricities e_x and e_y was developed by Bresler (Ref. 15.9). It has been found to give satisfactory results. The method is based on an approximate representation of a point on the failure surface, plotted in terms of variables $1/P_n$, e_x, and e_y. Bresler has shown that the nominal load capacity, P_n, at eccentricities of e_x and e_y, can be approximately represented by the equation:

$$\frac{1}{P_n} = \frac{1}{P_{nx}} + \frac{1}{P_{ny}} + \frac{1}{P_{no}} \quad (15.15)$$

where P_n = nominal axial load capacity at eccentricities of e_x and e_y
P_{nx} = nominal axial load capacity for load applied at eccentricity e_x only, with $e_y = 0$
P_{ny} = nominal axial load capacity for load applied at eccentricity e_y, with $e_x = 0$
P_{no} = nominal axial load capacity with concentric loading, that is, with $e_x = e_y = 0$

Note that *nominal capacity* here means the resistance computed with the material resistance factors, ϕ_s and ϕ_c, taken as unity. Introducing the material resistance factors, for design purposes, Eq. 15.15 can be rewritten, in terms of the factored axial load resistances, as:

$$\frac{1}{P_r} = \frac{1}{P_{rx}} + \frac{1}{P_{ry}} + \frac{1}{P_{ro}} \quad (15.16)$$

where P_r = factored axial load resistance with eccentricities e_x and e_y
P_{rx} = factored axial load resistance with e_x only and $e_y = 0$
P_{ry} = factored axial load resistance with e_y only and $e_x = 0$
P_{ro} = factored axial load resistance for concentric loading

To facilitate the use of the design charts for uniaxial bending (Figs. 15.16 to 15.19), Eq. 15.16 may be multiplied by the gross area of section, A_g, to give Eq. 15.17.

$$\frac{A_g}{P_r} = \frac{A_g}{P_{rx}} + \frac{A_g}{P_{ry}} + \frac{A_g}{P_{ro}} \tag{15.17}$$

Equation 15.17 can be used to compute the resistance, P_r, of a given section, loaded with given eccentricities e_x and e_y. The three terms on the right-hand side of Eq. 15.17 can be obtained from the interaction diagrams for uniaxial bending, as indicated in Fig. 15.29. This gives the value of P_r/A_g, and hence P_r, the strength of the section with biaxial eccentricity.

For the design of a biaxially loaded column, a trial section is first selected and its adequacy checked using Eq. 15.17. One approach to select the trial section is to design it as though it is bent uniaxially with a moment slightly greater than the resultant of the applied moments (that is, $M_f = \sqrt{M_{fx}^2 + M_{fy}^2}$) acting about the principal axis about which the larger of the two components acts. The accuracy of the method decreases when the axial force is smaller than the lesser of P_{rb} or $0.1\,\phi_c f_c' A_g$. However, in this range, it is sufficiently accurate and usually conservative to ignore the axial force and design the section for biaxial bending only.

The design of a biaxially loaded column using the method described above is illustrated in Example 15.7. An alternative design procedure based on an approximate relationship between M_{rx} and M_{ry} for a constant load, P_n, (which defines a load contour on the failure surface) is described in Refs. 15.7 and 15.10.

Fig. 15.29 Components of Eq. 15.15

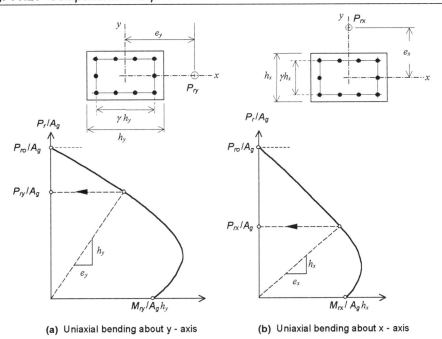

(a) Uniaxial bending about y - axis

(b) Uniaxial bending about x - axis

EXAMPLE 15.7

A corner column in the second floor of a braced frame is subjected to factored loads of $P_f = 1320$ kN, $M_{fx} = 112$ kN·m, and $M_{fy} = 80$ kN·m. The effective length of the column is $kl = 3000$ mm. Design the column, using a relatively high reinforcement ratio.

$f_c' = 30$ MPa and $f_y = 400$ MPa

SOLUTION

1. The eccentricities are:
   $e_x = M_{fx}/P_f = 112 \times 10^6 / 1320 \times 10^3 = 84.8$ mm
   $e_y = 80 \times 10^6 / 1320 \times 10^3 = 60.6$ mm
   Resultant moment is: $M_{fr} = (112^2 + 80^2)^{1/2} = 138$ kN·m

2. Estimate column size
   In practice, the approximate column size may be known from the preliminary designs, and this step may not be necessary.
   　　A reinforcement ratio of about 4 percent is selected. For 4 percent steel, the maximum value of P_r/A_g for minimum eccentricity is 22 MPa as seen from Fig. 15.16. Hence, the minimum area required to carry $P_f = 1320$ kN, with no moment, is: $1320 \times 10^3 / 22 = 60\,000$ mm²
   　　A column section having this area will have the eccentricities about both axes greater than $0.1h$. Hence, a much larger column area will be required. For trial a 350 mm square section is selected.

3. Check slenderness effects
   The loading causing maximum moment in exterior columns of braced frames bends the column in reverse curvature (see Fig. 10.4). In computing the limiting slenderness ratio for such columns, the value of M_1/M_2 in Eq. 15.1 may be taken conservatively as -0.5. Hence,
   $$\left(\frac{kl_u}{r}\right)_{lim} = \frac{25 - 10(-0.5)}{\sqrt{1320 \times 10^3 /(30 \times 350 \times 350)}} = 50.1$$
   $(kl_u)_{lim} = 50.1 \times 0.3 \times 350 = 5261$ mm
   The column is not slender.

4. Estimate reinforcement required
   Reinforcement may be selected assuming the column to be bent uniaxially with a moment equal to the resultant moment, increased by 10 to 20 percent. On

this basis, taking a 15 percent increase in moment:
$P_f/A_g = 1320 \times 10^3/350^2 = 10.8$ MPa
$M_f/A_g h = 1.15 \times 138 \times 10^6/350^3 = 3.70$ MPa

For these values, Fig. 15.16 gives:
$\rho_g = 3.6$ percent, and $A_s = 0.036 \times 350^2 = 4410$ mm^2

Select four No. 35 bars. Therefore, with corner bars only, Fig. 15.17 should be used and $\rho_g = 2.74$ and $A_s = 0.0274 \times 350^2 = 3357$ mm^2

Providing four No. 35 bars, $A_s = 4000$ mm^2, which gives
$\rho_g = 4000/350^2 = 0.033$

5. Check the adequacy of section for biaxial bending
With No. 10 ties and 40 mm cover:
$\gamma h = 350 - 2 \times (40 + 11.3) - 35.7 = 212$ mm; $\gamma = 212/350 = 0.60$

Figure 15.17 is applicable:
$e_x/h = 84.8/350 = 0.242$
$e_y/h = 60.6/350 = 0.173$

Biaxial bending strength is represented by the Eq. 15.17, as:

$$\frac{A_g}{P_r} = \frac{A_g}{P_{rx}} + \frac{A_g}{P_{ry}} + \frac{A_g}{P_{ro}} \qquad (15.17)$$

The reciprocal of the first term on the right-hand side, giving the strength under uniaxial bending about the x-axis with $e_y = 0$, is read from Fig. 15.17 off a radial line $e/h = e_x/h = 0.242$ at $\rho_g = 3.3$ percent as:

$P_{rx}/A_g = 14.1$ MPa
Similarly, for $e/h = e_y/h = 0.173$ and $\rho_g = 3.3$ percent
$P_{ry}/A_g = 16.5$ MPa

The reciprocal of the last term, P_{ro}/A_g which is the intercept of curve for $p_g = 3.3$ percent with the vertical axis in Fig. 15.17, may be calculated from Eq. 15.2 as:
$$\frac{P_{ro}}{A_g} = \alpha_1 \phi_c f_c'(1 - \rho_g) + \phi_s f_y \rho_g$$
$$= 0.805 \times 0.6 \times 30(1 - 0.033) + 0.85 \times 400 \times 0.033 = 25.2$$
Substituting in Eq. 15.17,

$$\frac{P_r}{A_g} = \frac{1}{\dfrac{A_g}{P_{rx}} + \dfrac{A_g}{P_{ry}} + \dfrac{A_g}{P_{ro}}} = \frac{1}{\dfrac{1}{14.1} + \dfrac{1}{16.5} + \dfrac{1}{25.2}} = 10.9 \text{ MPa}$$

$P_r = 10.9 \times 350^2 \times 10^{-3} = 1335$ kN

This being greater than biaxially eccentric load applied of 1320 kN, the design is adequate.

PROBLEMS

1. A column in a braced frame, with an effective length of 3 m, has the cross section shown in Fig. 15.23. Compute the factored axial load resistance of the column, the load being applied at an eccentricity of: (a) 150 mm about the major axis, and (b) 700 mm about the major axis. The material strengths are $f_c' = 40$ MPa and $f_y = 300$ MPa.

2. Figure 15.30 shows a circular spirally reinforced column at the foundation level in a braced frame. The effective length of the column is 2.7 m. Compute the factored axial load resistance of the column if the load is applied at an eccentricity of 120 mm. Take $f_c' = 30$ MPa, $f_y = 400$ MPa. Also, design the spiral reinforcement.

3. Design the cross section for a short rectangular column to carry a factored axial load of 2000 kN and a factored moment of 150 kN·m. Use a reinforcement ratio of 2 percent. Take $f_c' = 30$ MPa, $f_y = 400$ MPa.

4. A rectangular column in a braced frame has an effective length of 3 m. The column has to carry an unfactored dead load of 800 kN at an eccentricity of 180 mm and a specified live load of 1200 kN at an eccentricity of 250 mm, both eccentricities being about the major axis. Design a tied column. Take $f_c' = 30$ MPa and $f_y = 400$ MPa. Use a reinforcement ratio of about 4 percent.

5. A rectangular corner column at the ground floor level in a braced frame is subjected to a factored axial load of 1600 kN, and factored moments about the two principal axes of 140 kN·m and 190 kN·m, respectively. The slenderness effects can be neglected. Design a tied column. $f_c' = 30$ MPa and $f_y = 400$ MPa.

6. For the column in Problem 4, design a circular spirally reinforced column.

Fig. 15.30 Problem No. 2

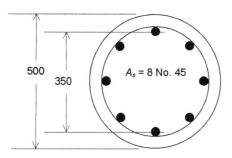

7. The biaxially loaded column, designed in Example 15.7, is now subjected to a factored axial load of P_f = 1400 kN and biaxial moments of $M_{fx} = M_{fy}$ = 80 kN·m. Check the adequacy of the design.

REFERENCES

15.1 ACI Committee 315, *Proposed ACI Standard: Details and Detailing of Concrete Reinforcement*, Concrete International, American Concrete Institute, Detroit, Vol. 2, No. 2, Feb. 1980, pp. 73-108.

15.2 *Reinforcing Steel - Manual of Standard Practice*, Reinforcing Steel Institute of Canada, Richmond Hill, 1996.

15.3 MacGregor, J.G., Breen, J.E., and Pfrang, E.O., *Design of Slender Concrete Columns*, ACI Journal, Proc., Vol. 67, No. 1, Jan. 1970, pp. 6-28.

15.4 ACI Standard 318-95, *Building Code Requirements for Structural Concrete, and Commentary*(ACI 318R-95), American Concrete Institute, Detroit, Michigan, 1995, 369 pp.

15.5 *Reinforced Concrete Column Investigation - Tentative Final Report of Committee 105*, ACI Journal, Proc. Vol. 29, No. 5, Feb. 1933, pp. 275-282.

15.6 *Concrete Design Handbook*, Canadian Portland Cement Association, Ottawa, 1995.

15.7 *Design Handbook, Vol. 2 - Columns*, ACI Special Publication SP-17 A (78), American Concrete Institute, Detroit, 1978.

15.8 *CRSI Handbook*, Concrete Reinforcing Steel Institute, Chicago, 1980.

15.9 Bresler, B., *Design Criteria for Reinforced Concrete Columns Under Axial Load and Biaxial Bending*, ACI Journal, Vol. 57, 1960, pp. 481-490.

15.10 Gouwens, A.J., *Biaxial Bending Simplified*, Reinforced Concrete Columns, ACI Special Publication SP-50, American Concrete Institute, Detroit, 1975, pp. 223-261.

CHAPTER 16 Slender Columns

16.1 BEHAVIOUR OF SLENDER COLUMNS

Slender columns carrying axial compression and bending may have their strength reduced, because of the increased bending moment caused by the transverse deflection. Figure 16.1 shows a column loaded at an eccentricity, e, at both ends, causing single curvature bending. The applied end moments, $M_o = Pe$ (which may be termed the primary moment), cause bending of the column, which increases the effective eccentricity of the load at sections away from the ends of the column. If Δ is the deflection of the column at its mid-height, the maximum bending moment in the column at this section (Fig. 16.1b) is:

$$M_{max} = P(e + \Delta) \qquad (16.1a)$$
$$= M_n + P\Delta \qquad (16.1b)$$

The additional moment due to the deflection of the column is $P\Delta$, and may be termed the "P-Δ moment". In Eq. 16.1, the deflection, Δ, itself is dependent on the moment. Therefore, the variation of M_{max} with P is nonlinear, with M_{max} increasing at a faster rate as the load, P, increases.

Figure 16.1c shows the maximum axial load-moment interaction diagram for the column cross section. If the column is very short (to be exact, of zero length), its lateral deflection, Δ, is negligibly small, and the maximum moment $M_{max} = Pe$, increases linearly with P. In this case, the P-M path, superimposed on the interaction diagram, will be along line OA in Fig. 16.1c. When the load and moment combination reaches $P = P_s$ and $M_{max} = P_s e$ corresponding to the point A on the interaction diagram, the column falls by crushing of the concrete at the section of maximum moment. If, on the other hand, the column is longer, with increasing load P, the deflection, Δ, is no longer negligible, and the moment $M_{max} = P_1(e + \Delta_1)$ will vary along the curve OB in Fig. 16.1c. Failure occurs at a load P_1, the moment $P_1(e + \Delta_1)$, represented by point B. Thus, although the column section and applied load eccentricity, e, are the same, the slenderness has caused a decrease in the load-carrying capacity of the column from P_s for the short column to P_1 for the longer one. In both cases, failure of the column occurs due to the maximum compressive strain in the concrete reaching the limiting value (this may or may not be preceded by tensile yielding of the reinforcement on the tension side, depending on which part of the interaction diagram the points A and B fall (Fig. 15.6)). Most columns in practical building frames will have this type of failure.

Fig. 16.1 Behaviour of slender columns

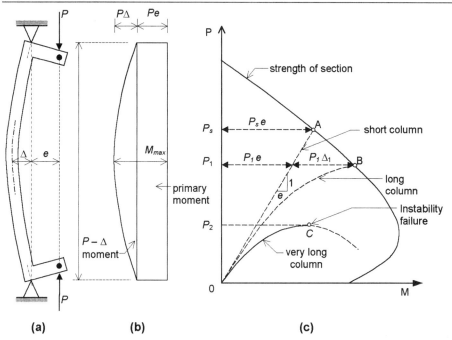

If the column in Fig. 16.1 is very long, the increase in deflection may be such that the variation of maximum moment with increasing P follows the path OC, with dP/dM reaching zero at point C. In this case the column falls by instability, and the capacity of the column is given by the peak load attained, equal to P_2. This type of failure may occur in very slender columns in unbraced frames.

The moment magnification due to the P-Δ effect is also dependent on the ratio of applied end moments, M_1/M_2, (or e_1/e_2), where M_1 and M_2 are the smaller and larger applied end moments, respectively (Fig. 16.2); and the ratio M_1/M_2 is considered positive if the column is bent in single curvature and negative if bent in double curvature. In braced columns bent in symmetrical single curvature (that is, $M_1/M_2 = +1.0$), as shown in Fig. 16.1a, the moment is always magnified by the deflection. In columns braced against sidesway, when the moments are unequal ($M_1/M_2 \neq 1.0$), and particularly when the column is bent in double curvature, the P-Δ effect due to column deflection may not always increase the moment above the maximum moment in the undeflected column. This is shown in Fig. 16.2. For the columns in Fig. 16.2, the maximum applied moment (primary moment) is at the ends of the columns, whereas the lateral deflections and related increase in moments occur away from the ends. If the deflection is relatively small, so that the additional moment is $P_y = P_{y1}$, as indicated by curve 1, the maximum moment is not affected by the P-Δ effect. If, however, the column is very slender, the deflections may be much larger,

Fig. 16.2 Columns braced against sidesway, unsymmetrical end moments – member stability effect

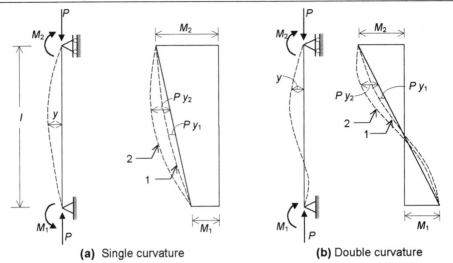

(a) Single curvature (b) Double curvature

causing an additional moment $P_y = P_{y2}$, as indicated by curve 2 in Fig. 16.2. In this case, the maximum moment is influenced by P-Δ effects.

Creep resulting from sustained loads increases column deflection. Therefore, the latter may also cause an increase in the P-Δ moment and, thereby, a decrease in the column capacity. Another factor that affects the column moment is the changes that may occur in the stiffnesses of either the column or the restraining beams framing into the column. Changes in stiffness may result from axial load, cracking, yielding, and creep.

For a column bent in symmetrical single curvature, with applied end moments of $M_o = Pe$, as in Fig. 16.1a, it can be shown (Ref. 16.1) that under elastic behaviour, the maximum moment is very nearly equal to:

$$M_{max} = M_o + P\Delta$$
$$\cong M_o \times \frac{1 \times \psi P / P_{cr}}{1 - P / P_{cr}} \qquad (16.2)$$

where ψ = a coefficient, = 0.23 for the loadings in Fig. 16.1a (ψ has slightly different, but smaller, values when the applied column moment is due to transverse loading on the column)
  P_{cr} = Euler critical load for the column

The values of both ψ and P/P_{cr} are very small in the practical range of columns. Therefore, their product is negligible compared to 1.0. With this approximation, Eq. 16.2 can be simplified to:

$$M_{max} = M_o \times \frac{1}{1-P/P_{cr}} \qquad (16.3)$$

Equations 16.2 and 16.3 are applicable for symmetrical single curvature bending only. In the derivation of Eq. 16.2, it is assumed that the maximum primary moment is at the mid-height of the column where the maximum P-Δ moment also occurs. The magnification of moment was shown to depend also on the ratio M_1/M_2 (Fig. 16.2). With unequal end moments, in columns braced against sidesway, the maximum values of the primary moment and of the P-Δ moment occur at different sections and cannot be directly added. For such columns with unequal end moments M_1 and M_2, Eq. 16.3 can be used by replacing M_o with an equivalent uniform moment equal to $C_m M_2$ (Ref. 16.1). The factor, C_m, is derived such that equal end moments of $C_m M_2$ causing symmetrical single curvature bending of the column would give the same maximum moment, M_{max}, as occurs under the actual moments, M_1 and M_2. Thus, replacing the actual moment by the equivalent uniform moment, for columns braced against sidesway.

$$M_{max} = M_2 \times \frac{C_m}{1-P/P_{cr}} \geq M_2 \qquad (16.4)$$

where
$$C_m = 0.6 + 0.4 M_1/M_2 \geq 0.4 \qquad (16.5)$$

For actual symmetrical angle curvature bending, $M_1/M_2 = 1.0$, and Eq. 16.4 reduces to Eq. 16.3.

The foregoing discussion, and the column examples in Figs. 16.1 and 16.2, dealt with braced columns where sidesway is prevented. In this case, the amplification of moment by the factor $C_m/(1 - P/P_{cr})$ is due to the axial load, P, acting on the lateral deflection due to the member curvature relative to the chord line joining the ends of the column. This is termed the *member stability effect*.

A frame which is not braced against sidesway may displace laterally resulting in a relative lateral displacement (lateral drift or sway) between the ends of the column, leading to magnification of column end moments, which is discussed in the next paragraph. A frame may be considered as braced against sidesway, or as non-sway, if the lateral loads in the direction under consideration are resisted by lateral load carrying member or members (such as shear walls, elevator shafts, stairwells, shear trusses, or other types of lateral bracing) extending from the foundation to the upper end of the frame. In such a frame, the lateral drift in the column can be negligibly small. A frame is unbraced if the frame itself resists the lateral load effects. Real structures are seldom fully braced or completely unbraced. To determine if a frame is non-sway for a

550 REINFORCED CONCRETE DESIGN

particular storey, CSA A23.3-94(10.14.4) defines the stability index, Q, as

$$Q = \frac{\sum P_f \Delta_o}{V_f l_c} \qquad (16.5a)$$

for which, if $Q \leq 0.05$, the frame may be considered as non-sway, and the columns of the frame as braced. In Eq. 16.5a, $\sum P_f$ and V_f are, respectively, the total factored vertical load and the factored storey shear in the storey, and Δ_o is the first order relative deflection of the top and bottom of the storey due to V_f. The deflection, Δ_o, is calculated using a modified value of EI, described in Section 16.2. The length of the columns, l_c, in the storey is measured from centre to centre of the horizontal framing members at the top and bottom of the storey.

An unbraced frame may sway laterally due to the effects of lateral and gravity loading, if the loading and/or the frame are unsymmetrical. The lateral sway or drift of a column is the relative transverse displacement, Δ, between the ends of the column (Fig. 16.3a). The additional moment resulting from the vertical loads acting on the

Fig. 16.3 Column subjected to sidesway — lateral drift effect

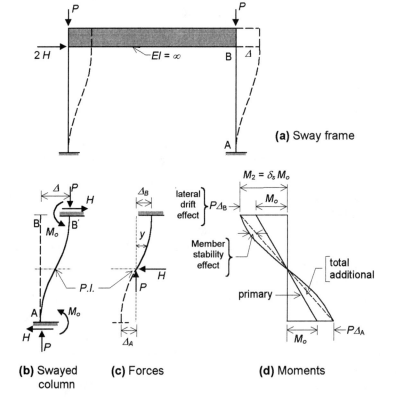

(a) Sway frame

(b) Swayed column

(c) Forces

(d) Moments

laterally displaced structure, $P\Delta_A$ and $P\Delta_B$ at column ends in Fig. 16.3d, is termed the *lateral drift effect*. In columns in unbraced frames, the magnifications of the moment may be more significant. Here, the maximum values of the primary moment, M_o, and the maximum values of the lateral drift effect both occur at the column ends and are additive, as shown in Fig. 16.3. Thus, the primary moment, amplified by the lateral drift effect, gives the end moment M_1 and M_2 in the column. In addition, there is a curvature in the member, and resulting transverse displacements at points along the length of the column relative to the chord joining the column ends. This gives rise to a further amplification of the moment over the length of the column due to the *member stability effect* described earlier (Fig. 16.2). However, usually for columns in unbraced frames, the lateral drift effect is the more dominant one and the amplification factor due to the member stability effect (Eq. 16.4) may be unity. Figure 16.3 shows a column completely restrained against rotation at the ends and subjected to sidesway. The final moments, shown in (d), are obtained by combining the primary moments with the lateral drift and member stability effects. The maximum applied (primary) moment is always increased by the lateral drift effect, while the increase due to member stability effect may not be significant. Practical frames will have beams of finite flexural stiffness. Therefore, in general, column ends are only partially restrained against rotation (Fig. 16.4). In such cases, the sidesway displacement and the additional P-Δ moment will be even greater. Furthermore, as explained in Section 15.1.3, columns in unbraced frames have larger slenderness ratios, kl/r ($k = 1$ to 4), than identical columns in braced frames. Therefore, columns in unbraced frames are weaker than similar columns in braced frames.

Fig. 16.4 Column sidesway in unbraced building frames

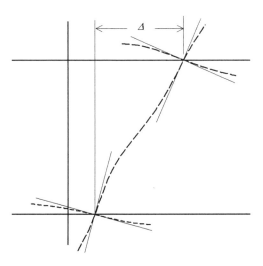

Briefly, the behaviour of slender columns can be summarised as follows:
1. The lateral deflection of the column results in significant additional moments, which, in general, reduce the load-carrying capacity of the column.
2. The moment magnification factor due to member stability effect depends mainly on the ratio of applied axial load to the critical load of the column and the end moment ratio M_1/M_2, as is evident from the second term on the right hand side of Eq. 16.4.
3. Columns free to sway are weaker than braced columns, and have the primary moment increased by both lateral drift effect and member stability effect. While lateral drift effects apply to columns in a frame with appreciable sway, member stability effects apply to all columns.

What is referred to in this section as *applied moment* or *primary moment* is the moment without inclusion of the P-Δ effects, In the case of a column in a frame, this is the moment determined by a conventional (first order) frame analysis, where the influence of the change in geometry of the frame due to deflections is neglected. If a second order analysis is made, including the effects of change in geometry of the frame due to deflections, as well as the effects of axial loads, variable moment of inertia and duration of loads on member stiffnesses, a better estimate of the real moments including slenderness effects is obtained. In lieu of such a detailed second order analysis, slenderness effects in compression members are accounted for approximately by magnifying the primary moment for lateral drift effects and member stability effects. The fundamental equations for moment magnifiers are based on the behaviour of columns with hinged ends, and these must be modified to account for the effect of end restraints. This is done by using an *effective column length*, kl_u, in the computation of slenderness effects. The effective length factor, k, was briefly discussed in Section 15.1.3b. A method for determining k for columns in frames is described in the next section. The radius of gyration is $r = (I/A)^2$, where I is the moment of inertia and A the area of the column section. The value of I is influenced by the extent of cracking and creep of concrete. These effects are considered in Section 16.3.

16.2 EFFECTIVE LENGTH OF COLUMNS IN FRAMES

Section 15.1.3 discusses how the k-factor of a column depends on the flexural stiffnesses of the beams framing into it, relative to the stiffness of the column itself. There is an accepted procedure (Ref. 16.2) for determining k for braced and unbraced frames, that includes alignment charts, shown in Figs. 16.5 and 16.6 for the graphical determination of k. These charts are in terms of two parameters, ψ_A and ψ_B, representing the degree of restraint at ends A and B, respectively, of the column under consideration. The charts are based on a simplified elastic analysis of a partial frame, consisting of the column being considered, plus all the members framing into the

Fig. 16.5 *Effective length factors for columns in braced frames*

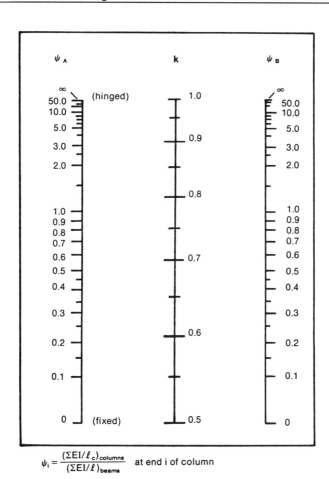

$$\psi_i = \frac{(\Sigma EI/\ell_c)_{columns}}{(\Sigma EI/\ell)_{beams}} \quad \text{at end i of column}$$

Reproduced, with permission, from Ref. 15.6

column ends and lying in the plane of bending (Fig. 16.7). Details of the analysis are also presented in Ref. 16.3.

The parameter ψ is given by
$$\psi = \frac{\Sigma\left(\dfrac{EI}{l} \text{ of columns}\right)}{\Sigma\left(\dfrac{EI}{l} \text{ of beams}\right)} \tag{16.6}$$

where Σ indicates summation for all members rigidly connected to that joint and lying in the plane of buckling of the column. For members braced against sidesway, k may

554 REINFORCED CONCRETE DESIGN

be taken as unity, since the actual value lies in the range of 0.5 to 1.0. For unbraced columns, k will be greater than 1.0.

The EI values used in Eq. 16.6 should reflect concrete cracking and the effects of sustained loading (CSA A23.3-94(10.14.1)). The Code recommends taking the modulus of elasticity of the member as that of concrete, E_c (Section 1.5), and the moment of inertia values for the members in the structure given in Table 16.1. In Table 16.1, I_g is the moment of inertia of the gross concrete section about the centroidal axis, neglecting the reinforcement. The values of EI, calculated using E_c for the modulus of elasticity of the member and the values in Table 16.1 for I, need to be divided by $(1+\beta_d)$ for braced columns and magnified sway moments (Section 16.3.3), and when sustained

Fig. 16.6 *Effective length factors for columns in unbraced frames*

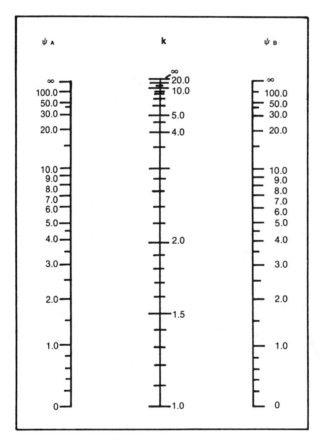

$$\psi_i = \frac{(\Sigma EI/\ell_c)_{columns}}{(\Sigma EI/\ell)_{beams}} \quad \text{at end i of column}$$

Reproduced, with permission, from Ref. 15.6

Fig. 16.7 Members considered in computing k of column

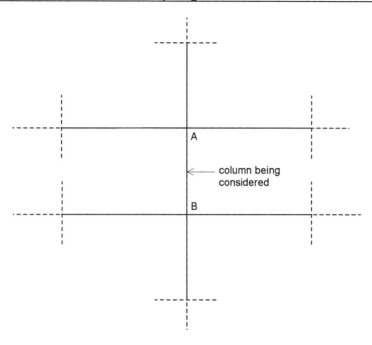

Table 16.1 *Effective Length Factors for Columns in Braced Frames (CSA A23.3-94)*

Top			k			
	Hinged		0.81	0.91	0.95	1.00
	Elastic		0.77	0.86	0.90	0.95
	Elastic		0.74	0.83	0.86	0.91
	Stiff		0.67	0.74	0.77	0.81
Member stability effects			Stiff	Elastic	Elastic	Hinged
			Bottom			

lateral loads act, or when structural stability is calculated (CSA A23.3(10.16.5)). For braced frames, except as required in CSA A23.3-94(10.16.5), β_d is the ratio of the maximum factored sustained shear within a storey to the maximum total factored shear in that storey.

The explanatory notes to CSA A23.3-94 also recommend the use of the effective length factors, k, given in Table 16.2 for braced compression members, as an alternative to the use of the monogram in Fig. 16.5 (Ref. 16.6). For braced columns and for computing member stability effects, the Code also suggests that k may be taken, conservatively, as unity.

Table 16.2 Moment of Inertia Values According to CSA A23.3-94

Beams		$0.35I_g$
Columns		$0.70I_g$
Walls	-uncracked	$0.70I_g$
	-cracked	$0.35I_g$
Flat plates and flat slabs		$0.25I_g$

16.3 DESIGN OF SLENDER COLUMNS

16.3.1 General

A detailed description of the background to the provisions in the Code for the design of slender columns is contained in Ref. 16.7. The Code requires consideration of slenderness effects for all compression members not braced against lateral deflection, and for braced compression members when the ratio kl_u/r of the member exceeds the limit set by Eq. 15.1.

The design of a slender column involves two parts:
1. The analysis of the structure to determine the forces and moments on the column, allowing for all the slenderness effects discussed in the last two sections.
2. The proportioning of the column cross section to resist the forces and moments determined in part 1 above.

The second part involves the design of the cross-sectional shape, size, and reinforcement details, necessary to resist a given axial load and moment. The procedure for this is the same as for the design of short columns, described in Chapter 15, and is not repeated here.

For the first part, CSA A23.3-94 presents two alternative approaches, namely: (1) a detailed second-order structural analysis; and (2) an approximate evaluation of slenderness effects using moment magnifiers. Procedure (1) is preferable in general, particularly for columns in unbraced frames. For all compression members with $kl_u/r > 100$, procedure (1) is mandatory. The moment magnifiers used in procedure (2) are to account for both lateral drift effects, if applicable, and the member stability effects. Comparisons with analytical and test results have shown that procedure (2) gives satisfactory results.

16.3.2 Detailed Analysis

In this approach (CSA A23.3-94(10.13.1)), the forces and moments in the compression members are determined from a second order structural analysis, which takes into account all slenderness effects, such as the influence of column and frame deflections on moments, effects of axial loads and variations in moment of inertia on member stiffnesses, and effects of sustained loads. Reference 16.5 lists the minimum required considerations for a satisfactory analysis procedure as follows:

1. use of realistic moment-curvature relationships for members,
2. effect of foundation rotations on displacements,
3. effect of axial loads on the stiffness and carry-over factors for very slender columns ($l_u/r > 45$),
4. effects of creep when sustained lateral loads are present,
5. effects of the lateral deflections of the frame, and of the column itself.

To satisfy item 1, it is acceptable to use a linear moment-curvature relation, with the flexural stiffness, EI, computed as (Ref. 16.5):

For columns, $(EI)_{col} = E_c I_g (0.2 + 1.2 \rho_t E_s/E_c)$ (16.7)

For beams, $(EI)_{beam} = 0.5 E_c I_g$ (16.8)

where ρ_t = ratio of total area of reinforcement to area of column. Procedures for second order analyses are presented in Refs. 16.8-16.10.

16.3.3 Approximate Evaluation of Slenderness Effects

In this procedure (CSA A23.3-94(10.14-10.16)), the factored axial loads and moments acting on the column are determined from a conventional elastic frame analysis. To account for the slenderness effects, the moment so obtained is magnified for the member stability effects and lateral drift effects, respectively. The column section is then designed for the factored axial load, P_f, obtained from the analysis, and the magnified factored moment, M_c.

Member Stability Effect

(a) The magnified factored moment, M_c, allowing for member stability effect, is given by:

$$M_c = \frac{C_m M_2}{1 - \frac{P_f}{\phi_m P_c}} \geq M_2 \qquad (16.9)$$

$$P_c = \frac{\pi^2 EI}{(kl_u)^2} \qquad (16.10)$$

where $C_m M_2$ is not taken as less than the moment associated with the minimum eccentricity, $P_c(15 + 0.03h)$, calculated about each axis separately, and for which h is the cross-sectional dimension in mm, in the directions of the eccentricity. Also, where

M_c = magnified factored moment, used in the design of section
M_2 = larger factored end moment on the column, considered positive
P_f = factored axial load
P_c = critical column load (Euler buckling load)
k = effective length factor
l_u = unsupported length of column
C_m = factor relating actual moment diagram to an equivalent uniform moment diagram
ϕ_m = member resistance factor, = 0.75

For columns with no transverse loads between supports, C_m is given by Eq. 16.11. For all other cases, C_m must be taken as unity.

$$C_m = 0.6 + 0.4\, M_1/M_2 \geq 0.4 \qquad (16.11)$$

where M_1 is the smaller factored end moment on the column associated with the same loading case as M_2, considered positive if the member is bent in single curvature and negative if bent in double curvature. Equation 16.9 is identical to Eq. 16.4 described earlier, except for the introduction of the member resistance factor to express the load ratio, P/P_{cr}, in terms of factored loads.

(b) Calculation of end moments M_1 and M_2.

The magnified factored moment, M_c, is calculated for both braced and unbraced frames. For this, the applicable end moments in the column, M_1 and M_2, are required.

For columns in braced frames, the drift of the frame is small even under lateral loads. Therefore, the magnification of primary end moments due to the lateral drift effect (Fig. 16.3) can be neglected. Hence, in braced frames, the column end moments, M_1 and M_2, to be used in Eq. 16.9 for computing M_c are the smaller and larger end moments calculated by a conventional elastic frame analysis.

For columns in unbraced frames, there are two conditions to be considered. For combinations of vertical loads, which do not produce *appreciable lateral deflections*, the lateral drift effect can be neglected. Therefore, the moments M_1 and M_2 are taken as those calculated by a conventional elastic frame analysis under these non-sway loads. The Code denotes these moments due to non-sway loads by M_{ns}, so that M_2 is the larger M_{ns} in this case. For this classification, the loads can be assumed to cause no *appreciable lateral deflections* if Q calculated using Eq. 16.5a is less than 0.05. Typically in a symmetric building frame, the load combination $1.25D + 1.5L$ applied uniformly may not cause appreciable lateral deflection, and the end moments obtained from a conventional elastic analysis can be directly used in Eq. 16.9 to compute the associated magnified factored moment for design. This case is illustrated in Fig. 16.8a.

The second case to be considered for columns in unbraced frames is the loading conditions that result in appreciable lateral deflections. This may occur under combinations of vertical loads (unsymmetrical pattern of loading, and/or unsymmetrical frame) or combinations of vertical and lateral loads. With lateral deflections, the column end moments are magnified due to the lateral drift effect (Fig. 16.3). In this case, the Code prescribes that the moment, M_s, due to the loads which result in appreciable lateral deflection (calculated by conventional elastic analysis), be first magnified by the factor δ_s to allow for the lateral drift effect (see Section 16.3.3(b)) and then combined with the part of moment, M_{ns}, which does not result in sway, and the so combined moments taken as M_1 and M_2 for computing M_c. This case is illustrated in Fig. 16.8b for the load combination $\alpha_D D + \psi(\alpha_L L + \alpha_w W)$. Thus, in this case, M_2 is the larger value of $(M_{ns} + \delta_s M_s)$ at either end of the column.

In columns subjected to transverse loading between supports, the maximum primary moment may occur at a section away from the supports. In such cases, the value of the largest calculated moment occurring anywhere along the member must be used as M_2 in Eq. 16.9.

If calculations show that the factored moments at both ends of a column are zero, or so small that the corresponding end eccentricities, $e = M/P$, are less than $(15 + 0.03h)$ mm, then the slender column must be designed for a moment M_2 in Eq. 16.9 corresponding to a minimum eccentricity of load of $(15 + 0.03h)$ mm, about each principal axis applied separately. (This minimum eccentricity is not applied about both axes simultaneously.) The ratio M_1/M_2 in Eq. 16.11 for this case is the actual ratio of the calculated end moments if they are not zero, and 1.0 when calculated end moments are zero.

(c) *EI* values and P_c.

In calculating the critical load using the Euler buckling formula in Eq. 16.10, the choice of an *EI* value must reflect the effects of cracking, creep, and the nonlinear concrete stress-strain relationship. Equations 16.12 and 16.13, specified in the Code

Fig. 16.8 Calculation of column end moments, M_1, and M_2 — columns in unbraced

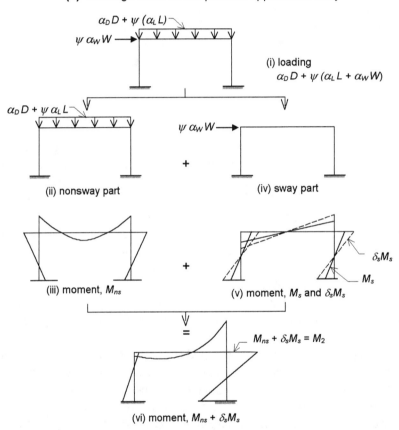

(b) Load combinations which result in appreciable sway

in lieu of a more accurate calculation represent lower bound values obtained from an evaluation of theoretical and experimental results (Refs. 16.5, 16.7). Equation 16.12 generally gives more accurate values, while the simpler Eq. 16.13 gives more conservative results. The accuracy of Eq. 16.13 improves for lightly reinforced columns. More accurate values of EI can be obtained from moment-curvature relationships for the section.

$$EI = \frac{0.2E_cI_g + E_sI_{st}}{1+\beta_d} = EI_{18} \qquad (16.12)$$

or more conservatively,

$$EI = 0.25E_cI_g = EI_{19} \qquad (16.13)$$

where notations EI_{18} and EI_{19} for the two values, with subscripts 18 and 19, identifying the respective equation numbers in Clause 10 of CSA A23.3-94, are introduced here for convenience in calculations

I_{se} = moment of inertia of reinforcement about the centroidal axis of the member cross section, and
β_d = as defined in Section 16.

Equation 16.13 is simpler to evaluate. The second term in the numerator of Eq. 16.12 depends on the amount and disposition of reinforcement. For design computations, it is convenient to express the more accurate value EI_{18} as $kbh^3/(1 + \beta_d)$, where k is a function of concrete strength, reinforcement location and steel ratio. Values of k for the various percentages and arrangements of reinforcement are presented in the interaction diagrams in Ref. 15.6 (see also Figs. 15.16 to 15.22). In terms of k,

$$EI_{18} = kbh^3/(1 + \beta_d) \qquad (16.14)$$

In calculating the critical column load, P_c, the effective length factor, k, in Eq. 16.10 may be determined using the alignment charts given in Section 16.2, or using Table 16.1. Alternatively, k may be taken as 1.0 in the computation of member stability effects.

Lateral Drift Effect

Frames not braced against lateral deflection may undergo appreciable drift, which causes magnification of column end moment (Figs. 16.3, 16.8b(v)). The magnified end moments, $\delta_s M_s$, in such a frame may be calculated by a second-order analysis (Section 16.3.2). In lieu of such an analysis, for columns with $kl_u/r \leq 100$, CSA A23.3-94(10.16.3) permits the approximate evaluation of lateral drift effect by multiplying the factored end moments, M_s, due to loads resulting in appreciable drift and calculated by conventional elastic frame analysis by a factor, δ_s, given by Eq. 16.15.

$$\delta_s = \frac{1}{1 - \dfrac{\sum P_f}{\phi_m \sum P_c}} \geq 1.0 \qquad (16.15)$$

where $\phi_m = 0.75$ and the summation applies to all columns in the storey that are rigidly attached to beams or footings and offers lateral restraint. In a multi-storey building structure, the door system acts as a stiff diaphragm, equalising the sway deflections of all the columns in a storey. Therefore, the moment magnifications due to lateral drift effect for all the columns in a storey are related. The sway modification factor, δ_s, given by Eq. 16.15 is applicable to all columns in the storey.

Alternatively, the sway modification factor, δ_s, can be calculated using Eq. 16.16, provided that the stability index, Q, calculated from Eq. 16.5a, is less than or equal to 1/3.

$$\delta_s = \frac{1}{1-Q} \tag{16.16}$$

In the calculation of $P_c = \pi^2 EI/(kl_u)^2$ in Eq. 16.15, the factor k must be computed as for an unbraced frame (Fig. 16.6) and must be greater than 1.0. The stiffness, EI, is calculated using Eq. 16.4 or Eq. 16.5. If Eq. 16.4 is used, β_d is the ratio of the maximum factored sustained shear within the storey to the maximum total factored shear in the storey. The column end moments magnified for the lateral drift effect, $\delta_s M_s$, must be used in Eq. 16.9 to calculate the final magnified factored moment, M_c, used for the design of the column (see Section 16.3.3(a) item 2 and Fig. 16.8b).

Figures 16.3 and 16.8b(v) show that, in frames not braced against sidesway, the beams at a joint are subjected to the amplified column moment. To avoid the formation of plastic hinges in the restraining beams, and to maintain the flexural restraint provided by the beams. CSA A23.3-94(10.16.2) requires flexural members to be designed for the total magnified end moments, $M_{ns} + \delta_s M_s$, from all of the columns at the joint.

Design Procedure

The column section must be designed to resist the factored axial load, P_f, and the magnified factored moment, M_c. However, the computation of M_c using Eqs. 16.9 to 16.15 requires knowing the cross-sectional dimensions, as both EI and k depend on them. (It may be recalled that member properties, at least in relative terms, are needed even for the initial structural analysis.) Thus, an iterative procedure is inevitable in the design process.

For the design of slender column sections using the moment magnifier method, two approaches may be used:
1. Start with a trial section, compute δ_s and M_c using the properties of the trial section, design the section for P_f and M_c so computed, compare with trial section, and revise if necessary.
2. Assume an initial value for M_c based on M_2 and δ_s, design the section for P_f and M_c, compute actual M_c and δ_s for the section designed, compare with initial

values, and revise if necessary.
The design of slender columns is illustrated in Examples 16.1 and 16.2.

EXAMPLE 16.1

A critical loading pattern in a braced building frame produces the following forces in an interior column: $P_f = 3000$ kN; and, at both ends, $M_f = 290$ kN·m. The moments are all due to live load, and cause symmetrical single curvature bending. The unsupported length of the column is 6 m. Design the column. Use $f_c' = 30$ MPa and $f_y = 400$ MPa.

SOLUTION

A tied rectangular section column will be used, so that the interaction diagrams in Figs. 15.17 to 15.19 can be used.

1. Trial section
   Assume about 3 percent reinforcement. For this, Fig. 15.17 gives the maximum value of $P_f/A_g = 19.5$ MPa, for a column with minimum eccentricity.
   Minimum $A_g = 3000 \times 10^3/19.5 = 153.8 \times 10^3$ mm^2
   The column, being relatively long and subjected to a large moment, may need a somewhat larger section. Select a section 400 mm wide and 500 mm deep, overall.

2. Check slenderness

$$\left(\frac{kl_u}{r}\right)_{\lim} = \frac{25 - 10(M_1/M_2)}{\sqrt{P_f/(f_c'A_g)}} \qquad (15.1)$$

$$= \frac{25 - 10}{\sqrt{(3000 \times 10^3)/(30 \times 200 \times 10^3)}} = 21.2$$

Since the frame is braced, $0.5 < k < 1.0$. The properties of the beams framing into the columns are not given so that the alignment chart (Fig. 16.5) cannot be used. Therefore, conservatively, k may be taken as unity (Code Cl. 10.15.1)).
Actual $kl_u/r = 1 \times 6000/(0.3 \times 500) = 40 > 21.2$
The column is *long* and slenderness effects must be considered.

3. Magnified moment
Using the simpler Eq. 16.13:
$EI_{19} = 0.25 E_c I_g$

$$= 0.25 \times 4500\sqrt{30} \times \frac{bh^3}{12} = 513.5 bh^3 \quad \text{(see Fig. 15.17)}$$

If the more detailed Eq. 16.14 is to be used:

$$EI_{18} = \frac{kbh^3}{1+\beta_d}$$

From Fig. 15.17, for 3 percent steel, $k = 984$

Since $\beta_d = M_D/M_{(D+L)} = 0$,
$EI_{18} = 984 \times 400 \times 500^3 = 4.98 \times 10^{13}$ N·mm²

$$P_c = \frac{\pi^2 EI}{(kl_u)^2} = \frac{\pi^2 \times 49.2 \times 10^{12}}{(1 \times 6000)^2} \times 10^{-3} = 13.4 \times 10^3 \text{ kN}$$

The magnified factored moment is (Eq. 16.9):

$$M_c = \frac{C_m M_2}{\left(1 - \dfrac{P_f}{\phi_m P_c}\right)}$$

For symmetrical single curvature bending, $M_1/M_2 = 1$, $M_2 = 290$ kN·m, and Eq. 16.11 gives $C_m = 1.0$.

$$M_c = \frac{1 \times 290}{1 - \dfrac{3000}{0.75 \times 13400}} = 413.4 \text{ kN·m}$$

4. Compute area of steel

$P_f/A_g = 3000 \times 10^3/(400 \times 500) = 15$ MPa

$$\frac{M_c}{A_g h} = \frac{413.4 \times 10^6}{400 \times 500^2} = 4.13 \text{ MPa}$$

With 40 mm cover, No. 10 ties and No. 35 bars,

$\gamma = [500 - (2 \times 40 + 2 \times 11.3 + 35.7)]/500 = 0.72$

Fig. 16.9 Example 16.1

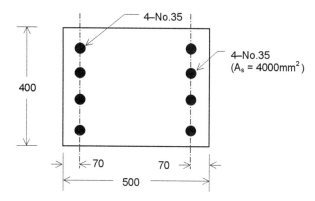

For $P_r/A_g = 15$ MPa, and $M_r/A_g h = 4.13$ MPa, Fig. 15.18 gives:

$\gamma = 0.7$, $\rho_g = 3.8$ percent
$A_s = 0.038 \times 400 \times 500 = 7600$ mm^2

Select eight No. 35 bars, giving area $A_s = 8000$ mm^2, as shown in Fig. 16.9.

Actual $\rho_g = \dfrac{8000 \times 100}{400 \times 500} = 4$ percent

The section initially selected is adequate, and the reinforcement is within normal limits. Therefore, no revision is needed. The design is slightly on the conservative side because:
(i) $k = 1$ has been used, while actual k should be less than 1.0;
(ii) for $\rho_g = 4$ percent and $\gamma = 0.70$, Fig. 15.18 shows $k = 1424$ compared to the 984 used in the EI computation; and
(iii) actual $\gamma = 0.72$, while curves for $\gamma = 0.70$ are used.

Items (i) and (ii) have the effect of increasing the critical load P_c, and hence decreasing M_c. Use of the lower γ value also gives conservative results. However, the difference is not significant.

EXAMPLE 16.2

The frame dimensions, and column sizes based on preliminary calculations for an unbraced frame are shown in Fig. 16.10. The column forces and moments obtained from a conventional frame analysis for separate loadings of $1.25D$, $1.5L$, and $1.5W$ (W = wind load) are given in Table 16.3 below.

TABLE 16.3 Example 16.2

Column Loading	AB			CD		
	1.25D	1.5L	1.5W	1.25D	1.5L	1.5W
Mom. at top, kN·m	-56	-59	-51	-8	54*	-72
Mom. at bottom, kN·m	59	61	75	9	56*	93
Axial load, kN	590	612	67	1100	1130	8

*single curvature case considered

For interior columns, live load is placed on all floors above, but in the immediate neighbouring floors, the loading causing single curvature bending is considered here. Design the columns. Use $f_c' = 30$ MPa and $f_y = 400$ MPa.

SOLUTION

1. Effective length factors
   Values of $0.7I_g$ of columns and $0.35I_g$ of beams will be used in the calculation of k factors using the alignment charts.

   For beams:
   $$\frac{0.35E_c I_g}{l_u} = 0.35 \times 4500\sqrt{30} \times \frac{350 \times 500^3}{12 \times (800-450)} = 4.17 \times 10^9 \text{ N·mm}$$

   For columns:
   $$\frac{0.7E_c I_g}{l_u} = 0.7 \times 4500\sqrt{30} \times \frac{350 \times 450^3}{12 \times (3300-500)} = 16.4 \times 10^9 \text{ N·mm}$$

Fig. 16.10 Example 16.2

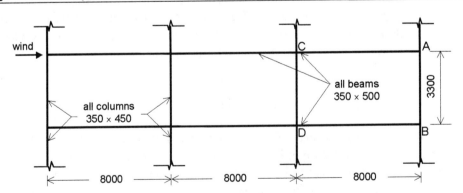

$$\psi = \frac{\sum (EI/l)_{col}}{\sum (EI/l)_{beam}}$$

For exterior columns:
$$\psi_A = \psi_B = \frac{2 \times 16.4 \times 10^9}{4.17 \times 10^9} = 7.87, \text{ from Fig. 16.6, } k = 2.68$$

For interior columns:
$$\psi_A = \psi_B = \frac{2 \times 16.4 \times 10^9}{2 \times 4.17 \times 10^9} = 3.93, \text{ from Fig. 16.6, } k = 2.02$$

The slenderness ratio is:
$$\frac{kl_u}{r} = \frac{2.02 \times 2800}{0.3 \times 450} = 41.9$$

This is less than 100 and, therefore, slenderness effects may be evaluated by the approximate method.

2. Loading cases and corresponding moments
   The load combinations to be considered for applying slenderness effects are:
   (i) $1.25D + 1.5L$
   (ii) $1.25D + 1.5W$
   (iii) $1.25D + 0.7(1.5L + 1.5W)$

   For each of the above cases of loading, the parts of moment due to loads causing appreciable sway, M_s, must be magnified by the sway magnification factor, δ_s, and combined with the non-sway part of the moment, M_s, before calculating M_c. For the three-load combinations, the moments M_s and M_{ns}, and the axial load are tabulated separately for exterior and interior columns in lines 1 to 3 in Table 16.4. 3.

3. Sway magnification factor, δ_s
   Assuming 40 mm clear cover, No. 10 ties and No. 35 bars,

$$\gamma = \frac{450 - (2 \times 40 + 2 \times 11.3 + 35.7)}{450} = 0.7$$

The value of EI will be computed using Eq. 16.12. Assuming $\rho_g = 0.02$, for $\gamma = 0.7$, Table 15.18 gives $k = 934$, so that

$$EI_{18} = \frac{kbh^3}{1+\beta_d} = \frac{934 \times 350 \times 450^3}{1+\beta_d} = \frac{29.8 \times 10^{12}}{1+\beta_d} \text{N} \cdot \text{mm}^2$$

568 REINFORCED CONCRETE DESIGN

The Euler load, P_c, is

$$P_c = \frac{\pi^2 EI}{(kl_u)^2} \qquad (16.10)$$

$$= \pi^2 \times \frac{29.8 \times 10^{12}}{1+\beta_d} \times \frac{1}{(k \times 2800)^2} \times 10^{-3} = \frac{37497}{k^2(1+\beta_d)} \text{ kN}$$

Values of the effective length factor k already determined, factor β_d giving the ratio of maximum factored dead load moment to the maximum factored total load moment, and the Euler load P_c given by Eq. 16.16 are tabulated in Table 16.4 at lines 4 to 6.

Sway magnification factor, δ_s, is applicable for loading cases (ii) and (iii). Using Eq. 16.15,

$$\delta_s = \frac{1}{1 - \dfrac{\sum P_f}{\phi_m \sum P_c}} \geq 1.0$$

For loading case (ii) (note that for wind load alone, $\sum P_f$ for a storey $= 0$),

$$\delta_s = \frac{1}{1 - \dfrac{2 \times 590 + 2 \times 1100}{0.75(2 \times 3434 + 2 \times 8358)}} = 1.24$$

For loading case (iii),

$$\delta_s = \frac{1}{1 - \dfrac{2 \times 1018 + 2 \times 1891}{0.75(2 \times 3681 + 2 \times 6611)}} = 1.60$$

The column end moments for computation of member stability effects are $M_{ns} + \delta_s M_s$, and these are tabulated in lines 8 and 9 of Table 16.4. Of these, the numerically larger end moment is taken as M_2 and the smaller as M_1. All columns are bent in double curvature, except column CD under loading case (i).

4. Moment magnification for member stability effect
   The magnified factored moment considering member stability is, by Eq. 16.9,

$$M_c = \frac{C_m M_2}{1 - \dfrac{P_f}{\phi_m P_c}} \geq M_2 \qquad (16.17a)$$

where, $C_m = 0.6 + 0.4\, M_1/M_2 \geq 0.4$ \qquad (16.17b)

CHAPTER 16 SLENDER COLUMNS 569

Table 16.4 Example 16.2

	Column AB						Column CD					
	(i) $1.25D+$ $1.5L$	(ii) $1.25D+1.5W$		(iii) $1.25D+0.7(1.5L$ $+1.5W)$			(i) $1.25D+$ $1.5L$	(ii) $1.25D+1.5W$		(iii) $1.25D+0.7(1.5L$ $+1.5W)$		
	M_{ns}	M_{ns} $1.25D$	M_s $1.5W$	M_{ns} $1.25D+$ $1.05L$	M_s $1.05W$		M_{ns}	M_{ns} $1.25D$	M_s $1.5W$	M_{ns} $1.25D+$ $1.05L$	M_s $1.05W$	
1 Mom. at top	-115	-56	-51	-97.3	-35.7		46	-8	-72	29.8	-50.4	
2 Mom. at bottom	120	59	75	102	52.5		65	9	93	48.2	65.1	
3 Axial load, P_f	1202	590	67	1018	47		2230	1100	8	1891	6	
4 Eff. length factor, k		2.68						2.02				
5 Ratio, β_d	0.49	0.52		0.42			0.17	0.10		0.39		
6 $P_c = 38\,000/[k^2(1+\beta_d)]$	3503	3434		3681			7855	8358		6611		
7 δ_s		1.24		1.60				1.24		1.60		
8 $M_{ns}+\delta_s M_s$, at top, $=M_1$	-115	-119		-154			46	-97.3		-50.8		
9 $M_{ns}+\delta_s M_s$, at bottom, $=M_2$	120	152		186			65	124		152		
10 $C_m = 0.6+0.4M_1/M_2 \geq 0.4$	0.4	0.4		0.4			0.88	0.4		0.47		
11 $P_c = 38\,000/(1+\beta_d)$	25 160	24 664		26 438			32 052	34 104		26 976		
12 $M_c \geq M_2$, (kN·m)	120	152		186			65	124		152		
13 P_f, (kN)	1202	657		1065			2230	1108		1897		
14 $M_c/(A_gh)$, MPa	1.69	2.14		2.62			0.92	1.75		2.14		
15 P_f/A_g, MPa	7.63	4.17		6.76			14.2	7.03		12.0		
16 $C_m M_2$	48	60.8		74.4			26.0*	49.6		60.8		
17 $P_f(15+0.03h)$	34.3	18.7		30.4			63.6*	31.6		54.1		

* here $C_m M_2 < P_f(15+0.03h)$

In computing M_c, the critical load, P_c, may be computed conservatively with the effective length factor, k, taken as 1.0, so that

$$P_c = \frac{\pi^2 EI}{l_u^2} = \pi^2 \times \frac{29.8 \times 10^{12}}{1+\beta_d} \times \frac{1}{2800^2} \times 10^{-3} = \frac{37497}{1+\beta_d} \text{ kN} \qquad (16.19c)$$

Values C_m, P_c, and M_c calculated using Eqs. 16.17b, c, a are tabulated at lines 10 to 12 in Table 16.4. In all cases, $M_c = M_2$, occurs at the column ends. The value of the factored axial load, P_f, to be used for design is given in line 12 of Table 16.4.

5. Design of columns

The design of the column for the moment, M_c, and load, P_f, is made as for short columns, illustrated in Chapter 15. Here use will be made of the interaction diagrams in Figs. 15.17 to 15.19.

The stress parameters $M_c/(A_g h)$ and P_f/A_g for the columns are given at lines 14 and 15 in Table 16.4. From the combinations of these parameters for various loading cases and entering chart in Fig. 15.19 for $\gamma = 0.7$, the required reinforcement ratios are:

for column, AB, $\rho_g = 0.01$ (for loading case (iii)), and

for column CD, $\rho_g = 0.011$ (for loading case (iii)).

$A_s = 0.011 \times 350 \times 450 = 1733 \text{ mm}^2$

Four No. 25 bars give an area of 2000 mm² and $\rho_g = 0.0127$, and may be used uniformly for all columns. Note that EI value was calculated using Eq. 16.12 with the factor k taken as 934 for an assumed steel ratio of 0.02. For $\rho_g = 0.0127$, k and EI will be slightly lower and the computation may be revised for this.

PROBLEMS

1. For the problem in Example 16.1, check whether a column 350 mm × 500 mm with four No. 35 bars is adequate.

2. Design the columns in the problem in Example 16.2 (Fig. 16.10) if the storey height is 3.5 m, assuming the moments to be the same as given in Table 16.3.

REFERENCES

16.1 Johnston, B.G., (ed) *SSRC Guide to Stability Design Criteria for Metal Structures*, 3rd ed., John Wiley and Sons, Inc., New York, 1976.

16.2 Kavanagh, T.C., *Effective Length of Framed Columns*, Transactions, ASCE, Vol. 127, Part II, 1962, pp. 81-101.

16.3 Salmon, C.G., and Johnson, J.E., *Steel Structures C Design and Behaviour*, 2nd ed., Harper and Row, New York, 1980.

16.4 Breen J.E., MacGregor, J.G., and Pfrang, E.O., *Determination of Effective Length Factors for Slender Concrete Columns*, ACI Journal, Proc. Vol. 69, No. 11, Nov. 1972, pp. 669-672.

16.5 FIP Recommendations, *Practical Design of Reinforced and Prestressed Concrete Structures*, Thomas Telford Ltd., London, 1984.

16.6 Collins, M.P., Dilger, W., Loov, R., MacGregor, J.G., MacGrath, R., Mitchell, D., Mutrie, J., and Simmonds, S., *Explanatory Notes on CSA Standard A23.3-94*, Canadian Portland Cement Association, Ottawa, Ontario, Canada, 1995.

16.7 MacGregor, J.G., Breen, J.E., and Pfrang, E.O., *Design of Slender Concrete Columns*, ACI Journal, Vol. 67, No. 1, Jan. 1970, pp. 6-28.

16.8 Wood, B.R., Beaulieu, D., and Adams, P.F., *Column Design by P-Delta Method*, Proc., ASCE, Vol. 102, No. ST2, Feb. 1976, pp. 411-427.

16.9 Wood, B.R., Beaulieu, D., and Adams, P.F., *Further Aspects of Design by P-Delta Model*, Proc., ASCE, Vol. 102, No. ST3, Mar. 1976, pp. 487-500.

16.10 MacGregor, J.G., and Hage, S.E., *Stability Analysis and Design of Concrete*, Proc., ASCE, Vol. 103, No. ST10, Oct. 1977, pp. 19-53.

CHAPTER 17 Footings

17.1 GENERAL

The loads acting on all structures erected on the ground have to be transmitted to the supporting soil. The part of the structure usually built below the ground surface, which interacts with the soil base and effects the transfer of loads from the parts above (superstructure) to the soil is identified as the foundation structure or substructure. In reinforced concrete structures, the loads from the superstructure are brought to the foundation level mostly through compression members such as columns, walls, and piers. The average specified (service) load stresses in these compression members are usually in the range of 7-20 MPa, whereas the allowable bearing pressures on the soil base may range from 200 to 500 kPa, and may be even less. Thus, the safe transfer of the high intensity loads to the relatively weaker soil is the function of the foundation structure. When soil of sufficient strength is available within a short vertical distance below the ground surface, the load transfer can be effected by spreading the load over a sufficiently large area around the local point. This type of foundation structure is generally classified as a spread foundation, and consists of different types of footings (Fig. 17.1) and mats. Mat (also called raft) foundations are used when the bearing capacity of the soil is low, or when the columns loads are so large that spread footings, if used, will cover most of the area of the building. They consist of solid two-way stabs (flat plates or flat slabs, upside down), inverted slab-beam-girder systems, or other variations of these, that usually extend over the entire foundation area (Fig. 17.2). Because of their high rigidity, mat foundations reduce both total and differential settlement. In the case of very weak surface and near-surface soils, resort is made to deep foundations, such as piles and caissons.

    Footings are the most widely used type of foundation in usual situations and are relatively cheap to build. This chapter is primarily concerned with the design of simple types of footings. The design of more complex forms of foundations is beyond the scope of this text, and for these reference may be made to books on foundation engineering (Refs. 17.1, 17.2).

17.2 TYPES OF FOOTINGS

Different types of common footings are shown in Fig. 17.1. A footing supporting a single column is called an isolated footing or an individual column footing. Such footings may be of uniform thickness, or they may be stepped or sloped, varying in

Fig. 17.1 Types of footings

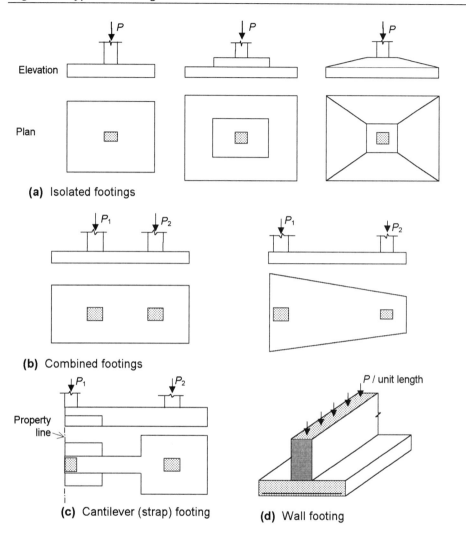

(a) Isolated footings

(b) Combined footings

(c) Cantilever (strap) footing

(d) Wall footing

thickness to suit the bending moment variation. In plan, the shape is usually square or rectangular, although other shapes may be used in special situations. Footings supporting two or more columns are termed combined footings (Section 17.6, Fig. 17.1). A variation of this type is the cantilever (or strap) footing.

The soil reaction from below causes a column footing to bend into a saucer-like shape. Therefore, the footing is generally reinforced in two directions, perpendicular to each other and parallel to the edges, similar to a flat slab supported on a column. Wall footings, similar to column footings, distribute the load from the wall to a wider area and are continuous throughout the length of the wall. However, they

Fig. 17.2 *Two types of mat foundation (flat plate, flat slab)*

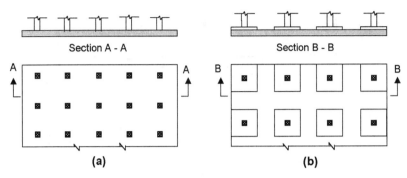

(a) (b)

bend essentially in one direction only, transverse to the length of the wall. Therefore, wall footings are reinforced mainly in the transverse direction.

17.3 ALLOWABLE SOIL PRESSURE

The base area of a footing is selected so as to limit the maximum soil pressure below it, due to the applied loads, to within safe limits. The safe soil pressure that can be allowed at the foundation level is determined using the principles of soil mechanics. The primary considerations in the determination of the allowable soil pressure are (1) that the soil does not fail under the applied loads; and (2) that the settlements, both overall and differential, are within the limits permissible for the structure. In general, structural damage is caused more by differential settlement than by a uniform vertical settlement or a slight rigid body rotation. Determination of allowable soil pressure is described in detail in Refs. 17.1 and 17.2.

The allowable soil pressure, q_a, given to the structural designer includes a suitable factor of safety. The safety factor ranges from 2 to 6, depending on the type of soil, and related uncertainties and approximations. Since the factor of safety is included in q_a, it is applicable to specified (service) load conditions. Therefore, the calculation for the required area of footing must be made with q_a, and applied specified loads, or, alternatively, with an increased soil pressure and the factored loads.

For computing required areas of footings under combinations of specified loads including dead plus live plus wind or earthquake, etc., the load combination factor, ψ, given in CSA A23.3-94(8.3.2.2) may be taken into account in computing load effects.

The allowable pressure for a given depth, D, of foundation may be expressed as the net or the gross value. The net pressure is the value that can be safely carried at the foundation depth, in excess of the existing overburden pressure. The gross pressure is the total pressure that can be carried, including the pressure due to existing overburden. In the design examples in this chapter, it is assumed that the allowable bearing pressure, q_a, indicated is the gross value. With the gross soil pressure known, the required area of footing must be determined from the total load (including the weight of the footing itself and of the backfill, if any), which is applied at the base of the footing.

If, instead of an allowable soil pressure, q_a, applicable for service load conditions, a factored soil bearing capacity, q_r, corresponding to ultimate limit state is given, the area of footing can be computed for the ultimate limit state under factored loads. In the examples in this chapter it will be assumed that the bearing capacity given corresponds to service load conditions.

17.4 SOIL REACTION UNDER FOOTING

The distribution of the soil reaction acting at the base of the footing is somewhat non-uniform, and is dependent on the rigidity of the footing itself and the properties of the soil. However, it is usual practice in structural design, and generally satisfactory, to assume a linear distribution of pressure at the base of the footing.

In a symmetrically loaded footing, where the resultant vertical load passes through the centroid of the footing, as shown in Fig. 17.3a, the distribution of the soil pressure is assumed uniform, and its magnitude, q, is given by

$$q = \frac{P}{A} \leq q_a$$

where P is the total vertical load above the foundation level under combinations of specified loads, and A is the base area of the footing. Limiting q to the allowable soil pressure, q_a, will give the minimum required area of footing:

$$\text{required } A \geq \frac{P}{q_a} \tag{17.1}$$

The load combination for computing P under service load conditions is (CSA A23.3-94(8.3.2)):

$$P = D + \psi(L + W + T)$$

where the load combination factor, ψ, has the values given in Section 3.6.

When the resultant vertical load is eccentric with respect to the centroid of the footing, the soil pressure is non-uniform. The eccentricity may result from one or more of the following effects: (1) the column transmitting a moment in addition to the

Fig. 17.3 Assumed distribution of soil reaction

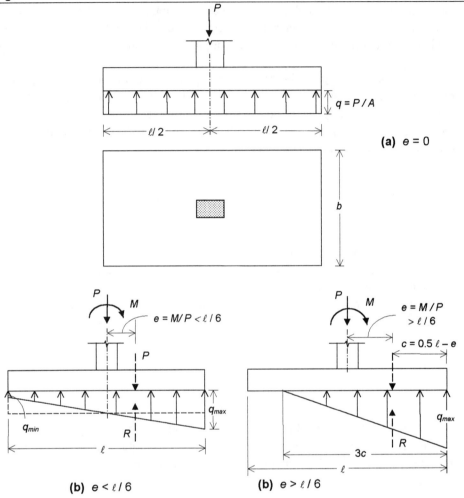

vertical load (Fig. 17.4a); (2) the column carrying a vertical load being offset from the centre (Fig. 17.4b); and (3) the column transmitting a lateral force located above the foundation level, in addition to the vertical load (Fig. 17.4c). In cases (1) and (3), if the moment and shear are fixed in direction and proportional to the load in magnitude, it is possible to offset the footing so that the resultant line of thrust passes through the centroid of the footing, and the pressure distribution is uniform (Fig. 17.4d). However, in practice, the moment (and lateral force) may vary in magnitude and alternate in direction (for example, the moments due to wind load), in which case offsetting the footing to obtain a uniform pressure distribution is not possible.

The general case of an eccentric loading is shown in Fig. 17.3b. Here, the

Fig. 17.4 Eccentric loading on footing

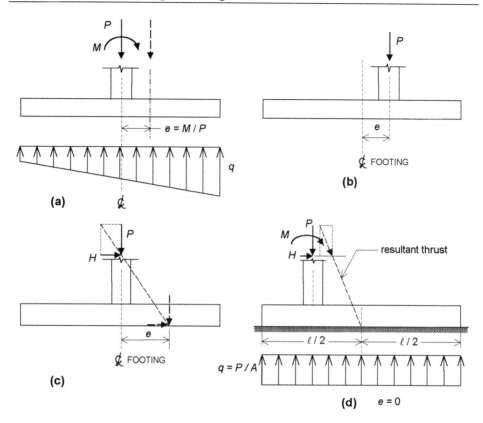

resultant vertical load, P, has an eccentricity, e, about one of the principal centroidal axes of the contact area between the footing and the soil below. Assuming the footing to be rigid, and a linear distribution of soil pressure, the latter may be computed by combining the pressures due to the vertical load and due to the moment. Thus, the extreme values of the pressure, q_{min} and q_{max}, are given by

$$q_{min,max} = \frac{P}{A}\left(1 \pm \frac{6e}{l}\right) \qquad (17.2)$$

where A is the area of the footing ($b \times l$ in Fig. 17.3a), and l is the length of the footing in the direction of bending. Equation 17.2 is valid so long as q_{min} is not tensile, that is $e \leq l/6$, or the resultant load acts within the middle third of the length, l.

When $e > l/6$, Eq. 17.2 will indicate that q_{min} is tensile. Since little tension can be developed between the footing and the soil below, it is common practice to assume no tensile force at the interface, and then the pressure distribution is triangular as shown in Fig. 17.3c. The resultant R of the soil pressure must coincide with the resultant vertical load of eccentricity e. This condition gives the effective width of contact with the soil base as $3c$, as shown in Fig. 17.3c. The maximum soil pressure is twice the average value and is readily computed as:

$$q_{max} = \frac{2P}{3bc} \qquad (17.3)$$

In all cases considered above, the dimensions of the footing must be selected so that q_{max} does not exceed the allowable soil pressure, q_a. Differential soil pressure at the base of the footing, such as shown in Fig. 17.3c, may lead to tilting of the footing, which in turn will reduce the degree of fixity at the column base. In general, the footing geometry must be selected to minimise differential soil pressure across the base, and particularly so if the soil is compressible.

17.5 GENERAL DESIGN CONSIDERATIONS FOR FOOTINGS AND CODE REQUIREMENTS

17.5.1 General

The base area required for a footing is determined from the forces and moments due to unfactored loads, in whatever combinations they may occur, and the allowable soil pressure, q_a (Eq. 17.1). Once the area has been determined, the subsequent structural design of the footing is done for the factored loads, following the Strength Design procedure of CSA A23.3-94. In order to compute the factored moments, shears, etc., acting at critical sections of the footing, a fictitious factored soil pressure, q_f, corresponding to the factored loads, is calculated using the factored loads.

The major considerations in the structural design of the footing are flexure, shear (both one-way and two-way action), bearing, and development of reinforcement. In these aspects, the design procedures are similar to those for beams and two-way slabs supported on columns. Additional considerations for footings involve the transfer of force from the column to the footing, and in some cases where horizontal forces are involved, safety against sliding and overturning.

CSA A23.3-94(15.7) specifies a minimum depth of the footing above the bottom reinforcement of 150 mm for footings on soil and 300 mm for footings on piles. For round and regular polygon shaped columns, the critical sections for moment, shear, and development of reinforcement in the footing may be located by replacing the actual column with an equivalent square column having the same area as the actual one.

17.5.2 Bending Moment

The factored moment at any section, such as *AB* in Fig. 17.5a, is determined from the net upward soil pressure, q_f, acting on the part *ABCD* of the footing. Based on results of extensive tests, the section of maximum design moment is taken as: (1) at the face of the column, pedestal, or wall for footings supporting a concrete column, pedestal, or wall (Fig. 17.5a); and (2) halfway between the face and centreline of the wall for footings supporting masonry walls (Fig. 17.5b).

In one-way reinforced footings (for example, wall footings) and in two-way reinforced square footings, the flexural reinforcement is placed at a uniform spacing across the full width of the footing. In two-way reinforced rectangular footings, again the reinforcement in the long direction is uniformly spaced across the full width of the footing. However, in the short direction, the Code requires a larger concentration of reinforcement to be provided within a band of width, *b*, centred on the column centreline. The amount of reinforcement so required in this band is given by:

Fig. 17.5 Critical sections for moments and shear

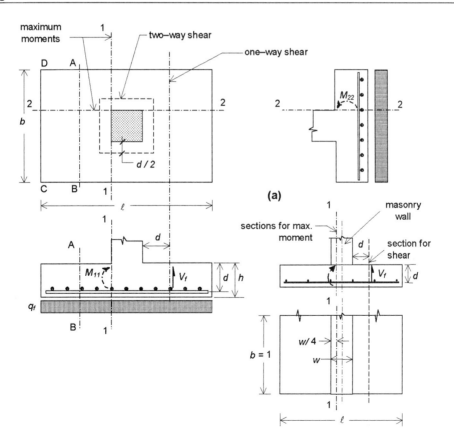

Reinforcement in bandwidth $\quad b = A_{s,short} \times 2/(\beta + 1)$ (17.4)

where b is the length of the short side of the footing, or the length of the supported wall or column, whichever is greater; $A_{s,\,short}$ is the total reinforcement in the short direction; and β is the ratio of the long side to the short side of the footing. The above portion of the reinforcement must be uniformly distributed within the width, b, and the remainder distributed uniformly in the outer portions of the footing. This approximately accounts for the non-uniform variation of the transverse bending moment along the length of the footing. In reinforced concrete footings, at all sections where flexural tension reinforcement is required, the minimum area of reinforcement must not be less than $2\sqrt{f'_c}b_t h / f_y$.

17.5.3 Shear and Development Length

Shear in footings, in general, involves both one-way and two-way shear. Wall footings are loaded along their full length, and under the soil pressure bend only one way (into a cylindrical surface) like an inverted double cantilever slab. Such footings need be designed only for one-way or beam shear. The critical section for one-way shear is taken, as for beams, at a distance, d, from the face of the wall or column (Fig. 17.5). Column footings may fail in one-way shear or two-way (punching) shear, and have to be designed for both. The behaviour of footings in two-way shear is identical to that of a two-way slab supported on columns (Chapter 14). The critical section for two-way shear is taken at a distance $d/2$ from the periphery of the column (Fig. 17.5, also Fig. 14.16). The design procedures for one- and two-way shear are identical to those discussed in Chapters 6 and 14, respectively. However, shear reinforcing is generally avoided in footings, and the factored shear force, V_f, is kept below the factored shear resistance provided by concrete, V_c, by providing the necessary depth. An exception to this is the situation where the thickness of footing that can be provided is restricted. The factored shear resistance of concrete, V_c, as given in Eqs. 6.11, 6.11a, 14.14, 14.15 and 14.16, is taken as:

For one-way (beam) shear in sections having an effective depth not exceeding 300 mm,

$$V_c = 0.2 \lambda \phi_c \sqrt{f'_c} b_w d$$

For one-way (beam) shear in sections having an effective depth greater than 300 mm,

$$V_c = \left(\frac{260}{1000 + d}\right) \lambda \phi_c \sqrt{f'_c} b_w d < 0.1 \lambda \phi_c \sqrt{f'_c} b_w d$$

For two-way shear, V_c is the smallest of

(a) $\quad V_c = (1 + 2/\beta_c)0.2\lambda\phi_c \sqrt{f'_c}\, b_o d$

(b) $\quad V_c = (\alpha_s d/b_o + 0.2)\lambda\phi_c \sqrt{f'_c}\, b_o d$

(c) $\quad V_c = 0.4\lambda\phi_c \sqrt{f'_c}\, b_o d$

where, β_c, is the ratio of the longer to the shorter dimension of the loaded area, and $\alpha_s = 4$ for loaded areas concentrically placed on a footing and 3 for loaded areas at the edge of a footing, usually owing to the adjacent location of a property line. Because of this need to limit the shear stress, the depth of footing is often dictated by shear considerations, and the design for shear usually precedes the flexural design.

The critical sections for development of flexural reinforcement are the sections of maximum moment, and all other sections where there is a change in section or in reinforcement. Furthermore, where the transfer of force at the base of the column, from the column to the footing, is accomplished by reinforcement, such reinforcement must also have adequate development length on each side (Section 17.5.4).

17.5.4 Transfer of Forces at Base of Column

All forces (axial, shear and moment) applied at the base of the column (or pedestal) must be transferred to the footing by compression in concrete or by reinforcement, dowels, or mechanical connectors between column and footing. The compressive force in the contact surface between the column and the footing at the top of the footing is limited to the factored bearing resistance, F_{br}, for either surface, given by Eq. 17.5. If the actual compressive force exceeds F_{br}, reinforcement, dowels, or mechanical connectors carry the excess. In the case of tensile forces in the column, the entire force must be transferred to the footing by reinforcement, dowels, or mechanical connectors. For transferring the column moment at its base to the footing, it may be necessary to use the same amount of reinforcement in the footing as in the column. All reinforcement, provided across the interface to facilitate transfer of column loads, must have adequate development length (in compression) on each side of the interface. This reinforcement may consist of the longitudinal reinforcement in the column extended into the footing or may be separate dowel bars (Fig. 17.6a). The diameter of dowels must not exceed that of the column bars by more than one bar size. The reinforcement across the interface must have an area not less than 0.5 percent of the area of the supported column (or pedestal).

The factored bearing resistance, F_{br}, is specified in CSA A23.3-94(10.8) as:

$$F_{br} = 0.85\phi_c f'_c A_1 \sqrt{A_2/A_1}$$
$$\leq 1.7\phi_c f'_c A_1 \quad (17.5)$$

Fig. 17.6 Transfer of stresses at column base

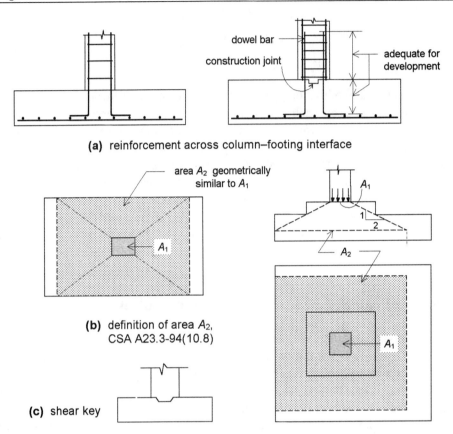

(a) reinforcement across column–footing interface

(b) definition of area A_2, CSA A23.3-94(10.8)

(c) shear key

where A_1 is the loaded area; A_2 is the maximum area of the portion of the supporting surface that is geometrically similar to and concentric with the loaded area; and $\sqrt{A_2/A_1}$ must not be taken larger than 2. The definition of area A_2 is illustrated in Fig. 17.6b. The factor $\sqrt{A_2/A_1}$ allows for the increase in strength of the concrete in the bearing area resulting from the confinement offered by the surrounding concrete. In bearing zones, the high axial compressive stresses give rise to transverse tensile stresses, which may lead to lateral splitting, bursting, or spalling of concrete. Where this possibility exists, transverse and confinement reinforcement must be provided to resist these effects.

Capacity to transfer horizontal shear forces between column and footing may be evaluated using the shear-friction method (Chapter 6). Additional capacity, if required, must be provided by shear keys (Fig. 17.6c) or other mechanical means.

Overturning and sliding of foundations may be significant considerations only when the horizontal loads are high and the gravity loads relatively low, as for foundations of retaining walls. This is seldom the case with footings of columns in usual building structures. However, where these effects are significant, the critical loading condition should maximise the horizontal loads, which tend to destabilise the foundation, and minimising the gravity loads, which stabilise it. Thus, all live gravity loads (except those that contribute to the sliding or overturning) are excluded and the dead loads that resist overturning are multiplied by a load factor of 0.85.

Design of wall footings and isolated column footings are illustrated in Examples 17.1 to 17.3. Combined footings are described in Section 17.6.

EXAMPLE 17.1

Design of Wall Footing
A concrete wall, 325 mm thick, carries unfactored loads per running metre of 80 kN dead load and 220 kN live load. Allowable soil pressure at a depth of 1.25 m below grade is 210 kN/m^2. Design the footing. Take $f_c' = 25$ MPa, $f_y = 400$ MPa.

SOLUTION

The design is made for a typical 1 m length of wall (see Fig. 17.7).

1. Width of footing
    It is assumed that the allowable soil pressure is the gross pressure at a depth of 1.25 m. Estimating the thickness of footing to be 0.3 m, and assuming the unit weight of concrete and soil as 24 kN/m^3 and 16 kN/m^3, respectively (it is

Fig. 17.7 Example 17.1

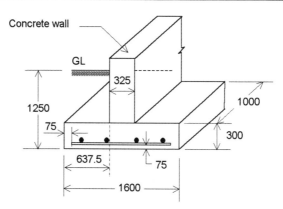

assumed that the soil will be backfilled to grade), the net soil pressure allowable at a depth 1.25 m is:

$$q_a = 210 - 0.3 \times 24 - (1.25 - 0.3) \times 16 = 188 \text{ kN/m}^2$$

Required width of footing is:
$$b = \frac{80 + 220}{188} = 1.60 \text{ m}$$

2. Factored pressure
The weight of footing and weight of overburden are uniformly distributed loads and are balanced by the uniform soil pressure. Therefore, they do not cause any bending moment or shear forces in the footing. The soil pressure for structural design of the footing is that due to the load transmitted through the wall. (The footing, by itself, buried in the soil is not subjected to moments and shears.)

Factored load = $1.25 \times 80 + 1.5 \times 220 = 430$ kN
Soil pressure corresponding to this load is:

$$q_f = \frac{430}{1.6 \times 1} = 269 \text{ kN/m}^2 \text{ (kPa)}$$

For design computations, it may be convenient to take $q_f = 0.269$ MPa (N/mm^2), and to use all dimensions in mm. This will be followed in the examples here.

3. Depth for shear
Critical section for shear is at a distance d (mm) from the wall face. The factored shear is:

$$V_f = 0.269 \times 1000 \times \left(\frac{1600 - 325}{2} - d \right) = 269(638 - d)$$

If d does not exceed 300 mm,
$$V_c = 0.2 \lambda \phi_c \sqrt{f'_c} b_w d = 0.2 \times 1.0 \times 0.6 \times \sqrt{25} \times 1000 d = 600 d$$

The minimum depth required to limit $V_f \leq V_c$ is given by:
$$269(638 - d) \leq 600 d$$

Solving, $d = 197$ mm < 300 mm OK

An overall thickness of 300 mm, with a cover of 75 mm and No. 15 bars will give $d = 300 - 75 - 16/2 = 217$ mm, which is adequate.

4. Design for moment
Maximum moment at wall face is:

$$M_f = 0.269 \times 1000 \times \left(\frac{1600-325}{2}\right)^2 \times \frac{1}{2} = 54.7 \times 10^6 \text{ N}\cdot\text{mm}$$

$$K_r = \frac{M_f}{bd^2} = \frac{54.7 \times 10^6}{1000 \times 217^2} = 1.16 \text{ MPa}$$

From Table 5.4,
$\rho = 0.00359 \qquad \rho_{max} = 0.0207$
$A_{smin} = 0.2 \sqrt{25} \times 1000 \times 300/400 = 750 \text{ mm}^2$
$A_s = 0.00359 \times 1000 \times 217 = 779 \text{ mm}^2$ OK

No. 15 bars at 250 mm gives $A_s = 800 \text{ mm}^2$.
Required development length for No. 15 bars (using Table 9.1) is 460 mm. Actual length available, assuming an end cover of 75 mm is:

$$\frac{1}{2}(1600-325) - 75 = 563 \text{ mm} > 460 \text{ mm} \qquad \text{OK}$$

The footing dimensions are given in Fig. 17.7.

The Code does not specify any temperature and shrinkage reinforcement for footings. Since it is covered by earth, the temperature stresses may not be significant in a footing. In a wall footing which is in the form of a long narrow strip, some secondary bending moment may develop as a result of slight differential settlements along the length of the wall, and it will be good practice to provide some secondary reinforcement in the longitudinal directions.

For the footing in this example, four No. 15 bars may be provided. (This gives a steel ratio of 0.0017.)

Factored bearing resistance in interface is (without factor $\sqrt{A_2/A_1}$), OK

$$\begin{aligned} F_{br} &= 0.85\phi_c f'_c A_1 \\ &= 0.85 \times 0.6 \times 25 \times 325 \times 1000 \times 10^{-3} = 4144 \text{ kN} \\ &> P_f = 430 \text{ kN} \end{aligned}$$

EXAMPLE 17.2

Design of Square Isolated Footing
A tied column, 450 mm square, and reinforced with eight No. 35 bars carries an unfactored dead load of 1300 kN and an unfactored live load of 1000 kN. Suitable soil with an allowable soil pressure of 300 kN/m^2 is available at a depth of 1.5 m. Design a square footing. The concrete compression strength, f_c', is 30 MPa for the column and 25 MPa for the footing. All steel has f_y = 400 MPa.

SOLUTION

1. Size of footing
   Assuming a depth of 750 mm for the footing, and unit weights of 24 kN/m^3 and 16 kN/m^3 for concrete and soil, respectively, the net allowable soil pressure at depth 1.5 m is:
   q_a = 300 - 0.75 × 24 - 0.75 × 16 = 270 kN/m^2

   Required area of footing = $\dfrac{1300+1000}{270}$ = 8.52 m^2

   Minimum size = $\sqrt{8.52}$ = 2.92 m
   Select a 3 m × 3 m footing.

2. Design for shear
   The factored load = 1.25 × 1300 + 1.5 × 1000 = 3125 kN
   Factored soil pressure q_f = 3125/(3 × 3) = 347 kN/m^2 = 0.347 N/mm^2

   (a) One-way or beam shear
   The critical section is at a distance d (in mm) from the column face. Referring to Fig. 17.8b:
   $$V_f = 0.347 \times 3000 \left(\dfrac{3000-450}{2} - d \right) = 1041(1275 - d)$$

   If d does not exceed 300 mm, the factored shear resistance of concrete, $V_c = 0.2 \times 1 \times 0.6 \times \sqrt{25} \times 3000d = 1800d$

   To limit the shear force, V_f, to less than V_c,
   1041(1275 - d) ≤ 1800d
   Solving, d ≥ 467 mm > 300 mm   not OK

Fig. 17.8 Example 17.2

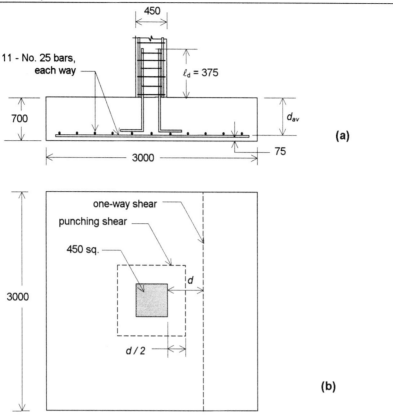

Since $d > 300$ mm,
$$V_c = \frac{260}{1000+d} \times 1.0 \times 0.6 \times \sqrt{25} \times 3000 \times d = \frac{2340 \times 10^3}{1000+d}d$$

To limit the shear force, V_f, to less than V_c,
$$1041(1275-d) \leq \frac{2340 \times 10^3}{1000+d}d$$
Solving, $d \geq 513$ mm

With $d = 513$ mm,
$V_c = (260/1513) \times 1.0 \times 0.6 \times \sqrt{25} \times 3000 \times 513$
$= 793$ N $\not< 0.1 \times 1.0 \times 0.6 \times \sqrt{25} \times 3000 \times 513 = 462$ OK

(b) Two-way shear
Critical section is at $d/2$ (in mm) from the periphery of the column. Factored shear force is:
$$V_f = 0.347\,[3000^2 - (450+d)^2]\ \text{N}$$
V_c is the smallest of:
$$(1 + 2/\beta_c)0.2\lambda\phi_c f'_c b_o d = 0.6\lambda\phi_c f'_c d\text{, since }\beta_c = 1\,(\text{square column})$$
or
$$(\alpha_s d/b_o + 0.2)\lambda\phi_c \sqrt{f'_c} b_o d,\text{ for which }\alpha_s = 4$$
or
$$0.4\lambda\phi_c \sqrt{f'_c} b_o d\text{ , where } b_o = 4(450+d)$$

Therefore, for $V_f \le V_c$, the solution is the smaller of either
$$0.347[3000^2 - (450+d)^2] \le$$
$$\left(\frac{4 \times d}{4(450+d)} + 0.2\right) \times 1.0 \times 0.6 \times \sqrt{25} \times 4(450+d)d$$

Solving, $d \ge 410$ mm
or
$$0.347[3000^2 - (450+d)^2] \le 0.4 \times 1.0 \times 0.6 \times \sqrt{25} \times 4(450+d)d$$

Solving, $d \ge 567$ mm
Therefore, use $d \ge 567$ mm

(It is usually satisfactory, and always conservative, to neglect the upward pressure in the punching failure block, that is, the area $(450 + d) \times (450 + d)$, and assume $V_f = P_f = 3125$ kN in this case. Also, instead of getting an exact solution as above, a depth may be assumed and then the computed factored shear resistance, V_c, compared with V_f. This can simplify the calculations.)

The depth for two-way shear controls in this example. Taking the average d as 567 mm, and assuming a cover of 75 mm and No. 20 bars, the thickness of footing required is:

$$h = 567 + 75 + 19.5/2 = 652\text{ mm}$$

Select $h = 700$ mm. The depth to the centre of reinforcement will differ by one bar diameter for the reinforcement along the two directions. For relatively deep square footings, it is adequate to use an

average depth, measured to the interface between the two layers, for the reinforcement computations, and provide the reinforcement as computed in both directions (see Fig. 17.8).
$d_{av} = 700 - 75 - 19.5 = 605$ mm

3. Design for flexure
Factored moment, M_f, at column face is:
$M_f = 0.347 \times 3000 (1275)^2/2 = 846 \times 10^6$ N·mm
$$K_r = \frac{M_f}{bd^2} = \frac{846 \times 10^6}{3000 \times 605^2} = 0.770 \text{ MPa}$$
$\rho = 0.00234 < \rho_{max} = 0.0207$
$A_{smin} = 0.2 \times \sqrt{25} \times 3000 \times 700/400 = 5250$ mm^2
$A_s = 0.00234 \times 3000 \times 605 = 4247$ mm^2 $< A_{smin}$
Therefore, $A_s = 5250$ mm^2

Provide eighteen No. 20 bars at uniform spacing in each direction giving $A_s = 5400$ mm^2 (spacing 166 mm) or, alternatively, eleven No. 25 bars.

Required development length for No. 20 bars (using Table 9.1) = 563 mm, and for No. 25 bars = 908 mm.
Length available = 1275 - 75 = 1200 mm, which is adequate with either bar size.

4. Transfer of force at column base
Actual factored compressive force is:
$P_f = 3125$ kN

The factored bearing resistance is given by
$$F_{br} = 0.85 \phi_c f'_c A_1 \sqrt{A_2/A_1}$$

For column face,
$F_{br} = 0.85 \times 0.6 \times 30 \times 450^2 \times 1 \times 10^{-3} = 3098$ kN < 3125 kN

For footing, face, $\sqrt{A_2/A_1} = \sqrt{3000^2/450^2} = 6.67$, limited to 2
$F_{br} = 0.85 \times 0.6 \times 25 \times 450^2 \times 2 \times 10^{-3} = 5164$ kN > 3125 kN   OK
The compressive force exceeds the factored bearing resistance of concrete at column face by 27 kN, that must be transferred by reinforcement, dowels, or mechanical connectors. Here, the minimum reinforcement provided across the

interface is more than adequate to transfer this excess force of 27 kN.

The minimum reinforcement required to be provided across the interface, at 0.5 percent of column area, is:
$A_s = 0.005 \times 450^2 = 1013 \text{ mm}^2$

Provide four No. 20 bars giving $A_s = 1200 \text{ mm}^2$.
Development length required for these bars in compression is:

$$l_d = \frac{0.24 f_y d_b}{\sqrt{f_c'}} = \frac{0.24 \times 400 \times 19.5}{\sqrt{25}} = 374 \text{ mm}$$

or, $\not\geq 0.044 f_y d_b = 0.044 \times 400 \times 19.5 = 343 \text{ mm}$

Available vertical embedment length within the footing is
$d - d_b = 605 - 20 = 585 \text{ mm} > 374 \text{ mm}$ OK

These bars must also be extended into the column for a length equal to their development length, say $l_d = 375$ mm. The details of the footing are given in Fig. 17.8.

EXAMPLE 17.3

Design of a Rectangular Footing
Redesign the footing for the column in Example 17.2, if one of the dimensions of the footing is limited to a maximum of 2.5 m due to space restrictions. Use $f_y = 400$ MPa and $f_c' = 35$ MPa.

SOLUTION

1. Size of footing
   From step 1 of Example 17.2, the required area of footing = 8.52 m^2. With a width of 2.5 m, required length $l = 8.52/2.5 = 3.41$ m. Select a rectangular footing, 3.5 m × 2.5 m, as shown in Fig. 17.9.

2. Depth required for shear An exact solution for required depth for shear may be obtained using the condition $V_f = V_c$, as was done in the previous two examples. However, in this example a trial and error procedure will be used.

Fig. 17.9 Example 17.3

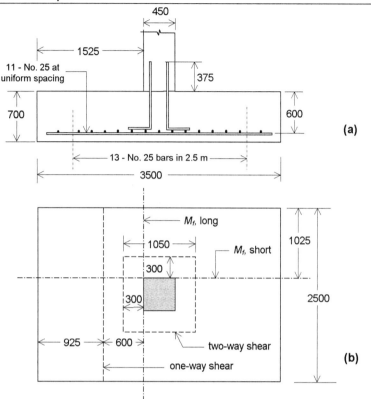

Estimate required footing thickness as 700 mm. Assume cover of 75 mm and No. 25 bars as flexural reinforcement. The bottom layer of reinforcement will be placed in the long direction in which the moment is greater. The effective depths are:

In long direction, $d = 700 - 75 - 25.2/2 = 612$ mm.
In short direction, $d = 700 - 75 - 1.5 \times 25.2 = 587$ mm.

Average d for shear calculation = 600 mm.
Factored soil pressure $q_f = 3125/(3.5 \times 2.5) = 357$ kN/m^2 = 0.357 MPa.

Critical section for two-way shear is at $d/2 = 300$ mm from column perimeter, as shown in Fig. 17.9b. Factored shear force, V_f, is:
$V_f = 0.357 \, [3500 \times 2500 - 1050^2] = 2.73 \times 10^6$ N

V_c is the smaller of
$(4 \times 600/(4 \times 1050) + 0.2) \times 1.0 \times 0.6 \times \sqrt{35} \times 4 \times 1050 \times 600 = 6.90 \times 10^6$ N

or
$$0.4 \times 1.0 \times 0.6 \times \sqrt{35} \times 4 \times 1050 \times 600 = 3.58 \times 10^6 \text{ N}$$

Take $V_c = 3.58 \times 10^6$ N, $V_c > V_f$ OK

Critical section for one-way shear is at $d = 600$ mm from column face, as shown in Fig. 17.9b.
$$V_f = 0.357 \times 2500 \times 925 = 826 \times 10^3 \text{ N}$$

Since $d > 300$ mm,
$$\begin{aligned}V_c &= [260/(1000+612)] \times 1.0 \times 0.6 \times \sqrt{35} \times 2500 \times 612 \\ &= 876 \times 10^3 \text{ N}, \not< 0.1 \times 1.0 \times 0.6 \times \sqrt{35} \times 2500 \times 612 = 543 \times 10^3 \text{ N}\end{aligned}$$

$V_c > V_f$ OK

The depth is adequate for shear.

3. Design for flexure
   In the long direction,
   $$M_f = 0.357 \times 2500 \times 1525^2/2 = 1.04 \times 10^9 \text{ N·mm}$$
   $$K_r = \frac{1.04 \times 10^9}{2500 \times 612^2} = 1.11 \text{ MPa}$$
   $$\rho = 0.00338 < \rho_{max} = 0.0277$$

   $A_{smin} = 0.2 \times \sqrt{35} \times 2500 \times 700 / 400 = 5177 \text{ mm}^2$
   $A_{s,long} = 0.00338 \times 2500 \times 612 = 5171 \text{ mm}^2$

   Therefore, $A_{s,long} = 5177 \text{ mm}^2$

   Provide eleven No. 25 bars at uniform spacing. Spacing is within permissible limits.

   In the short direction,
   $$M_f = 0.357 \times 3500 \times 1025^2/2 = 656 \times 10^6 \text{ N·mm}$$
   $$K_r = \frac{656 \times 10^6}{3500 \times 587^2} = 0.544 \text{ MPa}$$
   $$\rho = 0.00163$$

$A_{smin} = 0.2 \times \sqrt{35} \times 3500 \times 700 / 400 = 7247 \text{ mm}^2$

$A_{s,short} = 0.00163 \times 3500 \times 587 = 3349 \text{ mm}^2$

Therefore, $A_{s,short} = 7247 \text{ mm}^2$

Fifteen No. 25 bars give an area of 7500 mm^2.
The area of reinforcement to be provided within a central band of width $b = 2.5$ m is:

$$A_s \times \frac{2}{\beta+1} = 7247 \times \frac{2}{(3.5/2.5)+1} = 6039 \text{ mm}^2$$

that is, thirteen No. 25 bars.

In this case, provide thirteen bars within the middle band of width 2.5 m, and one bar each on the two outer segments making up a total of fifteen bars (Fig. 17.9). The spacings are within limits ($< 3h$ and 500 mm).

Required development length for No. 25 bars is 768 mm (using Table 9.1). Available least length (in the short direction) is 1025 - 75 (cover) = 950 mm which is greater than required 768 mm.

4. Transfer of force at base of column
Actual factored compressive force at interface = 3125 kN. For footing,

$\sqrt{A_2/A_1} = \sqrt{2500^2/450^2} = 5.5$, limited to 2

$F_{br} = 0.85 \times 0.6 \times 35 \times 450 \times 450 \times 2 \times 10^{-3} = 7229 \text{ kN} > 3125 \text{ kN}$ OK

As in Example 17.2, the minimum requirement of four No. 20 bars across the interface will suffice, and this will have adequate length of embedment for development.

17.6 COMBINED FOOTINGS

A footing supporting more than a single column or wall is called a combined footing (Fig. 17.1). In soils having very low allowable soil pressures, various forms of mat foundations supporting several columns are used (Refs. 17.1-17.3). However, even in soils having sufficient bearing capacity for the use of individual footings, some combined footings may become necessary under the following situations:

594 REINFORCED CONCRETE DESIGN

Fig. 17.10 Two-column combined footing

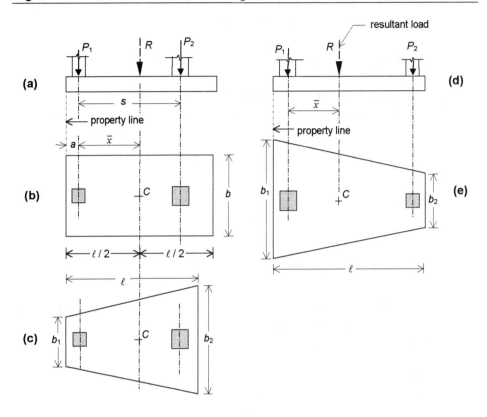

1. The columns are so closely spaced that individual footings, if used, will overlap in areas.
2. An exterior column along the periphery of the building is so close to the property line that an individual footing, symmetrically placed, will project beyond the boundary.

The soil pressure under a combined footing may be determined assuming a straight line distribution or by analysis as a beam on an elastic foundation. The former method is approximately true for relatively rigid footings, while for flexible footings the latter method is more appropriate (Refs. 17.1, 17.4). Two-column footings can be considered rigid, and the soil pressure distribution determined assuming a straight line variation.

Figure 17.10 shows examples of two-column footings. Non-uniform soil pressure may lead to the tilting of the footing. To avoid such a situation, the footing geometry is so selected that the centroid of the footing area coincides with the resultant of the column loads (including consideration of the moments and lateral forces, if any). With this arrangement, the soil pressure under the footing may be assumed to be uniform. The combined two-column footing may be rectangular (or trapezoidal with the wider end near the heavier load), when the exterior column, which has the space limitation for an independent footing, carries a lesser load than the interior column (that is, $\bar{x} > s/2$, as shown in Fig. 17.10a,b,c). When the exterior column carries the heavier load, $\bar{x} < s/2$, a trapezoidal footing, as shown in Fig. 17.10e, is needed to align the load resultant, R, and the centroid, c, of the footing.

17.7 DESIGN OF TWO-COLUMN COMBINED FOOTINGS

Referring to Fig. 17.10, in usual situations, the edge distance, a, and the distance, $\bar{x}$, to the resultant column load are known. The plan dimensions of the footing can be selected to satisfy the two requirements:

1. Area of footing, A = Total load/q_a.
2. Centroid of footing area must coincide with resultant of column loads.

When a rectangular footing is used, as in Fig. 17.10a, the second requirement gives $l = 2(\bar{x} + a)$, and then $b = A/l$. In a trapezoidal footing (Fig. 17.10c,e) usually the length l is selected first, and the dimensions b_1 and b_2 computed using the two requirements above.

Once the dimensions of the footing area are determined, the uniform factored soil pressure may be computed as q_f = *factored load/area*. A simplified and usually conservative load transfer arrangement, useful in the structural design of the footing, is shown in Fig. 17.11. The footing is considered as a uniformly loaded wide longitudinal beam supported by transverse beam segments at the column line. The longitudinal distribution of bending moments and shear forces may be determined from statics, as shown in Fig. 17.11c,d. The critical sections for one-way or beam shear are at a distance d from the column face, and for two-way shear at $d/2$ from the column perimeter. The footing depth, d, is determined from shear strength considerations. The flexural reinforcement in the longitudinal direction is designed for the positive moment at the face of the column (reinforcement placed at the bottom) and for the maximum negative moment (top reinforcement) between the columns. Since the design sections for both shear and moment are outside the column width, the shear and moment diagrams may be obtained assuming the column loads as concentrated at their centrelines, as shown by the dashed lines in Fig. 17.11c,d. The values of the design shear and moments are unaffected by this simplification.

596 REINFORCED CONCRETE DESIGN

In order for the footing to act as a wide beam supported at the column lines, two narrow transverse bands of the footing below the columns act as transverse beams (Fig. 17.11b). These two bands, acting as transverse beams cantilevered from the columns and supporting the longitudinal wide beam, must be designed for the associated bending moment and shear force. The effective width of footing acting as a transverse beam may be taken approximately as the width of column plus $0.75d$ on either side of the column (Ref. 17.1).

Fig. 17.11 *Assumed load transfer in two-column footing*

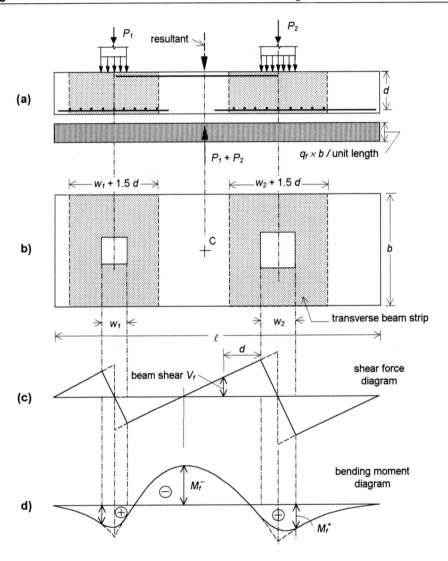

The design of a rectangular combined footing is illustrated in Example 17.4. The procedure just described is the conventional approach that assumes the footing as rigid. Slightly more economical designs will result by the use of the beam-on-elastic foundation method (Refs. 17.1, 17.4). More detailed discussions of various types of spread foundations are contained in Refs. 17.1 and 17.2.

EXAMPLE 17.4

The layout and relevant data for an exterior and adjacent interior column are given in Fig. 17.12. The allowable soil pressure at the base of the footing, 1.5 m below grade, is 250 kN/m². Concrete for the footing has $f_c' = 25$ MPa, and $f_y = 400$ MPa. Design a combined footing.

SOLUTION

1. Footing dimensions
   Assuming a depth of footing of 750 mm, and unit weights of concrete at 24 kN/m and of soil at 16 kN/m³, net allowable soil pressure is:
   $q_a = 250 - 0.75 \times 24 - 0.75 \times 16 = 220$ kN/m²

   Area of footing required is given by:
   $A = (600 + 400 + 1000 + 800)/220 = 12.7$ m²

   In arranging footing dimensions to obtain a uniform soil pressure distribution, q_f, at factored loads, the distance $\bar{x}$ (see Fig. 17.10b) to be used must be the value corresponding to factored loads (rather than the value of $\bar{x}$ at service loads. $\bar{x}$ at service loads and at factored loads may be somewhat

Fig. 17.12 Example 17.4

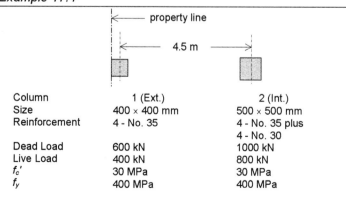

different if the dead load to live load ratios are different for the two columns).

The factored loads are:
For exterior column, $P_{f1} = 1.25 \times 600 + 1.5 \times 400 = 1350$ kN
For interior column, $P_{f2} = 1.25 \times 1000 + 1.5 \times 800 = 2450$ kN

$$\bar{x} = \frac{2450 \times 4.5}{(1350 + 2450)} = 2.90 \text{ m}$$

Since $\bar{x} > s/2 = 2.25$ m, a rectangular combined footing can be used. To locate the centroid of the rectangular footing at a distance $\bar{x}$ from the centre of the exterior column, the footing length must be:

$l = 2 \times (2.90 + 0.20) = 6.2$ m

Selecting a length of 6.2 m, the required width of footing, b, for providing an area of 12.7 m^2 is:

$$b = \frac{12.7}{6.2} = 2.05 \text{ m}$$

Select a footing size 6.2 m × 2.10 m, as shown in Fig. 17.13a. The actual factored soil pressure at factored load is:

$$q_f = \frac{1350 + 2450}{6.2 \times 2.1} = 292 \text{ kN/m}^2 = 0.292 \text{ MPa}$$

2. Shear force and bending moment in the longitudinal direction

Considering the footing in the longitudinal direction as a wide beam, supported on the columns, the distributed upward load per unit length of this beam is (Fig. 17.13b):
292 × 2.1 = 613 kN/m

The distribution of shear and moment are shown in Fig. 17.13c,d. The section of zero shear, a distance x from the exterior edge is obtained as
613 x = 1350, x = 2.20 m

The maximum moment at this section is:

$$M = \frac{613 \times 2.2^2}{2} - 1350 \times (2.2 - 0.2) = -1217 \text{ kN} \cdot \text{m}$$

3. Design depth for shear
   (a) Wide-beam action
   The maximum shear is at the inside face of an interior column, and equal to 1377 kN. The shear at the critical section a distance d (mm) from the face of the column is (Fig. 17.13c):

   $$V_f = 1377 - 613 \times \frac{d}{1000} = (1377 - 0.613d)$$

   If d does not exceed 300 mm,
   $$V_c = 0.2\lambda\phi_c\sqrt{f_c'}b_w d = 0.2 \times 1.0 \times 0.6 \times \sqrt{25} \times 2100 d \times 10^{-3}$$
   $$= 1.26d \text{ kN}$$

   Substituting in $V_f \leq V_c$,
   $1377 - 0.613d \leq 1.26d$
   Solving, $d \geq 735$ mm   not OK

   Since $d > 300$ mm,
   $$V_c = \frac{260}{(1000+d)}\lambda\phi_c\sqrt{f_c'}b_w d \not< 0.1\lambda\phi_c\sqrt{f_c'}b_w d$$
   Substituting in $V_f \leq V_c$,
   $$1377 - 0.613d \leq \frac{260}{1000+d} \times 1.0 \times 0.6 \times \sqrt{25} \times 2100 \times d \times 10^{-3}$$
   Solving, $d = 947$ mm

   Check
   $$V_c = 260/1947 \times 1.0 \times 0.6 \times \sqrt{25} \times b_w d \not< 0.1 \times 1.0 \times 0.6 \times \sqrt{25} \times b_w d$$
   $$V_c = 0.4 b_w d \not< 0.3 b_w d$$ OK

   (b) Two-way shear at exterior column
   An exact value for the required d can be obtained by setting up the equation $V_f \leq V_c$ and solving for d, as was illustrated in Example 17.2. Here, the stress will be checked using the value $d = 947$ mm determined for one-way shear.

   The critical section is at $d/2 = 474$ mm from the column perimeter (Fig. 17.13a). The shear is:

Fig. 17.13 Design example 17.4

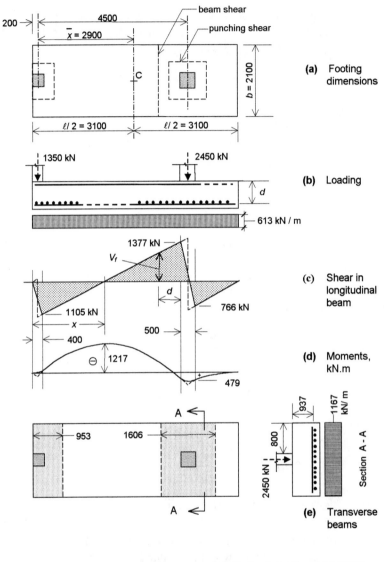

$$V_f = 1350 - 292 \times (0.4 + 0.474)(0.4 + 0.947) = 1006 \text{ kN}$$

For square column, $\beta_c = 1$, and V_c is the smallest of

$(1 + 2/\beta_c) 0.2\lambda\phi_c\sqrt{f_c'} b_o d$

$= (1 + 2/1) \times 0.2 \times 1.0 \times 0.6 \times \sqrt{25} \times (874 \times 2 + 1347) \times 947$

$= 5275 \text{ kN}$

or

$$(\alpha_s d / b_o + 0.2)\lambda\phi_c \sqrt{f_c'} b_o d$$
$$= [3 \times 947/(874 \times 2 + 1347) + 0.2] \times 1.0 \times 0.6 \times \sqrt{25}$$
$$\times (874 \times 2 + 1347) \times 947$$
$$= 9830 \text{ kN}$$

or

$$0.4\lambda\phi_c \sqrt{f_c'} b_o d$$
$$= 0.4 \times 1.0 \times 0.6 \times \sqrt{25} \times (874 \times 2 + 1347) \times 947 = 3517 \text{ kN}$$
Therefore, $V_c = 3517$, $V_f < V_c$ OK

(c) Two-way shear at interior column
Using $d = 947$ mm at the critical section (Fig. 17.13a):
$V_f = 2450 - 292 (0.5 + 0.947)^2 = 1839$ kN

V_f is obviously $< V_c$ at interior column since even for the exterior column $V_c = 3517$kN.

One-way shear controls in this example, and the effective depth required is 947 mm. Select an overall thickness of 1050 mm. With a clear cover of 75 mm and assuming No. 25 bars, the depth provided is: $d = 1050 - 75 - 25/2 = 962$ mm

4. Design of longitudinal flexural steel
Maximum negative moment is:

$M_f = 1217$ kN·m
$$K_r = \frac{M_f}{bd^2} = \frac{1217 \times 10^6}{2100 \times 962^2} = 0.626 \text{ MPa}$$
$\rho = 0.00189 < \rho_{max}$ (Table 5.4)

and $A_{smin} = 0.2 \times \sqrt{25} \times 2100 \times 1050/400 = 5513$ mm^2
$A_s = 0.00189 \times 2100 \times 962 = 3818$ mm^2. A_{smin} governs, so $A_s = 5513$ mm^2.
Provide twelve No. 25 bars giving $A_s = 6000$ mm^2, and a spacing of 175 mm which is satisfactory.

    Development length for No. 25 top bars is (using Table 9.1) 1181 mm. The length available beyond the peak moment section (Fig. 17.13b,d) is in excess of this on both sides, and hence OK.

602 REINFORCED CONCRETE DESIGN

The positive moment in the zone near the interior column is far less than the peak negative moment and, therefore, here also the minimum reinforcement requirement governs. Therefore, provide twelve No. 25 bars here also. Required development length for bottom bars is 908 mm, and the length provided is in excess of this (Fig. 17.13b).

5. Design of transverse beams (Fig. 17.13e)

   (a) Transverse beam under exterior column The total load on the beam is equal to the load transferred to the column, 1350 kN. Load per unit length of beam is:

   $$\frac{1350}{2.1} = 643 \text{ kN/m}$$

   Projection of beam beyond column face:

   $$\frac{2.1 - 0.4}{2} = 0.85 \text{ m}$$

   Moment in transverse beam at column face is:

   $$M_f = 643 \times 0.85^2 / 2 = 232 \text{ kN} \cdot \text{m}$$

   Effective depth for transverse beam (reinforcement placed above longitudinal reinforcement) = 1050 - 75 - 1.5 × 25.2 = 937 mm.
   Width of beam = $w + 0.75d$ = 400 + 0.75 × 937 = 1103 mm.

   $$K_r = \frac{232 \times 10^6}{1103 \times 937^2} = 0.240 \text{ MPa}$$

   $\rho = 0.00071$ (Table 5.4)
   $A_s = A_{smin} = 0.2 \times \sqrt{25} \times 1103 \times 1050/400 = 2895 \text{ mm}^2$
   Provide ten No. 20 bars giving $A_s = 3000 \text{ mm}^2$.
   Required development length is 563 mm.
   Available length = 850 - cover = 775 mm   OK

   The beam shear in the transverse beam must be checked at a distance d = 937 mm from the column face. This is outside the edge of the beam. Hence, there is no problem with shear.

   (b) Transverse beam under interior column

   Load per unit length = $\frac{2450}{2.1}$ = 1167 kN/m

   Beam projection from column face = 0.8 m

Moment at column face, $M_f = 1167 \times \dfrac{0.8^2}{2} = 373 \, \text{kN} \cdot \text{m}$

Width of beam = 500 + 2 × 0.75 × 937 = 1906 mm

$K_r = \dfrac{373 \times 10^6}{1906 \times 937^2} = 0.223 \, \text{MPa}$

Again the minimum reinforcement requirement controls.
$A_s = 0.2 \times \sqrt{25} \times 1906 \times 1050/400 = 5003 \, \text{mm}^2$
Provide seventeen No. 20 bars.
Development length available = 800 - 75 = 725 mm > 563 mm   OK
Once again, the critical section for beam-shear is at d = 937 mm from column face, which is outside the edge of the footing.

6. Transfer of force at base of column
For exterior column, factored bearing resistance of footing is:
$F_{br} = 0.85 \times 0.6 \times 25 \times 400^2 \times 10^{-3} = 2040 \, \text{kN} > 1350 \, \text{kN}$ OK

At base of interior column, factored bearing resistance (even without the factor $\sqrt{A_2/A_1}$) is:
$F_{br} = 0.85 \times 0.6 \times 25 \times 500^2 \times 10^{-3} = 3188 \, \text{kN}$

Factored compressive force = 2450 kN < F_{br} OK

In both cases, provide the minimum reinforcement as follows:
Exterior column, $0.005 \times 400^2 = 800 \, \text{mm}^2$;
provide four No. 25 bars.
Interior column, $0.005 \times 500^2 = 1250 \, \text{mm}^2$
provide four No. 25 bars.

Vertical development length required for the No. 25 bars, in compression, is:
$0.24 f_y d_b / \sqrt{f_c'} \geq 0.044 f_y d_b$
The depth available in the footing is greater than this, and hence adequate.

7. Details of reinforcement
Figure 17.14 gives a sketch of the reinforcement details. Some of the longitudinal reinforcement is (arbitrarily) extended to the full length of the footing to: (1) provide support for the transverse reinforcement under the exterior column; and (2) provide some reinforcement in the large (otherwise unreinforced) area of concrete at the top outside of the interior column.

Fig. 17.14 Details of reinforcement, Example 17.4

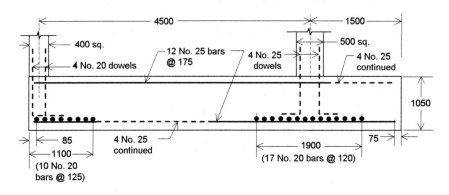

Fig. 17.15 Problem 3

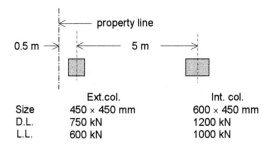

	Ext.col.	Int. col.
Size	450 × 450 mm	600 × 450 mm
D.L.	750 kN	1200 kN
L.L.	600 kN	1000 kN

PROBLEMS

1. For two columns in Example 17.4, assuming that there is no property line limitation, design and detail independent square footings.

2. A circular column, 650 mm in diameter, carries unfactored loads of 100 kN dead load and 800 kN live load. The column is reinforced with eight No. 30 bars, and has $f_c' = 30$ MPa and $f_y = 300$ MPa. The column centreline is 1.2 m from the property line. Allowable soil pressure at a depth of 1.2 m below grade is 240 kN/m². Design a rectangular footing for the column. For footing, assume $f_c' = 25$ MPa and $f_y = 300$ MPa.

3. The layout of an exterior and adjacent interior column in a building is shown in Fig. 17.15. Design a combined footing for the columns. Allowable soil pressure at base of footing, 2 m below grade, is 300 kN/m². Use $f_c' = 25$ MPa, $f_y = 400$ MPa.

REFERENCES

17.1 Bowles, J.E., *Foundation Analysis and Design*, 5th ed., McGraw-Hill Book Co., New York, 1996, 1,168 pp.
17.2 Peck, R.R., Hanson, W.E., and Thornburn, T.H., *Foundation Engineering*, 2nd ed., John Wiley & Sons, Inc., 1974, 514 pp.
17.3 ACI Committee 336, *Suggested Design Procedures for Combined Footings and Mats*, ACI Journal, Proc. Vol. 63, No. 10, Oct. 1966, pp. 1041-1057.
17.4 Kramisch, F., and Rogers, P., *Simplified Design of Combined Footings*, ASCE J., Soil Mech. Div., Proc. Vol. 87, No. SM5, Oct. 1961, pp. 19-44.

CHAPTER 18 Special Provisions for Seismic Design

18.1 GENERAL

The National Building Code of Canada (Refs. 18.1, 18.2) presents the minimum design requirements for earthquake-resistant design in order to ensure an acceptable level of safety. These requirements take into consideration the characteristics and probability of occurrence of earthquakes, characteristics of the structure and the foundation, and the amount of damage that is considered tolerable.

The minimum lateral seismic force at the base of a structure specified in NBC 1995 is given by Eq. 2.3. In this equation, the force modification factor R reflects the material and type of construction, damping, ductility and/or energy absorptive capacity of the structure. Types of construction that have ductility and damping characteristics and have been shown to perform satisfactorily in earthquakes are assigned higher values of R. Values for R are given in NBC 1995 (Sentence 4.1.9.1(8)), and range from 1.0 to 4.0. By providing ductility and energy-absorption capability in the structure, the seismic forces can be minimised. For cast-in-place, continuously reinforced concrete structures, the value of the factor R ranges from 1.5 to 4.0, depending on the ductility and damping that are built-in. To meet the ductility requirements associated with these reduced seismic forces, the structure must be designed and detailed to meet the special provisions specified in CSA A23.3-94 (Clause 21). For reinforced concrete structures without special provisions for ductility in the lateral load resisting system (for example, a conventional reinforced concrete building not more than 60 m in height and conforming to the requirements in the Code Clauses 1 to 20 and 21.9 and having nominal ductility), the value for R is 2.0. The lateral force resisting systems specifically considered in the Code are ductile moment-resisting space frames and/or ductile flexural walls.

Ductility may be defined as the ability of a structure or member to undergo inelastic deformations beyond the initial yield deformation, with no decrease in the load resistance. Yielding, in this context, is flexural. A *ductile moment-resisting frame* is a frame of continuous construction consisting of flexural members and columns designed and detailed to accommodate reversible lateral displacements after the formation of plastic hinges, without decrease in strength. A *flexural wall* is a reinforced concrete structural wall cantilevering from the foundation and designed and detailed to be ductile, and to resist seismic forces and dissipate energy through flexural yielding at one or more plastic hinges.

18.2 ROLE OF DUCTILITY IN SEISMIC DESIGN

Although a general qualitative definition of ductility was given in the preceding Section, it is difficult in practice to quantify the degree of ductility of a structure. For an under-reinforced beam section in flexure, a typical moment-curvature ($M - \psi$) relation is shown in Fig. 18.1a. Based on the idealised $M - \psi$ relationship, a *ductility factor* in terms of curvature may be expressed as the ratio ψ_u/ψ_y, where ψ_y is the curvature at first yield (idealised) and ψ_u is the maximum curvature at the section. The value of ψ_u/ψ_y is a section property, and can be computed relatively easily using the principles described in Chapter 4. In the case of a single beam member (Fig. 18.1b), the definition of a ductility factor is more difficult, as it could be in terms of the curvature at a section, the rotation (θ) at a joint, or the displacement (Δ) (Fig. 18.1b) at a selected point. The ductility factor obtained by the three methods will differ, and

***Fig. 18.1** Concept of ductility*

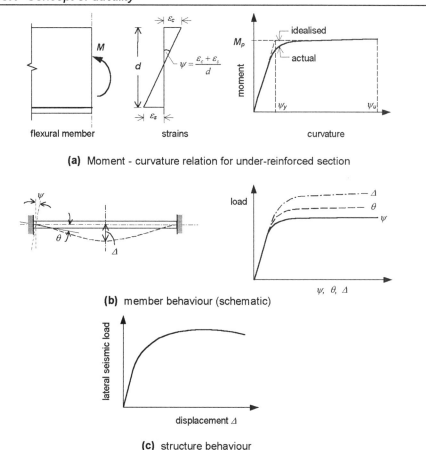

(a) Moment - curvature relation for under-reinforced section

(b) member behaviour (schematic)

(c) structure behaviour

furthermore, the rotations and displacements will depend on several factors, such as the span dimensions, shape of moment diagram, support restraint, etc. The problem is more complex when it comes to the ductility factor of an entire structure. However, qualitatively, a ductile structure will have a load-displacement response as shown in Fig. 18.1c. This ductility is achieved by ensuring member section responses similar to that in Fig. 18.1a, so that an adequate number of plastic hinges (Section 10.7.1) would develop under extreme lateral seismic forces.

Ductility helps reduce the forces in members and dissipate energy due to seismic effects. This is illustrated schematically in Fig. 18.2, for the simple case of one cycle of static loading and unloading. For the same energy input, the maximum force, P_u, induced in a system responding elastically is greater than that in a system responding in an elastic-plastic manner. Furthermore, the latter system dissipates a substantial part of the energy by inelastic deformation. However, the maximum displacement, δ_u, of the elastic-plastic (ductile) system is more than that in the elastic (linear) system. Thus, while ductility helps in reducing induced forces and in dissipating energy, it also demands that the deformations that the system can accommodate without damage be larger.

Forces and moments in the elements of the lateral force resisting system due to earthquake effects are usually determined by a linear elastic analysis of the system under static loads equivalent to the earthquake. The equivalent static loads are determined as prescribed in NBC 1995 (Section 2.2 and Eq. 2.3), and the magnitude of these forces is inversely proportional to the force modification factor R. Thus, for the same ground motion, a ductile moment-resisting space frame ($R = 4.0$) will have induced lateral forces which are only 25 percent of those if the structure has little ductility and damping ($R = 1.0$). However, the ductile frame has to sustain inelastic

Fig. 18.2 Influence of ductile structural response on member forces and energy dissipation

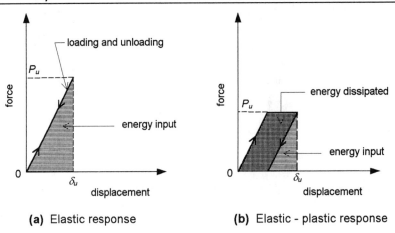

(a) Elastic response (b) Elastic - plastic response

deformations that are several times the elastic deformations calculated for these reduced forces. Since reinforced concrete is relatively less ductile in compression and shear, dissipation of seismic energy is best achieved by flexural yielding. Hence, weakness in compression and shear in relation to flexure is avoided.

In a structure composed of a ductile frame and/or a ductile flexural wall conforming to CSA A23.3-94 (Clause 21), the inelastic response is developed by the formation of plastic hinges (flexural yielding) in the members, as shown in Fig. 18.3. For ductile frames, hinges may form in the beams or in the columns, as shown in Fig. 18.3a. Plastic hinges in beams have larger rotation capacities than column hinges,

Fig. 18.3 *Formation of inelastic deformations (plastic hinges) in ductile structure*

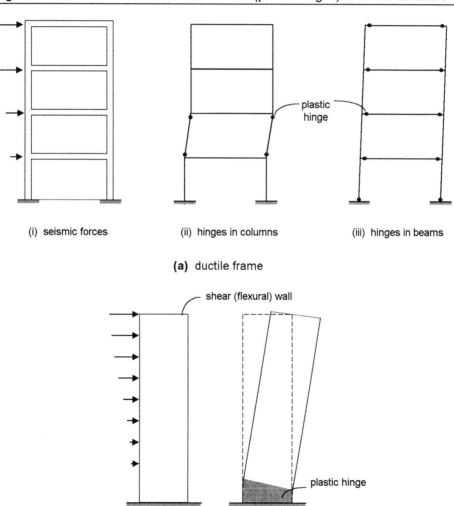

and mechanisms involving beam hinges have larger energy-absorptive capacity on account of the larger number of beam hinges. Furthermore, in the event of residual deformation and damage, a column is more difficult to straighten and repair than a beam. Therefore, it is preferable to design the frame so as to form the inelastic deformations in the beams rather than the columns.

18.3 MAJOR DESIGN CONSIDERATIONS

18.3.1 General

The objective of the special seismic design provision in CSA A23.3-94 (Clause 21) is to accomplish a satisfactory inelastic structural response, similar to that shown in Fig. 18.1c. This ensures adequate ductile flexural behaviour, with ability to undergo large inelastic reversible deformations, for individual members, such as beams, flexural walls and columns, and prevents other non-ductile types of failure. This leads, in relative terms, to combinations of strong columns and weak beams, members strong in shear and weak in flexure, and stronger foundations and weaker superstructure. From earlier chapters, especially Chapters 4, 6 and 15, some of the major factors which contribute to member (flexural) ductility are: (1) a low tensile reinforcement ratio and/or the use of compression reinforcement; (2) adequate shear reinforcement to provide a shear strength in excess of the flexural strength; (3) confinement of concrete and compression reinforcement by closely spaced hoops or spirals; and (4) proper detailing with regard to anchorage of reinforcement, splicing, minimum reinforcement ratios. Similarly, continuity assists in development of more inelastic response and, thereby, moment redistribution and energy dissipation at several hinges.

18.3.2 Code Requirements

The design and detailing requirements for beams, columns and their joints, forming part of the lateral (seismic) force resisting system in ductile moment-resisting frames are given in Clauses 21.3, 21.4, 21.6 and 21.7 of CSA A23.3-94, and for frame members not considered as part of the lateral force resisting system in Clause 21.8. Requirements for ductile flexural walls are given in Clauses 21.5 and 21.7. Clause 21.9 gives the requirements for building members with nominal ductility, so that they can be classified as cast-in-place continuously reinforced concrete construction with an $R = 2.0$. These Code requirements are presently considered to be the minimum requirements necessary to enable the structure to sustain a series of oscillations into the inelastic range of response without critical deterioration in strength. Some of the pertinent considerations and the application of these requirements are briefly presented below (also Ref. 18.3).

CHAPTER 18 SPECIAL PROVISIONS FOR SEISMIC DESIGN 611

18.3.3 Structural Framing, Analysis and Proportioning

In a structure designed to resist earthquake forces in a ductile manner, it is expected that a severe earthquake will induce lateral deformations and oscillations sufficient to develop reversible plastic hinges at suitable locations in the ductile frame and the ductile flexural walls. The structural scheme should be such that these plastic hinge formations in members, or failure of an individual element, will not lead to instability or progressive collapse. Building redundancy into the structural system is desirable, as it assists in the development of alternative load paths, thereby helping redistribution of forces, avoidance of progressive collapse, and dissipation of energy.

In addition to adequate strength to resist the earthquake forces, the structure must have sufficient stiffness to limit the lateral deflection or drift. The distribution of forces to the elements of the lateral force resisting system is usually based on a linearly elastic model of the system acted on by the equivalent factored static loads. Allowing for some plastic deformations in the structural system, Ref. 18.1 recommends that the anticipated drift due to earthquake forces be taken as the deflection obtained from the elastic analysis under the equivalent static loads multiplied by R. Assuming at least all the horizontal structural members to be fully cracked is likely to result in better estimates of the possible drift than using the uncracked stiffness for all members (Ref. 18.3).

The limiting value of drift depends on the acceptable damage to the nonstructural components, such as plaster, glass panels, etc., and a recommended maximum for interstorey drift is 0.02 times the storey height under the specified earthquake loads (Ref. 18.1). The effect of drift on the vertical load-carrying capacity of the lateral force resisting system must also be taken into account. The analysis also needs to account for the effects of members not forming part of the lateral force resisting system and of nonstructural members on the response of the structure to earthquake motion.

18.3.4 Materials

The fact that yielding is permitted to occur at certain locations in the structure also necessitates the avoidance of too much over-strength at such locations. Over-strength will result in the member not yielding at the expected lateral load level as intended, and this may force adjoining elements and/or foundations to receive higher seismic load and suffer consequent damage. The plastic moment capacity of under-reinforced sections is primarily dependent on the yield stress of steel. Furthermore, yield strength far in excess of that specified may lead to excessive shear and bond stresses as the plastic moment is developed. For resisting earthquake-induced flexural and axial forces, the Code prescribes the use of reinforcing steel that is in conformance with CSA

Standard G30.18. This steel has more closely controlled chemical composition and mechanical properties, better elongation, and a specified upper limit on the yield stress (Table 1.3). It also has an ultimate tensile strength, f_u, which is at least 1.25 times the tensile yield strength, f_y. The length of the hinging region along the member axis, and thereby the inelastic rotation capacity of hinging regions, is found to increase with the ratio f_u/f_y. In addition, the reinforcing steel needs to be weldable grade for lateral load resisting systems designed with force modification factors, R, greater than 2.0, but need not be weldable grade for lateral load resisting systems designed with force modification factors of 2.0 or less.

High strength, low and normal density concretes have relatively low ultimate compressive strains, and this may adversely affect the rotation capacity of flexural members. In the absence of experimental and field data on behaviour of members with high strength concrete subjected to displacement reversals in the nonlinear range, the Code limits the maximum strengths of normal and low-density concrete for use in seismic resistant design to 55 MPa and 30 MPa, respectively.

18.3.5 Foundation

As plastic deformations are permitted to occur at suitable locations in the structure under severe earthquakes, the maximum earthquake loads transmitted to the foundation are not governed by the factored loads, but by the loads at which actual yielding takes place in the structural elements that transfer the lateral loads to the foundation, such as frame members and walls. Therefore, the Code (Cl. 21.2.2.3) requires that the *factored resistance* of the foundation system be sufficient to develop the nominal moment capacity (that is, the resistance computed using ϕ_c and $\phi_s = 1.0$) and the corresponding shears of the frames and walls. Under factored load conditions this means a relatively stronger foundation and weaker superstructure combination, ensuring ductile behaviour of the superstructure without serious damage to the foundation.

18.3.6 Flexural Members in Ductile Frames

As shown in Section 4.4 and Fig. 4.7, the ductility of flexural members decreases with increasing steel ratios. In order to ensure significant ductile behaviour, even under reversals of displacements into the inelastic range, to avoid congestion of steel, and to limit the shear stresses in beams of usual proportions, the Code limits the tensile reinforcement ratio to a maximum of 0.025. To avoid sudden brittle failure of a beam as the cracking moment of the section is reached, a minimum area of longitudinal reinforcement equal to $1.4 b_w d/f_y$ must be provided at both the top and bottom for the entire length of the member, with at least two bars being placed on each side. The reversible plastic hinges in flexural members usually develop near continuous supports such as at beam-column connections (Fig. 18.3a(iii)). A minimum positive moment capacity must be provided at such supports. Hinging regions are subjected to very large and reversible strains, and the concrete shell outside the hoop reinforcements may spall off. Splices of flexural reinforcement are not permitted in and near possible plastic hinge locations. Lap splices, when used elsewhere, must be enclosed within hoop or spiral reinforcement. Transverse reinforcement for confining concrete and to support longitudinal bars must also be provided in hinging regions, as explained below.

The bar extensions must provide for possible unpredictable shifts in the inflection points, which may occur under combined loadings, including seismic loads. Tests have shown that at interior beam-column connections, the continuation of top and bottom beam reinforcement through the column will develop the theoretical flexural strength of the beams. Therefore, such bars must be made continuous wherever possible. In other cases, both top and bottom bars must be extended to the far face of the confined column core, and anchored in accordance with Code Cl. 21.6.1.3. The confined core is the portion of the member cross section confined by the perimeter of the transverse reinforcement conforming to the requirement in Clause 21 of the Code, measured from out-to-out of the transverse reinforcement.

Flexural members of lateral force resisting ductile frames are assumed to yield at the design earthquake. Thus, the required factored shear resistance of flexural members is related to the flexural strength of the members, rather than to the factored shear force obtained from the lateral-load analysis. The members must be capable of developing reversible plastic hinges without premature non-ductile failure in shear. Therefore, flexural members must have a factored shear resistance exceeding the maximum probable shear force in the member, as plastic hinges develop at both ends of the beam, as in Fig. 18.3a(iii). Since these beams are expected to develop large inelastic rotations at these hinges, the strains in the flexural reinforcement at the face of the joint may be well in excess of the yield strain and may go into the strain-hardening range. To allow for this possibility of strain-hardening, and to get a conservative estimate of the factored shear resistance required, the Code specifies that the moment at the plastic hinges be taken as the *probable moment resistance*, M_{pr}, which is the moment resistance of the section calculated using $1.25f_y$ as the stress in the tension reinforcement, and f_c' as the strength of the concrete, and with the material resistance factors, ϕ_c and ϕ_s, taken as unity. In addition to these moments, M_{pr}, at each end of the beam, the factored gravity loads acting on the beam also cause beam shear. Thus, the required factored shear resistance of the beam under the load combination, for occupancies other than storage or assembly, $1.0D + \gamma(0.5L + 1.0E)$, with $\psi = 1.0$, is as shown in Fig. 18.4. This over-estimation of the shear strength requirement of the beam in relation to flexural strength ensures that the member develops plastic hinges before shear failure. However, the factored shear need not exceed that determined from factored load combinations with load effects calculated using $R = 1.0$.

614 REINFORCED CONCRETE DESIGN

Fig. 18.4 *Required factored shear resistance for beams* *

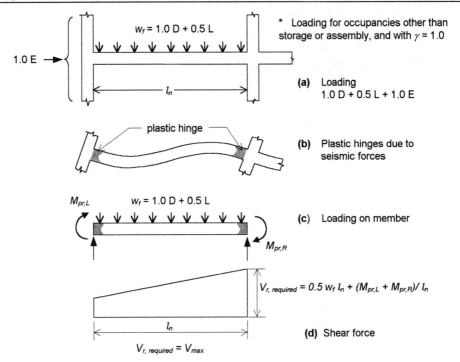

Fig. 18.5 *Type of web reinforcement for reversed shear*

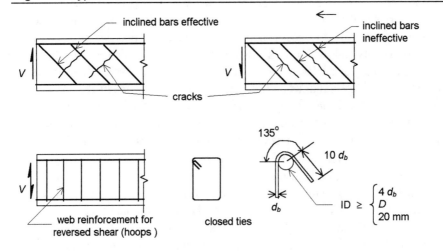

Because of the alternating nature of the shear force due to seismic effects, the direction of the associated diagonal tensile stress also alternates, as shown in Fig. 18.5. Therefore, inclined web reinforcement, which is effective for shear in one direction, will not be effective when the shear is reversed. Web reinforcement for seismic design must be placed perpendicular to the longitudinal reinforcement, and must be provided throughout the length of the member. Since plastic hinges are likely to form at the ends of flexural members near the beam-column connections, in order to confine the concrete and to prevent buckling of the longitudinal bars, such regions must be provided with closely spaced hoops or spirals enclosing the main flexural reinforcement. At sections where inelastic rotations occur alternately in opposite directions, the reinforcement on the opposite faces also alternates in tension and compression. Furthermore, after a few cycles of inelastic rotations, the concrete cover may spall off. Therefore, the web reinforcement in such regions must be in the form of closed hoops with their free ends bent at 135° and adequately anchored into the core concrete (Fig. 18.5). Finally, in the shear design of flexural members of lateral force resisting frames by the Simplified Method (Chapter 6), the shear resistance of concrete, V_c, is ignored.

18.3.7 Columns

The Code requires that at beam-column joints, the sum of the *factored* moment resistance (with $\phi_c = 0.6$ and $\phi_s = 0.85$) of the columns framing into the joint be at least 1.1 times the sum of the *nominal* moment resistance (that is, with $\phi_c = \phi_s = 1.0$) of the beams and girders framing into the joint. The nominal resistance of the beams must be computed, including the effects of the slab reinforcement within a distance of three times slab thickness on either side of the beam. This requirement, shown in Fig. 18.6a, must be satisfied for beam moments acting in either direction. The requirement is intended to ensure the formation of plastic hinges in the beams rather than in the columns (strong column-weak beam combination). The column resistances are computed conservatively (with inclusion of ϕ-factors) compared to the beam resistances. Again, lap splices are not permitted near the ends of the column where spalling of the concrete shell is likely to occur.

The factored shear resistance of columns must be greater than the larger of (1) the shear force due to the factored loads, and (2) the shear force in the column corresponding to the development of the *probable* moment resistances in the beams framing into it. The latter condition and corresponding column shears are shown in Fig. 18.6b. This allows for possible over-strength of the beams, thereby avoiding shear failure of the columns.

Apart from the normal transverse reinforcement required for columns, special transverse confining reinforcement must be provided near the joints (high moment regions) and on both sides of any section, where flexural yielding may occur due to

616 REINFORCED CONCRETE DESIGN

Fig. 18.6 Column resistance requirements

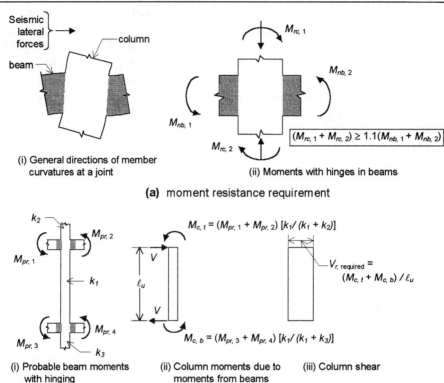

inelastic lateral displacement of the frame. In columns, which may develop plastic hinges, such as at the base of the building, and columns supporting discontinued stiff members like walls or trusses, this special transverse confining reinforcement is required for the full height of the column.

18.3.8 Joints in Ductile Frames

The design requirements for beam-column joints in ductile frames given in the Code (Cl. 21.6) are based on Ref. 18.4. The joints are zones of heavy concentration of stresses. Joints must have adequate strength to facilitate the development of large inelastic reversible rotations in the beams at the face of the joints. The flexural stresses transferred to the faces of the joint by the beams generate high shear forces within the joint on horizontal shear planes as shown in Fig. 18.7. As mentioned in Section 18.3.6, to allow for the very large strains in and the possibility of strain-hardening of flexural reinforcement in the beams at the faces of the joint, the Code specifies that the factored

CHAPTER 18 SPECIAL PROVISIONS FOR SEISMIC DESIGN 617

Fig. 18.7 Horizontal factored shearing forces on an interior joint

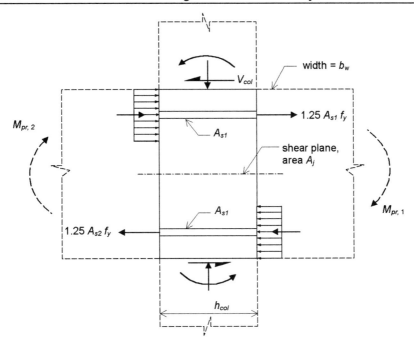

shear force in the joint be determined assuming a tensile stress of $1.25f_y$ in the beam reinforcement (Fig. 18.7).

Tests have indicated that the shear strength of joints is not sensitive to the amount of shear reinforcement, and may be taken as a function of only the compressive strength of concrete (Ref. 18.3). On this basis, the Code recommends that the factored shear resistance of the joint be taken as:

1) for confined joints,

$$V_r = 2.4\lambda\phi_c\sqrt{f'_c}A_j$$

2) for joints confined on three faces or on two opposite faces,

$$V_r = 1.8\lambda\phi_c\sqrt{f'_c}A_j$$

3) and for others,

$$V_r = 1.5\lambda\phi_c\sqrt{f'_c}A_j$$

where A_j is the minimum cross-sectional area within a joint in a plane parallel to the axis of the reinforcement generating the shear in the joint (equal to the lesser of A_g of the column or $2b_w h_{col}$).

Irrespective of the magnitude of the shear force in the joint, transverse confining (hoop) reinforcement must be provided through the joint around the column reinforcement. Where the joint is externally confined, namely if members frame into all vertical faces of the joint, and if at least 3/4 of each face of the joint is covered by the framing members, the amount of transverse hoop reinforcement may be reduced in that portion of the joint. At a column where the width of beam exceeds the width of column, the longitudinal beam reinforcements located outside the column core must also be provided with lateral confinement, either by beams framing into the joint laterally, or by transverse hoop reinforcement through the joint. Since flexural reinforcement is expected to develop stresses of the order of $1.25f_y$ at the face of the joint and is furthermore subjected to stress reversals, the development length requirements within the joint are also very critical.

The connection zone is an area of high concentration of reinforcement, consisting of beam bars, column bars and connection hoops. Many failures in structures under seismic loading can be traced to poor detailing, particularly of beam-column joints. Extreme care is needed in detailing the reinforcement at the connection to provide for proper stress transfer and to avoid congestion and placing difficulties for both reinforcement and concrete.

18.3.9 Ductile Flexural Walls

Ductile flexural walls are vertical members cantilevering from the foundation and subjected to bending moment, horizontal shear, and axial load. In ductile walls, inelastic deformation usually involves plastic hinges at the base, as shown in Fig. 18.3b, and/or at locations of abrupt changes in strength and stiffness of the lateral load resisting system. In the absence of such abrupt changes, plastic hinges will form within the lower half of the wall height. Unlike a beam, a flexural wall is relatively thin and deep, and is subjected to substantial axial forces. These must be designed as axially loaded beams capable of forming reversible plastic hinges with sufficient rotation capacity. The wall being thin, the regions of high compressive strain must be checked for instability (buckling). Stability of the compressed zone can be improved by thickening this zone or by providing lateral support, such as by a flange or cross wall. Flanged walls, having I or channel sections, have higher bending resistance and ductility, and less likelihood of lateral buckling at the compressed flange, may be used in elements such as elevator shafts.

CSA A23.3-94 (21.5) gives minimum design and detailing requirements for ductile flexural walls. The wall is reinforced with distributed reinforcement, provided in both vertical and horizontal directions, and with vertical reinforcement concentrated at each end of the wall. The distributed reinforcement provides the shear resistance, controls the cracking, inhibits local breakdown in the event of severe cracking during an earthquake, and also resists temperature and shrinkage stresses. The concentrated and the distributed vertical reinforcement provide the flexural and axial load resistance.

The concentrated vertical flexural reinforcement near the ends must be tied together by transverse ties, as in a column, to provide confinement of the concrete and to ensure the yielding, without buckling, of the flexural bars on the compression side, when a plastic hinge is formed. A minimum ductility is ensured by placing an upper limit on the depth to the neutral axis.

Owing to the minimum reinforcement requirements, and the possibility of strain-hardening in the steel, the actual moment capacity of a flexural wall may be greater than the design moment, M_f. To prevent a premature brittle shear failure of the wall, before the development of its full plastic resistance in bending, the Code requires the wall to be designed to resist a shear force greater than the shear that is present when the wall develops a plastic hinge with the probable moment resistance, M_{pr}. M_{pr} is computed using $1.25 f_y$ in the tension reinforcement and f'_c for the concrete, without the application of any resistance factors. Because of possible severe shear cracking under cyclic loads, the shear carried by the concrete (Cl. 11.3) is neglected in the region of the plastic hinge.

18.4   RECENT ADVANCES

The purpose of this chapter is to explain the background to the seismic design provisions of the Code. A detailed discussion of seismic design of reinforced concrete structures is beyond the scope of this book. Rapid advances are being made in this area and recent publications on this topic (for example, Refs. 18.5 to 18.10) may be consulted for more details.

REFERENCES

18.1   *National Building Code of Canada* 1995, Part 4: Structural Design, National Research Council of Canada, Ottawa, 1995, pp. 137-159.

18.2   *User's Guide - NBC 1995 Structural Commentaries (Part 4)*, Canadian Commission on Building and Fire Codes, Commentary J, National Research Council of Canada, 1996, pp. 75-91.

18.3   ACI Standard 318-95, *Building Code Requirements for Structural Concrete* and *Commentary* (ACI 318R-95), American Concrete Institute, Detroit, 1995, 369 pp.

18.4   ACI-ASCE Committee 352, *Recommendations for Design of Beam-Column Joints in Monolithic Reinforced Concrete Structures*, (ACI 352 R-76, Reaffirmed 1981), American Concrete Institute, Detroit, 1976, 19 pp.
18.5   Arnold, C. and Reitherman, R., *Building Configuration and Seismic Design*, John Wiley & Sons, Inc., New York, N.Y., 1982, 296 pp.
18.6   Paulay, T., and Priestley, M.J.N., *Seismic Design of Reinforced Concrete and Masonry Buildings*, John Wiley & Sons, New York, 1992, 744 pp.
18.7   Dowrick, D.J., *Earthquake Resistant Design*, John Wiley & Sons, Chichester, U.K., 1977.
18.8   NEHRP, *Recommended Provisions for the Development of Seismic Regulations for New Buildings*, Federal Emergency Management Agency, Washington, D.C., 1994.
18.9   *Reinforced Concrete Structures Subjected to Wind and Earthquake Forces*, ACI Publication SP-63, American Concrete Institute, Detroit, 1980.
18.10 *Earthquake Effects on Reinforced Concrete Structure*, ACI Publication SP-84, American Concrete Institute, Detroit, 1985.

APPENDIX A Bookkeeping

Bookkeeping refers to the art or practice of keeping a systematic record of design correspondence, data, and calculations. Because the design of a project may span several months or even years, and may be worked on by many designers, and because calculations may cover thousands of pages, a good system of bookkeeping is mandatory if designers are to work efficiently. Benefits derived from the use of a sound, uniform bookkeeping system are as follows:

1. Assist the designers to proceed with the design in a systematic and previously proven manner.
2. Ensures the organisation and effective indexing of design calculations.
3. Facilitates the checking of calculations.
4. Facilitates the transfer of information to drawings.
5. Minimises repetition of calculations and wasted design effort.

Some basic guidelines or bookkeeping hints which should be followed by designers in designing a structure such as a multi-storey building are the following:

1. Do all calculations on special paper provided for the purpose.
2. Ensure that every page is indexed and that all data such as project number, date, designer, etc., is recorded. Each major division should have a "General" subdivision which summarises codes, design and analysis methods, loads, material properties and other data required for the design of the components or elements in that division.
3. Keep calculations in three ring binders. Binders ensure that pages do not get lost and facilitate amending of design files.
4. Where possible, calculations to the applicable code provisions, design aids, or technical publications.
5. Reference all calculations to the applicable code provisions, design aids, or technical publications.
6. For each structural system or component being designed, clearly define the geometry by referring to sketches or drawings or provide a sketch showing span, cross section, tributary width, stiffness, restraint conditions, openings or variations in bent widths.
7. For each structural system or component being designed, identify the source of all loading data and illustrate the loads by sketches if appropriate.
8. The derivation of all forces should be shown on a sketch, if necessary.
9. Cross reference calculations if previous results are used.
10. Identify each member being designed by grid references or member marks.

Fig. A-1 *Format A for recording vertical information on the "Model Building"*

Floor Level	Fl-Fl Height (m)	Elev. (m)	Ht. Ab. Ground (m)	Ht. Ab. Bsmt (m)
PARA'PT ROOF	2.40	174.39	97.78	99.56
25	5.66	168.73	92.12	93.90
24	3.60	165.13	88.52	90.30
23	3.60	161.53	84.92	86.70
22	3.60	157.93	81.32	83.10
21	3.60	154.33	77.72	79.50
20	3.60	150.73	74.12	75.90
19	3.60	147.13	70.52	72.30
18	3.60	143.53	66.92	68.70
17	3.60	139.93	63.32	65.10
16	3.60	136.33	59.72	61.50
15	3.60	132.73	56.12	57.90
14	3.60	129.13	52.52	54.30
13	3.60	125.53	48.92	50.70
12	3.60	121.93	45.32	47.10
11	3.60	118.33	41.72	43.50
10	3.60	114.73	38.12	39.90
9	3.60	111.13	34.52	36.30
8	3.60	107.53	30.92	32.70
7	3.60	103.93	27.32	29.10
6	3.60	100.33	23.72	25.50
5	4.50	95.83	19.22	21.00
4	4.50	91.33	14.72	16.50
3	4.50	86.83	10.22	12.00
2	4.50	82.33	5.72	7.50
1	4.50	77.83	1.22	3.00
BSMT. CAISSON	3.00	74.83		

11. Separate by underlining or highlighting final components dimensions, reinforcement, etc., and transfer applicable results (wall, beams, column, footing designs), directly to schedules.
12. Utilise reusable formats to present dates, identify members or display results. A vertical format, (Fig. A.1), which represents, to scale, the number of floors, floor to floor heights, elevations, and distances above the basement floor is convenient for:
    a. Column rundowns.
    b. Column designs.
    c. Lateral force calculations.

d. Manual lateral load analyses.
e. Displaying, graphically, column loads.
f. Displaying, graphically, lateral forces.

A horizontal formats, (Fig. A.2), which consists of a floor plan containing only grid lines and basic geometry is convenient for showing:

a. Floor loads on loading drawings.
b. Slab and beam identification marks.
c. Column identifications groupings, and tributary areas.
d. Wall and core tributary areas
e. Slab reinforcement results.
f. Opening, framing details.
g. Foundation layouts.

Both these formats are used repeatedly throughout the various stages of design.

Fig. A.2 *Format B for recording horizontal information on the "Model Building"*

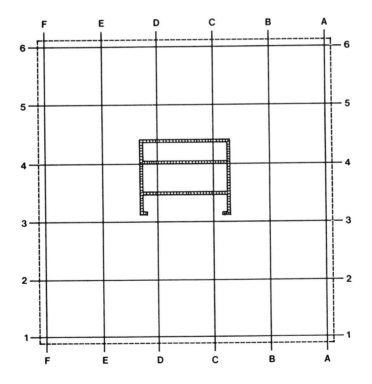

APPENDIX B Design Aids

Table 1.2	Metric Deformed Bar Size and Designation Numbers – CSA G30.18	20
Table 1.3	Standard Reinforcing Steel Data	23
Table 2.2	Minimum Specified Distributed Live Loads for Floors and Roofs (Abstracted from National Building Code of Canada – 1995, Sentence 4.1.6)	31-32
Table 2.3	Minimum Specified Concentrated Live Loads for Floors and Roofs (Ref. 2.1)	32
Table 4.1	Reinforcement Ratios ρ and Resistance Factors K_r for Rectangular Sections	96-97
Table 5.1	Minimum Specified Clear Concrete Cover for Reinforcement in Cast-in-place Concrete, CSA A23.3-94	117
Table 5.2	Thickness Below Which Deflections Must Be Computed for Non-prestressed Beams or One-Way Slabs	120
Table 5.3	Recommended Tension Reinforcement for Non-prestressed One-Way Construction So That Deflections Will Normally Be Within Acceptable Limits	121
Table 5.4	Reinforcement Percentage 100ρ for Resistance Factor K_r	131-132
Table 5.5	Area Reinforcement Bars in Slabs, in mm^2 per Metre Width	134
Figure 5.6	Recommended Typical Bending Details	139
Table 9.1	Basic Development Length l_{bd} in mm, for Bars and Deformed Wires in Tension	279
Table 9.2	Basic Development Length l_{hb} for Standard Hooks in Tension	283
Table 9.3	Tension Lap Splice Classification and Specified Minimum Lap Lengths, l_s	298
Table 10.1	Moments and Shears in Continuous Beams Using CSA Coefficients (CSA A23.3-94(9.3.3))	308

Figure 10.6	Non-dimensional Moment Diagrams for CSA Moment Coefficients	311
Table 13.1	Coefficients for Negative Moments in Slabs	405
Table 13.2	Coefficients for Dead Load Positive Moments in Slabs	406
Table 13.3	Coefficients for Live Load Positive Moments in Slabs	407
Table 13.4	Ratio of Load w in A and B Directions for Shear in Slab and Load on Supports	408
Table 14.2	Moment Distribution Constants for Slab-Beam Elements	468
Table 14.3	Moment Distribution Constants for Slab-Beam Elements, Drop Thickness = $0.5h$	469
Table 14.4	Stiffness and Carry-Over Factors for Columns	472
Figure 14.28	Minimum Length of Reinforcement – Slabs without Beams	476
Figure 15.16	Interaction Diagram for Axial Load and Moment Resistance for Rectangular Columns with an Equal Number of Bars on all Four Faces	523
Figure 15.17	Interaction Diagram for Axial Load and Moment Resistance for Rectangular Columns	524
Figure 15.18	Interaction Diagram for Axial Load and Moment Resistance for Rectangular Columns	524
Figure 15.19	Interaction Diagram for Axial Load and Moment Resistance for Rectangular Columns	525
Figure 15.20	Interaction Diagram for Axial Load and Moment Resistance for Circular Columns	525
Figure 15.21	Interaction Diagram for Axial Load and Moment Resistance for Circular Columns	526
Figure 15.22	Interaction Diagram for Axial Load and Moment Resistance for Circular Columns	526
Figure 16.5	Effective Length Factors for Columns in Braced Frames	553
Figure 16.6	Effective Length Factors for Columns in Unbraced Frames	554
Table 16.1	Effective Length Factors for Columns in Braced Frames (CSA A23.3-94)	555

INDEX

Aggregates:
  coarse, 2
  fine, 2
  high density, 3
  low density, 3
Analysis, 306, 309, 315
Anchorage, 190, 267
Axial compression, 504

Beams:
  analysis:
    compression reinforcement, 87
    rectangular sections, 86, 94, 108
    slabs, 99
    T-beams, 90, 106
  balanced strain condition, 68, 83
  bar spacing, 116, 117
  behaviour in flexure, 65-70
  bond stresses, (See Bond)
  code provisions, 85, 94
  compression reinforcement, 87, 100
  cover, 116, 117
  crack control, 119, 365
  deep, 254
  deflection of, (see Deflections)
  design procedure:
    tension reinforcement, 121-128
    aids, 128, APPENDIX B, 624
    compression reinforcement, 138
    T-beams, 142
  doubly reinforced, 87, 100
  development length, (see Bond, development length)
  flexural behaviour, 65
  flexural strength, 79-85
  moment-curvature, 70
  over-reinforced, 69, 108
  service load stresses, 71-79
  shear strength, (see Shear)
  singly reinforced, 86
  size guidelines, 119
  steel area:
    maximum, 87
    minimum, 93
  straight-line theory, 71
  stress block, 84
  T-beam analysis, 90, 106
  T-beam design, 142
  tension reinforcement, 86
    CSA limitations, 94
  torsion, 212-234
  transformed section, 61, 66
  under-reinforced, 67
Beam and girder floors, 373-397
Bent bars, 138, 148
Bond:
  anchorage, 190, 267
  bundled bars, 280
  compression bars, 279
  development length, 275, 277, 281
  failure, 270
  flexural, 266
  hooks, 281
  tension bars, 277
  tests, 274
  welded wire fabric, 280
Buildings, reinforced concrete, 27-45
Bundled bars, 116, 280

Cement, 2
  types of, 2
  hydration of, 1
Codes and specifications, 22
Columns:
  axial load and bending, 508
  axial load capacity, 504
  balanced strain conditions, 509
  biaxial bending, 537
  circular, 533
  compression failure, 512
  design aids, 521, APPENDIX B, 624
  effective length, 502, 552
  interaction diagrams, 518, 521, 523, 528
  lateral drift effect, 562
  magnified factored moment, 558
  member stability effect, 557
  P-Δ effect, 546
  plastic centroid, 505

reinforced requirements, 499
short, 504
slender, 546-571
  behaviour, 546
  design of, 556
slenderness effects, 502
  lateral drift effect, 561
  member stability effect, 557
spiral, 499, 501, 534
tension failure, 515
tied, 499, 501
types, 499
Combined footings, (see Footings, combined)
Compatibility torsion, 212, 225
Compression field theory, 241
Compression members, (see Columns)
Compression steel, 87, 100
Compressive strength, 5
Continuous structures, 300-339
  loading patterns on, 301
Concrete:
  admixtures, 4
  air-entraining, 4
  biaxial strength, 11-13
  coefficient of thermal expansion, 19
  combined stresses, 11
  compressive strength, 5
  constituents, 1
  creep, 14
  diagonal compressive stress, 164, 173, 239
  diagonal crushing strength, 241
  factor for low density, 10
  high density, 3
  low density, 3
  modulus of elasticity, 7, 8
  Poisson's ratio, 8
  proportioning of, 3
  resistance factor for, 55
  shear strength, 10
  shrinkage and temperature effects, 17, 19
  stress-strain curves, 6
  tensile strength, 9
    modulus of rupture, 9, 10

split cylinder, 9
triaxial strength, 13, 14
water-cement ratio, 2
Corbels, 204, 255
Cracking:
  control of, 119, 365
  height of, 349
  moment, 345
  section properties, 349
Creep, 14, 357
Cut-off of reinforcement, 148-152, 286

Deflections, 341
  allowable, 366
  creep, 358
  effective moment of inertia, 345-348
  elastic theory, 342
  immediate, 344
  long-time, 355, 360
  shrinkage, 356, 358
Design aids, APPENDIX B, 624
Development length, (see Bond, development length)
Diagonal tension, 166, 168, 169, 239
Direct design method for slabs, 431, 452
Doubly reinforced beams, 87, 100, 138
Drop panels, 38, 445, 449, 447-479
Ductility, 70, 604

Equilibrium torsion, 212

Factored resistance, 53
  moment, 79, 82, 285
  shear, 184
  torsion, 222
Flange of T-beams, 91
Footings:
  allowable soil pressure, 572
  code requirements, 576
  combined, 573, 591, 597
  design:
    bearing at column base, 581
    bending moment, 579
    combined, 573, 591, 597
    considerations, 576
    development length, 580

628 *INDEX*

    rectangular, 590
    shear, 580
    square, 586
    transfer of forces, 581
    wall, 583
  dowels, 581
  net soil pressure, 573
  rectangular, 590
  square, 573, 586
  types of, 570
  wall, 573, 583
Frame analysis:
  code approximations, 306
  procedures, 309
  stiffness of members, 311

Hooks, 281
Hydration, 1
Horizontal loads on buildings, 30, 32, 34, 606-619

Inelastic redistribution of moments, 331
Interaction diagrams, (see Columns, interaction diagrams)

Joint detail for ductility, 616
Joists, 39, 396

Lateral load resisting systems, 42
Limit analysis, 331
  code provisions, 335
Limit states design, 49-51
  code provisions, 51
Loads:
  building materials, 29
  earthquake, 34
  factors, 55
  live, 30, 31, 32
  wind, 30, 45-46
Load and resistance factor design, 52
Load factors, 55

Material resistance factors, 55
Mechanisms, collapse, 333
Modular ratio, 64
Modulus of elasticity, 7, 8

Modulus of rupture, 9, 10
Moment coefficients CSA, 307-309
Moment of inertia:
  for beams, 311
  for columns, 311
Moment-curvature relations, 70
Moment redistribution, 331-335
Moment resistance, 72
  factored, 79, 82, 285
  nominal, 79

One-way slabs, (see Slabs, one-way)

P-Δ effect, 546
Pattern loading, 301
Plastic hinge, 333
Poisson's ratio, 8

Redistribution of moments, 331, 335
Reinforcement:
  bend points, 138, 148
  cover, 116, 117
  cut-off locations, 148-161
  design aids, 23, 134, APPENDIX B, 624
  development length, 154
  grades, 20
  resistance factors for, 55
  sizes, 20, 21
  skin reinforcement, 93
  spacing, 116, 117
  specifications, 20, 21, 23
  splices, 296
  stirrups, 177, 181, 183, 192, 251
  stress-strain curves, 24
  temperature and shrinkage, 93, 133, 379
  types, 20

Safety provisions:
  CSA, 52, 54
  factored load effects, 53
  factored resistance, 53
  importance factor, 55
  load combination factor, 55
  load factors, 55
  material resistance factors, 55
  member resistance factors, 56

resistance factors, 49, 52
tolerances, 57
Seismic design:
  code requirements, 610
  columns, 615
  design considerations, 610
  ductility, 607
  flexural members, 612
  foundation, 612
  joints, 616
  materials, 611
  structural framing, 611
  walls, 618
Serviceability limits, 54, 341-372
Shear:
  beams with axial loads, 202
  beams with varying depth, 203
  bracket and corbels, 204, 255
  code requirements for design:
    compression field theory, 241
    general method, 237, 241
    simplified method, 183
  coefficients, CSA, 308
  concrete:
    cracking shear resistance, 175
    factored shear resistance, 176
  deep beams, 254
  design procedure:
    compression field theory, 241
    general method, 237, 241
    simplified method, 183
  diagonal cracking, 168, 169, 174, 239-241
  failure modes, 172
  nominal stresses, 168
  reinforcement:
    anchorage, 190
    behaviour, 177
    factored shear 183istance of,
    minimum, 185
    spacing, 188, 259
  strength of plain beams, 174-176
  strength with web reinforcement, 181-183
    stirrup area, 183-185
    stirrup spacing, 188

Shear-friction, 204, 581
Shrinkage, 17-18
  curvature, 357
  deflection due to, 356, 358
Shrinkage and temperature effects, 17-18
  reinforcement, 93, 136, 379
Slabs:
  behaviour, 400
  one-way, 99, 132, 138
    flexural reinforcement, 138
  one-way floor systems, 318-339
    design of beams, 381
    design of girders, 387
    design of slabs, 376
    joist floors, 396
  reinforcement details, 138, 464
  systems:
    beam and slab, 40
    joists, 38
    flat, 37
    flat plate, 37
    wall supported, 35
  two-way:
    code provisions, 425
    deflection control, 429
    direct design method, 433, 454
    equivalent frame method, 464, 477
    shear design, 443-454
    supported on stiff beams or walls, 400-421
    transfer of moments to columns, 431
Slenderness (see Columns, slenderness effects)
Space-truss theory, 221, 253
Specifications and codes, 22
Splices, 296
Stiffness of members, 311
Stirrups:
  design of, 183, 241
  spacing of, 183, 253
Strength design, 79-85
Stress block, 84
Stress-strain curves, 6, 22, 24
Structural building systems:
  load transfer, 35
  floors, 35

lateral load resisting, 42
vertical elements, 40
Strut-and-tie model, 254
Structural safety, 47-57

T-beams, 90-93
Temperature and shrinkage steel, 57, 133, 379
Ties, 501
Tolerances, 57
Torsion:
  combined loading, 223
  compatibility, 213, 225
  cracking, resistance in, 217
  detailing of reinforcement, 226
  equilibrium, 212
  equivalent tube concept, 218
  factored resistance in, 222
  formulas, 213
  reinforcement for, 220, 226
  space truss theory, 221, 254-259
Transformed area, 61, 64
Two-way slabs (see Slabs, two-way)

Water-cement ratio, 2
Web reinforcement, 170, 180, 183, 192, 254